MARKOV CHAIN MONTE CARLO IN PRACTICE

MARKOV CHAIN MONTE CARLO IN PRACTICE

Edited by

W.R. Gilks

Medical Research Council Biostatistics Unit
Cambridge, UK

S. Richardson

French National Institute for Health and Medical Research
Villejuif, France

and

D.J. Spiegelhalter

Medical Research Council Biostatistics Unit
Cambridge, UK

CHAPMAN & HALL/CRC

Boca Raton   London   New York   Washington, D.C.

Library of Congress Cataloging-in-Publication Data

Catalog record is available from the Library of Congress.

© 1996 by Chapman & Hall

First Edition 1996
Reprinted 1996
First CRC Press reprint 1998
No claim to original U.S. Government works
International Standard Book Number 0-412-05551-1
Printed in the United States of America   3 4 5 6 7 8 9 0
Printed on acid-free paper

Contents

Contributors

James E Bennett   Department of Mathematics, Imperial College,
London, UK.

Carlo Berzuini   Dipartimento di Informatica e Sistemistica,
University of Pavia, Italy.

Nicola G Best   Medical Research Council Biostatistics Unit,
Institute of Public Health, Cambridge, UK.

Caitlin Buck   School of History and Archaeology,
University of Wales, Cardiff, UK.

Bradley P Carlin   Division of Biostatistics,
School of Public Health,
University of Minnesota, Minneapolis, USA.

David G Clayton   Medical Research Council Biostatistics Unit,
Institute of Public Health, Cambridge, UK.

Jean Diebolt   Département de Statistique
et Modèles Aléatoires, CNRS,
Université Paris 6, France.

W James Gauderman   Department of Preventive Medicine,
University of Southern California,
Los Angeles, USA.

Alan E Gelfand   Department of Statistics,
University of Connecticut, USA.

Andrew Gelman   Department of Statistics,
University of California, Berkeley,
California, USA.

Edward I George   MSIS Department,
University of Texas at Austin, USA.

| Charles J Geyer | School of Statistics, University of Minnesota, Minneapolis, USA. |

Charles J Geyer — School of Statistics, University of Minnesota, Minneapolis, USA.

Walter R Gilks — Medical Research Council Biostatistics Unit, Institute of Public Health, Cambridge, UK.

Peter J Green — Department of Mathematics, University of Bristol, UK.

Hazel Inskip — Medical Research Council Environmental Epidemiology Unit, Southampton, UK.

Eddie H S Ip — Educational Testing Service, Princeton, USA.

Steven M Lewis — School of Social Work, University of Washington, Seattle, USA.

Cliff Litton — Department of Mathematics, University of Nottingham, UK.

Robert E McCulloch — Graduate School of Business, University of Chicago, USA.

Xiao-Li Meng — Department of Statistics, University of Chicago, USA.

Annie Mollié — INSERM Unité 351, Villejuif, France.

David B Phillips — NatWest Markets, London, UK.

Amy Racine-Poon — Ciba Geigy, Basle, Switzerland.

Adrian E Raftery — Department of Statistics, University of Washington, Seattle, USA.

Sylvia Richardson — INSERM Unité 170, Villejuif, France.

Christian P Robert — Laboratoire de Statistique, University of Rouen, France.

Gareth O Roberts — Statistical Laboratory, University of Cambridge, UK.

Adrian F M Smith — Department of Mathematics, Imperial College, London, UK.

David J Spiegelhalter — Medical Research Council Biostatistics Unit, Institute of Public Health, Cambridge, UK.

Duncan C Thomas   Department of Preventive Medicine,
University of Southern California,
Los Angeles, USA.

Luke Tierney   School of Statistics, University of Minnesota,
Minneapolis, USA.

Jon C Wakefield   Department of Mathematics, Imperial College,
London, UK.

1

Introducing Markov chain Monte Carlo

Walter R Gilks
Sylvia Richardson
David J Spiegelhalter

1.1 Introduction

Markov chain Monte Carlo (MCMC) methodology provides enormous scope for realistic statistical modelling. Until recently, acknowledging the full complexity and structure in many applications was difficult and required the development of specific methodology and purpose-built software. The alternative was to coerce the problem into the over-simple framework of an available method. Now, MCMC methods provide a unifying framework within which many complex problems can be analysed using generic software.

MCMC is essentially Monte Carlo integration using Markov chains. Bayesians, and sometimes also frequentists, need to integrate over possibly high-dimensional probability distributions to make inference about model parameters or to make predictions. Bayesians need to integrate over the posterior distribution of model parameters given the data, and frequentists may need to integrate over the distribution of observables given parameter values. As described below, *Monte Carlo* integration draws samples from the the required distribution, and then forms sample averages to approximate expectations. *Markov chain* Monte Carlo draws these samples by running a cleverly constructed Markov chain for a long time. There are many ways of constructing these chains, but all of them, including the Gibbs sampler (Geman and Geman, 1984), are special cases of the general framework of Metropolis *et al.* (1953) and Hastings (1970).

It took nearly 40 years for MCMC to penetrate mainstream statistical practice. It originated in the statistical physics literature, and has been used for a decade in spatial statistics and image analysis. In the last few years, MCMC has had a profound effect on Bayesian statistics, and has also found applications in classical statistics. Recent research has added considerably to its breadth of application, the richness of its methodology, and its theoretical underpinnings.

The purpose of this book is to introduce MCMC methods and their applications, and to provide pointers to the literature for further details. Having in mind principally an applied readership, our role as editors has been to keep the technical content of the book to a minimum and to concentrate on methods which have been shown to help in real applications. However, some theoretical background is also provided. The applications featured in this volume draw from a wide range of statistical practice, but to some extent reflect our own biostatistical bias. The chapters have been written by researchers who have made key contributions in the recent development of MCMC methodology and its application. Regrettably, we were not able to include all leading researchers in our list of contributors, nor were we able to cover all areas of theory, methods and application in the depth they deserve.

Our aim has been to keep each chapter self-contained, including notation and references, although chapters may assume knowledge of the basics described in this chapter. This chapter contains enough information to allow the reader to start applying MCMC in a basic way. In it we describe the Metropolis–Hastings algorithm, the Gibbs sampler, and the main issues arising in implementing MCMC methods. We also give a brief introduction to Bayesian inference, since many of the following chapters assume a basic knowledge. Chapter 2 illustrates many of the main issues in a worked example. Chapters 3 and 4 give an introduction to important concepts and results in discrete and general state-space Markov chain theory. Chapters 5 through 8 give more information on techniques for implementing MCMC or improving its performance. Chapters 9 through 13 describe methods for assessing model adequacy and choosing between models, using MCMC. Chapters 14 and 15 describe MCMC methods for non-Bayesian inference, and Chapters 16 through 25 describe applications or summarize application domains.

1.2 The problem

1.2.1 Bayesian inference

Most applications of MCMC to date, including the majority of those described in the following chapters, are oriented towards Bayesian inference. From a Bayesian perspective, there is no fundamental distinction between observables and parameters of a statistical model: all are considered random quantities. Let D denote the observed data, and θ denote model parameters and missing data. Formal inference then requires setting up a joint probability distribution $P(D, \theta)$ over all random quantities. This joint distribution comprises two parts: a *prior* distribution $P(\theta)$ and a *likelihood* $P(D|\theta)$. Specifying $P(\theta)$ and $P(D|\theta)$ gives a *full probability model*, in which

$$P(D, \theta) = P(D|\theta) P(\theta).$$

Having observed D, Bayes theorem is used to determine the distribution of θ conditional on D:

$$P(\theta|D) = \frac{P(\theta)P(D|\theta)}{\int P(\theta)P(D|\theta)d\theta}.$$

This is called the *posterior* distribution of θ, and is the object of all Bayesian inference.

Any features of the posterior distribution are legitimate for Bayesian inference: moments, quantiles, highest posterior density regions, etc. All these quantities can be expressed in terms of posterior expectations of functions of θ. The posterior expectation of a function $f(\theta)$ is

$$E[f(\theta)|D] = \frac{\int f(\theta)P(\theta)P(D|\theta)d\theta}{\int P(\theta)P(D|\theta)d\theta}. \qquad (1.0)$$

The integrations in this expression have until recently been the source of most of the practical difficulties in Bayesian inference, especially in high dimensions. In most applications, analytic evaluation of $E[f(\theta)|D]$ is impossible. Alternative approaches include numerical evaluation, which is difficult and inaccurate in greater than about 20 dimensions; analytic approximation such as the Laplace approximation (Kass *et al.*, 1988), which is sometimes appropriate; and Monte Carlo integration, including MCMC.

1.2.2 Calculating expectations

The problem of calculating expectations in high-dimensional distributions also occurs in some areas of frequentist inference; see Geyer (1995) and Diebolt and Ip (1995) in this volume. To avoid an unnecessarily Bayesian flavour in the following discussion, we restate the problem in more general terms. Let X be a vector of k random variables, with distribution $\pi(.)$. In Bayesian applications, X will comprise model parameters and missing data; in frequentist applications, it may comprise data or random effects. For Bayesians, $\pi(.)$ will be a posterior distribution, and for frequentists it will be a likelihood. Either way, the task is to evaluate the expectation

$$E[f(X)] = \frac{\int f(x)\pi(x)dx}{\int \pi(x)dx} \qquad (1.1)$$

Here: $\pi(\theta) = P(\theta)P(D/\theta)$ in (1.0)

$\int \pi(x) dx = 1$

for some function of interest $f(.)$. Here we allow for the possibility that the distribution of X is known only up to a constant of normalization. That is, $\int \pi(x)dx$ is unknown. This is a common situation in practice, for example in Bayesian inference we know $P(\theta|D) \propto P(\theta)P(D|\theta)$, but we cannot easily evaluate the normalization constant $\int P(\theta)P(D|\theta)d\theta$. For simplicity, we assume that X takes values in k-dimensional Euclidean space, i.e. that X comprises k continuous random variables. However, the methods described here are quite general. For example, X could consist of discrete random variables, so then the integrals in (1.1) would be replaced by summations. Alternatively, X could be a mixture of discrete and continuous random variables, or indeed a collection of random variables on any probability space. Indeed, k can itself be variable: see Section 1.3.3. Measure theoretic notation in (1.1) would of course concisely accommodate all these possibilities, but the essential message can be expressed without it. We use the terms *distribution* and *density* interchangeably.

← Bad style!

1.3 Markov chain Monte Carlo

In this section, we introduce MCMC as a method for evaluating expressions of the form of (1.1). We begin by describing its constituent parts: Monte Carlo integration and Markov chains. We then describe the general form of MCMC given by the Metropolis–Hastings algorithm, and a special case: the Gibbs sampler.

generated by Monte Carlo

1.3.1 Monte Carlo integration

Monte Carlo integration evaluates $E[f(X)]$ by drawing samples $\{X_t, t = 1, \ldots, n\}$ from $\pi(.)$ and then approximating

$$E[f(X)] \approx \frac{1}{n}\sum_{t=1}^{n} f(X_t).$$

So the population mean of $f(X)$ is estimated by a sample mean. When the samples $\{X_t\}$ are independent, laws of large numbers ensure that the approximation can be made as accurate as desired by increasing the sample size n. Note that here n is under the control of the analyst: it is not the size of a fixed data sample.

In general, drawing samples $\{X_t\}$ independently from $\pi(.)$ is not feasible, since $\pi(.)$ can be quite non-standard. However the $\{X_t\}$ need not necessarily be independent. The $\{X_t\}$ can be generated by any process which, loosely speaking, draws samples throughout the support of $\pi(.)$ in the correct proportions. One way of doing this is through a Markov chain having $\pi(.)$ as its stationary distribution. This is then *Markov chain* Monte Carlo.

1.3.2 Markov chains

Suppose we generate a sequence of random variables, $\{X_0, X_1, X_2, \ldots\}$, such that at each time $t \geq 0$, the next state X_{t+1} is sampled from a distribution $P(X_{t+1}|X_t)$ which depends only on the current state of the chain, X_t. That is, *given* X_t, the next state X_{t+1} does not depend further on the history of the chain $\{X_0, X_1, \ldots, X_{t-1}\}$. This sequence is called a *Markov chain*, and $P(.|.)$ is called the *transition kernel* of the chain. We will assume that the chain is time-homogenous: that is, $P(.|.)$ does not depend on t.

How does the starting state X_0 affect X_t? This question concerns the distribution of X_t given X_0, which we denote $P^{(t)}(X_t|X_0)$. Here we are not given the intervening variables $\{X_1, X_2, \ldots, X_{t-1}\}$, so X_t depends directly on X_0. Subject to regularity conditions, the chain will gradually 'forget' its initial state and $P^{(t)}(.|X_0)$ will eventually converge to a unique *stationary* (or *invariant*) distribution, which does not depend on t or X_0. For the moment, we denote the stationary distribution by $\phi(.)$. Thus as t increases, the sampled points $\{X_t\}$ will look increasingly like dependent samples from $\phi(.)$. This is illustrated in Figure 1.1, where $\phi(.)$ is univariate standard normal. Note that convergence is much quicker in Figure 1.1(a) than in Figures 1.1(b) or 1.1(c).

Thus, after a sufficiently long *burn-in* of say m iterations, points $\{X_t; t = m+1, \ldots, n\}$ will be dependent samples approximately from $\phi(.)$. We discuss methods for determining m in Section 1.4.6. We can now use the output from the Markov chain to estimate the expectation $E[f(X)]$, where X has distribution $\phi(.)$. Burn-in samples are usually discarded for this calculation, giving an estimator

$$\overline{f} = \frac{1}{n-m} \sum_{t=m+1}^{n} f(X_t). \tag{1.2}$$

This is called an *ergodic average*. Convergence to the required expectation is ensured by the ergodic theorem.

See Roberts (1995) and Tierney (1995) in this volume for more technical discussion of several of the issues raised here.

1.3.3 The Metropolis–Hastings algorithm

Equation (1.2) shows how a Markov chain can be used to estimate $E[f(X)]$, where the expectation is taken over its stationary distribution $\phi(.)$. This would seem to provide the solution to our problem, but first we need to discover how to construct a Markov chain such that its stationary distribution $\phi(.)$ is precisely our distribution of interest $\pi(.)$.

Constructing such a Markov chain is surprisingly easy. We describe the form due to Hastings (1970), which is a generalization of the method

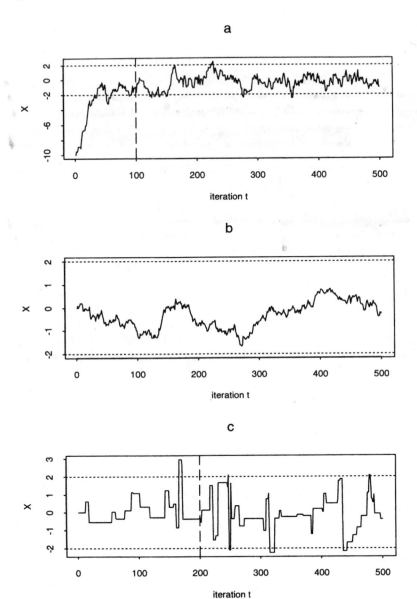

Figure 1.1 *500 iterations from Metropolis algorithms with stationary distribution*
$N(0,1)$ *and proposal distributions (a)* $q(.|X) = N(X, 0.5)$; *(b)* $q(.|X) = N(X, 0.1)$;
and (c) $q(.|X) = N(X, 10.0)$. *The burn-in is taken to be to the left of the vertical
broken line.*

first proposed by Metropolis *et al.* (1953). For the *Metropolis–Hastings* (or *Hastings–Metropolis*) algorithm, at each time t, the next state X_{t+1} is chosen by first sampling a *candidate* point Y from a *proposal* distribution $q(.|X_t)$. Note that the proposal distribution may depend on the current point X_t. For example, $q(.|X)$ might be a multivariate normal distribution with mean X and a fixed covariance matrix. The candidate point Y is then *accepted* with probability $\alpha(X_t, Y)$ where

$$\alpha(X, Y) = \min\left(1, \frac{\pi(Y)q(X|Y)}{\pi(X)q(Y|X)}\right). \qquad (1.3)$$

If the candidate point is accepted, the next state becomes $X_{t+1} = Y$. If the candidate is rejected, the chain does not move, i.e. $X_{t+1} = X_t$. Figure 1.1 illustrates this for univariate normal proposal and target distributions; Figure 1.1(c) showing many instances where the chain did not move for several iterations.

Thus the Metropolis–Hastings algorithm is extremely simple:

```
Initialize X_0; set t = 0.
Repeat {
    Sample a point Y from q(.|X_t)
    Sample a Uniform(0,1) random variable U
    If U ≤ α(X_t, Y) set X_{t+1} = Y
        otherwise set X_{t+1} = X_t
    Increment t
}.
```

Remarkably, the proposal distribution $q(.|.)$ can have any form and the stationary distribution of the chain will be $\pi(.)$. (For regularity conditions see Roberts, 1995: this volume.) This can be seen from the following argument. The transition kernel for the Metropolis–Hastings algorithm is

$$P(X_{t+1}|X_t) = q(X_{t+1}|X_t)\alpha(X_t, X_{t+1})$$
$$+I(X_{t+1} = X_t)[1 - \int q(Y|X_t)\alpha(X_t, Y)dY], \qquad (1.4)$$

where $I(.)$ denotes the indicator function (taking the value 1 when its argument is true, and 0 otherwise). The first term in (1.4) arises from acceptance of a candidate $Y = X_{t+1}$, and the second term arises from rejection, for all possible candidates Y. Using the fact that

$$\pi(X_t)q(X_{t+1}|X_t)\alpha(X_t, X_{t+1}) = \pi(X_{t+1})q(X_t|X_{t+1})\alpha(X_{t+1}, X_t)$$

which follows from (1.3), we obtain the *detailed balance* equation:

$$\pi(X_t)P(X_{t+1}|X_t) = \pi(X_{t+1})P(X_t|X_{t+1}). \qquad (1.5)$$

Integrating both sides of (1.5) with respect to X_t gives:

$$\int \pi(X_t)P(X_{t+1}|X_t)dX_t = \pi(X_{t+1}). \qquad (1.6)$$

The left-hand side of equation (1.6) gives the marginal distribution of X_{t+1} under the assumption that X_t is from $\pi(.)$. Therefore (1.6) says that if X_t is from $\pi(.)$, then X_{t+1} will be also. Thus, once a sample from the stationary distribution has been obtained, all subsequent samples will be from that distribution. This only proves that the stationary distribution is $\pi(.)$, and is not a complete justification for the Metropolis–Hastings algorithm. A full justification requires a proof that $P^{(t)}(X_t|X_0)$ will converge to the stationary distribution. See Roberts (1995) and Tierney (1995) in this volume for further details.

So far we have assumed that X is a fixed-length vector of k continuous random variables. As noted in Section 1.2, there are many other possibilities, in particular X can be of *variable dimension*. For example, in a Bayesian mixture model, the number of mixture components may be variable: each component possessing its own scale and location parameters. In this situation, $\pi(.)$ must specify the joint distribution of k and X, and $q(Y|X)$ must be able to propose moves between spaces of differing dimension. Then Metropolis–Hastings is as described above, with formally the same expression (1.3) for the acceptance probability, but where dimension-matching conditions for moves between spaces of differing dimension must be carefully considered (Green, 1994a,b). See also Geyer and Møller (1993), Grenander and Miller (1994), and Phillips and Smith (1995: this volume) for MCMC methodology in variably dimensioned problems.

1.4 Implementation

There are several issues which arise when implementing MCMC. We discuss these briefly here. Further details can be found throughout this volume, and in particular in Chapters 5–8. The most immediate issue is the choice of proposal distribution $q(.|.)$.

1.4.1 Canonical forms of proposal distribution

As already noted, any proposal distribution will ultimately deliver samples from the target distribution $\pi(.)$. However, the rate of convergence to the stationary distribution will depend crucially on the relationship between $q(.|.)$ and $\pi(.)$. Moreover, having 'converged', the chain may still *mix* slowly (i.e. move slowly around the support of $\pi(.)$). These phenomena are illustrated in Figure 1.1. Figure 1.1(a) shows rapid convergence from a somewhat extreme starting value: thereafter the chain mixes rapidly. Figure 1.1(b),(c) shows slow mixing chains: these would have to be run much longer to obtain reliable estimates from (1.2), despite having been started at the mode of $\pi(.)$.

In high-dimensional problems with little symmetry, it is often necessary to perform exploratory analyses to determine roughly the shape and orientation of $\pi(.)$. This will help in constructing a proposal $q(.|.)$ which leads to rapid mixing. Progress in practice often depends on experimentation and craftmanship, although untuned canonical forms for $q(.|.)$ often work surprisingly well. For computational efficiency, $q(.|.)$ should be chosen so that it can be easily sampled and evaluated.

Here we describe some canonical forms for $q(.|.)$. Roberts (1995), Tierney (1995) and Gilks and Roberts (1995) in this volume discuss rates of convergence and strategies for choosing $q(.|.)$ in more detail.

The Metropolis Algorithm

The *Metropolis algorithm* (Metropolis *et al.*, 1953) considers only symmetric proposals, having the form $q(Y|X) = q(X|Y)$ for all X and Y. For example, when X is continuous, $q(.|X)$ might be a multivariate normal distribution with mean X and constant covariance matrix Σ. Often it is convenient to choose a proposal which generates each component of Y conditionally independently, given X_t. For the Metropolis algorithm, the acceptance probability (1.3) reduces to

$$\alpha(X, Y) = \min\left(1, \frac{\pi(Y)}{\pi(X)}\right) \tag{1.7}$$

A special case of the Metropolis algorithm is *random-walk Metropolis*, for which $q(Y|X) = q(|X - Y|)$. The data in Figure 1.1 were generated by random-walk Metropolis algorithms.

When choosing a proposal distribution, its scale (for example Σ) may need to be chosen carefully. A cautious proposal distribution generating small steps $Y - X_t$ will generally have a high acceptance rate (1.7), but will nevertheless mix slowly. This is illustrated in Figure 1.1(b). A bold proposal distribution generating large steps will often propose moves from the body to the tails of the distribution, giving small values of $\pi(Y)/\pi(X_t)$ and a low probability of acceptance. Such a chain will frequently not move, again resulting in slow mixing as illustrated in Figure 1.1(c). Ideally, the proposal distribution should be scaled to avoid both these extremes.

The independence sampler

The *independence sampler* (Tierney, 1994) is a Metropolis–Hastings algorithm whose proposal $q(Y|X) = q(Y)$ does not depend on X. For this, the acceptance probability (1.3) can be written in the form

$$\alpha(X, Y) = \min\left(1, \frac{w(Y)}{w(X)}\right), \tag{1.8}$$

where $w(X) = \pi(X)/q(X)$.

In general, the independence sampler can work very well or very badly (see Roberts, 1995: this volume). For the independence sampler to work well, $q(.)$ should be a good approximation to $\pi(.)$, but it is safest if $q(.)$ is heavier-tailed than $\pi(.)$. To see this, suppose $q(.)$ is lighter-tailed than $\pi(.)$, and that X_t is currently in the tails of $\pi(.)$. Most candidates will not be in the tails, so $w(X_t)$ will be much larger than $w(Y)$ giving a low acceptance probability (1.8). Thus heavy-tailed independence proposals help to avoid long periods stuck in the tails, at the expense of an increased overall rate of candidate rejection.

In some situations, in particular where it is thought that large-sample theory might be operating, a multivariate normal proposal might be tried, with mean at the mode of $\pi(.)$ and covariance matrix somewhat greater than the inverse Hessian matrix

$$\left[-\frac{d^2 \log \pi(x)}{dx^\mathrm{T} dx} \right]^{-1}$$

evaluated at the mode.

Single-component Metropolis–Hastings

Instead of updating the whole of X *en bloc*, it is often more convenient and computationally efficient to divide X into components $\{X_{.1}, X_{.2}, \ldots, X_{.h}\}$ of possibly differing dimension, and then update these components one by one. This was the framework for MCMC originally proposed by Metropolis *et al.* (1953), and we refer to it as *single-component Metropolis–Hastings*. Let $X_{.-i} = \{X_{.1}, \ldots, X_{.i-1}, X_{.i+1}, \ldots, X_{.h}\}$, so $X_{.-i}$ comprises all of X except $X_{.i}$.

An iteration of the single-component Metropolis–Hastings algorithm comprises h updating steps, as follows. Let $X_{t.i}$ denote the state of $X_{.i}$ at the end of iteration t. For step i of iteration $t + 1$, $X_{.i}$ is updated using Metropolis–Hastings. The candidate $Y_{.i}$ is generated from a proposal distribution $q_i(Y_{.i}|X_{t.i}, X_{t.-i})$, where $X_{t.-i}$ denotes the value of $X_{.-i}$ after completing step $i - 1$ of iteration $t + 1$:

$$X_{t.-i} = \{X_{t+1.1}, \ldots, X_{t+1.i-1}, X_{t.i+1}, \ldots, X_{t.h}\},$$

where components $1, 2, \ldots, i - 1$ have already been updated. Thus the ith proposal distribution $q_i(.|., .)$ generates a candidate only for the ith component of X, and may depend on the *current* values of any of the components of X. The candidate is accepted with probability $\alpha(X_{t.-i}, X_{t.i}, Y_{.i})$ where

$$\alpha(X_{.-i}, X_{.i}, Y_{.i}) = \min\left(1, \frac{\pi(Y_{.i}|X_{.-i})q_i(X_{.i}|Y_{.i}, X_{.-i})}{\pi(X_{.i}|X_{.-i})q_i(Y_{.i}|X_{.i}, X_{.-i})} \right). \qquad (1.9)$$

Here $\pi(X_{.i}|X_{.-i})$ is the *full conditional* distribution for $X_{.i}$ under $\pi(.)$ (see below). If $Y_{.i}$ is accepted, we set $X_{t+1.i} = Y_{.i}$; otherwise, we set $X_{t+1.i} = X_{t.i}$. The remaining components are not changed at step i.

Thus each updating step produces a move in the direction of a coordinate axis (if the candidate is accepted), as illustrated in Figure 1.2. The proposal distribution $q_i(.|.,.)$ can be chosen in any of the ways discussed earlier in this section.

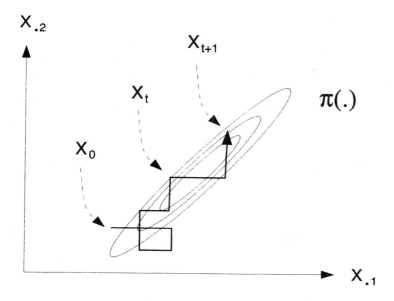

Figure 1.2 *Illustrating a single-component Metropolis–Hastings algorithm for a bivariate target distribution $\pi(.)$. Components 1 and 2 are updated alternately, producing alternate moves in horizontal and vertical directions.*

The full conditional distribution $\pi(X_{.i}|X_{.-i})$ is the distribution of the ith component of X conditioning on all the remaining components, where X has distribution $\pi(.)$:

$$\pi(X_{.i}|X_{.-i}) = \frac{\pi(X)}{\int \pi(X)dX_{.i}}. \tag{1.10}$$

Full conditional distributions play a prominent role in many of the applications in this volume, and are considered in detail by Gilks (1995: this volume). That the single-component Metropolis–Hastings algorithm with acceptance probability given by (1.9) does indeed generate samples from the target distribution $\pi(.)$ results from the fact that $\pi(.)$ is uniquely determined by the set of its full conditional distributions (Besag, 1974).

In applications, (1.9) often simplifies considerably, particularly when $\pi(.)$ derives from a conditional independence model: see Spiegelhalter *et al.* (1995) and Gilks (1995) in this volume. This provides an important computational advantage. Another important advantage of single-component updating occurs when the target distribution $\pi(.)$ is naturally specified in terms of its full conditional distributions, as commonly occurs in spatial models; see Besag (1974), Besag *et al.* (1995) and Green (1995: this volume).

Gibbs sampling

A special case of single-component Metropolis–Hastings is the *Gibbs sampler*. The Gibbs sampler was given its name by Geman and Geman (1984), who used it for analysing Gibbs distributions on lattices. However, its applicability is not limited to Gibbs distributions, so 'Gibbs sampling' is really a misnoma. Moreover, the same method was already in use in statistical physics, and was known there as the *heat bath algorithm*. Nevertheless, the work of Geman and Geman (1984) led to the introduction of MCMC into mainstream statistics via the articles by Gelfand and Smith (1990) and Gelfand *et al.* (1990). To date, most statistical applications of MCMC have used Gibbs sampling.

For the Gibbs sampler, the proposal distribution for updating the ith component of X is

$$q_i(Y_{.i}|X_{.i}, X_{.-i}) = \pi(Y_{.i}|X_{.-i}) \qquad (1.11)$$

where $\pi(Y_{.i}|X_{.-i})$ is the full conditional distribution (1.10). Substituting (1.11) into (1.9) gives an acceptance probability of 1; that is, Gibbs sampler candidates are always accepted. Thus Gibbs sampling consists purely in sampling from full conditional distributions. Methods for sampling from full conditional distributions are described in Gilks (1995: this volume).

1.4.2 Blocking

Our description of single-component samplers in Section 1.4.1 said nothing about how the components should be chosen. Typically, low-dimensional or scalar components are used. In some situations, multivariate components are natural. For example, in a Bayesian random-effects model, an entire precision matrix would usually comprise a single component. When components are highly correlated in the stationary distribution $\pi(.)$, mixing can be slow; see Gilks and Roberts (1995: this volume). Blocking highly correlated components into a higher-dimensional component may improve mixing, but this depends on the choice of proposal.

1.4.3 Updating order

In the above description of the single-component Metropolis–Hastings algorithm and Gibbs sampling, we assumed a fixed updating order for the components of X_t. Although this is usual, a fixed order is not necessary: random permutations of the updating order are quite acceptable. Moreover, not all components need be updated in each iteration. For example, we could instead update only one component per iteration, selecting component i with some fixed probability $s(i)$. A natural choice would be to set $s(i) = \frac{1}{h}$. Zeger and Karim (1991) suggest updating highly correlated components more frequently than other components, to improve mixing. Note that if $s(i)$ is allowed to depend on X_t then the acceptance probability (1.9) should be modified, otherwise the stationary distribution of the chain may no longer be the target distribution $\pi(.)$. Specifically, the acceptance probability becomes

$$\min\left(1, \frac{\pi(Y_{.i}|X_{.-i})s(i|Y_{.i}, X_{.-i})q_i(X_{.i}|Y_{.i}, X_{.-i})}{\pi(X_{.i}|X_{.-i})s(i|X_{.i}, X_{.-i})q_i(Y_{.i}|X_{.i}, X_{.-i})}\right).$$

1.4.4 Number of chains

So far we have considered running only one chain, but multiple chains are permissible. Recommendations in the literature have been conflicting, ranging from many short chains (Gelfand and Smith, 1990), to several long ones (Gelman and Rubin, 1992a,b), to one very long one (Geyer, 1992). It is now generally agreed that running many short chains, motivated by a desire to obtain independent samples from $\pi(.)$, is misguided unless there is some special reason for needing independent samples. Certainly, independent samples are not required for ergodic averaging in (1.2). The debate between the several-long-runs school and the one-very-long-run school seems set to continue. The latter maintains that one very long run has the best chance of finding new modes, and comparison between chains can never prove convergence, whilst the former maintains that comparing several seemingly converged chains might reveal genuine differences if the chains have not yet approached stationarity; see Gelman (1995: this volume). If several processors are available, running one chain on each will generally be worthwhile.

1.4.5 Starting values

Not much has been written on this topic. If the chain is irreducible, the choice of starting values X_0 will not affect the stationary distribution. A rapidly mixing chain, such as in Figure 1.1(a), will quickly find its way from extreme starting values. Starting values may need to be chosen more carefully for slow-mixing chains, to avoid a lengthy burn-in. However, it is seldom necessary to expend much effort in choosing starting values. Gelman and Rubin (1992a,b) suggest using 'over-dispersed' starting values in multiple chains, to assist in assessing convergence; see below and Gelman (1995: this volume).

1.4.6 Determining burn-in

The length of burn-in m depends on X_0, on the rate of convergence of $P^{(t)}(X_t|X_0)$ to $\pi(X_t)$ and on how similar $P^{(t)}(.|.)$ and $\pi(.)$ are required to be. Theoretically, having specified a criterion of 'similar enough', m can be determined analytically. However, this calculation is far from computationally feasible in most situations (see Roberts, 1995: this volume). Visual inspection of plots of (functions of) the Monte-Carlo output $\{X_t, t = 1, \ldots, n\}$ is the most obvious and commonly used method for determining burn-in, as in Figure 1.1. Starting the chain close to the mode of $\pi(.)$ does not remove the need for a burn-in, as the chain should still be run long enough for it to 'forget' its starting position. For example, in Figure 1.1(b) the chain has not wandered far from its starting position in 500 iterations. In this case, m should be set greater than 500.

More formal tools for determining m, called *convergence diagnostics*, have been proposed. Convergence diagnostics use a variety of theoretical methods and approximations, but all make use of the Monte Carlo output in some way. By now, at least 10 convergence diagnostics have been proposed; for a recent review, see Cowles and Carlin (1994). Some of these diagnostics are also suited to determining run length n (see below).

Convergence diagnostics can be classified by whether or not they are based on an arbitrary function $f(X)$ of the Monte Carlo output; whether they use output from a single chain or from multiple chains; and whether they can be based purely on the Monte Carlo output.

Methods which rely on monitoring $\{f(X_t), t = 1, \ldots, n\}$ (e.g. Gelman and Rubin, 1992b; Raftery and Lewis, 1992; Geweke, 1992) are easy to apply, but may be misleading since $f(X_t)$ may appear to have converged in distribution by iteration m, whilst another unmonitored function $g(X_t)$ may not have. Whatever functions $f(.)$ are monitored, there may be others which behave differently.

From a theoretical perspective, it is better to compare globally the full joint distribution $P^{(t)}(.)$ with $\pi(.)$. To avoid having to deal with $P^{(t)}(.)$ directly, several methods obtain samples from it by running multiple parallel chains (Ritter and Tanner, 1992; Roberts, 1992; Liu and Liu, 1993), and make use of the transition kernel $P(.|.)$. However, for stability in the procedures, it may be necessary to run many parallel chains. When convergence is slow, this is a serious practical limitation.

Running parallel chains obviously increases the computational burden, but can be useful, even informally, to diagnose slow convergence. For example, several parallel chains might individually appear to have converged, but comparisons between them may reveal marked differences in the apparent stationary distributions (Gelman and Rubin, 1992a).

From a practical perspective, methods which are based purely on the Monte Carlo output are particularly convenient, allowing assessment of convergence without recourse to the transition kernel $P(.|.)$, and hence without model-specific coding.

This volume does not contain a review of convergence diagnostics. This is still an active area of research, and much remains to be learnt about the behaviour of existing methods in real applications, particularly in high dimensions and when convergence is slow. Instead, the chapters by Raftery and Lewis (1995) and Gelman (1995) in this volume contain descriptions of two of the most popular methods. Both methods monitor an arbitrary function $f(.)$, and are based purely on the Monte Carlo output. The former uses a single chain and the latter multiple chains.

Geyer (1992) suggests that calculation of the length of burn-in is unnecessary, as it is likely to be less than 1% of the total length of a run sufficiently long to obtain adequate precision in the estimator $\overline{f}$ in (1.2), (see below). If extreme starting values are avoided, Geyer suggests setting m to between 1% and 2% of the run length n.

1.4.7 Determining stopping time

Deciding when to stop the chain is an important practical matter. The aim is to run the chain long enough to obtain adequate precision in the estimator $\overline{f}$ in (1.2). Estimation of the variance of $\overline{f}$ (called the *Monte Carlo* variance) is complicated by lack of independence in the iterates $\{X_t\}$.

The most obvious informal method for determining run length n is to run several chains in parallel, with different starting values, and compare the estimates $\overline{f}$ from (1.2). If they do not agree adequately, n must be increased. More formal methods which aim to estimate the variance of $\overline{f}$ have been proposed: see Roberts (1995) and Raftery and Lewis (1995) in this volume for further details.

1.4.8 Output analysis

In Bayesian inference, it is usual to summarize the posterior distribution $\pi(.)$ in terms of means, standard deviations, correlations, credible intervals and marginal distributions for components $X_{.i}$ of interest. Means, standard deviations and correlations can all be estimated by their sample equivalents in the Monte Carlo output $\{X_{t.i}, t = m+1, \ldots, n\}$, according to (1.2). For example, the marginal mean and variance of $X_{.i}$ are estimated by

$$\overline{X}_{.i} = \frac{1}{n-m} \sum_{t=m+1}^{n} X_{t.i}$$

and

$$S_{.i}^2 = \frac{1}{n-m-1} \sum_{t=m+1}^{n} (X_{t.i} - \overline{X}_{.i})^2.$$

Note that these estimates simply ignore other components in the Monte Carlo output.

A $100(1 - 2p)\%$ credible interval $[c_p, c_{1-p}]$ for a scalar component $X_{.i}$ can be estimated by setting c_p equal to the pth quantile of $\{X_{t.i}, t = m+1, \ldots, n\}$, and c_{1-p} equal to the $(1 - p)^{th}$ quantile. Besag *et al.* (1995) give a procedure for calculating rectangular credible regions in two or more dimensions.

Marginal distributions can be estimated by kernel density estimation. For the marginal distribution of $X_{.i}$, this is

$$\pi(X_{.i}) \approx \frac{1}{n - m} \sum_{t=m+1}^{n} K(X_{.i}|X_t),$$

where $K(.|X_t)$ is a density concentrated around $X_{t.i}$. A natural choice for $K(X_{.i}|X_t)$ is the full conditional distribution $\pi(X_{.i}|X_{t.-i})$. Gelfand and Smith (1990) use this construction to estimate expectations under $\pi(.)$. Thus their *Rao–Blackwellized* estimator of $E[f(X_{.i})]$ is

$$\overline{f}_{RB} = \frac{1}{n - m} \sum_{t=m+1}^{n} E[f(X_{.i})|X_{t.-i}], \tag{1.12}$$

where the expectation is with respect to the full conditional $\pi(X_{.i}|X_{t.-i})$. With reasonably long runs, the improvement from using (1.12) instead of (1.2) is usually slight, and in any case (1.12) requires a closed form for the full conditional expectation.

1.5 Discussion

This chapter provides a brief introduction to MCMC. We hope we have convinced readers that MCMC is a simple idea with enormous potential. The following chapters fill out many of the ideas sketched here, and in particular give some indication of where the methods work well and where they need some tuning or further development.

MCMC methodology and Bayesian estimation go together naturally, as many of the chapters in this volume testify. However, Bayesian model validation is still a difficult area. Some techniques for Bayesian model validation using MCMC are described in Chapters 9–13.

The philosophical debate between Bayesians and non-Bayesians has continued for decades and has largely been sterile from a practical perspective. For many applied statisticians, the most persuasive argument is the availability of robust methods and software. For many years, Bayesians had difficulty solving problems which were straightforward for non-Bayesians, so it is not surprising that most applied statisticians today are non-Bayesian. With the arrival of MCMC and related software, notably the Gibbs sampling program **BUGS** (see Spiegelhalter *et al.*, 1995: this volume), we hope more applied statisticians will become familiar and comfortable with Bayesian ideas, and apply them.

References

Besag, J. (1974) Spatial interaction and the statistical analysis of lattice systems (with discussion). *J. R. Statist. Soc.* B, **36**, 192–236.

Besag, J., Green, P., Higdon, D. and Mengersen, K. (1995) Bayesian computation and stochastic systems. *Statist. Sci.* (in press).

Cowles, M. K. and Carlin, B. P. (1994) Markov chain Monte Carlo convergence diagnostics: a comparative review. *Technical Report 94-008*, Division of Biostatistics, School of Public Health, University of Minnesota.

Diebolt, J. and Ip, E. H. S. (1995) Stochastic EM: methods and application. In *Markov Chain Monte Carlo in Practice* (eds W. R. Gilks, S. Richardson and D. J. Spiegelhalter), pp. 259–273. London: Chapman & Hall.

Gelfand, A. E. and Smith, A. F. M. (1990) Sampling-based approaches to calculating marginal densities. *J. Am. Statist. Ass.*, **85**, 398–409.

Gelfand, A. E., Hills, S. E., Racine-Poon, A. and Smith, A. F. M. (1990) Illustration of Bayesian inference in normal data models using Gibbs sampling. *J. Am. Statist. Ass.*, **85**, 972–985.

Gelman, A. (1995) Inference and monitoring convergence. In *Markov Chain Monte Carlo in Practice* (eds W. R. Gilks, S. Richardson and D. J. Spiegelhalter), pp. 131–143. London: Chapman & Hall.

Gelman, A. and Rubin, D. B. (1992a) A single series from the Gibbs sampler provides a false sense of security. In *Bayesian Statistics 4* (eds J. M. Bernardo, J. Berger, A. P. Dawid and A. F. M. Smith), pp. 625–631. Oxford: Oxford University Press.

Gelman, A. and Rubin, D. B. (1992b) Inference from iterative simulation using multiple sequences. *Statist. Sci.*, **7**, 457–472.

Geman, S. and Geman, D. (1984) Stochastic relaxation, Gibbs distributions and the Bayesian restoration of images. *IEEE Trans. Pattn. Anal. Mach. Intel.*, **6**, 721–741.

Geweke, J. (1992) Evaluating the accuracy of sampling-based approaches to the calculation of posterior moments. In *Bayesian Statistics 4* (eds J. M. Bernardo, J. Berger, A. P. Dawid and A. F. M. Smith), pp. 169–193. Oxford: Oxford University Press.

Geyer, C. J. (1992) Practical Markov chain Monte Carlo. *Statist. Sci.*, **7**, 473–511.

Geyer, C. J. (1995) Estimation and optimization of functions. In *Markov Chain Monte Carlo in Practice* (eds W. R. Gilks, S. Richardson and D. J. Spiegelhalter), pp. 241–258. London: Chapman & Hall.

Geyer, C. J. and Møller, J. (1993) Simulation procedures and likelihood inference for spatial point processes. *Technical Report*, University of Aarhus.

Gilks, W. R. (1995) Full conditional distributions. In *Markov Chain Monte Carlo in Practice* (eds W. R. Gilks, S. Richardson and D. J. Spiegelhalter), pp. 75–88. London: Chapman & Hall.

Gilks, W. R. and Roberts, G. O. (1995) Strategies for improving MCMC. In *Markov Chain Monte Carlo in Practice* (eds W. R. Gilks, S. Richardson and D. J. Spiegelhalter), pp. 89–114. London: Chapman & Hall.

Green, P. J. (1994a) Discussion on Representations of knowledge in complex systems (by U. Grenander and M. I. Miller). *J. R. Statist. Soc.* B, **56**, 589–590.

Green, P. J. (1994b) Reversible jump MCMC computation and Bayesian model determination. *Technical Report*, Department of Mathematics, University of Bristol.

Green, P. J. (1995) MCMC in image analysis. In *Markov Chain Monte Carlo in Practice* (eds W. R. Gilks, S. Richardson and D. J. Spiegelhalter), pp. 381–399. London: Chapman & Hall.

Grenander, U. and Miller, M. I. (1994) Representations of knowledge in complex systems. *J. R. Statist. Soc.* B, **56**, 549–603.

Hastings, W. K. (1970) Monte Carlo sampling methods using Markov chains and their applications. *Biometrika*, **57**, 97–109.

Kass, R. E., Tierney, L. and Kadane, J. B. (1988) Asymptotics in Bayesian computation (with discussion). In *Bayesian Statistics 3* (eds J. M. Bernardo, M. H. DeGroot, D. V. Lindley and A. F. M. Smith), pp. 261–278. Oxford: Oxford University Press.

Liu, C. and Liu, J. (1993) Discussion on the meeting on the Gibbs sampler and other Markov chain Monte Carlo methods. *J. R. Statist. Soc.* B, **55**, 82–83.

Metropolis, N., Rosenbluth, A. W., Rosenbluth, M. N., Teller, A. H. and Teller, E. (1953) Equations of state calculations by fast computing machine. *J. Chem. Phys.*, **21**, 1087–1091.

Phillips, D. B. and Smith, A. F. M. (1995) Bayesian model comparison via jump diffusions. In *Markov Chain Monte Carlo in Practice* (eds W. R. Gilks, S. Richardson and D. J. Spiegelhalter), pp. 215–239. London: Chapman & Hall.

Raftery, A. E. and Lewis, S. M. (1992) How many iterations of the Gibbs sampler? In *Bayesian Statistics 4* (eds J. M. Bernardo, J. Berger, A. P. Dawid and A. F. M. Smith), pp. 641–649. Oxford: Oxford University Press.

Raftery, A. E. and Lewis, S. M. (1995) Implementing MCMC. In *Markov Chain Monte Carlo in Practice* (eds W. R. Gilks, S. Richardson and D. J. Spiegelhalter), pp. 115–130. London: Chapman & Hall.

Ritter, C. and Tanner, M. A. (1992) Facilitating the Gibbs sampler: the Gibbs stopper and the Griddy-Gibbs sampler. *J. Am. Statist. Ass.*, **87**, 861–868.

Roberts, G. O. (1992) Convergence diagnostics of the Gibbs sampler. In *Bayesian Statistics 4* (eds J. M. Bernardo, J. Berger, A. P. Dawid and A. F. M. Smith), pp. 775–782. Oxford: Oxford University Press.

Roberts, G. O. (1995) Markov chain concepts related to samping algorithms. In *Markov Chain Monte Carlo in Practice* (eds W. R. Gilks, S. Richardson and D. J. Spiegelhalter), pp. 45–57. London: Chapman & Hall.

Spiegelhalter, D. J., Best, N. G., Gilks, W. R. and Inskip, H. (1995) Hepatitis B: a case study in MCMC methods. In *Markov Chain Monte Carlo in Practice* (eds W. R. Gilks, S. Richardson and D. J. Spiegelhalter), pp. 21–43. London: Chapman & Hall.

Tierney, L. (1994) Markov chains for exploring posterior distributions (with discussion). *Ann. Statist.*, **22**, 1701–1762.

Tierney, L. (1995) Introduction to general state-space Markov chain theory. In *Markov Chain Monte Carlo in Practice* (eds W. R. Gilks, S. Richardson and D. J. Spiegelhalter), pp. 59–74. London: Chapman & Hall.

Zeger, S. L. and Karim, M. R. (1991) Generalized linear models with random effects: a Gibbs sampling approach. *J. Am. Statist. Ass.*, **86**, 79–86.

Hepatitis B: a case study in MCMC methods

David J Spiegelhalter

Nicola G Best

Walter R Gilks

Hazel Inskip

2.1 Introduction

This chapter features a worked example using the most basic of MCMC techniques, the Gibbs sampler, and serves to introduce ideas that will be developed more fully in later chapters. Our data for this exercise are serial antibody-titre measurements, obtained from Gambian infants after hepatitis B immunization.

We begin our analysis with an initial statistical model, and describe the use of the Gibbs sampler to obtain inferences from it, briefly touching upon issues of convergence, presentation of results, model checking and model criticism. We then step through some elaborations of the initial model, emphasizing the comparative ease of adding realistic complexity to the traditional, rather simplistic, statistical assumptions; in particular, we illustrate the accommodation of covariate measurement error. The Appendix contains some details of a freely available software package (BUGS, Spiegelhalter *et al.*, 1994), within which all the analyses in this chapter were carried out.

We emphasize that the analyses presented here cannot be considered the definitive approach to this or any other dataset, but merely illustrate some of the possibilities afforded by computer-intensive MCMC methods. Further details are provided in other chapters in this volume.

2.2 Hepatitis B immunization

2.2.1 Background

Hepatitis B (HB) is endemic in many parts of the world. In highly endemic areas such as West Africa, almost everyone is infected with the HB virus during childhood. About 20% of those infected, and particularly those who acquire the infection very early in life, do not completely clear the infection and go on to become chronic carriers of the virus. Such carriers are at increased risk of chronic liver disease in adult life and liver cancer is a major cause of death in this region.

The Gambian Hepatitis Intervention Study (GHIS) is a national programme of vaccination against HB, designed to reduce the incidence of HB carriage (Whittle *et al.*, 1991). The effectiveness of this programme will depend on the duration of immunity that HB vaccination affords. To study this, a cohort of vaccinated GHIS infants was followed up. Blood samples were periodically collected from each infant, and the amount of surface-antibody was measured. This measurement is called the *anti-HBs titre*, and is measured in milli-International Units (mIU). A similar study in neighbouring Senegal concluded that for all infants

$$\text{anti-HBs titre} \propto \frac{1}{t}, \tag{2.1}$$

where t denotes time since the infant's final vaccination, and the constant of proportionality may vary between infants (Coursaget *et al.*, 1991). This is equivalent to a linear relationship between log titre and log time:

$$y = \alpha_i - 1 \times \log t, \tag{2.2}$$

where y denotes log anti-HBs titre and α_i is constant after the final dose of vaccine for each infant i.

Here we analyse the GHIS data to validate the findings of Coursaget *et al.* (1991). In particular, we investigate the plausibility of individuals having a common gradient of minus 1, as in (2.2). This relationship, if true, would provide a simple tool for predicting individual protection against HB, via (2.1).

2.2.2 Preliminary analysis

Figure 2.1 shows the raw data, plotted on a log-log scale, for a subset of 106 infants from the GHIS follow-up study. These infants each had a *baseline* anti-HBs titre measurement taken at the time of the final vaccination, and at least two titre measurements taken subsequently. For these infants, a total of 288 post-baseline measurements were made (30 infants with two measurements, 76 with three) at approximately six-monthly intervals after final vaccination.

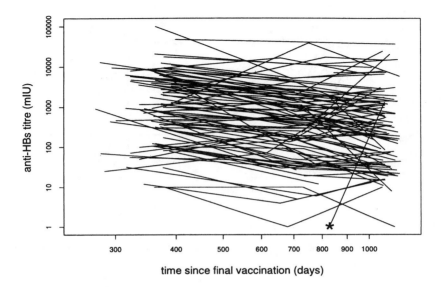

Figure 2.1 *Raw data for a subset of 106 GHIS infants: straight lines connect anti-HBS measurements for each infant.*

Initial examination of the data in Figure 2.1 suggests that it might be reasonable to fit straight lines to the data for each infant, but that these lines should be allowed to have different intercepts and possibly different gradients. Of particular note is the infant labelled with a '*' in Figure 2.1, whose titre apparently rose from 1 mIU at 826 days to 1329 mIU at day 1077. This somewhat atypical behaviour could be thought of as an outlier with respect to the change of titre over time, i.e. an outlying gradient; or due to one or both of the measurements being subject to extraneous error, i.e. outlying observations.

As a preliminary exploratory analysis, for each infant in Figure 2.1 we fitted a straight line:

$$\mu_{ij} = E[y_{ij}] = \alpha_i + \beta_i (\log t_{ij} - \log 730),\qquad(2.3)$$

where E denotes expectation and subscripts ij index the jth post-baseline observation for infant i. We standardized $\log t$ around $\log 730$ for numerical stability; thus the intercept α_i represents estimated log titre at two years post-baseline. The regressions were performed independently for each infant using ordinary least squares, and the results are shown in Figure 2.2.

The distribution of the 106 estimated intercepts $\{\hat{\alpha}_i\}$ in Figure 2.2 appears reasonably Gaussian apart from the single negative value associated

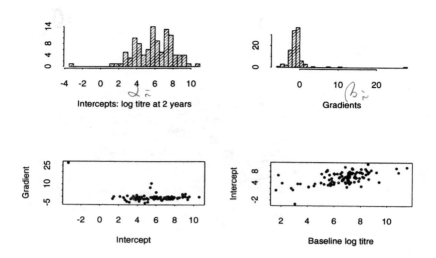

Figure 2.2 *Results of independently fitting straight lines to the data for each of the infants in Figure 2.1.*

with infant '∗' mentioned above. The distribution of the estimated gradients $\{\hat{\beta}_i\}$ also appears Gaussian apart from a few high estimates, particularly that for infant '∗'. Thirteen (12%) of the infants have a positive estimated gradient, while four (4%) have a 'high' estimated gradient greater than 2.0. Plotting estimated intercepts against gradients suggests independence of α_i and β_i, apart from the clear outlier for infant '∗'. This analysis did not explicitly take account of baseline log titre, y_{i0}: the final plot in Figure 2.2 suggests a positive relationship between y_{i0} and α_i, indicating that a high baseline titre predisposes towards high subsequent titres.

Our primary interest is in the population from which these 106 infants were drawn, rather than in the 106 infants themselves. Independently applying the linear regression model (2.3) to each infant does not provide a basis for inference about the population; for this, we must build into our model assumptions about the underlying population distribution of α_i and β_i. Thus we are concerned with 'random-effects growth-curve' models. If we are willing to make certain simplifying assumptions and asymptotic approximations, then a variety of techniques are available for fitting such models, such as restricted maximum likelihood or penalized quasi-likelihood (Breslow and Clayton, 1993). Alternatively, we can take the more general approach of simulating 'exact' solutions, where the accuracy of the solution depends only the computational care taken. This volume is concerned with the latter approach.

2.3 Modelling

This section identifies three distinct components in the construction of a full probability model, and applies them in the analysis of the GHIS data:

- Specification of model quantities and their qualitative conditional independence structure: we and several other authors in this volume find it convenient to use a *graphical* representation at this stage.

- Specification of the parametric form of the direct relationships between these quantities: this provides the likelihood terms in the model. Each of these terms may have a standard form but, by connecting them together according to the specified conditional-independence structure, models of arbitrary complexity may be constructed.

- Specification of prior distributions for parameters: see Gilks *et al.* (1995: this volume) for a brief introduction to Bayesian inference.

2.3.1 Structural modelling

We make the following minimal structural assumptions based on the exploratory analysis above. The y_{ij} are independent conditional on their mean μ_{ij} and on a parameter σ that governs the sampling error. For an individual i, each mean lies on a 'growth curve' such that μ_{ij} is a deterministic function of time t_{ij} and of intercept and gradient parameters α_i and β_i. The α_i are independently drawn from a distribution parameterized by α_0 and σ_α, while the β_i are independently drawn from a distribution parameterized by β_0 and σ_β.

Figure 2.3 shows a *directed acyclic graph* (DAG) representing these assumptions (*directed* because each link between nodes is an arrow; *acyclic* because, by following the directions of the arrows, it is impossible to return to a node after leaving it). Each quantity in the model appears as a node in the graph, and directed links correspond to direct dependencies as specified above: solid arrows are probabilistic dependencies, while dashed arrows show functional (deterministic) relationships. The latter are included to simplify the graph but are collapsed over when identifying probabilistic relationships. Repetitive structures, of blood-samples within infants for example, are shown as stacked 'sheets'. There is no essential difference between any node in the graph in that each is considered a random quantity, but it is convenient to use some graphical notation: here we use a double rectangle to denote quantities assumed fixed by the design (i.e. sampling times t_{ij}), single rectangles to indicate observed data, and circles to represent all unknown quantities.

To interpret the graph, it will help to introduce some fairly self-explanatory definitions. Let v be a node in the graph, and V be the set of all nodes. We define a 'parent' of v to be any node with an arrow emanating from it pointing to v, and a 'descendant' of v to be any node on a directed path

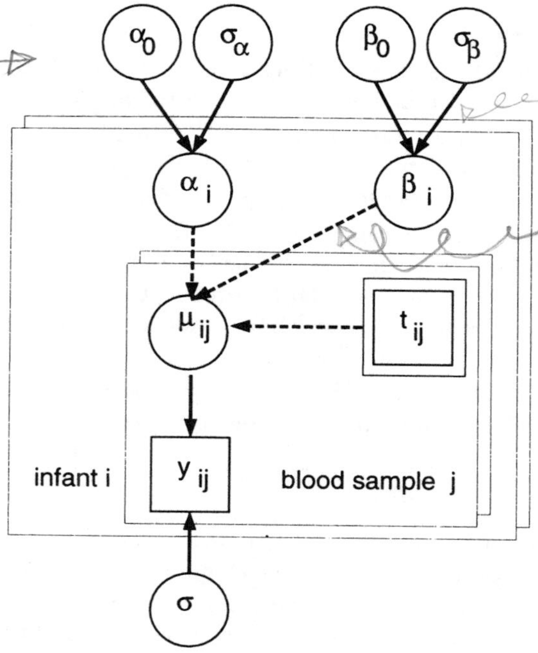

Prior distribution

free Probabilistic dependency

free Deterministic dependency

Figure 2.3 *Graphical model for hepatitis B data.*

starting from v. In identifying parents and descendants, deterministic links are collapsed so that, for example, the parents of y_{ij} are α_i, β_i and σ. The graph represents the following formal assumption: for any node v, if we know the value of its parents, then no other nodes would be informative concerning v except descendants of v. The genetic analogy is clear: if we know your parents' genetic structure, then no other individual will give any additional information concerning your genes except one of your descendants. Thomas and Gauderman (1995: this volume) illustrate the use of graphical models in genetics.

Although no probabilistic model has yet been specified, the conditional independencies expressed by the above assumptions permit many properties of the model to be derived; see for example Lauritzen *et al.* (1990), Whittaker (1990) or Spiegelhalter *et al.* (1993) for discussion of how to read off independence properties from DAGs. It is important to understand that the graph represents properties of the full model before any data is observed, and the independence properties will change when conditioning on data. For example, although nodes that have no common 'ancestor' will be initially marginally independent, such as α_i and β_i, this independence will not necessarily be retained when conditioning upon other quantities. For example, when y_{i1}, y_{i2}, y_{i3} are observed, dependence between α_i and β_i may be induced.

Our use of a graph in this example is primarily to facilitate communication of the essentials of the model without needing algebra. However, as we now show, it forms a convenient basis for the specification of the full joint distribution of all model quantities.

2.3.2 Probability modelling

The preceding discussion of graphical models has been in terms of conditional independence properties without necessarily a probabilistic interpretation. If we wish to construct a full probability model, it can be shown (Lauritzen *et al.*, 1990) that a DAG model is equivalent to assuming that the joint distribution of all the random quantities is fully specified in terms of the conditional distribution of each node given its parents:

$$P(V) = \prod_{v \in V} P(v \mid \text{parents}[v]), \qquad (2.4)$$

where $P(.)$ denotes a probability distribution. This factorization not only allows extremely complex models to be built up from local components, but also provides an efficient basis for the implementation of some forms of MCMC methods.

For our example, we therefore need to specify exact forms of 'parent–child' relationships on the graph shown in Figure 2.3. We shall make the initial assumption of normality both for within- and between-infant variability, although this will be relaxed in later sections. We shall also assume a simple linear relationship between expected log titre and log time, as in (2.3). The likelihood terms in the model are therefore

$$y_{ij} \sim \text{N}(\mu_{ij}, \sigma^2), \qquad (2.5)$$
$$\mu_{ij} = \alpha_i + \beta_i(\log t_{ij} - \log 730), \qquad (2.6)$$
$$\alpha_i \sim \text{N}(\alpha_0, \sigma_\alpha^2), \qquad (2.7)$$
$$\beta_i \sim \text{N}(\beta_0, \sigma_\beta^2), \qquad (2.8)$$

where '$\sim$' means 'distributed as', and $\text{N}(a, b)$ generically denotes a normal distribution with mean a and variance b. Scaling $\log t$ around $\log 730$ makes the assumed prior independence of gradient and intercept more plausible, as suggested in Figure 2.2.

2.3.3 Prior distributions

To complete the specification of a full probability model, we require prior distributions on the nodes without parents: σ^2, α_0, σ_α^2, β_0 and σ_β^2. These nodes are known as 'founders' in genetics. In a scientific context, we would often like these priors to be not too influential in the final conclusions, although if there is only weak evidence from the data concerning some secondary aspects of a model, such as the degree of smoothness to be expected in a set of adjacent observations, it may be very useful to be able to include external information in the form of fairly informative prior distributions. In hierarchical models such as ours, it is particularly important to avoid casual use of standard improper priors since these may result in improper posterior distributions (DuMouchel and Waternaux, 1992); see also Clayton (1995) and Carlin (1995) in this volume.

The priors chosen for our analysis are

$$\alpha_0, \ \beta_0 \ \sim \ N(0, \ 10\,000), \qquad (2.9)$$

$$\sigma^{-2}, \ \sigma_\alpha^{-2}, \ \sigma_\beta^{-2} \ \sim \ Ga(0.01, \ 0.01), \qquad (2.10)$$

where $Ga(a, b)$ generically denotes a gamma distribution with mean a/b and variance a/b^2. Although these are proper probability distributions, we might expect them to have minimal effect on the analysis since α_0 and β_0 have standard deviation 100, and the inverse of the variance components (the precisions) all have prior standard deviation 10. Examination of the final results shows these prior standard deviations are at least an order of magnitude greater than the corresponding posterior standard deviations.

2.4 Fitting a model using Gibbs sampling

We estimate our model by Gibbs sampling using the BUGS software (Gilks *et al.*, 1994; Spiegelhalter *et al.*, 1995). See Gilks *et al.* (1995: this volume) for a description of Gibbs sampling.

In general, four steps are required to implement Gibbs sampling:

- starting values must be provided for all unobserved nodes (parameters and any missing data);

- full conditional distributions for each unobserved node must be constructed and methods for sampling from them decided upon;

- the output must be monitored to decide on the length of the 'burn-in' and the total run length, or perhaps to identify whether a more computationally efficient parameterization or MCMC algorithm is required;

- summary statistics for quantities of interest must be calculated from the output, for inference about the true values of the unobserved nodes.

For a satisfactory implementation of Gibbs sampling, a fifth step should also be added: to examine summary statistics for evidence of lack of fit of the model.

We now discuss each of these steps briefly; further details are provided elsewhere in this volume.

2.4.1 Initialization

In principle, the choice of starting values is unimportant since the Gibbs sampler (or any other MCMC sampler) should be run long enough for it to 'forget' its initial states. It is useful to perform a number of runs with widely dispersed starting values, to check that the conclusions are not sensitive to the choice of starting values (Gelman, 1995: this volume). However, very extreme starting values could lead to a very long burn-in (Raftery, 1995: this volume). In severe cases, the sampler may fail to converge towards the main support of the posterior distribution, this possibility being aggravated by numerical instability in the extreme tails of the posterior. On the other hand, starting the simulation at the mode of the posterior is no guarantee of success if the sampler is not mixing well, i.e. if it is not moving fluidly around the support of the posterior.

We performed three runs with starting values shown in Table 2.1. The first run starts at values considered plausible in the light of Figure 2.2, while the second and third represent substantial deviations in initial values. In particular, run 2 is intended to represent a situation in which there is low measurement error but large between-individual variability, while run 3 represents very similar individuals with very high measurement error.

From these parameters, initial values for for α_i and β_i were independently generated from (2.7) and (2.8). Such 'forwards sampling' is the default strategy in the BUGS software.

Parameter	Run 1	Run 2	Run 3
α_0	5.0	20.0	-10.00
β_0	-1.0	-5.0	5.00
σ_α	2.0	20.0	0.20
σ_β	0.5	5.0	0.05
σ	1.0	0.1	10.00

Table 2.1 *Starting values for parameters in three runs of the Gibbs sampler*

2.4.2 Sampling from full conditional distributions

Gibbs sampling works by iteratively drawing samples from the full conditional distributions of unobserved nodes in the graph. The full conditional distribution for a node is the distribution of that node given current or known values for all the other nodes in the graph. For a directed graphical model, we can exploit the structure of the joint distribution given in (2.4). For any node v, we may denote the remaining nodes by V_{-v}, and from (2.4) it follows that the full conditional distribution $P(v|V_{-v})$ has the form

$$
\begin{aligned}
P(v \mid V_{-v}) \quad &\propto \quad P(v, V_{-v}) \\
&\propto \quad \text{terms in } P(V) \text{ containing } v \\
&= \quad P(v \mid \text{parents}[v]) \times \\
&\qquad \prod_{w \in children[v]} P(w \mid \text{parents}[w]), \qquad (2.11)
\end{aligned}
$$

where $\propto$ means 'proportional to'. (The proportionality constant, which ensures that the distribution integrates to 1, will in general be a function of the remaining nodes V_{-v}.) We see from (2.11) that the full conditional distribution for v contains a *prior* component $P(v \mid \text{parents}[v])$ and *likelihood* components arising from each child of v. Thus the full conditional for any node depends only on the values of its parents, children, and co-parents, where 'co-parents' are other parents of the children of v.

For example, consider the intercept term α_i. The general prescription of (2.11) tells us that the full conditional distribution for α_i is proportional to the product of the prior for α_i, given by (2.7), and n_i likelihood terms, given by (2.5, 2.6), where n_i is the number of observations on the ith infant. Thus

$$
\begin{aligned}
P(\alpha_i \mid \cdot) \quad &\propto \quad \exp\left\{-\frac{(\alpha_i - \alpha_0)^2}{2\sigma_\alpha^2}\right\} \times \\
&\qquad \prod_{j=1}^{n_i} \exp\left\{-\frac{[y_{ij} - \alpha_i - \beta_i(\log t_{ij} - \log 730)]^2}{2\sigma^2}\right\}, \quad (2.12)
\end{aligned}
$$

where the '$\cdot$' in $P(\alpha_i \mid \cdot)$ denotes all data nodes and all parameter nodes except α_i, (i.e. $V_{-\alpha_i}$). By completing the square for α_i in the exponent of (2.12), it can be shown that $P(\alpha_i \mid \cdot)$ is a normal distribution with mean

$$
\frac{\frac{\alpha_0}{\sigma_\alpha^2} + \frac{1}{\sigma^2}\sum_{j=1}^{n_i} y_{ij} - \beta_i(\log t_{ij} - \log 730)}{\frac{1}{\sigma_\alpha^2} + \frac{n_i}{\sigma^2}}
$$

and variance

$$
\frac{1}{\frac{1}{\sigma_\alpha^2} + \frac{n_i}{\sigma^2}}.
$$

The full conditionals for β_i, α_0 and β_0 can similarly be shown to be normal distributions.

The full conditional distribution for the precision parameter σ_α^{-2} can also be easily worked out. Let τ_α denote σ_α^{-2}. The general prescription (2.11) tells us that the full conditional for τ_α is proportional to the product of the prior for τ_α, given by (2.10), and the 'likelihood' terms for τ_α, given by (2.7) for each i. These are the likelihood terms for τ_α because the α_i

note: $X \sim \Gamma(\alpha, \beta)$

parameters are the only children of τ_α. Thus we have

$EX = \dfrac{\alpha}{\beta}$, $D(X) = \dfrac{\alpha}{\beta^2}$

$$
\begin{aligned}
P(\tau_\alpha \mid \cdot) \;&\propto\; \tau_\alpha^{0.01-1} e^{-0.01\tau_\alpha} \prod_{i=1}^{106} \tau_\alpha^{\frac{1}{2}} \exp\left\{-\frac{1}{2}\tau_\alpha (\alpha_i - \alpha_0)^2\right\} \\
&=\; \tau_\alpha^{0.01+\frac{106}{2}-1} \exp\left\{-\tau_\alpha \left(0.01 + \frac{1}{2}\sum_{i=1}^{106}(\alpha_i - \alpha_0)^2\right)\right\} \\
&\propto\; \mathrm{Ga}\left(0.01 + \frac{106}{2},\; 0.01 + \frac{1}{2}\sum_{i=1}^{106}(\alpha_i - \alpha_0)^2\right).
\end{aligned}
$$

Thus the full conditional distribution for τ_α is another gamma distribution. The full conditional distributions for σ_β^{-2} and σ^{-2} can similarly be shown to be gamma distributions.

In this example, all full conditionals reduce to normal or gamma distributions, from which sampling is straightforward (see for example Ripley, 1987). In many applications, full conditional distributions do not simplify so conveniently. However, several techniques are available for efficiently sampling from such distributions; see Gilks (1995: this volume).

2.4.3 Monitoring the output

The values for the unknown quantities generated by the Gibbs sampler must be graphically and statistically summarized to check mixing and convergence. Here we illustrate the use of Gelman and Rubin (1992) statistics on three runs of length 5 000 iterations, started as in Table 2.1. Details of the method of Gelman and Rubin are given by Gelman (1995: this volume). Each run took around two minutes on a SPARCstation 10 using BUGS, and the runs were monitored using the suite of S-functions called CODA (Best *et al.*, 1995). Figure 2.4 shows the trace of the sampled values of β_0 for the three runs: while runs 1 and 2 quickly settled down, run 3 took around 700 iterations to stabilize.

Table 2.2 shows Gelman–Rubin statistics for four parameters being monitored (b_4 is defined below). For each parameter, the statistic estimates the reduction in the pooled estimate of its variance if the runs were continued indefinitely. The estimates and their 97.5% points are near 1, indicating that reasonable convergence has occurred for these parameters.

The Gelman–Rubin statistics can be calculated sequentially as the runs proceed, and plotted as in Figure 2.5: such a display would be a valuable tool in parallel implementation of a Gibbs sampler. These plots suggest discarding the first 1 000 iterations of each run and then pooling the remaining $3 \times 4\,000$ samples.

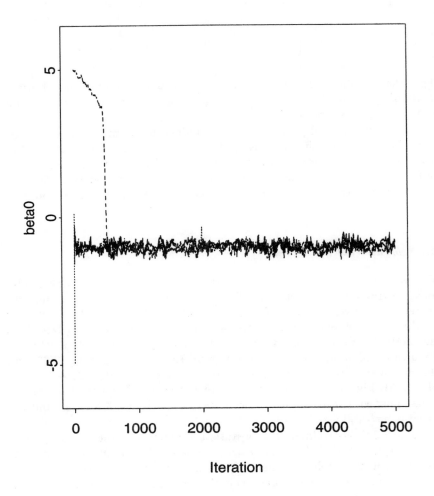

Figure 2.4 *Sampled values for β_0 from three runs of the Gibbs sampler applied to the model of Section 2.3; starting values are given in Table 2.1: run 1, solid line; run 2, dotted line; run 3, broken line.*

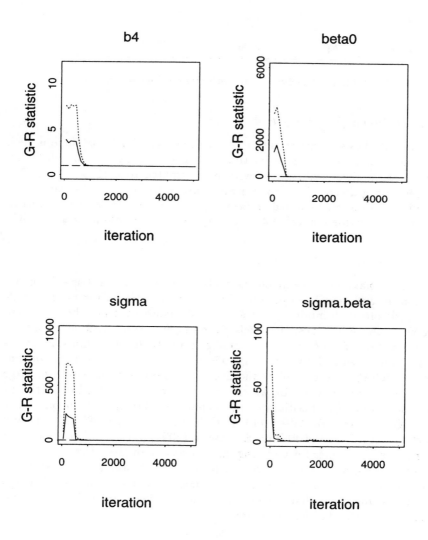

Figure 2.5 *Gelman–Rubin statistics for four parameters from three parallel runs. At each iteration the median and 97.5 centile of the statistic are calculated and plotted, based on the sampled values of the parameter up to that iteration. Solid lines: medians; broken lines: 97.5 centiles. Convergence is suggested when the plotted values closely approach 1.*

Parameter	Estimate	97.5% quantile
β_0	1.03	1.11
σ_β	1.01	1.02
σ	1.00	1.00
b_4	1.00	1.00

Table 2.2 *Gelman–Rubin statistics for four parameters*

2.4.4 Inference from the output

Figure 2.6 shows kernel density plots of the sampled values. There is clear evidence of variability between the gradients, although the absolute size of the variability is not great, with σ_β estimated to be around 0.3. The underlying gradient β_0 is concentrated around minus 1: this value is of particular interest, as noted in Section 2.2.1. Summary statistics for this model are provided in Table 2.3 (see page 37), in the column headed 'GG'.

2.4.5 Assessing goodness-of-fit

Standard maximum likelihood methods provide a natural basis for good-ness-of-fit and model comparison, since the parameter estimates are specif-ically designed to minimize measures of deviance that may be compared between alternative nested models. Cox and Solomon (1986) also consider classical tests for detecting departures from standard within-subject as-sumptions in such data, assuming independence between subjects. In con-trast, MCMC methods allow the fitting of multi-level models with large numbers of parameters for which standard asymptotic likelihood theory does not hold. Model criticism and comparison therefore require particular care. Gelfand (1995), Raftery (1995), Gelman and Meng (1995), George and McCulloch (1995) and Phillips and Smith (1995) in this volume describe a variety of techniques for assessing and improving model adequacy.

Here, we simply illustrate the manner in which standard statistics for measuring departures from an assumed model may be calculated, although we again emphasize that our example is by no means a definitive analysis. We define a standardized residual

$$r_{ij} = \frac{y_{ij} - \mu_{ij}}{\sigma}$$

with mean zero and variance 1 under the assumed error model. These residuals can be calculated at each iteration (using current values of μ_{ij} and σ), and can be used to construct summary statistics. For example, we can calculate a statistic that is intended to detect deviations from the

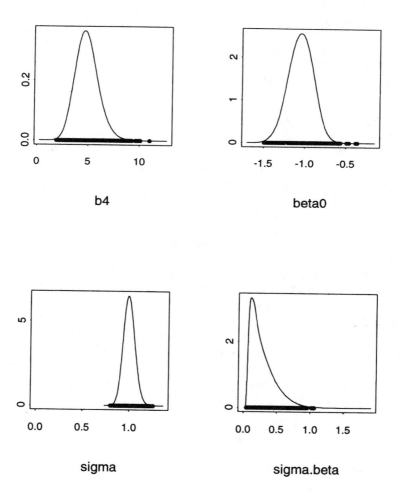

Figure 2.6 *Kernel density plots of sampled values for parameters of the model of Section 2.3 based on three pooled runs, each of 4 000 iterations after 1 000 iterations burn-in. Results are shown for β_0, the mean gradient in the population; σ_β, the standard deviation of gradients in the population; σ, the sampling error; and b_4, the standardized fourth moment of the residuals.*

assumed normal error model. Various functions of standardized residuals
could be considered, and here we calculate

$$b_4 = \frac{1}{288} \sum_{ij} r_{ij}^4,$$

the mean fourth moment of the standardized residual. If the error distribution is truly normal then this statistic should be close to 3; see Gelman
and Meng (1995: this volume) for more formal assessment of such summary
statistics. Figure 2.6 clearly shows that sampled values of b_4 are substantially greater than 3 (mean = 4.9; 95% interval from 3.2 to 7.2). This
strongly indicates that the residuals are not normally distributed.

2.5 Model elaboration

2.5.1 Heavy-tailed distributions

The data and discussion in Section 2.2 suggest we should take account
of apparent outlying observations. One approach is to use heavy-tailed
distributions, for example t distributions, instead of Gaussian distributions
for the intercepts α_i, gradients β_i and sampling errors $y_{ij} - \mu_{ij}$.

Many researchers have shown how t distributions can be easily introduced within a Gibbs sampling framework by representing the precision
(the inverse of the variance) of each Gaussian observation as itself being
a random quantity with a suitable gamma distribution; see for example
Gelfand *et al.* (1992). However, in the BUGS program, a t distribution on
ν degrees of freedom can be specified directly as a sampling distribution,
with priors being specified for its scale and location, as for the Gaussian
distributions in Section 2.3. Which value of ν should we use? In BUGS, a
prior distribution can be placed over ν so that the data can indicate the
degree of support for a heavy- or Gaussian-tailed distribution. In the results shown below, we have assumed a discrete uniform prior distribution
for ν in the set

$$\{\ 1,\ 1.5,\ 2,\ 2.5,\ \ldots,\ 20,\ 21,\ 22,\ \ldots,\ 30,\ 35,\ 40,\ 45,\ 50,\ 75,$$
$$100,\ 200,\ \ldots,\ 500,\ 750,\ 1000\ \}.$$

We fitted the following models:

GG   Gaussian sampling errors $y_{ij} - \mu_{ij}$; Gaussian intercepts α_i
and gradients β_i, as in Section 2.3;

GT   Gaussian sampling errors; t-distributed intercepts and gradients;

TG t-distributed samping errors; Gaussian intercepts and gradients;

TT t-distributed samping errors; t-distributed intercepts and gradients.

Results for these models are given in Table 2.3, each based on 5 000 iterations after a 1 000-iteration burn-in. Strong auto-correlations and cross-correlations in the parameters of the t distributions were observed, but the β_0 sequence was quite stable in each model (results not shown).

Parameter		GG	GT	TG	TT
β_0	mean	-1.05	-1.13	-1.06	-1.11
	95% c.i.	$-1.33, -0.80$	$-1.35, -0.93$	$-1.24, -0.88$	$-1.26, -0.93$
σ_β	mean	0.274	0.028	0.033	0.065
	95% c.i.	*0.070, 0.698*	*0.007, 0.084*	*0.004, 0.111*	*0.007, 0.176*
ν	mean	∞	∞	3.5	2.5
	95% c.i.			*2.5, 3.5*	*2, 3.5*
ν_α	mean	∞	12	∞	19
	95% c.i.		*4, 20*		*5, 30*
ν_β	mean	∞	1	∞	16
	95% c.i.		*1, 1*		*8.5, 26*

Table 2.3 *Results of fitting models GG, GT, TG and TT to the GHIS data: posterior means and 95% credible intervals (c.i). Parameters ν, ν_α and ν_β are the degrees of freedom in t distributions for sampling errors, intercepts and gradients, respectively. Degrees of freedom $= \infty$ corresponds to a Gaussian distribution.*

We note that the point estimate of β_0 is robust to secondary assumptions about distributional shape, although the width of the interval estimate is reduced by 35% when allowing t distributions of unknown degrees of freedom for both population and sampling distributions. Gaussian sampling errors and t distributions for regression coefficients (model GT) leads to overwhelming belief in very heavy (Cauchy) tails for the distribution of gradients, due to the outlying individuals ($\hat{\nu}_\beta \approx 1$). Allowing the sampling error alone to have heavy tails (model TG) leads to a confident judgement of a heavy-tailed sampling distribution ($\hat{\nu} \approx 3.5$), while allowing t distributions at all levels (model TT) supports the assumption of a heavy-tailed sampling distribution ($\hat{\nu} \approx 2.5$) and a fairly Gaussian shape for intercepts and gradients ($\hat{\nu}_\alpha \approx 19$, $\hat{\nu}_\beta \approx 16$).

2.5.2 Introducing a covariate

As noted in Section 2.2.2, the observed baseline log titre measurement, y_{i0}, is correlated with subsequent titres. The obvious way to adjust for this is to replace the regression equation (2.6) with

$$\mu_{ij} = \alpha_i + \gamma(y_{i0} - y_{.0}) + \beta_i(\log t_{ij} - \log 730), \qquad (2.13)$$

where $y_{.0}$ is the mean of the observations $\{y_{i0}\}$. In (2.13), the covariate y_{i0} is 'centred' by subtracting $y_{.0}$: this will help to reduce posterior correlations between γ and other parameters, and consequently to improve mixing in the Gibbs sampler. See Gilks and Roberts (1995: this volume) for further elaboration of this point.

As for all anti-HBS titre measurements, y_{i0} is subject to measurement error. We are scientifically interested in the relationship between the 'true' underlying log titres μ_{i0} and μ_{ij}, where μ_{i0} is the unobserved 'true' log titre on the ith infant at baseline. Therefore, instead of the obvious regression model (2.13), we should use the 'errors-in-variables' regression model

$$\mu_{ij} = \alpha_i + \gamma(\mu_{i0} - y_{.0}) + \beta_i(\log t_{ij} - \log 730). \qquad (2.14)$$

Information about the unknown μ_{i0} in (2.14) is provided by the measurement y_{i0}. We model this with

$$y_{i0} \sim N(\mu_{i0}, \sigma^2). \qquad (2.15)$$

Note that we have assigned the same variance σ^2 to both y_{ij} in (2.5) and y_{i0} in (2.15), because we believe that y_{ij} and y_{i0} are subject to the same sources of measurement (sampling) error.

We must also specify a prior for μ_{i0}. We choose

$$\mu_{i0} \sim N(\theta, \phi^2), \qquad (2.16)$$

where the hyperparameters θ and ϕ^{-2} are assigned vague but proper Gaussian and gamma prior distributions.

Equations (2.14–2.16) constitute a measurement-error model, as discussed further by Richardson (1995: this volume). The measurement-error model (2.14–2.16) forms one component of our complete model, which includes equations (2.5) and (2.7–2.10). The graph for the complete model is shown in Figure 2.7, and the results from fitting both this model and the simpler model with a fixed baseline (2.13) instead of (2.14–2.16), are shown in Table 2.4.

We note the expected result: the coefficient γ attached to the covariate measured with error increases dramatically when that error is properly taken into account. Indeed, the 95% credible interval for γ under the errors-in-baseline model does not contain the estimate for γ under the fixed-baseline model. The estimate for σ_β from the errors-in-baseline model suggests that population variation in gradients probably does not have a major impact on the rate of loss of antibody, and results (not shown) from an analysis of a much larger subsample of the GHIS data confirm this. Setting $\sigma_\beta = 0$, with the plausible values of $\gamma = 1$, $\beta_0 = -1$, gives the satisfyingly

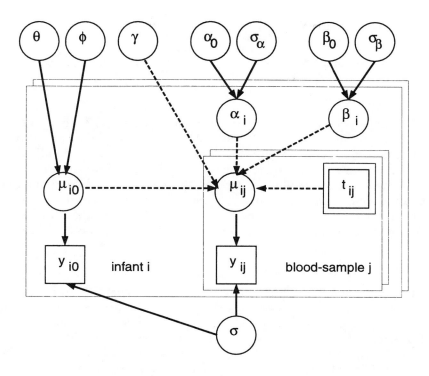

Figure 2.7 *Graphical model for the GHIS data, showing dependence on baseline titre, measured with error.*

Parameter		Fixed baseline (2.13)	Errors in baseline (2.14–2.16)
β_0	mean	−1.06	−1.08
	95% c.i.	*−1.32, −0.80*	*−1.35, −0.81*
σ_β	mean	0.31	0.24
	95% c.i.	*0.07, 0.76*	*0.07, 0.62*
γ	mean	0.68	1.04
	95% c.i.	*0.51, 0.85*	*0.76, 1.42*

Table 2.4 *Results of fitting alternative regression models to the GHIS data: posterior means and 95% credible intervals*

simple model:
$$\frac{\text{titre at time } t}{\text{titre at time } 0} \propto \frac{1}{t},$$
which is a useful elaboration of the simpler model given by (2.1).

2.6 Conclusion

We have provided a brief overview of the issues involved in applying MCMC to full probability modelling. In particular, we emphasize the possibility for constructing increasingly elaborate statistical models using 'local' associations which can be expressed graphically and which allow straightforward implementation using Gibbs sampling. However, this possibility for complex modelling brings associated dangers and difficulties; we refer the reader to other chapters in this volume for deeper discussion of issues such as convergence monitoring and improvement, model checking and model choice.

Acknowledgements

The BUGS project is supported by a grant from the UK Economic and Social Science Research Council's initiative for the Analysis of Large and Complex Datasets. The GHIS was generously funded by a grant from the Direzione Generale per la Cooperazione allo Sviluppo of the Ministry of Foreign Affairs of Italy. The GHIS was conducted at the Medical Research Council Laboratories, The Gambia, under the auspicies of the International Agency for Research on Cancer.

References

Best, N. G., Cowles, M. K. and Vines, S. K. (1995) *CODA: Convergence Diagnosis and Output Analysis software for Gibbs Sampler output: Version 0.3*. Cambridge: Medical Research Council Biostatistics Unit.

Breslow, N. E. and Clayton, D. G. (1993) Approximate inference in generalized linear mixed models. *J. Am. Statist. Ass.*, **88**, 9–25.

Carlin, B. P. (1995) Hierarchical longitudinal modelling. In *Markov Chain Monte Carlo in Practice* (eds W. R. Gilks, S. Richardson and D. J. Spiegelhalter), pp. 303–319. London: Chapman & Hall.

Clayton, D. G. (1995) Generalized linear mixed models. In *Markov Chain Monte Carlo in Practice* (eds W. R. Gilks, S. Richardson and D. J. Spiegelhalter), pp. 275–301. London: Chapman & Hall.

Coursaget, P., Yvonnet, B., Gilks, W. R., Wang, C. C., Day, N. E., Chiron, J. P. and Diop-Mar, I. (1991) Scheduling of revaccinations against Hepatitis B virus. *Lancet*, **337**, 1180–3.

Cox, D. R. and Solomon, P. J. (1986) Analysis of variability with large numbers of small samples. *Biometrika*, **73**, 543–54.

DuMouchel, W. and Waternaux, C. (1992) Discussion on hierarchical models for combining information and for meta-analyses (by C. N. Morris and S. L. Normand). In *Bayesian Statistics 4* (eds J. M. Bernardo, J. O. Berger, A. P. Dawid and A. F. M. Smith), pp. 338–341. Oxford: Oxford University Press.

Gelfand, A. E. (1995) Model determination using sampling-based methods. In *Markov Chain Monte Carlo in Practice* (eds W. R. Gilks, S. Richardson and D. J. Spiegelhalter), pp. 145–161. London: Chapman & Hall.

Gelfand, A. E., Smith, A. F. M. and Lee, T.-M. (1992) Bayesian analysis of constrained parameter and truncated data problems using Gibbs sampling. *J. Am. Statist. Ass.*, **87**, 523–32.

Gelman, A. (1995) Inference and monitoring convergence. In *Markov Chain Monte Carlo in Practice* (eds W. R. Gilks, S. Richardson and D. J. Spiegelhalter), pp. 131–143. London: Chapman & Hall.

Gelman, A. and Meng, X.-L. (1995) Model checking and model improvement. In *Markov Chain Monte Carlo in Practice* (eds W. R. Gilks, S. Richardson and D. J. Spiegelhalter), pp. 189–201. London: Chapman & Hall.

Gelman, A. and Rubin, D. B. (1992) Inference from iterative simulation using multiple sequences (with discussion). *Statist. Sci.*, **7**, 457–511.

George, E. I. and McCulloch, R. E. (1995) Stochastic search variable selection. In *Markov Chain Monte Carlo in Practice* (eds W. R. Gilks, S. Richardson and D. J. Spiegelhalter), pp. 203–214. London: Chapman & Hall.

Gilks, W. R. (1995) Full conditional distributions. In *Markov Chain Monte Carlo in Practice* (eds W. R. Gilks, S. Richardson and D. J. Spiegelhalter), pp. 75–88. London: Chapman & Hall.

Gilks, W. R. and Roberts, G. O. (1995) Strategies for improving MCMC. In *Markov Chain Monte Carlo in Practice* (eds W. R. Gilks, S. Richardson and D. J. Spiegelhalter), pp. 89–114. London: Chapman & Hall.

Gilks, W. R., Richardson, S. and Spiegelhalter, D. J. (1995) Introducing Markov chain Monte Carlo. In *Markov Chain Monte Carlo in Practice* (eds W. R. Gilks, S. Richardson and D. J. Spiegelhalter), pp. 1–19. London: Chapman & Hall.

Gilks, W. R., Thomas, A. and Spiegelhalter, D. J. (1994) A language and program for complex Bayesian modelling. *The Statistician*, **43**, 169–78.

Lauritzen, S. L., Dawid, A. P., Larsen, B. N. and Leimer, H.-G. (1990) Independence properties of directed Markov fields. *Networks*, **20**, 491–505.

Phillips, D. B. and Smith, A. F. M. (1995) Bayesian model comparison via jump diffusions. In *Markov Chain Monte Carlo in Practice* (eds W. R. Gilks, S. Richardson and D. J. Spiegelhalter), pp. 215–239. London: Chapman & Hall.

Raftery, A. E. (1995) Hypothesis testing and model selection. In *Markov Chain Monte Carlo in Practice* (eds W. R. Gilks, S. Richardson and D. J. Spiegelhalter), pp. 163–187. London: Chapman & Hall.

Richardson, S. (1995) Measurement error. In *Markov Chain Monte Carlo in Practice* (eds W. R. Gilks, S. Richardson and D. J. Spiegelhalter), pp. 401–417. London: Chapman & Hall.

Ripley, B. D. (1987) *Stochastic Simulation*. New York: Wiley.

Spiegelhalter, D. J., Dawid, A. P., Lauritzen, S. L. and Cowell, R. G. (1993) Bayesian analysis in expert systems (with discussion) *Statist. Sci.*, **8**, 219–83.

Spiegelhalter, D. J., Thomas, A. and Best, N. G. (1995) Computation on Bayesian graphical models. In *Bayesian Statistics 5*, (eds J. M. Bernardo, J. O. Berger, A. P. Dawid and A. F. M. Smith). Oxford: Oxford University Press, (in press).

Spiegelhalter, D. J., Thomas, A., Best, N. G. and Gilks, W. R. (1994) *BUGS: Bayesian inference Using Gibbs Sampling, Version 0.30.* Cambridge: Medical Research Council Biostatistics Unit.

Thomas, D. C. and Gauderman, W. J. (1995) Gibbs sampling methods in genetics. In *Markov Chain Monte Carlo in Practice* (eds W. R. Gilks, S. Richardson and D. J. Spiegelhalter), pp. 419–440. London: Chapman & Hall.

Whittaker, J. (1990) *Graphical Models in Applied Multivariate Analysis.* Chichester: Wiley.

Whittle, H. C., Inskip, H., Hall, A. J., Mendy, M., Downes, R. and Hoare, S. (1991) Vaccination against Hepatitis B and protection against viral carriage in The Gambia. *Lancet*, **337**, 747–750.

Appendix: BUGS

BUGS is a program which provides a syntax for specifying graphical models and a command language for running Gibbs sampling sessions. An idea of the syntax can be obtained from the model description shown below, corresponding to the errors-in-variables model described in equations (2.5, 2.7–2.10, 2.14–2.16) and shown in Figure 2.7.

```
{
    for(i in 1:I){

        for(j in 1:n[i]){
            y[i,j]   ~ dnorm(mu[i,j],tau);
            mu[i,j] <- alpha[i] + gamma * (mu0[i] - mean(y0[]))
                                + beta[i] * (log.time[j] - log(730));
        }

    #  covariate with measurement error
        y0[i]    ~ dnorm(mu0[i],tau);
        mu0[i]   ~ dnorm(theta, phi);

    # random lines
        beta[i]  ~ dnorm(beta0, tau.beta);
        alpha[i] ~ dnorm(alpha0, tau.alpha);

    }


    # prior distributions
        tau          ~ dgamma(0.01,0.01);
```

```
gamma         ~ dnorm(0,0.0001);
alpha0        ~ dnorm(0,0.0001);
beta0         ~ dnorm(0,0.0001);
tau.beta      ~ dgamma(0.01,0.01);
tau.alpha     ~ dgamma(0.01,0.01);
theta         ~ dnorm(0.0, 0.0001);
phi           ~ dgamma(0.01, 0.01);


sigma         <- 1/sqrt(tau);
sigma.beta    <- 1/sqrt(tau.beta);
sigma.alpha   <- 1/sqrt(tau.alpha);
}
```

The essential correspondence between the syntax and the graphical representation should be clear: the relational operator $\sim$ corresponds to 'is distributed as' and <- to 'is logically defined by'. Note that BUGS parameterizes the Gaussian distribution in terms of mean and precision (= 1/variance). The program then interprets this declarative model description and constructs an internal representation of the graph, identifying relevant prior and likelihood terms and selecting a sampling method. Further details of the program are given in Gilks *et al.* (1994) and Spiegelhalter *et al.* (1995).

The software will run under UNIX and DOS, is available for a number of computer platforms, and can be freely obtained by anonymous *ftp* together with a manual and extensive examples. Contact the authors by post or at bugs@mrc-bsu.cam.ac.uk.

3

Markov chain concepts related to sampling algorithms

Gareth O Roberts

3.1 Introduction

The purpose of this chapter is to give an introduction to some of the theoretical ideas from Markov chain theory and time series relevant for MCMC. We begin with a brief review of basic concepts and rates of convergence for Markov chains. An account of estimation from ergodic averages follows, and finally we discuss the Gibbs sampler and Metropolis–Hastings algorithm in the context of the theory outlined below. For a more extensive introduction to the basic ideas of Markov chain theory, see Grimmett and Stirzaker (1992), and for an excellent treatment of the general theory, see Meyn and Tweedie (1993). Priestley (1981) contains a thorough treatment of some of the time series ideas we discuss in Section 3.4. See also Tierney (1995: this volume) for a more mathematical treatment of some of the material presented here.

3.2 Markov chains

A Markov chain X is a discrete time stochastic process $\{X_0, X_1, \ldots\}$ with the property that the distribution of X_t given all previous values of the process, $X_0, X_1, \ldots, X_{t-1}$ only depends upon X_{t-1}. Mathematically, we write,

$$P[X_t \in A | X_0, X_1, \ldots, X_{t-1}] = P[X_t \in A | X_{t-1}]$$

for any set A, where $P[.|.]$ denotes a conditional probability. Typically (but not always) for MCMC, the Markov chain takes values in $\mathbb{R}^d$ (d-dimensional Euclidean space). However, to illustrate the main ideas, for most of this chapter we shall restrict attention to discrete state-spaces.

Extensions to general state-spaces are more technical, but do not require any major new concepts (see Tierney, 1995: this volume). Therefore we consider transition probabilities of the form $P_{ij}(t) = P[X_t = j | X_0 = i]$.

For the distribution of X_t to converge to a *stationary distribution*, the chain needs to satisfy three important properties. First, it has to be *irreducible*. That is, from all starting points, the Markov chain can reach any non-empty set with positive probability, in some number of iterations. This is essentially a probabilistic connectedness condition. Second, the chain needs to be *aperiodic*. This stops the Markov chain from oscillating between different sets of states in a regular periodic movement. Finally, and most importantly, the chain must be positive recurrent. This can be expressed in terms of the existence of a stationary distribution $\pi(.)$, say, such that if the initial value X_0 is sampled from $\pi(.)$, then all subsequent iterates will also be distributed according to $\pi(.)$. There are also various equivalent definitions. These ideas are made precise in the following definition. Let τ_{ii} be the time of the first return to state i, $(\tau_{ii} = \min\{t > 0 : X_t = i | X_0 = i\})$.

Definition 3.1

(i) X is called irreducible if for all i, j, there exists a $t > 0$ such that $P_{ij}(t) > 0$.

(ii) An irreducible chain X is recurrent if $P[\tau_{ii} < \infty] = 1$ for some (and hence for all) i. Otherwise, X is transient. Another equivalent condition for recurrence is

$$\sum_t P_{ij}(t) = \infty$$

for all i, j.

(iii) An irreducible recurrent chain X is called positive recurrent if $E[\tau_{ii}] < \infty$ for some (and hence for all) i. Otherwise, it is called null-recurrent. Another equivalent condition for positive recurrence is the existence of a stationary probability distribution for X, that is there exists $\pi(\cdot)$ such that

$$\sum_i \pi(i) P_{ij}(t) = \pi(j) \tag{3.1}$$

for all j and $t \geq 0$.

(iv) An irreducible chain X is called aperiodic if for some (and hence for all) i,

greatest common divider $\{t > 0 : P_{ii}(t) > 0\} = 1$.

In MCMC, we already have a target distribution $\pi(.)$, so that by (iii) above, X will be positive recurrent if we can demonstrate irreducibility.

A good account of the generalizations of the above definition to general state spaces is given in Tierney (1995: this volume), and full details appear in Meyn and Tweedie (1993). See also Roberts and Smith (1994) and Roberts and Tweedie (1994).

In practice, output from MCMC is summarized in terms of *ergodic averages* of the form

$$\bar{f}_N = \frac{\sum_{t=1}^{N} f(X_t)}{N},$$

where $f(.)$ is a real valued function. Therefore asymptotic properties of $\bar{f}_N$ are very important.

It turns out that for an aperiodic positive-recurrent Markov chain, the stationary distribution (which is our target distribution of course), is also the *limiting* distribution of successive iterates from the chain. Importantly, this is true regardless of the starting value of the chain. Moreover, under these conditions, ergodic averages converge to their expectations under the stationary distribution. These limit results are expressed more precisely in the following theorem.

Theorem 3.1 *If X is positive recurrent and aperiodic then its stationary distribution $\pi(.)$ is the unique probability distribution satisfying (3.1). We then say that X is ergodic and the following consequences hold:*

(i) $P_{ij}(t) \to \pi(j)$ *as* $t \to \infty$ *for all* i, j.

(ii) *(Ergodic theorem) If* $E_\pi[|f(X)|] < \infty$, *then*

$$P\left[\bar{f}_N \to E_\pi[f(X)]\right] = 1,$$

where $E_\pi[f(X)] = \sum_i f(i)\pi(i)$, *the expectation of $f(X)$ with respect to $\pi(.)$.*

Part *(ii)* of Theorem 3.1 is clearly very important in practice for MCMC, although it does not offer any reassurance as to how long we need to run the Markov chain before its iterations are distributed approximately according to $\pi(.)$, and it offers no estimate as to the size of the error of any Monte Carlo estimate $\bar{f}_N$.

More sophisticated estimation procedures based on regenerative techniques are discussed in Tierney (1995: this volume).

Most of the Markov chains produced in MCMC are *reversible*, are derived from reversible components, or have reversible versions. A Markov chain is said to be reversible if it is positive recurrent with stationary distribution $\pi(.)$, and

$$\pi(i)P_{ij} = \pi(j)P_{ji}.$$

We shall assume that the Markov chains we consider are reversible unless otherwise stated.

3.3 Rates of convergence

We say that X is geometrically ergodic (in total variation norm), if it is ergodic (positive recurrent and aperiodic) and there exists $0 \leq \lambda < 1$ and a function $V(.) > 1$ such that

$$\sum_j |P_{ij}(t) - \pi(j)| \leq V(i)\lambda^t \tag{3.2}$$

for all i. The smallest λ for which there exists a function V satisfying (3.2) is called the *rate of convergence*. We shall denote this by λ^*. (Formally, we define λ^* as $\inf\{\lambda : \exists V$ such that (3.2) holds$\}$.)

To understand more closely the implications of geometric convergence, we need to consider the spectral analysis of Markov chains. For reversible Markov chains, spectral theory provides an extremely powerful method for analysing chains.

Sufficiently regular problems have transition probabilities described by a sequence of eigenvalues $\{\lambda_0, \lambda_1, \ldots\}$, where $\lambda_0 = 1$, and corresponding left eigenvectors $\{e_0, e_1, \ldots\}$, that is,

$$\sum_i e_k(i)P_{ij}(t) = \lambda_k e_k(j)$$

for all j and for each k, such that

$$P_{ij}(t) = \sum_k e_k(i)e_k(j)\lambda_k^t. \tag{3.3}$$

Here $e_0(.) = \pi(.)$. In general, the eigenvalues can be complex with modulus bounded by unity. Reversibility of the Markov chain ensures that the eigenvalues and eigenvectors are real. In general, for infinite state-spaces, there are an infinite number of eigenvalues. However, for geometrically ergodic chains, all but the principal eigenvalue, $\lambda_0 = 1$, are uniformly bounded away from ± 1. Chains which fail to be geometrically ergodic have an infinite number of eigenvalues in any open interval containing either -1 or 1.

For large t, the dominant term in (3.3) is $\pi(j) = e_0(j)$. However, the speed at which convergence is achieved, depends on the second largest eigenvalue in absolute value, which is just the rate of convergence of the Markov chain. Therefore, an alternative definition of λ^* is

$$\lambda^* = \sup_{k > 0} |\lambda_k|$$

Of course, eigen-analysis is not the whole story; for instance, the constants in the eigenfunction expansion (3.3) could be very large. However, given reasonable starting points (or starting distributions), it is often possible to bound these also, thus obtaining an upper bound for λ^*; see, for example, Roberts (1995) and Polson (1995).

In practice, it is usually too difficult to obtain useful upper bounds on λ^*. However, see Rosenthal (1993) and Polson (1995) for notable exceptions.

3.4 Estimation

One of the most important consequences of geometric convergence is that it allows the existence of central limit theorems for ergodic averages, that is, results of the form

$$N^{1/2}(\bar{f}_N - E_\pi[f(X)]) \rightarrow N(0, \sigma^2) \tag{3.4}$$

for some positive constant σ, as $N \rightarrow \infty$, where the convergence is in distribution. Such results are essential in order to put inference from MCMC output on a sound footing. Even when central limit theorems do exist, algorithms can often be extremely inefficient in cases where σ is large in comparison with the variance (under π) of $f(X)$. An extensive treatment of geometric convergence and central limit theorems for Markov chains can be found in Meyn and Tweedie (1993), applications to Metropolis–Hastings algorithms appear in Roberts and Tweedie (1994), and to the Gibbs sampler in Chan (1993), Schervish and Carlin (1992) and Roberts and Polson (1994). Tierney (1995: this volume) discusses these results further, and Geyer (1992) provides a review of the use of central limit theorems for MCMC.

In this section, we will assume that (3.4) holds for the function $f(.)$. See Priestley (1981) and Geyer (1992) for further details. The following result gives equivalent expressions for σ^2.

Theorem 3.2

$$\sigma^2 = \text{var}_\pi(f(X_0)) + 2\sum_{i=1}^\infty \text{cov}(X_0, X_i) \tag{3.5}$$

$$= \sum_{j=1}^\infty \frac{1 + \lambda_j}{1 - \lambda_j} a_i \, \text{var}_\pi(f(X_0))$$

$$\leq \frac{1 + \lambda^*}{1 - \lambda^*}$$

for some non-negative constants a_i, where X_0 is distributed according to $\pi(.)$, and $\sum_{i=0}^\infty a_i = 1$.

The ratio

$$\text{eff}_{\bar{f}} = \frac{\text{var}_\pi(f(X_0))}{\sigma^2}$$

is a measure of the efficiency of the Markov chain for estimating $E_\pi[f(X)]$.

To assess the accuracy of our estimate of $E_\pi[f(X)]$, $\bar{f}_N$, it is essential to be able to estimate σ^2. This problem is reviewed extensively in Geyer

(1992). We content ourselves here with a short description of two of the simplest and most commonly used methods.

3.4.1 Batch means

Run the Markov chain for $N = mn$ iterations, where n is assumed sufficiently large that

$$Y_k = \frac{1}{n} \sum_{i=(k-1)n+1}^{kn} f(X_i) \qquad (3.6)$$

are approximately independently $N(E_\pi[f(X)], \sigma^2/n)$. Therefore σ^2 can be approximated by

$$\frac{n}{m-1} \sum_{k=1}^{m} (Y_k - \bar{f}_N)^2,$$

or alternatively a t-test can be used to give bounds on the accuracy of $\bar{f}_N$.

This method is extremely attractive due to its ease of implementation. However, care has to be taken to choose n large enough for the approximations involved to be valid.

3.4.2 Window estimators

From (3.5), an obvious way to try to approximate σ^2 is to estimate $\gamma_i \equiv \text{cov}_\pi(f(X_0), f(X_i))$ by the empirical covariance function

$$\hat{\gamma}_i = \frac{1}{n} \sum_{j=1}^{N-i} (f(X_j) - \bar{f}_N)(f(X_{j+i}) - \bar{f}_N). \qquad (3.7)$$

Unfortunately, this approach runs into trouble since the estimates become progressively worse as i increases (there are less and less terms in the average (3.7)). In fact, the estimator produced by this approach is not even consistent. Instead, it is necessary to use a truncated *window estimator* of σ^2,

$$\hat{\sigma}_N^2 = \hat{\gamma}_0 + 2 \sum_{i=1}^{\infty} w_N(i)\hat{\gamma}_i \qquad (3.8)$$

where $0 \leq w_N(i) \leq 1$. Typically, the $w_N(i)$ are chosen to be unity within a certain range (which can depend upon N), and zero outside this range.

Geyer (1992) suggests an appealing data-dependent window which is based on the heuristic that it makes no sense to continue summing the lag correlations when they are dominated by noise. For a reversible Markov chain, $\Gamma_m = \gamma_{2m} + \gamma_{2m+1}$ is a non-negative, non-increasing convex function of m. Geyer defines the *initial convex sequence estimator* by examining the empirical estimates given by $\hat{\Gamma}_m = \hat{\gamma}_{2m} + \hat{\gamma}_{2m+1}$. Now define m^* to be the largest integer such that the function $\{\hat{\Gamma}_m, \ 1 \leq m \leq m^*\}$ is a non-negative, non-increasing convex function of m. The $w_N(i)$ are then chosen to be unity for $i \leq 2m^*$ and zero otherwise.

Geyer (1992) introduces other similar estimators, and provides some comparisons of their performance.

3.5 The Gibbs sampler and Metropolis–Hastings algorithm

Many MCMC algorithms are hybrids or generalizations of the two simplest methods: the Gibbs sampler and the Metropolis–Hastings algorithm. Therefore if we are to design more sophisticated techniques, it is first of all essential to understand the strong and weak points of these basic elements. For the remainder of this section, we give a brief review of convergence results and heuristics for the Gibbs sampler and the Metropolis–Hastings algorithm.

From now on, for notational convenience, we shall revert to the continuous state-space case, assuming that our target densities have Lebesgue densities. The discrete case follows in almost identical fashion and all the ideas, that we will discuss, can be assumed to carry over to this case.

3.5.1 The Gibbs sampler

The Gibbs sampler is best described mathematically by its transition *kernel*. Let $X = \{X_t; \ t \geq 0\}$ be the Markov chain induced by a d-dimensional Gibbs sampler, as described in Gilks *et al.* (1995: this volume), where we write $X_t = (X_{t.1}, \ldots, X_{t.d})$ in component form. Given the full conditional densities (usually with respect to Lebesgue measure), $\pi(X_{.i}|X_{.j}, j \neq i)$, the transition kernel describes the density of going from one point, X, to another, Y:

$$K(X,Y) = \prod_{i=1}^{d} \pi(Y_{.i}|\{Y_{.j}, j < i\}, \{X_{.j}, j > i\}) \qquad (3.9)$$

This is a product of the conditional densities of the individual steps required to produce an iteration of the Gibbs sampler.

Given (3.9), we can express the Gibbs sampler transition probabilities as follows. For any set A:

$$P[X_1 \in A|X_0 = X] = \int_A K(X,Y)dY.$$

Note that the Gibbs sampler in this form is not reversible, although it consists of d consecutive reversible components. Reversible versions of the Gibbs sampler are easily produced (see for example Roberts, 1995), and include a random scan sampler – at each iteration, this sampler picks a random component to update. There is some theoretical evidence to suggest that the random scan Gibbs sampler can be more efficient that fixed cycle ones (Amit and Grenander, 1991).

Conditions that ensure irreducibility, aperiodicity (and hence ergodicity) of the Gibbs sampler are given (for example) in Roberts and Smith (1994). Chan (1993), Schervish and Carlin (1992), and Roberts and Polson (1994) all give geometric convergence results without offering any upper bounds on λ^*. Rates of convergence estimates in general are difficult to obtain, but see Rosenthal (1993) and its references for techniques and upper bounds on convergence rates in certain problems.

Often, a natural parameterization of a d-dimensional space is suggested by the form of π. For example, there may be a conditional independence structure which allows a number of the 1-dimensional conditional densities to have particularly simple forms. This occurs for example in hierarchical models; see for example Smith and Roberts (1993) and Spiegelhalter *et al.* (1995: this volume). Therefore, in many problems, an artificial change in the parameterization of the problem in order to speed up the Gibbs sampler may be inappropriate due to the greatly increased computational burden. However, the efficiency of the algorithm can vary radically with the choice of parameterization. Therefore, it is natural to ask how λ^* varies with this choice.

It is clear that when $X_{.1}, \ldots, X_{.d}$ are independent under π, then the Gibbs sampler will output independent identically distributed (iid) random variables. In the context of the discussion above, this corresponds to $\lambda^* = 0$. There is a collection of function analytic results which generalizes this idea to support the heuristic:

If $X_{.1}, \ldots, X_{.d}$ are approximately independent under π, then λ^* will be close to 0 and the Gibbs sampler will output almost iid random variables.

These results were first noticed in the context of MCMC by Amit (1991). The simplest such result is the following.

Theorem 3.3 *Suppose $d = 2$, then for the Gibbs sampler,*
$$\lambda^* = \{\sup \, \mathrm{corr}_\pi(f(X_{.1}), g(X_{.2}))\}^2 ,$$
where the supremum is taken over all possible functions f and g.

As an example, we shall look at a Gibbs sampler which is in fact reducible (see Figure 3.1). Consider the target distribution $\pi(.)$ which is uniform on $([-1, 0] \times [-1, 0]) \cup ([0, 1] \times [0, 1])$ using the coordinate directions as sampling directions. It is easy to see in this case that the Gibbs sampler is reducible. Now let $f(X) = g(X) = I[X \geq 0]$, taking the value 1 if and only if

$X \geq 0$. In this case, it is easy to show that $\text{corr}_\pi(f(X_{.1}), g(X_{.2})) = 1$. Then from the above theorem, $\lambda^* = 1$, i.e. the chain fails to be geometrically ergodic. It is easy to see that the Gibbs sampler fails to be irreducible in this example; a chain starting in the positive quadrant will never visit the negative quadrant, and vice versa. So that in fact in this example, the chain is not even ergodic.

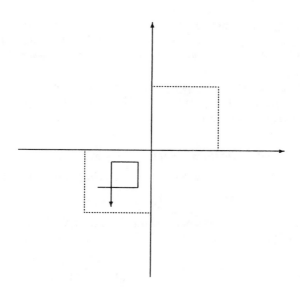

Figure 3.1 *A reducible Gibbs sampler.*

Clearly these results cannot be used directly; calculating the correlation structure for all possible functions f and g, let alone carrying out the supremum operation, will not be possible in practice. However, in the common situation where $\pi(.)$ is a posterior distribution from a Bayesian analysis, it is very commonly approximately Gaussian. In practice, posteriors are frequently well approximated by Gaussian distributions even for relatively small data sets. For the multivariate normal distribution,

$$\sup \text{corr}_\pi(f(X_{.1}), g(X_{.2}))$$

simplifies considerably, since the supremum is always attained by linear functions, so that the convergence rate of a two-dimensional Gibbs sampler on a bivariate normal distribution has $\lambda^* = \rho^2$, where ρ is the correlation of the bivariate normal. The idea generalizes (if not as explicitly) to the case $d \geq 3$, and lends support to the following practical strategy.

Try to choose a parameterization scheme to make the correlations between coordinate directions (according to $\pi(.)$) as small as possible.

This could be achieved by monitoring the sample variance-covariance matrix Σ of the output from a pilot sample from $\pi(.)$, possibly obtained from an initial Gibbs sampler run. Assuming that Σ well approximates the true variance-covariance matrix of $\pi(.)$ according to the chosen parameterization, a simple linear transformation can be used to produce an approximately uncorrelated parameterization. See Wakefield (1993) for details. More sophisticated transformation techniques are discussed in Hills and Smith (1992).

It is, however, worth emphasizing that for highly non-normal target distributions, an uncorrelated parameterization scheme may not necessarily lead to an efficient Gibbs sampler; see, for example, Roberts (1992).

3.5.2 The Metropolis–Hastings algorithm

The Metropolis–Hastings algorithm has transition kernel density,

$$K(X, Y) = q(X, Y)\alpha(X, Y),$$

where $q(X, .)$ is the proposal density from X, and $\alpha(X, Y)$ is the acceptance probability for the move from X to Y:

$$\alpha(X, Y) = \min\left\{1, \frac{\pi(Y)q(Y, X)}{\pi(X)q(X, Y)}\right\}.$$

K does not describe all possible transitions from X; there is also the possibility of a rejected move. For continuous distributions, we can write,

$$r(X) = 1 - \int_Y q(X, Y)\alpha(X, Y)dY$$

for the rejection probability from a point X. Thus we can write

$$P[X_1 \in A | X_0 = X] = \int_A K(X, Y)dY + r(X)I[X \in A].$$

Unfortunately, the links between efficiency of the Metropolis–Hastings algorithm and the statistical properties of $\pi(.)$ are not nearly as well understood as for the Gibbs sampler. We shall look at two simple (but widely used) examples where some progress has been made.

(1) For the *independence sampler*, we write $q(X, Y) = q(Y)$. In this case, an explicit expression for the iterated transition probabilities is available; see Liu (1995) and Smith (1994). Here, we merely state that the spectral gap $1 - \lambda^*$ for this algorithm is given by $\inf_Y \frac{q(Y)}{\pi(Y)}$. (Note that $\pi(.)$ here represents the *normalized* target density.)

Therefore, when the independence proposal density well approximates the target density, the spectral gap is large, leading to fast convergence and efficient estimation. Note that it is common for $\inf_Y \frac{q(Y)}{\pi(Y)}$ to be zero, so that the algorithm in this case is not geometrically convergent and is hence liable to have 'sticky' points and be sensitive to starting values. When this expression is not zero, simple rejection sampling is always possible, (see Gilks, 1995: this volume), and this produces an independent identically distributed sample. In either case, it is rare for the independence sampler to be useful as a *stand-alone* algorithm. However, within a hybrid strategy which combines and mixes different MCMC methods, the method is extremely easy to implement and often very effective; see Gilks and Roberts (1995: this volume).

(2) *The random-walk Metropolis algorithm* can be defined by the relation $q(X, Y) = q(Y - X)$. Without a rejection step, this Markov chain would carry out a random walk, where its iterations were independent and identically distributed. This is a very popular algorithm which can be shown to be geometrically ergodic under reasonably general conditions assuming that the tails of the target density $\pi(.)$ are exponential or lighter, and that its contours are sufficiently regular; see Roberts and Tweedie (1994). Frequently, the proposed random walk jump has a form indexed by a scale parameter (for example, Y could be distributed $N(X, \sigma^2)$), and it is necessary to choose the scale parameter (σ) reasonably carefully for any particular target density. If σ is too large, an extremely large proportion of iterations will be rejected, and the algorithm will therefore be extremely inefficient. Conversely, if σ is too small, the random walk will accept nearly all proposed moves but will move around the space slowly, again leading to inefficiency.

An appealing approach is to monitor overall acceptance rates for the algorithm, i.e. the proportion of proposed moves which are subsequently accepted. By the ergodic theorem (Theorem 3.1), for large samples, this should normally be a good estimate of the theoretical expected acceptance rate:

$$a(\sigma) = \int_X (1 - r(X))\pi(X)dX.$$

The empirical overall acceptance rate is an easy to monitor, non-parametric quantity, which can be tuned by changing σ. Roberts *et al.* (1994) and Gelman *et al.* (1995) provide some theoretical justification for aiming to have acceptance rates in the range [0.15, 0.5], using a limiting argument for high-dimensional target densities. Simulation results for relatively well-behaved densities support these ideas; see Gelman *et al.* (1995).

Armed with a basic understanding of the Gibbs sampler and Metropolis–Hastings algorithms, we are now in a position to study more sophisticated methods. Gilks and Roberts (1995: this volume) is devoted to a survey of existing techniques, many of which are modifications, generalizations, or hybrids of Gibbs and Metropolis–Hastings methods.

References

Amit, Y. (1991) On rates of convergence of stochastic relaxation for Gaussian and non-Gaussian distributions. *J. Mult. Anal.*, **38**, 82–99.

Amit, Y. and Grenander, U. (1991) Comparing sweep strategies for stochastic relaxation. *J. Mult. Anal.*, **37**, 197–222.

Chan, K. S. (1993) Asymptotic behaviour of the Gibbs sampler. *J. Am. Statist. Ass.*, **88**, 320–326.

Gelman, A., Roberts, G. O. and Gilks, W. R. (1995) Efficient Metropolis jumping rules. In *Bayesian Statistics 5* (eds J. M. Bernardo, J. O. Berger, A. P. Dawid and A. F. M. Smith). Oxford: Oxford University Press (in press).

Geyer, C. J. (1992) Practical Markov chain Monte Carlo (with discussion), *Statist. Sci.*, **7**, 473–511.

Gilks, W. R. (1995) Full conditional distributions. In *Markov Chain Monte Carlo in Practice* (eds W. R. Gilks, S. Richardson and D. J. Spiegelhalter), pp. 75–88. London: Chapman & Hall.

Gilks, W. R. and Roberts, G. O. (1995) Strategies for improving MCMC. In *Markov Chain Monte Carlo in Practice* (eds W. R. Gilks, S. Richardson and D. J. Spiegelhalter), pp. 89–114. London: Chapman & Hall.

Gilks, W. R., Richardson, S. and Spiegelhalter, D. J. (1995) Introducing Markov chain Monte Carlo. In *Markov Chain Monte Carlo in Practice* (eds W. R. Gilks, S. Richardson and D. J. Spiegelhalter), pp. 1–19. London: Chapman & Hall.

Grimmett, G. R. and Stirzaker, D. R. (1992) *Probability and Random Processes, 2nd edn.*. Oxford: Oxford University Press.

Hills, S. E. and Smith, A. F. M. (1992) Parameterization issues in Bayesian inference (with discussion). In *Bayesian Statistics 4* (eds J. M. Bernardo, J. Berger, A. P. Dawid and A. F. M. Smith), pp. 227–246. Oxford: Oxford University Press.

Liu, J. (1995) Metropolized independent sampling with comparison to rejection sampling and importance sampling. *Statist. Comput.*, (in press).

Meyn, S. P. and Tweedie, R. L. (1993) *Markov Chains and Stochastic Stability*. London: Springer-Verlag.

Polson, N. G. (1995). Convergence of Markov chain Monte Carlo algorithms. In *Bayesian Statistics 5* (eds J. M. Bernardo, J. Berger, A. P. Dawid and A. F. M. Smith). Oxford: Oxford University Press (in press).

Priestley, M. B. (1981) *Spectral Analysis and Time Series*. London: Academic Press.

Roberts, G. O. (1992) Discussion on parameterisation issues in Bayesian inference (by S. E. Hills and A. F. M. Smith). In *Bayesian Statistics 4* (eds J. M. Bernardo, J. Berger, A. P. Dawid and A. F. M. Smith), pp. 241. Oxford: Oxford University Press.

Roberts, G. O. (1995) Methods for estimating L^2 convergence of Markov chain Monte Carlo. In *Bayesian Statistics and Econometrics: Essays in Honor of Arnold Zellner* (eds D. Berry, K. Chaloner and J. Geweke). New York: Wiley, (in press).

Roberts, G. O. and Polson, N. G. (1994) On the geometric convergence of the Gibbs sampler. *J. R. Statist. Soc.* B, **56**, 377–384.

Roberts, G. O. and Smith, A. F. M. (1994) Simple conditions for the convergence of the Gibbs sampler and Hastings-Metropolis algorithms. *Stoch. Proc. Appl.*, **49**, 207-216.

Roberts, G. O. and Tweedie, R. L. (1994) Geometric convergence and central limit theorems for multidimensional Hastings and Metropolis algorithms. *Research Report 94.9*, Statistical Laboratory, University of Cambridge, UK.

Roberts, G. O., Gelman, A. and Gilks, W. R. (1994) Weak convergence and optimal scaling of random walk Metropolis algorithms. *Research Report 94.16*, Statistical Laboratory, University of Cambridge.

Rosenthal, J. S. (1993) Rates of convergence for data augmentation on finite sample spaces. *Ann. Appl. Prob.*, **3**, 819–839.

Schervish, M. J. and Carlin, B. P. (1992) On the convergence of successive substitution sampling. *J. Comp. Graph. Statist.*, **1**, 111-127.

Smith, A. F. M. and Roberts, G. O. (1993) Bayesian computation via the Gibbs sampler and related Markov Chain Monte Carlo methods (with discussion). *J. R. Statist. Soc.* B, **55**, 3–24.

Smith, R. (1994) Exact transition probabilities for Metropolized independent sampling. *Research Report*, Statistical Laboratory, University of Cambridge, UK

Spiegelhalter, D. J., Best, N. G., Gilks, W. R. and Inskip, H. (1995) Hepatitis B: a case study in MCMC methods. In *Markov Chain Monte Carlo in Practice* (eds W. R. Gilks, S. Richardson and D. J. Spiegelhalter), pp. 21–43. London: Chapman & Hall.

Tierney, L. (1995) Introduction to general state-space Markov chain theory. In *Markov Chain Monte Carlo in Practice* (eds W. R. Gilks, S. Richardson and D. J. Spiegelhalter), pp. 59–74. London: Chapman & Hall.

Wakefield, J. C. (1993) Discussion on the meeting on the Gibbs sampler and other Markov chain Monte Carlo methods. *J. R. Statist. Soc.* B, **55**, 56–57.

4

Introduction to general state-space Markov chain theory

Luke Tierney

4.1 Introduction

Markov chain Monte Carlo is a method for exploring a distribution π in which a Markov chain with invariant distribution π is constructed and sample path averages of this Markov chain are used to estimate characteristics of π. How well this strategy works in a particular problem depends on certain characteristics of the Markov chain.

Problems arising in Bayesian inference and other areas often lead to distributions π that are continuous, and thus to MCMC samplers that have continuous state-spaces. It is therefore natural to base a theoretical analysis of these samplers on theory for general state-space Markov chains. This theory has recently seen several developments that have made it both more accessible and more powerful. Very minimal conditions turn out to be sufficient and essentially necessary to ensure convergence of the distribution of the sampler's state to the invariant distribution and to provide a law of large numbers for sample path averages. Additional conditions provide central limit results.

This chapter summarizes some of the results of general state-space Markov chain theory as described in Nummelin (1984) and Meyn and Tweedie (1993) as they apply to MCMC samplers. The only special feature of these samplers used is that they are known by construction to have a particular distribution π as an invariant distribution. Results designed to establish the existence of an invariant distribution are thus not needed. Other approaches to establishing properties of MCMC samplers, for example approaches based on operator theory, are also possible but are not addressed here. References to some of these approaches can be found in Smith and

Roberts (1993) and Tierney (1994). Many results are analogous to results for discrete state-space chains, but there are some differences. Some of these differences are illustrated with a simple random walk example. This chapter assumes that the reader is familiar with basic properties of discrete state-space Markov chains as presented, for example, in Taylor and Karlin (1984); see also Roberts (1995: this volume).

All results given in this chapter are derived from results presented in Nummelin (1984) and Meyn and Tweedie (1993). Many of these results are given in Tierney (1994) with proofs or references to proofs. To simplify the presentation, no proofs are provided in this chapter, and references are only given for results not directly available in these sources.

4.2 Notation and definitions

The distribution π is assumed to be defined on a set E called the state-space. Often E will be k-dimensional Euclidean space $\mathbb{R}^k$, but it can be quite general. The only technical requirement, which will not be used or mentioned explicitly in the remainder of this chapter, is that the collection $\mathcal{E}$ of subsets of E on which π is defined must be a countably generated sigma algebra. All subsets of E and all functions defined on E used in the remainder of this chapter are assumed to be $\mathcal{E}$-measurable. At times, π will be assumed to have a density, which will be denoted by $\pi(x)$.

For a probability distribution ν and a real valued function h on E, the expectation of h under ν is denoted by

$$\nu h = \int h(x)\nu(dx).^*$$

Thus if ν is discrete with probability mass function $f(x)$, then

$$\nu h = \int h(x)\nu(dx) = \sum h(x)f(x),$$

and if ν is continuous with density $f(x)$, then

$$\nu h = \int h(x)\nu(dx) = \int h(x)f(x)dx.$$

The distribution of a time-homogeneous Markov chain $\{X_n\} = \{X_n, n = 0, 1, \ldots\}$ on the state-space E is specified by its initial distribution and its transition kernel. The transition kernel is a function $P(x, A)$ such that for any $n \geq 0$

$$P\{X_{n+1} \in A | X_n = x\} = P(x, A)$$

for all $x \in E$ and $A \subset E$. That is, $P(x, \cdot)$ is the distribution of the Markov chain after one step given that it starts at x. Different strategies, such as the

---

* $\nu(dx)$ is the probability under ν of a small measurable subset $dx \in E$; $h(x)$ is the value of h in dx; and the integral is over such subsets dx.

Gibbs sampler, Metropolis–Hastings algorithms, etc., give rise to different transition kernels. Transition kernels are the general state-space versions of discrete chain transition matrices.

For a probability distribution ν on E, define the distribution νP by

$$\nu P(A) = \int P(x, A)\nu(dx).$$

The distribution νP is the distribution of the position of a Markov chain with transition kernel P and initial distribution ν after one step. A Markov chain has invariant distribution π if $\pi = \pi P$. Invariant distributions are also often called stationary distributions.

For a real-valued function h on E, the function Ph is defined as

$$Ph(x) = \int P(x, dy)h(y)^\dagger = E[h(X_1)|X_0 = x].$$

The product PQ of two transition kernels P and Q is the transition kernel defined by

$$PQ(x, A) = \int P(x, dy)Q(y, A)$$

for all $x \in E$ and $A \subset E$. The nth iterate P^n of P is defined recursively for $n \geq 1$ by $P^1 = P$ and $P^n = PP^{n-1}$ for $n \geq 2$. By convention, P^0 is the identity kernel that puts probability one on remaining at the initial value. Using this notation, we can write

$$P\{X_n \in A|X_0 = x\} = P^n(x, A)$$

for any $n \geq 0$.

Probabilities and expectations for a Markov chain started with $X_0 = x$ are denoted by P_x and E_x, respectively. Probabilities and expectations for a Markov chain with initial distribution ν are denoted by P_ν and E_ν.

A probability distribution ν_1 is absolutely continuous with respect to a probability distribution ν_2 if $\nu_2(A) = 0$ implies $\nu_1(A) = 0$ for any $A \subset E$, i.e. if every null set of ν_2 is a null set of ν_1. Two probability distributions are said to be equivalent if they are mutually absolutely continuous, or have the same null sets.

The total variation distance between two probability distributions ν_1 and ν_2 is defined as

$$\| \nu_1 - \nu_2 \| = 2 \sup_{A \subset E} |\nu_1(A) - \nu_2(A)|.$$

The first return time of a Markov chain to a set $A \subset E$ is denoted by τ_A. That is, $\tau_A = \inf\{n \geq 1 : X_n \in A\}$, with the convention that $\tau_A = \infty$

$\dagger$ $P(x, dy)$ is the probability of moving to a small measurable subset $dy \in E$ given that the move starts at x; $h(y)$ is the value of h in dy; and the integral is over such subsets dy.

if the chain never returns to A.

The indicator function of a set C is

$$I_C(x) = \left\{ \begin{array}{ll} 1 & \text{if } x \in C \\ 0 & \text{if } x \notin C. \end{array} \right.$$

For a probability distribution ν on E a statement holds for 'ν-almost all x' if ν gives probability zero to the set of points in E where the statement fails.

4.3 Irreducibility, recurrence and convergence

4.3.1 Irreducibility

Suppose we have a Markov chain with transition kernel P that has invariant distribution π. For a MCMC experiment using this chain to be successful, we need to ensure that the sample path average

$$\overline{f}_n = \frac{1}{n+1} \sum_{i=0}^{n} f(X_i) \tag{4.1}$$

converges to the expectation πf for any initial distribution whenever this expectation exists. A minimal requirement for this property is that the chain must be able to reach all interesting parts of the state-space, i.e. that the chain is *irreducible*.

For general state-space Markov chains, irreducibility is defined with respect to a distribution:

Definition 4.1 *A Markov chain is φ-irreducible for a probability distribution φ on E if $\varphi(A) > 0$ for a set $A \subset E$ implies that*

$$P_x\{\tau_A < \infty\} > 0$$

for all $x \in E$. A chain is irreducible if it is φ-irreducible for some probability distribution φ. If a chain is φ-irreducible, then φ is called an irreducibility distribution for the chain.

This definition is a bit more general than the traditional definition in the discrete case in that discrete chains that are not irreducible in the traditional sense can be irreducible in the sense of Definition 4.1. As an example, consider a discrete Markov chain with state-space $E = \{0, 1, \ldots\}$ and transition matrix

$$P = \begin{bmatrix} 1 & 0 & 0 & 0 & 0 & \cdots \\ q & 0 & p & 0 & 0 & \cdots \\ 0 & q & 0 & p & 0 & \cdots \\ \vdots & \ddots & \ddots & \ddots & \ddots & \ddots \end{bmatrix} \tag{4.2}$$

where $q + p = 1$, $q > 0$ and $p > 0$. This is a random walk on the non-negative half line with an absorbing barrier at the origin. Traditional state analysis would classify this chain as reducible with one absorbing class, the origin, and with transient states $\{1, 2, \ldots\}$. But this chain is irreducible under Definition 4.1 since it is φ-irreducible for the distribution φ that puts probability one on the origin.

Verifying irreducibility is often straightforward. A sufficient condition for a kernel P to be irreducible with respect to a distribution φ is that P^n has a positive density f with respect to φ for some $n \geq 1$, i.e. that there exist a positive function f such that $P^n(x, A) = \int_A f(x, y)\varphi(dy)$ for all $x \in E$ and $A \subset E$. This is often the case for Gibbs samplers. If P has both discrete and continuous components, as is often the case in Metropolis–Hastings samplers, then it is sufficient for the continuous component of P^n to have a positive density with respect to φ.

If a chain is irreducible, then it has many different irreducibility distributions. However, it is possible to show that any irreducible chain has a maximal irreducibility distribution ψ in the sense that all other irreducibility distributions are absolutely continuous with respect to ψ. Maximal irreducibility distributions are not unique but are equivalent, i.e. they have the same null sets.

4.3.2 Recurrence

Irreducibility means that all interesting sets can be reached. Recurrence is the property that all such sets will be reached infinitely often, at least from almost all starting points.

Definition 4.2 *An irreducible Markov chain with maximal irreducibility distribution ψ is recurrent if for any set $A \subset E$ with $\psi(A) > 0$ the conditions*

(i) $P_x\{X_n \in A \text{ infinitely often}\} > 0$ for all x

(ii) $P_x\{X_n \in A \text{ infinitely often}\} = 1$ for ψ-almost all x

are both satisfied. An irreducible recurrent chain is positive recurrent if it has an invariant probability distribution. Otherwise it is null recurrent.

This definition defines recurrence as a property of an entire irreducible chain. In contrast, for discrete chains one usually defines recurrence of individual states, and then defines a chain to be recurrent if it is irreducible and all its states are recurrent. The definition of positive recurrence given here is usually derived as a theorem in the discrete case. That is, it can be shown that a recurrent discrete chain is positive recurrent if and only if a proper invariant distribution exists.

For the random walk example with transition matrix (4.2), the maximal irreducibility distribution ψ cannot give positive probability to any of the states $1, 2, \ldots$, and is therefore the distribution that puts probability one on the origin. A set $A \subset E$ has $\psi(A) > 0$ if and only if $0 \in A$. Since $0 < p < 1$, the probability of reaching the absorbing origin, and therefore the probability of visiting the origin infinitely often, is positive for all starting points, and thus condition *(i)* holds. The probability of visiting the origin infinitely often starting from the origin is one; since $\psi(\{1, 2, \ldots\}) = 0$ this implies that condition *(ii)* holds. For $p \leq \frac{1}{2}$ the probability of reaching the origin from any state is one, and condition *(ii)* therefore holds for all starting points. But, for $\frac{1}{2} < p < 1$, the probability of reaching the origin from an initial state x is $(q/p)^x < 1$ and therefore condition *(ii)* does not hold for $x \in \{1, 2, \ldots\}$.

If a Markov chain is irreducible and is known to have a proper invariant distribution π, then it must be positive recurrent. This is easy to show for a discrete chain using the following argument. An irreducible discrete chain must be either transient, null recurrent or positive recurrent; see Roberts (1995: this volume). Transience and null recurrence both imply that $P\{X_n = y | X_0 = x\} \to 0$ as $n \to \infty$ for all $x, y \in E$. But this contradicts the fact that π is invariant, i.e. $P_\pi\{X_n = y\} = \pi(y)$ for all $y \in E$. So, by elimination, the chain must be positive recurrent. A similar argument applies even in the general state-space setting and yields the following result:

Theorem 4.1 *Suppose the Markov chain $\{X_n\}$ is irreducible and has invariant distribution π. Then the chain is π-irreducible, π is a maximal irreducibility distribution, π is the unique invariant distribution of the chain, and the chain is positive recurrent.*

4.3.3 Convergence

Recurrence is sufficient to imply convergence of averages of probabilities:

Theorem 4.2 *Suppose $\{X_n\}$ is an irreducible Markov chain with transition kernel P and invariant distribution π. Define the average transition kernel $\overline{P}^n$ by*

$$\overline{P}^n(x, A) = \frac{1}{n+1} \sum_{i=0}^{n} P^i(x, A)$$

for all $x \in E$ and $A \subset E$. Then

$$\| \overline{P}^n(x, \cdot) - \pi(\cdot) \| \to 0$$

for π-almost all x.

Recurrence also implies a strong law of large numbers:

Theorem 4.3 *Suppose $\{X_n\}$ is an irreducible Markov chain with transition kernel P and invariant distribution π, and let f be a real-valued function on E such that $\pi|f| < \infty$. Then $P_x\{\overline{f}_n \to \pi f\} = 1$ for π-almost all x, where $\overline{f}_n$ is given by (4.1).*

Theorems 4.2 and 4.3 show that the observed and expected proportion of time spent in a set A converge to $\pi(A)$. Stronger distributional results are possible but, as in the discrete case, it is necessary to rule out periodic or cyclic behavior. An m-cycle for an irreducible chain with transition kernel P is a collection $\{E_0, \ldots, E_{m-1}\}$ of disjoint sets such that $P(x, E_j) = 1$ for $j = i + 1 \bmod m$ and all $x \in E_i$. The period d of the chain is the largest m for which an m-cycle exists. The chain is aperiodic if $d = 1$.

Verifying aperiodicity can be rather difficult. A condition that is easier to verify is strong aperiodicity. An irreducible chain with invariant distribution π is strongly aperiodic if there exists a probability distribution ν on E, a constant $\beta > 0$ and a set $C \subset E$ such that $\nu(C) > 0$ and

$$P(x, A) \geq \beta\nu(A)$$

for all $x \in C$ and all $A \subset E$. Strong aperiodicity implies aperiodicity.

If a chain is aperiodic, then the result of Theorem 4.2 can be strengthened:

Theorem 4.4 *Suppose $\{X_n\}$ is an irreducible, aperiodic Markov chain with transition kernel P and invariant distribution π. Then*

$$\| P^n(x, \cdot) - \pi(\cdot) \| \to 0$$

for π-almost all x.

Even though most Markov chain theory emphasizes the aperiodic case, it is important to note that in MCMC experiments aperiodicity is of little importance since we are typically interested in results concerning sample path averages. It is also, at times, convenient to use chains that are periodic, for example when transition kernels are changed in a cyclic fashion.

4.4 Harris recurrence

The need to allow for a null set of initial values where the results of Theorems 4.1, 4.2, 4.3, and 4.4 might fail is a nuisance. It arises because condition *(ii)* in Definition 4.2 is allowed to fail on a null set as long as condition *(i)* holds everywhere. The concept of Harris recurrence is designed to eliminate this problem by requiring that condition *(ii)* should hold everywhere.

Definition 4.3 *An irreducible Markov chain with maximal irreducibility distribution ψ is Harris recurrent if for any $A \subset E$ with $\psi(A) > 0$ we have*

$$P_x\{X_n \in A \text{ infinitely often}\} = 1$$

for all $x \in E$.

The results of Theorems 4.1, 4.2, 4.3, and 4.4 hold for all initial values x if the chain is Harris recurrent. In fact, the convergence result in Theorem 4.4 holds for all x if and only if the chain is irreducible, aperiodic, and positive Harris recurrent with invariant distribution π.

Harris recurrence does not automatically follow from irreducibility. However, sufficient conditions that allow Harris recurrence to be verified are available for many important samplers. For example, an irreducible Markov chain with invariant distribution π is Harris recurrent if for some n the transition kernel P^n has a component that has a density with respect to π. This is satisfied by most Gibbs samplers and all pure Metropolis–Hastings samplers. A sufficient condition for a variable-at-a-time Metropolis–Hastings sampler to be Harris recurrent is given in Chan and Geyer (1994).

If it is not possible to ensure that a chain is Harris recurrent, then there is an absorbing subset H of the state-space E such that the restriction of the chain to H is Harris recurrent. The remainder is not only a π-null set, but is also dissipative – the chain essentially has to drift to infinity to avoid entering H. To state this precisely, we need the following definition.

Definition 4.4 *A set $A \subset E$ is transient for a Markov chain if the number of visits to A is finite with probability one for any starting point $x \in A$ of the chain. A set $A \subset E$ is dissipative if it is a countable union of transient sets.*

With this definition, we can state the decomposition result:

Theorem 4.5 *The state-space of an irreducible Markov chain with invariant distribution π can be decomposed as $E = H \cup N$, where H is absorbing, N is a π-null set and is dissipative, and the restriction of the chain to H is Harris recurrent. The set H is called a Harris set for the chain. The chain is Harris recurrent if and only if E is a Harris set.*

One consequence of this result is that the set N can be avoided by using an initial distribution that is absolutely continuous with respect to π, since this ensures that the starting point is in H with probability one. For a Harris set H, the results of Theorems 4.1, 4.2, 4.3, and 4.4 hold for all initial values $x \in H$.

For the random walk with transition matrix (4.2) the chain is Harris recurrent if $p \leq \frac{1}{2}$. But if $\frac{1}{2} < p < 1$, then the chain is recurrent but not

Harris recurrent; the Harris set is $H = \{0\}$ and the set $N = \{1, 2, \ldots\}$ is dissipative.

4.5 Mixing rates and central limit theorems

Recurrence alone is sufficient to ensure that a law of large numbers holds for a Markov chain. Stronger conditions are needed to provide a central limit theorem. Traditionally, stronger conditions are stated in terms of the rate of convergence in Theorem 4.4 for *ergodic* chains, i.e. chains that are irreducible, aperiodic and positive Harris recurrent. These convergence rate conditions are related to mixing conditions from stationary process theory. For a positive Harris recurrent chain, asymptotic results, such as laws of large numbers and central limit theorems, that do not depend on any initial finite portion of a sample path can be shown to hold for all initial distributions if they hold for any.

One convergence rate condition that is often considered is geometric ergodicity:

Definition 4.5 *An ergodic Markov chain with invariant distribution π is geometrically ergodic if there exist a non-negative extended real valued function M such that $\pi M < \infty$ and a positive constant $r < 1$ such that*

$$\| P^n(x, \cdot) - \pi(\cdot) \| \leq M(x)r^n$$

for all x and all n.

A geometrically ergodic Markov chain started with its invariant distribution is α-mixing at a geometric rate. That is,

$$\alpha(n) = \sup_{A,B \subset E} |P_\pi\{X_0 \in A, X_n \in B\} - \pi(A)\pi(B)| = O(r^n)$$

for some $r < 1$.

A stronger condition is uniform ergodicity:

Definition 4.6 *An ergodic Markov chain with invariant distribution π is uniformly ergodic if there exist a positive, finite constant M and a positive constant $r < 1$ such that*

$$\| P^n(x, \cdot) - \pi(\cdot) \| \leq Mr^n$$

for all x and all n.

Uniform ergodicity is equivalent to Doeblin's condition, and to exponential ϕ-mixing. That is, for a uniformly ergodic Markov chain

$$\phi(n) = \sup_{A,B \subset E} |P_\pi\{X_n \in B | X_0 \in A\} - \pi(B)| = O(r^n)$$

for some $r < 1$. Uniform ergodicity is the strongest convergence rate condition in common use.

A Markov chain that is geometrically or uniformly ergodic satisfies a central limit theorem:

Theorem 4.6 *Suppose an ergodic Markov chain $\{X_n\}$ with invariant distribution π and a real valued function f satisfy one of the following conditions:*

(i) The chain is geometrically ergodic and $\pi|f|^{2+\epsilon} < \infty$ for some $\epsilon > 0$.

(ii) The chain is uniformly ergodic and $\pi f^2 < \infty$.

Then

$$\sigma_f^2 = E_\pi[(f(X_0) - \pi f)^2] + 2 \sum_{k=1}^{\infty} E_\pi[(f(X_0) - \pi f)(f(X_k) - \pi f)]$$

is well defined, non-negative and finite, and $\sqrt{n}(\overline{f}_n - \pi f)$ converges in distribution to a $N(0, \sigma_f^2)$ random variable.

The result for geometric ergodicity is given in Chan and Geyer (1994).

As in the case of the law of large numbers, the assumption of aperiodicity is not needed for a central limit theorem to hold. For a periodic chain with period d, the chain with transition kernel P^d restricted to any of the components $\{E_0, \ldots, E_{m-1}\}$ of a d-cycle is aperiodic. One of these restricted chains is geometrically ergodic or uniformly ergodic if and only if all are. When this is the case, the chain is said to be geometrically recurrent or uniformly recurrent, respectively. The result of Theorem 4.6 remains valid if the assumptions of geometric or uniform ergodicity are replaced by geometric or uniform recurrence.

Geometric ergodicity or recurrence is usually verified using a drift condition. To give an example of such a condition, we need the concept of a small set:

Definition 4.7 *A set $C \subset E$ is small for an irreducible transition kernel P with maximal irreducibility distribution ψ if $\psi(C) > 0$ and there exist a probability distribution ν on E, a non-negative integer m, and a constant $\beta > 0$ such that $P^m(x, A) \geq \beta \nu(A)$ for all $x \in C$ and all $A \subset E$.*

In many problems compact sets are small; for general state-space chains, they play a role similar to individual states in discrete chain theory. A sufficient condition for geometric ergodicity is then given by the following theorem.

Theorem 4.7 *Suppose an ergodic Markov chain has the property that for some real valued function V with values in the interval $[1, \infty)$, some constants $\beta > 0$ and $b < \infty$ and a small set $C \subset E$*

$$PV(x) - V(x) \le -\beta V(x) + b I_C(x)$$

for all $x \in E$. Then the chain is geometrically ergodic.

The function V can often be thought of as an energy function that increases as x moves away from C. For points $x \notin C$, the inequality bounds the expected change in energy $PV(x) - V(x)$ above by $-\beta V(x)$, a negative value proportional to the current energy level. The negative bound means that the chain tends to move towards lower energy states, or to drift towards C. The proportionality of the bound to the current energy level ensures that the distribution of the return time to C has a geometric tail, which implies geometric ergodicity for an irreducible, aperiodic chain. Chan (1993), Rosenthal (1993), Mengersen and Tweedie (1994) and Roberts and Tweedie (1994) provide examples of using drift conditions such as this one to examine geometric recurrence of MCMC samplers.

A useful necessary and sufficient condition for uniform ergodicity is that a chain is uniformly ergodic if and only if the entire state-space is a small set for the chain. Roberts and Polson (1994) give sufficient conditions for uniform ergodicity based on continuity and compactness.

Even with results such as Theorem 4.7 at hand, verifying geometric ergodicity can be quite difficult; many reasonable samplers are conjectured to be geometrically ergodic but have not yet been shown to be so. As a stronger condition, uniform ergodicity is of course more restrictive. But the concept of uniform ergodicity interacts well with certain methods of combining samplers into hybrid chains. In particular, if P is constructed as a mixture in which at each step one of $P_1, \ldots, P_k$ is chosen according to positive probabilities $\alpha_1, \ldots, \alpha_k$, then the mixture hybrid kernel P is uniformly ergodic if any of the component kernels is uniformly ergodic. A slightly weaker result is available for cycle hybrid kernels of the form $P = P_1 \cdots P_k$. Thus if a MCMC sampler is not, or cannot be shown to be, uniformly ergodic, it may be possible to modify the sampler to make it so.

One example of a sampler where uniform ergodicity is easy to establish is an independence Metropolis–Hastings chain. Suppose candidates are generated from a density f and accepted or rejected with the usual Metropolis–Hastings rule. The probability of accepting a candidate value y when the current value of the chain is x is then

$$\alpha(x, y) = \min\left\{ \frac{w(y)}{w(x)}, 1 \right\}$$

where $w(x) = \pi(x)/f(x)$. The function w is the importance weight function that would be used in importance sampling to weight a sample from f

toward the distribution π. A chain using this kernel is uniformly ergodic if the function w is bounded. Boundedness is essential for this result; if the function w is not bounded, then Mengersen and Tweedie (1994) show that the chain is essentially not even geometrically ergodic.

Both mixture and cycle hybrid kernels that incorporate an independence component with a bounded weight function are uniformly ergodic. Experience from importance sampling can be helpful in finding candidate generating densities that produce bounded weight functions. For example, if π has a density that can be written as a product $\pi(x) = g(x)h(x)$ where g is bounded and h is a density that can be sampled, then taking $f = h$ produces a bounded weight function.

4.6 Regeneration

One area where there is an interesting interplay between theory and practice in MCMC is the use of regenerative methods. A process is regenerative if there is a sequence of random times at which the process starts over independently and identically. The tours of the process between these times are independent and identically distributed.

From a theoretical perspective, the existence of regeneration times provides an extremely powerful approach for deriving asymptotic results such as those of the previous section. From a practical point of view, the existence of regeneration times allows parallel generation of independent tours using a parallel computing environment and also allows the use of regenerative simulation output analysis methods (e.g. Ripley, 1987). If a regenerative simulation is run for a fixed number of tours, then initialization issues do not arise since the simulation is started at a regeneration, and variance estimation methods based on independent identically distributed random variables can be used to assess the accuracy of results. This alleviates some of the difficulties of analysing output from dependent simulations that are not started in equilibrium. Of course, this is only a method of analysis and does not by itself improve a simulation. If a regenerative simulation process mixes slowly, then this will be reflected in a very heavy-tailed tour length distribution.

For an ergodic discrete Markov chain, regeneration times are easily identified. One need only fix a particular state and consider the times at which the chain returns to this state. Each time the chain enters the distinguished state, it starts a new tour with the same distribution, regardless of the preceding sample path; this implies that successive tours are independent and identically distributed.

Recent theoretical results show that any ergodic general state-space Markov chain has, or can be modified to have, a series of regeneration times. These results establish the existence of regeneration times but do not necessarily lead to practical methods for computing them. Approaches for computing regeneration times for general state-space Markov chains are available, and some are outlined in Mykland *et al.* (1995). The easiest way to introduce regeneration into a MCMC sampler is to use a hybrid algorithm in which a basic sampler, a Gibbs sampler for example, is combined in a mixture or a cycle with a particular independence kernel where the candidate (proposal) density f is sampled by a rejection sampling algorithm.

To construct the rejection sampling independence chain, suppose π has a density $\pi(x)$, and that we can find a density h and a constant c such that the set $C = \{x \in E : \pi(x) \leq ch(x)\}$ has reasonably high probability under h. We can then generate independent variables Y from h and U uniform on $[0, 1]$ and continue to generate such pairs independently until we obtain a pair that satisfies $Uch(Y) \leq \pi(Y)$. The final Y produced by this rejection algorithm has density $f(x) \propto \min\{\pi(x), ch(x)\}$. Using this Y as a candidate for a Metropolis–Hastings algorithm produces an acceptance probability for the candidate of

$$
\alpha(x, y) = \begin{cases} 1 & \text{for } x \in C \\ 1/w(x) & \text{for } x \notin C, y \in C \\ \min\{w(y)/w(x), 1\} & \text{for } x \notin C, y \notin C. \end{cases}
$$

Whenever the current state of the chain is in C, any candidate is accepted. Thus, when the chain is in C, the next state is generated as a draw from the distribution f, no matter what the particular state within C is. The set C thus plays a role similar to the distinguished state in the discrete case mentioned above, and visits to C therefore represent regeneration times. When a kernel of this form is used in a mixture or a cycle, regenerations occur each time this kernel is used and the chain is in the set C.

The effectiveness of this hybrid/rejection approach depends on the choice of h and c as well as the frequency with which the rejection kernel is used. Some experimentation is needed in any particular problem to determine good choices.

Other approaches to introducing regeneration are also under investigation. The simulated tempering algorithm described by Geyer and Thompson (1995) produces regenerations each time the chain uses the 'hottest' chain if that chain produces independent draws from some distribution; see Gilks and Roberts (1995: this volume). Mengersen and Tweedie (1994) describe a lumping algorithm in which part of the state-space is collapsed into a single state.

4.7 Discussion

This chapter has outlined some of the tools provided by general state-space Markov chain theory that can be used to investigate properties of MCMC samplers. General state-space Markov chain theory is still an area of active development, and further work is likely to produce better characterizations of rates of convergence and weaker sufficient conditions for central limit theorems.

An important point which deserves repeating is that these samplers are not an end in themselves but merely a means to an end: that of exploring the distribution π. It is thus reasonable to allow the sampler to be modified to allow its properties to be improved or to make the analysis of its properties easier. At an exploratory stage, it is even reasonable to modify the distribution π to make it easier to examine and develop an understanding of parts of it, if the complete distribution is initially too hard to explore all at once. Mixture and cycle hybrid samplers are one useful approach for merging chains; methods such as simulated tempering are another. Some results for deriving properties of such compound strategies from their components are available, but more are needed.

There are a number of open issues that are not addressed by the theory outlined here. One such issue is the performance of samplers when the distribution π is improper. Whether there is any statistical use for samplers with an improper invariant distribution is not clear, as there is considerable doubt about the usefulness of improper posterior distributions. But the possibility of an improper invariant distribution π does raise some interesting theoretical questions. It is clear that an irreducible sampler with an improper invariant distribution cannot be positive recurrent. If it is null recurrent, ratio limit theorems are available and may be of some use. But it is far from obvious how to determine whether such a chain is null recurrent or not.

Another open issue concerns adaptive choice of sampling strategies. It is reasonable to attempt to choose parameters of a sampler to fit the distribution π. As the sampler is run and more information about π becomes available, it may be useful to adjust the parameter settings. If these adjustments are confined to an initial phase then standard theory can be used to analyse the subsequent behaviour of the chain. But if adjustment is allowed to continue indefinitely, then the sampler may no longer be a Markov chain and other methods will be needed to assess its behaviour. Similar issues arise in studying samplers used in simulated annealing, and some of the results from that literature may be applicable.

Finally, most results discussed in this chapter deal with long-run behaviour of a sampler. A sampler can only be run for a finite period, and it would be useful to know whether that period is long enough. Some information can be obtained using standard simulation variance estimation methods if it can be assumed that the sampler has been run long enough to have reached equilibrium in some reasonable sense. Clearly, a diagnostic that detects this would be very useful, and considerable effort has been expended in recent years to develop such diagnostics. Most proposals for universal diagnostics have been based on the observed sample path, the unnormalized values of the density of π along the sample path, or the conditional distributions of individual components with others held fixed at observed sample path points. Similar efforts can be found in the general simulation literature under the headings of initialization bias or initial transient detection. Recently, Asmussen *et al.* (1992) have give a formal version of a negative result: a diagnostic that will work for any simulation problem does not exist. A calculation in the same spirit is possible for MCMC. In particular, one can show that no diagnostic based only on the information mentioned above can be universally effective. Such negative theoretical results may be useful to retarget efforts towards designing classes of diagnostic methods to take advantage of more special structure of the distribution π or the sampler used in a particular problem.

References

Asmussen, S., Glynn, P. W. and Thorisson, H. (1992) Stationarity detection in the initial transient problem. *ACM Trans. Model. Comput. Simul.*, **2**, 130–157.

Chan, K. S. (1993) Asymptotic behavior of the Gibbs sampler. *J. Am. Statist. Ass.*, **88**, 320–326.

Chan, K. S. and Geyer C. J. (1994) Discussion on Markov chains for exploring posterior distributions (by L. Tierney). *Ann. Statist.*, **22**, 1747–1758.

Geyer, C. J. and Thompson, E. A. (1995) Annealing Markov chain Monte Carlo with applications to pedigree analysis. *J. Am. Statist. Ass.*, (in press).

Gilks, W. R. and Roberts, G. O. (1995) Strategies for improving MCMC. In *Markov Chain Monte Carlo in Practice* (eds W. R. Gilks, S. Richardson and D. J. Spiegelhalter), pp. 89–114. London: Chapman & Hall.

Mengersen, K. L. and Tweedie, R. L. (1994) Rates of convergence for the Hastings and Metropolis algorithms. *Technical Report*, Department of Statistics, Colorado State University.

Meyn, S. P. and Tweedie, R. L. (1993) *Markov Chains and Stochastic Stability.* New York: Springer-Verlag.

Mykland, P., Tierney, L. and Yu, B. (1995) Regeneration in Markov chain samplers. *J. Am. Statist. Ass.*, **90**, 233–241.

Nummelin, E. (1984) *General Irreducible Markov Chains and Non-negative Operators.* Cambridge: Cambridge University Press.

Ripley, B. D. (1987) *Stochastic Simulation.* New York: Wiley.

Roberts, G. O. (1995) Markov chain concepts related to sampling algorithms. In *Markov Chain Monte Carlo in Practice* (eds W. R. Gilks, S. Richardson and D. J. Spiegelhalter), pp. 45–57. London: Chapman & Hall.

Roberts, G. O. and Polson N. (1994) On the geometric convergence of the Gibbs sampler. *J. R. Statist. Soc.* B, **56**, 377-384.

Roberts, G. O. and Tweedie, R. L. (1994) Geometric convergence and central limit theorems for multidimensional Hastings and Metropolis algorithms. *Research Report 94.9*, Statistical Laboratory, University of Cambridge, UK.

Rosenthal, J. S. (1993) Minorization conditions and convergence rates for Markov chain Monte Carlo. *Technical Report*, School of Mathematics, University of Minnesota.

Smith, A. F. M. and Roberts, G. O. (1993) Bayesian computation via the Gibbs sampler and related Markov chain Monte Carlo methods. *J. R. Statist. Soc. B*, **55**, 3–24.

Taylor, H. M. and Karlin, S. (1984) *An Introduction to Stochastic Modeling.* Orlando: Academic Press.

Tierney, L. (1994) Markov chains for exploring posterior distributions (with discussion). *Ann. Statist.*, **22**, 1701–1762.

5

Full conditional distributions

Walter R Gilks

5.1 Introduction

As described in Gilks *et al.* (1995b: this volume), Gibbs sampling involves little more than sampling from full conditional distributions. This chapter shows how full conditional distributions are derived, and describes methods for sampling from them.

To establish notation, vector X denotes a point in the state-space of the Gibbs sampler and $\pi(X)$ denotes its stationary distribution. The elements of X are partitioned into k components $(X_{.1}, X_{.2}, \ldots, X_{.k})$. Each of the k components of X may be scalar or vector. We define an iteration of the Gibbs sampler to be an updating of one component of X; X_t denotes the state of X at iteration t. Vector X without component s is denoted $X_{.-s} = (X_{.1}, \ldots, X_{.s-1}, X_{.s+1}, \ldots, X_{.k})$. The full conditional distribution for $X_{.s}$ at iteration t is denoted $\pi(X_{.s}|X_{t.-s})$. To avoid measure-theoretic notation, all random variables are assumed real and continuous, although much of this chapter applies also to other kinds of variable. $P(.)$ generically denotes a probability density function.

5.2 Deriving full conditional distributions

Full conditional distributions are derived from the joint distribtion of the variables:

$$\pi(X_{.s}|X_{t.-s}) \quad = \quad \frac{\pi(X_{.s}, X_{t.-s})}{\int \pi(X_{.s}, X_{t.-s})dX_{.s}}. \qquad (5.1)$$

5.2.1 A simple example

Consider the following simple two-parameter Bayesian model:

$$y_i \sim \text{N}(\mu, \tau^{-1}), \qquad i = 1, \dots, n; \tag{5.2}$$
$$\mu \sim \text{N}(0, 1);$$
$$\tau \sim \text{Ga}(2, 1),$$

where $\text{N}(a, b)$ generically denotes a normal distribution with mean a and variance b, and $\text{Ga}(a, b)$ generically denotes a gamma distribution with mean a/b and variance a/b^2. Here we assume the $\{y_i\}$ are conditionally independent given μ and τ, and μ and τ are themselves independent. Let $y = \{y_i; \ i = 1, \dots, n\}$.

The joint distribution of y, μ and τ is

$$
\begin{aligned}
P(y, \mu, \tau) &= \prod_{i=1}^{n} P(y_i|\mu, \tau) P(\mu) P(\tau) \\
&= (2\pi)^{-\frac{n+1}{2}} \tau^{\frac{n}{2}} \exp\left\{-\frac{\tau}{2}\Sigma(y_i - \mu)^2\right\} \exp\left\{-\frac{1}{2}\mu^2\right\} \tau e^{-\tau}.
\end{aligned}
\tag{5.3}
$$

When y is observed, the joint posterior distribution of μ and τ is

$$\pi(\mu, \tau) = P(\mu, \tau|y) = \frac{P(y, \mu, \tau)}{\int P(y, \mu, \tau) d\mu d\tau}. \tag{5.4}$$

From (5.1) and (5.4), the full conditional for μ is

$$
\begin{aligned}
\pi(\mu|\tau) &= \frac{P(\mu, \tau|y)}{P(\tau|y)} \\
&= \frac{P(y, \mu, \tau)}{P(y, \tau)} \\
&\propto P(y, \mu, \tau).
\end{aligned}
\tag{5.5}
$$

Here, proportionality follows because $\pi(\mu|\tau)$ is a distribution for μ, and the denominator of (5.5) does not depend on μ. Thus, to construct the full conditional for μ, we need only pick out the terms in (5.3) which involve μ, giving:

$$
\begin{aligned}
\pi(\mu|\tau) &\propto \exp\left\{-\frac{\tau}{2}\Sigma(y_i - \mu)^2\right\} \exp\left\{-\frac{1}{2}\mu^2\right\} \\
&\propto \exp\left\{-\frac{1}{2}(1 + n\tau)\left(\mu - \frac{\tau\Sigma y_i}{1 + n\tau}\right)^2\right\}.
\end{aligned}
$$

Thus, the full conditional for μ is a normal distribution with mean $\frac{\tau\Sigma y_i}{1+n\tau}$ and variance $(1 + n\tau)^{-1}$. Similarly, the full conditional for τ depends only on the terms in (5.3) involving τ, giving:

$$\pi(\tau|\mu) \quad \propto \quad \tau^{\frac{n}{2}} \exp\left\{-\frac{\tau}{2}\Sigma(y_i - \mu)^2\right\} \tau e^{-\tau}$$

$$= \quad \tau^{1+\frac{n}{2}} \exp\left\{-\tau\left[1 + \frac{1}{2}\Sigma(y_i - \mu)^2\right]\right\},$$

which is the kernel of a gamma distribution with index $2 + \frac{n}{2}$ and scale $1 + \frac{1}{2}\Sigma(y_i - \mu)^2$.

In this simple example, prior distributions are conjugate to the likelihood (5.2), so full conditionals reduce analytically to closed-form distributions. Highly efficient sampling routines are available for these distributions; see for example Ripley (1987).

5.2.2 Graphical models

Full conditional distributions for complex models can also be constructed easily. In particular, for Bayesian directed acyclic graphical (DAG) models, the joint distribution of the data and parameters is a product of many terms, each involving only a subset of the parameters. For such models, the full conditional distribution for any given parameter can be constructed from those few terms of the joint distribution which depend on it; see Spiegelhalter *et al.* (1995b: this volume).

Normal random-effects model

For example, consider the random-effects model:

$$\begin{aligned}
y_{ij} &\sim N(\alpha_i, \tau^{-1}), & j = 1, \ldots, m_i, & \quad i = 1, \ldots, n; \\
\alpha_i &\sim N(\mu, \omega^{-1}), & i = 1, \ldots, n; \\
\mu &\sim N(0, 1); \\
\tau &\sim Ga(2, 1); \\
\omega &\sim Ga(1, 1),
\end{aligned}$$

where we assume independence between the $\{y_{ij}\}$ given all model parameters; between the $\{\alpha_i\}$ given the hyperparameters μ, τ and ω; and between the hyperparameters themselves. The joint distribution of the data and parameters for this model is:

$$P(y, \alpha, \mu, \tau, \omega) = \prod_{i=1}^{n}\left\{\prod_{j=1}^{m_i} P(y_{ij}|\alpha_i, \tau)P(\alpha_i|\mu, \omega)\right\} P(\mu)P(\tau)P(\omega).$$

Then the full conditional for α_i is

$$\pi(\alpha_i|y, \alpha_{-i}, \mu, \tau, \omega) \quad \propto \quad \prod_{j=1}^{m_i} P(y_{ij}|\alpha_i, \tau)P(\alpha_i|\mu, \omega) \tag{5.6}$$

$$\propto \quad \exp\left\{-\frac{1}{2}(\omega + m_i\tau)\left(\alpha_i - \frac{\omega\mu + \tau\sum_{j=1}^{m_i} y_{ij}}{\omega + m_i\tau}\right)^2\right\},$$

which is a normal distribution with mean

$$\frac{\omega\mu + \tau\Sigma_{j=1}^{m_i} y_{ij}}{\omega + m_i\tau}$$

and variance $(\omega + m_i\tau)^{-1}$.

Logistic regression model

Although for DAG models it is trivial to write down expressions for full conditionals, as in (5.6), it is often not possible to make further progress analytically. For example, consider the following Bayesian logistic regression model of y on covariate z:

$$y_i \quad \sim \quad \text{Bernoulli}\left(\frac{1}{1 + e^{-(\mu + \alpha z_i)}}\right), \qquad i = 1, \dots, n; \qquad (5.7)$$
$$\alpha \quad \sim \quad N(0, 1);$$
$$\mu \quad \sim \quad N(0, 1),$$

where we assume conditional independence between the $\{y_i\}$ given the model parameters and covariates, and independence between the parameters themselves. Here, the full conditional for α is

$$\pi(\alpha|\mu) \quad \propto \quad e^{-\frac{1}{2}\alpha^2} \prod_{i=1}^{n} \{1 + e^{-(\mu + \alpha z_i)}\}^{-y_i} \{1 + e^{\mu + \alpha z_i}\}^{y_i - 1}, \qquad (5.8)$$

which unfortunately does not simplify. Thus methods are required for sampling from arbitrarily complex full conditional distributions. This is the subject of the remainder of this chapter.

Undirected graphical models

For non-DAG models, full conditionals may be difficult to derive, although for some partially-DAG models the derivation is straightforward; see for example Mollié (1995: this volume).

5.3 Sampling from full conditional distributions

Full conditionals change from iteration to iteration as the conditioning $X_{t.-s}$ changes, so each full conditional is used only once and then disposed of. Thus it is essential that sampling from full conditional distributions is highly efficient computationally. When analytical reduction of a full conditional is not possible, it will be necessary to evaluate the full conditional function at a number of points, and in typical applications each function evaluation will be computationally expensive. Thus any method for sampling from full conditional distributions should aim to minimize the number of function evaluations. Sampling methods such as inversion (see Ripley, 1987), which require a large number of function evaluations, should be avoided if possible.

Two techniques for sampling from a general density $g(y)$ are rejection sampling and the ratio-of-uniforms method. A third method, which does not produce independent samples, is the Metropolis–Hastings algorithm. All three methods can be used for sampling multivariate distributions, and none require evaluation of the normalizing constant for g. This is an important practical point, since the normalizing constant for full conditional distributions is typically unavailable in closed form (as in (5.8), for example). We now describe these methods, and hybrids of them, for sampling from full conditional distributions. Below, Y represents $X_{t+1.s}$ and $g(Y)$ is proportional to the density of interest $\pi(X_{t+1.s}|X_{t.-s})$.

5.3.1 Rejection sampling

Rejection sampling requires an envelope function G of g (so $G(Y) \geq g(Y)$ for all Y: see Figure 5.1). Samples are drawn from the density proportional to G, and each sampled point Y is subjected to an accept/reject test. This test takes the form: accept point Y with probability $g(Y)/G(Y)$. If the point is not accepted, it is discarded. Sampling continues until the required number of points have been accepted: for Gibbs sampling just one point is required from each full conditional g. Accepted points are then exactly independent samples from the density proportional to g (see for example Ripley, 1987).

The algorithm then is:

```
Repeat {
    Sample a point Y from G(.);
    Sample a Uniform(0, 1) random variable U;
    If  U ≤ g(Y)/G(Y) accept Y; }
until one Y is accepted.
```

Several rejections may occur before an acceptance. Each accept/reject test involves evaluating $g(Y)$ and $G(Y)$, and typically the former will be computationally expensive. Marginally, the probability of accepting a point is $\int g(Y)dY / \int G(Y)dY$, so to reduce the number of rejections, it is essential that the envelope G be close to g. For computational efficiency, it is also essential that G be cheap to evaluate and sample from.

Some computational savings may result from using *squeezing functions* $a(Y)$ and $b(Y)$, where $a(Y) \geq g(Y) \geq b(Y)$ for all Y, and a and b are cheaper to evaluate than g (see Figure 5.1). The accept/reject test on line 4 of the above algorithm can then be replaced by

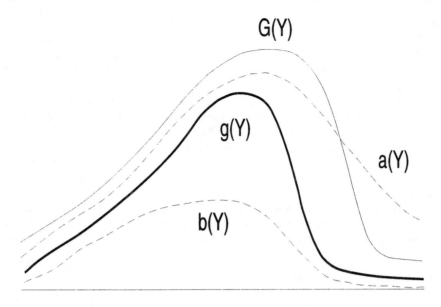

Figure 5.1 *Functions for rejection sampling. Thin line: envelope $G(Y)$; heavy line: density $g(Y)$; broken lines: squeezing functions $a(Y)$ and $b(Y)$.*

```
If  U > a(Y)/G(Y) reject Y;
   else if  U ≤ b(Y)/G(Y) accept Y;
   else if  U ≤ g(Y)/G(Y) accept Y.
```

The first two tests enable a decision to be made about Y without calculating $g(Y)$.

Zeger and Karim (1991) and Carlin and Gelfand (1991) propose rejection sampling for multivariate full conditional distributions, using multivariate normal and multivariate split-t distributions as envelopes. A difficulty with these methods is in establishing that the proposed envelopes are true envelopes. Bennett *et al.* (1995: this volume) use rejection sampling for multivariate full conditional distributions in nonlinear models. For the envelope function G, they use the prior distribution multiplied by the likelihood at the maximum likelihood estimate.

5.3.2 Ratio-of-uniforms method

Suppose Y is univariate. Let U and V be two real variables, and let $\mathcal{D}$ denote a region in U, V space defined by $0 \leq U \leq \sqrt{g(V/U)}$ (see Figure 5.2). Sample a point U, V uniformly from $\mathcal{D}$. This can be done by first determining an envelope region $\mathcal{E}$ which contains $\mathcal{D}$ and from which it is easy to sample uniformly. U and V can then be generated by rejection sampling from $\mathcal{E}$. Rather surprisingly, $Y = V/U$ is a sample from the density proportional to g (see for example Ripley, 1987).

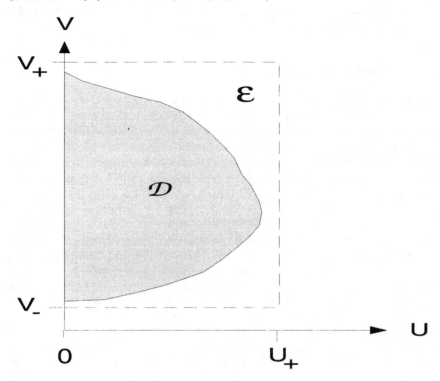

Figure 5.2 *An envelope $\mathcal{E}$ (broken line) for a region $\mathcal{D}$ defined by $0 \leq U \leq \sqrt{g(V/U)}$, for the ratio-of-uniforms method.*

Typically $\mathcal{E}$ is chosen to be a rectangle with vertices at $(0, v_-)$; (u_+, v_-); $(0, v_+)$; and (u_+, v_+), where constants u_+, v_- and v_+ are such that $\mathcal{E}$ contains $\mathcal{D}$. This leads to the following algorithm.

```
Determine constants u_+, v_-, v_+;
Repeat {
    Sample a Uniform(0, u_+) random variable U;
    Sample a Uniform(v_-, v_+) random variable V;
    If (U, V) is in D, accept Y = V/U; }
until one Y is accepted.
```

As in pure rejection sampling, it is important for computational efficiency to keep the number of rejections low. If squeezing regions can be found, efficiency may be improved. Wakefield *et al.* (1991) give a multivariate generalization of the ratio-of-uniforms method, and suggest variable transformations to improve its efficiency. Bennett *et al.* (1995: this volume) compare the ratio-of-uniforms method with other methods for sampling from full conditional distributions in nonlinear models.

5.3.3 Adaptive rejection sampling

The practical problem with both rejection sampling and the ratio-of-uniforms method is in finding a tight envelope function G or region $\mathcal{E}$. Often this will involve time-consuming maximizations, exploiting features peculiar to g. However, for the important class of log-concave univariate densities, efficient methods of envelope construction have been developed. A function $g(Y)$ is log-concave if the determinant of $\frac{d^2 \log g}{dY \, dY^T}$ is non-positive.

In many applications of Gibbs sampling, all full conditional densities $g(Y)$ are log-concave (Gilks and Wild, 1992). In particular, this is true for all generalized linear models with canonical link function (Dellaportas and Smith, 1993). For example, full conditional distributions in the logistic regression model (5.7) are log-concave. Gilks and Wild (1992) show that, for univariate Y, an envelope function $\log G_{\mathcal{S}}(Y)$ for $\log g(Y)$ can be constructed by drawing tangents to $\log g$ at each abscissa in a given set of abscissae $\mathcal{S}$. An envelope between any two adjacent abscissae is then constructed from the tangents at either end of that interval (Figure 5.3(a)). An alternative envelope construction which does not require evaluation of derivatives of $\log g$ is given by Gilks (1992). For this, secants are drawn through $\log g$ at adjacent abscissae, and the envelope between any two adjacent abscissae is constructed from the secants immediately to the left and right of that interval (Figure 5.3(b)). For both constructions, the envelope is piece-wise exponential, from which sampling is straightforward. Also, both constructions automatically provide a lower squeezing function $\log b_{\mathcal{S}}(Y)$.

Three or four starting abscissae usually suffice, unless the density is exceptionally concentrated. Both methods require starting abscissae to be placed on both sides of the mode if the support of g is unbounded. This does not involve locating the mode, since gradients of tangents or secants determine whether the starting abscissae are acceptable. If desired, starting abscissae can be set with reference to the envelope constructed at the previous Gibbs iteration.

The important feature of both of these envelope constructions is that they can be used *adaptively*. When a Y is sampled, $g(Y)$ must be evaluated to perform the rejection step. Then, with negligible computational cost, the point $(Y, g(Y))$ can be incorporated in the envelope, just as if Y had been among the initial abscissae. This is called *adaptive rejection sampling* (ARS):

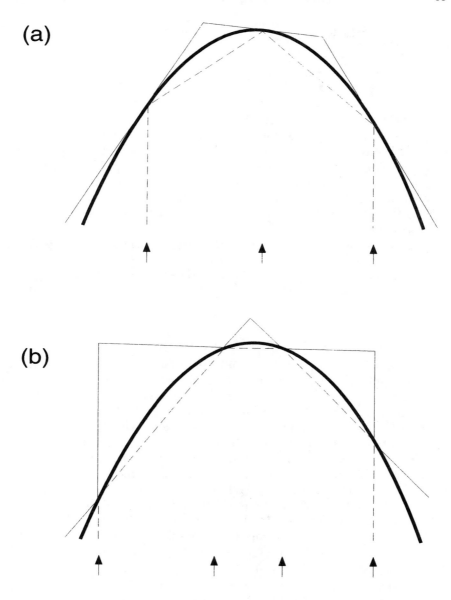

Figure 5.3 *Adaptive rejection sampling: (a) tangent method; (b) secant method. Heavy line:* $\log g(Y)$*; thin line: envelope* $\log G_S(Y)$*; broken line: squeezing function* $\log b_S(Y)$*; arrows: abscissae used in the construction.*

```
Initialize S
Repeat {
  Sample Y from G_S(.);
  Sample U from Uniform(0,1);
  If  U ≤ g(Y)/G_S(Y) accept Y;
  Include Y in S; }
until one Y is accepted.
```

At each iteration of ARS, the envelope $G_S(Y)$ is brought closer to g and the risk of further rejections and function evaluations is reduced. To accept one Y, the tangent version of adaptive rejection sampling typically involves about four function evaluations including those at the initial abscissae; for the secant version, five or six function evaluations are usually required. These performance figures are surprisingly robust to location of starting abscissae and to the form of g.

Multivariate generalizations of adaptive rejection sampling are possible, but have not yet been implemented. The amount of computation for such methods could be of order m^5, where m is the number of dimensions. Thus multivariate adaptive rejection sampling would probably be useful only in low dimensions.

5.3.4 Metropolis–Hastings algorithm

When an approximation $h(Y)$ to full conditional $g(Y)$ is available, from which sampling is easy, it is tempting to sample from h instead of from g. Then ergodic averages calculated from the output of the Gibbs sampler will not correspond exactly to π, no matter how long the chain is run. Ritter and Tanner (1992) propose grid-based methods for approximate sampling from full conditional distributions, successively refining the grid as the iterations proceed to reduce the element of approximation. Thus, approximation is improved at the cost of increasing computational burden.

Tierney (1991) and Gelman (1992) suggest a way to sample from approximate full conditional distributions whilst maintaining exactly the required stationary distribution of the Markov chain. This involves using the approximate full conditional h as a proposal distribution in an independence-type Metropolis–Hastings algorithm (see Gilks *et al.*, 1995b: this volume):

```
Sample a point Y from h(.);
Sample a Uniform(0,1) random variable U;
If  U ≤ min[1, g(Y)h(Y')/g(Y')h(Y)] accept Y;
  else set Y equal to Y';
```

where $Y' = X_{t,s}$ is the 'old' value of $X_{.s}$. Note that only one iteration of Metropolis–Hastings is required, because if X_t is from π, then so is $X_{t+1} = (X_{t+1.s}, X_{t.-s})$. Note also that multivariate full conditionals can be handled using this technique.

If $g(Y)$ is unimodal and not heavy-tailed, a convenient independence-type proposal $h(Y)$ might be a normal distribution whose scale and location are chosen to match g, perhaps via a least-squares fit of $\log h$ to $\log g$ at several well-spaced points. For more complex g, proposals could be mixtures of normals or scale- and location-shifted t-distributions. In general, if h approximates g well, there will be few Metropolis–Hastings rejections, and this will generally assist mixing in the Markov chain. However, there is clearly a trade-off between reducing the rejection rate and the computational burden of calculating good approximations to g.

The above algorithm is no longer purely Gibbs sampling: it produces a different Markov chain but with the same stationary distribution π. The proposal density h need not be an approximation to g, nor need it be of the independence type. Tierney (1991) and Besag and Green (1993) suggest that it can be advantageous to use an h which is distinctly different from g, to produce an antithetic variables effect in the output which will reduce Monte-Carlo standard errors in ergodic averages. Such chains have been called 'Metropolis–Hastings-within-Gibbs', but as the original algorithm described by Metropolis $et~al.$ (1953) uses single-component updating, the term 'single-component Metropolis–Hastings' is more appropriate (Besag and Green, 1993).

5.3.5 Hybrid adaptive rejection and Metropolis–Hastings

Tierney (1991) discusses the use of Metropolis–Hastings in conjunction with rejection sampling. Extending this idea, ARS can be used to sample adaptively from non-log-concave univariate full conditional distributions. For non-log-concave densities, the 'envelope' functions $G_{\mathcal{S}}(Y)$ calculated as described in Section 5.3.3 may not be true envelopes; in places the full conditional $g(Y)$ may protrude above $G_{\mathcal{S}}(Y)$. Then the sample delivered by ARS will be from the density proportional to $h(Y) = \min[g(Y), G_{\mathcal{S}}(Y)]$, where $\mathcal{S}$ is the set of abscissae used in the final accept/reject step of ARS. A sample Y from g can then be obtained by appending the following Metropolis–Hastings step to ARS:

> Sample U from Uniform$(0, 1)$;
> If $U \le \min\{1, \frac{g(Y)h(Y')}{g(Y')h(Y)}\}$ accept Y;
>   else set Y equal to Y'.

Here, as before, $Y' = X_{t.s}$. This is $adaptive~rejection~Metropolis~sampling$ (ARMS) (Gilks $et~al.$, 1995a).

ARMS works well when g is nearly log-concave, and reduces to ARS when g is exactly log-concave. When g is grossly non-log-concave, ARMS still delivers samples from g, but rejections at the Metropolis–Hastings step will be more frequent. As for ARS, the initial set of abscissae in $\mathcal{S}$ may be chosen to depend on $G_{\mathcal{S}}$ constructed at the previous Gibbs iteration.

Besag *et al.* (1995) note that the number of iterations in the repeat loop of ARS (or ARMS) is unbounded, and suggest using Metropolis–Hastings to curtail the number of these iterations. Roberts *et al.* (1995) suggest the following implementation of that idea. Let c denote the maximum permitted number of iterations of the repeat loop of ARS. If each of the c iterations result in rejection, perform a Metropolis–Hastings step as for ARMS above, but with $h(Y) = G_S(Y) - \min[g(Y), G_S(Y)]$, and using the value of Y generated at the cth step of ARS. It is unlikely that curtailment for log-concave g would offer computational advantages, since log-concavity ensures acceptance of a Y within very few iterations. However, curtailment for very non-log-concave g may sometimes be worthwhile.

5.4 Discussion

In general, the Gibbs sampler will be more efficient (better mixing) if the number of components k of X is small (and the dimensionality of the individual components is correspondingly large). However, sampling from complex multivariate distributions is generally not possible unless MCMC itself is used, as in Section 5.3.4. Why not therefore abandon Gibbs sampling in favour of Metropolis–Hastings applied to the whole of X simultaneously? Often this would be a sensible strategy, but Metropolis–Hastings requires finding a reasonably efficient proposal distribution, which can be difficult in problems where dimensions are scaled very differently to each other. In many problems, Gibbs sampling applied to univariate full conditional distributions works well, as demonstrated by the wealth of problems efficiently handled by the BUGS software (Spiegelhalter *et al.*, 1994, 1995a), but for difficult problems and for robust general-purpose software, hybrid methods are likely to be most powerful. See Gilks and Roberts (1995: this volume) for a discussion of techniques for improving the efficiency of MCMC, and Bennett *et al.* (1995: this volume) for a comparison of various methods for sampling from full conditional distributions in the context of nonlinear models.

FORTRAN code for ARS and C code for ARMS are available from the author (e-mail wally.gilks@mrc-bsu.cam.ac.uk).

References

Bennett, J. E., Racine-Poon, A. and Wakefield, J. C. (1995) MCMC for nonlinear hierarchical models. In *Markov Chain Monte Carlo in Practice* (eds W. R. Gilks, S. Richardson and D. J. Spiegelhalter), pp. 339–357. London: Chapman & Hall.

Besag, J. and Green, P. J. (1993) Spatial statistics and Bayesian computation. *J. R. Statist. Soc.* B, **55**, 25–37.

Besag, J., Green, P. J., Higdon, D. and Mengerson, K. (1995) Bayesian computation and stochastic systems. *Statist. Sci.*, **10**, 3–41.

Carlin, B. P. and Gelfand, A. E. (1991) An iterative Monte Carlo method for nonconjugate Bayesian analysis. *Statist. Comput.*, **1**, 119–128.

Dellaportas, P. and Smith, A. F. M. (1993) Bayesian inference for generalised linear and proportional hazards models via Gibbs sampling. *Appl. Statist.*, **42**, 443–460.

Gelman, A. (1992) Iterative and non-iterative simulation algorithms. In *Computing Science and Statistics* (ed. H. J. Newton), pp. 433–438. Fairfax Station: Interface Foundation of North America.

Gilks, W. R. (1992) Derivative-free adaptive rejection sampling for Gibbs sampling. In *Bayesian Statistics 4* (eds J. M. Bernardo, J. O. Berger, A. P. Dawid, and A. F. M. Smith), pp. 641–649. Oxford: Oxford University Press.

Gilks, W. R. and Roberts, G. O. (1995) Strategies for improving MCMC. In *Markov Chain Monte Carlo in Practice* (eds W. R. Gilks, S. Richardson and D. J. Spiegelhalter), pp. 89–114. London: Chapman & Hall.

Gilks, W. R. and Wild, P. (1992) Adaptive rejection sampling for Gibbs sampling. *Appl. Statist.*, **41**, 337–348.

Gilks, W. R., Best, N. G. and Tan, K. K. C. (1995a) Adaptive rejection Metropolis sampling within Gibbs sampling. *Appl. Statist.*, (in press).

Gilks, W. R., Richardson, S. and Spiegelhalter, D. J. (1995b) Introducing Markov chain Monte Carlo. In *Markov Chain Monte Carlo in Practice* (eds W. R. Gilks, S. Richardson and D. J. Spiegelhalter), pp. 1–19. London: Chapman & Hall.

Metropolis, N, Rosenbluth, A. W., Rosenbluth, M. N., Teller, A. H. and Teller, E. (1953) Equations of state calculations by fast computing machine. *J. Chem. Phys.*, **21**, 1087–1091.

Mollié, A. (1995) Bayesian mapping of disease. In *Markov Chain Monte Carlo in Practice* (eds W. R. Gilks, S. Richardson and D. J. Spiegelhalter), pp. 359–379. London: Chapman & Hall.

Ripley, B. D. (1987) *Stochastic Simulation*. New York: Wiley.

Ritter, C. and Tanner, M. A. (1992) Facilitating the Gibbs sampler: the Gibbs stopper and the griddy–Gibbs sampler. *J. Am. Statist. Ass.*, **87**, 861–868.

Roberts, G. O., Sahu, S. K. and Gilks, W. R. (1995) Discussion on Bayesian computation and stochastic systems (by J. Besag, P. J. Green, D. Higdon and K. Mengerson). *Statist. Sci.*, **10**, 49–51.

Spiegelhalter, D. J., Thomas, A. and Best, N. G. (1995a). Computation on Bayesian graphical models. In *Bayesian Statistics 5* (eds J. M. Bernardo, J. Berger, A. P. Dawid and A. F. M. Smith). Oxford: Oxford University Press (in press).

Spiegelhalter, D. J., Best, N. G., Gilks, W. R. and Inskip, H. (1995b) Hepatitis B: a case study in MCMC methods. In *Markov Chain Monte Carlo in Practice* (eds W. R. Gilks, S. Richardson and D. J. Spiegelhalter), pp. 21–43. London: Chapman & Hall.

Spiegelhalter, D. J., Thomas, A., Best, N. G. and Gilks, W. R. (1994) *BUGS: Bayesian inference Using Gibbs Sampling.* Cambridge: MRC Biostatistics Unit.

Tierney, L. (1991) Exploring posterior distributions using Markov chains. In *Computer Science and Statistics: Proc. 23rd Symp. Interface* (ed. E. Keramidas), pp. 563–570. Fairfax Station: Interface Foundation.

Wakefield, J. C., Gelfand, A. E. and Smith, A. F. M. (1991) Efficient generation of random variates via the ratio-of-uniforms method. *Statist. Comput.*, 1, 129–133.

Zeger, S. and Karim, M. R. (1991) Generalised linear models with random effects: a Gibbs sampling approach. *J. Am. Statist. Ass.*, 86, 79–86.

6

Strategies for improving MCMC

Walter R Gilks
Gareth O Roberts

6.1 Introduction

In many applications raw MCMC methods, in particular the Gibbs sampler, work surprisingly well. However, as models become more complex, it becomes increasingly likely that untuned methods will not *mix* rapidly. That is, the Markov chain will not move rapidly throughout the support of the target distribution. Consequently, unless the chain is run for very many iterations, Monte-Carlo standard errors in output sample averages will be large. See Roberts (1995) and Tierney (1995) in this volume for further discussion of Monte-Carlo standard errors and Markov chain mixing.

In almost any application of MCMC, many models must be explored and refined. Thus poor mixing can be severely inhibiting. Run times of the order of seconds or minutes are desirable, runs taking hours are tolerable, but longer run times are practically impossible to work with. As models become more ambitious, the practitioner must be prepared to experiment with strategies for improving mixing. Techniques for reducing the amount of computation per iteration are also important in reducing run times.

In this chapter, we review strategies for improving run times of MCMC. Our aim is to give sufficient detail for these strategies to be implemented: further information can be found in the original references. For readers who are new to MCMC methodology, we emphasize that familiarity with the material in this chapter is not a prerequisite for successful application of MCMC; Gilks *et al.* (1995b: this volume) provide enough information to permit application of MCMC in straightforward situations.

For simplicity, we will mostly assume that the Markov chain takes values in k-dimensional Euclidean space $\mathbb{R}^k$, although most of the techniques we discuss apply more generally. The target density (for example a posterior distribution) is denoted $\pi(.)$, and $X_t = (X_{t.1}, X_{t.2}, \ldots, X_{t.k})^{\mathrm{T}}$ denotes the state of the chain at iteration t. Note that throughout this chapter $X_{t.i}$ will denote a scalar. A generic point in the sample space will be denoted $X = (X_{.1}, X_{.2}, \ldots, X_{.k})^{\mathrm{T}}$.

6.2 Reparameterization

In this section, we consider simple techniques for reparameterizing commonly used models to improve mixing. We begin by examining the consequences of correlations among the $\{X_{.i}\}$.

6.2.1 Correlations and transformations

Both the Gibbs sampler and the Metropolis–Hastings algorithm may be sensitive to the choice of parameterization of the model. Consider for example the two-dimensional target density $\pi(.)$ illustrated in Figure 6.1(a). The ellipses represent contours of $\pi(.)$, and indicate that the probability in $\pi(.)$ is concentrated around the diagonal line $X_{.1} = X_{.2}$.

To sample from this target distribution, we might use a Gibbs sampler, updating one component of X at a time. Suppose that, at iteration t, the current point X_t is at the intersection of the arrows in Figure 6.1(a). Updating one of the components of X_t will produce a new point X_{t+1} lying in the direction of one of the arrows (depending on which component is updated). Since $\pi(.)$ is concentrated around the diagonal, the full conditional densities $\pi(X_{t+1.1}|X_{t.2})$ and $\pi(X_{t+1.2}|X_{t.1})$ will be concentrated near X_t (see Figure 6.1(c)) and so X_{t+1} will probably lie quite close to the current point. This will tend to happen at every iteration, so the Gibbs chain will move around rather slowly, in general taking small steps so as to stay close to the main diagonal. The consequence of this slow mixing is that a long simulation will be required to obtain adequate precision in the output analysis.

Alternatively, a Metropolis algorithm might be used for this problem. Suppose, for simplicity, that the Metropolis proposal comprises a uniform distribution on a disc centred on the current point X_t. This is indicated by the circle in Figure 6.1(a). As most of the disc lies away from the diagonal, the proposal distribution will tend to generate candidate points X' for which $\pi(X')$ is small compared to $\pi(X_t)$. Such candidates will probably be rejected (the acceptance probability being $\min[1, \pi(X')/\pi(X_t)]$) and then X_{t+1} will be identical to X_t. Thus the chain is likely to get stuck at X_t for several iterations. When a candiate point is eventually accepted, it will probably lie close to the main diagonal, and the same problem will recur at the next iteration. Consequently, this Metropolis algorithm will mix slowly.

A solution to the slow mixing of the Gibbs sampler and Metropolis algorithms in this example is to transform X to a new variable Y. Let

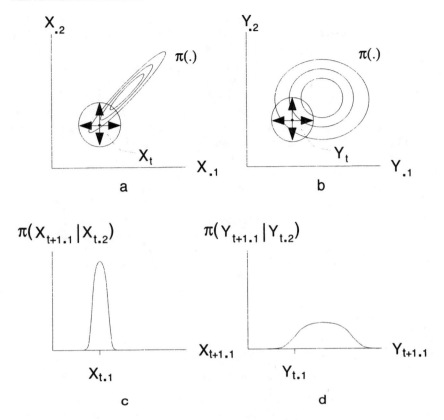

Figure 6.1 *Illustrating Gibbs sampling and Metropolis algorithms for a bivariate target density $\pi(.)$. Contours of $\pi(.)$: (a) before reparameterization; (b) after reparameterization. Full conditional densities at time t: (c) before reparameterization; (d) after reparameterization. See text for explanation.*

$Y_{.1} = X_{.1} + X_{.2}$ and $Y_{.2} = 3[X_{.1} - X_{.2}]$. Contours for $\pi(Y)$ are approximately drawn in Figure 6.1(b). These contours are much more open, and indicate absence of correlation between the components of Y under $\pi(.)$. The Gibbs sampler will now be able to move around freely since moves away from the diagonal are no longer inhibited (as indicated by the full conditional density drawn in Figure 6.1(d)). Moreover, candidate points generated using the original Metropolis proposal density will be rejected less often.

For the Metropolis algorithm, an alternative strategy to improve mixing would be to use a different proposal distribution. For example, for the target distribution shown in Figure 6.1(a), a uniform proposal distribution on a diagonally oriented elliptical disc would generate a more rapidly mixing chain. However, using a transformed proposal for the original variable X is equivalent to using the original proposal for a transformed variable $Y(X)$. For present purposes, it is convenient to think in terms of variable transformation, since thereby we can address the issue of mixing in both Gibbs samplers and Metropolis–Hastings algorithms simultaneously.

After running the MCMC for the transformed variable Y, the original variables X can be recovered simply by transforming back the values in the MCMC output; (we assume that a unique inverse transform $X(Y)$ exists). In the above example, this would involve calculating for each iteration t: $X_{t.1} = \frac{1}{2}Y_{t.1} + \frac{1}{6}Y_{t.2}$ and $X_{t.2} = \frac{1}{2}Y_{t.1} - \frac{1}{6}Y_{t.2}$.

Placing the above discussion in a Bayesian context, variable X represents a vector of parameters; $\pi(X)$ is a posterior distribution; Figure 6.1(a) illustrates a strong posterior correlation; and any one-to-one transformation $Y(X)$ represents a *reparameterization*. Below, we discuss reparameterization strategies designed to reduce posterior correlations in commonly used models. However, posterior correlations are not the only cause of slow mixing; we return to this point in Section 6.2.5.

6.2.2 Linear regression models

Problems of slow mixing can occur in the simplest of statistical models. Consider for example the linear regression model

$$y_i = \alpha + \beta x_i + \epsilon_i, \quad \epsilon_i \sim N(0, \sigma^2), \quad i = 1 \ldots n,$$

where $N(a, b)$ generically denotes a normal distribution with mean a and variance b. Here x_i and y_i are observed and, for simplicity, we assume that σ is known. For a Bayesian analysis of these data, we will assume flat priors on the parameters α and β. The posterior correlation between α and β can then be shown to be

$$\rho_{\alpha\beta} = -\frac{\bar{x}}{\sqrt{\bar{x}^2 + \frac{1}{n}\sum_1^n (x_i - \bar{x})^2}},$$

where $\bar{x} = \frac{1}{n}\sum_1^n x_i$. If $|\bar{x}|$ is large compared to the sample standard deviation of $\{x_i\}$, then $\rho_{\alpha\beta}$ will be close to 1 or minus 1. As discussed above and in Roberts (1995: this volume), high posterior correlations cause poor mixing in Gibbs samplers and in untuned Metropolis–Hastings algorithms.

A simple remedy is to work with centred covariates $x_i' = x_i - \bar{x}$. Then the regression equation becomes $y_i = \alpha' + \beta' x_i' + \epsilon_i$ where $\alpha' = \alpha + \beta\bar{x}$ and $\beta' = \beta$. Note that this has induced a reparameterization $(\alpha, \beta) \rightarrow (\alpha', \beta')$, for which the posterior correlation $\rho_{\alpha'\beta'} = 0$. In this parameterization, the Gibbs sampler will work well. In fact α' and β' are *a posteriori* independent, so the Gibbs sampler will produce samples immediately from the posterior distribution without any burn in. A simple Metropolis algorithm with independent normal proposals for α' and β' will also work well, provided that the proposals are scaled adequately; see Roberts (1995: this volume).

More generally, consider the multiple linear regression model

$$y_i = \theta^T x_i + \epsilon_i, \quad \epsilon_i \sim N(0, \sigma^2), \quad i = 1 \ldots n,$$

where x_i is a vector of covariates (the first covariate being unity); θ is a vector of parameters with a flat prior; and σ is assumed known. For this model, the posterior variance matrix of θ is $\sigma^2 \{x^T x\}^{-1}$, where $x = (x_1, x_2, \ldots, x_n)^T$ is the design matrix. If the elements of θ are to be uncorrelated in the posterior, the columns of x must be orthogonal. Orthogonalization with respect to the first column of x is achieved by centring, as described above. For orthogonalizing the whole of x, Gram–Schmidt orthogonalization might be used. However, adequate mixing can be achieved without perfect orthogonalization, since it is usually sufficient to avoid nearly collinear covariates (Hills and Smith, 1992). Also, scaling covariates to roughly equalize sample standard deviations can often help mixing in 'off-the-shelf' Metropolis–Hastings algorithms. The same general recommendations carry over to generalized linear models, with known or unknown scale parameters.

6.2.3 Random-effects models

Gibbs sampling has proved particularly successful for random-effects models, as several chapters in this volume testify. By comparison with fixed-effects models, random-effects models typically contain many parameters, but mixing is usually rapid. However, in some circumstances, mixing can be very poor. To see this, consider the simple random-effects model

$$
\begin{aligned}
y_{ij} &= \mu + \alpha_i + \epsilon_{ij}, & (6.1) \\
\alpha_i &\sim N(0, \sigma_\alpha^2),
\end{aligned}
$$

$i = 1, \ldots, m, j = 1, \ldots, n$; where $\epsilon_{ij} \sim N(0, \sigma_y^2)$. For simplicity we suppose σ_α and σ_y are known, and assume a flat prior on μ. Gelfand *et al.* (1995a) show that posterior correlations for this model depend on the relative sizes of the variance components. Posterior correlations for model (6.1) are

$$\rho_{\mu,\alpha_i} = -\left\{1 + \frac{m\sigma_y^2}{n\sigma_\alpha^2}\right\}^{-\frac{1}{2}}, \quad \rho_{\alpha_i,\alpha_j} = \left\{1 + \frac{m\sigma_y^2}{n\sigma_\alpha^2}\right\}^{-1},$$

for $i \neq j$. Large posterior correlations and poor mixing will be avoided if σ_y^2/n is not small in relation to σ_α^2/m. Thus a large number m of random effects or a small random-effects variance σ_α^2 will improve mixing, but a large number n of observations per random effect or small observational variance σ_y^2 will worsen mixing. In short, mixing is worst when the data are most informative!

We now consider two methods for reparameterizing random-effects models.

Reparameterization by hierarchical centring

Gelfand *et al.* (1995a) propose reparameterization to improve mixing in model (6.1). Let $\eta_i = \mu + \alpha_i$, then the reparameterized model is:

$$
\begin{aligned}
y_{ij} &= \eta_i + \epsilon_{ij}, & (6.2)\\
\eta_i &\sim \mathrm{N}(\mu, \sigma_\alpha^2).
\end{aligned}
$$

With this parameterization, Gelfand *et al.* calculate posterior correlations:

$$
\rho_{\mu,\eta_i} = \{1 + \tfrac{mn\sigma_\alpha^2}{\sigma_y^2}\}^{-\frac{1}{2}}, \quad \rho_{\eta_i,\eta_j} = \{1 + \tfrac{mn\sigma_\alpha^2}{\sigma_y^2}\}^{-1}, \tag{6.3}
$$

for $i \neq j$. Here, large m or n both improve mixing, and poor mixing will only result if σ_α^2 is very small. Thus, in general, parameterization (6.2) will be preferred since σ_α^2 will not be small if random effects are deemed necessary.

Gelfand *et al.* (1995a) call (6.2) a *centred* parameterization. Note that 'centring' is used here in a quite different sense than in Section 6.2.2; here we are centring parameters, and in Section 6.2.2 we were centring (in a different way) covariates. Gelfand *et al.* extend their approach to nested random-effects models. For example,

$$
\begin{aligned}
y_{ijk} &= \mu + \alpha_i + \beta_{ij} + \epsilon_{ijk}, & (6.4)\\
\beta_{ij} &\sim \mathrm{N}(0, \sigma_\beta^2),\\
\alpha_i &\sim \mathrm{N}(0, \sigma_\alpha^2),
\end{aligned}
$$

where $i = 1, \ldots, m$; $j = 1, \ldots, n$; $k = 1, \ldots, r$; and $\epsilon_{ijk} \sim \mathrm{N}(0, \sigma_y^2)$. As before, we assume a flat prior on μ and fixed variance components. For this model, the *hierarchically centred* parameterization is $\zeta_{ij} = \mu + \alpha_i + \beta_{ij}$; $\eta_i = \mu + \alpha_i$, so the reparameterized model is:

$$
\begin{aligned}
y_{ijk} &= \zeta_{ij} + \epsilon_{ijk},\\
\zeta_{ij} &\sim \mathrm{N}(\eta_i, \sigma_\beta^2),\\
\eta_i &\sim \mathrm{N}(\mu, \sigma_\alpha^2).
\end{aligned}
$$

Partial centrings are also possible, for example the β parameters might be centred but not the α, giving a (μ, α, ζ) parameterization; or the α parameters might be centered but not the β, giving a (μ, η, β) parameterization. As before, the best of these parameterizations will depend on variance components, and Gelfand *et al.* recommend that effects having large posterior variance relative to σ_y^2 should be centred, along with all effects lower in the hierarchy. If variance components are unknown, the authors suggest using posterior expectations of variance components estimated from a preliminary MCMC run.

Gelfand *et al.* (1995a, 1995b) further generalize their approach to linear and generalized linear mixed-effects models incorporating covariates. With these more general models exact calculations are more difficult, but the recommendation is still to try centred parameterizations.

Reparameterization by sweeping

Vines *et al.* (1995) propose a different approach to reparameterizing random-effects models. They note that model (6.1) is essentially overparameterized. For example, a quantity c could be added to μ and subtracted from each α_i without altering the likelihood of the data. Thus the data are unable to provide any information on the single degree of freedom $\mu - \bar{\alpha}$, where $\bar{\alpha} = \frac{1}{m} \sum_i \alpha_i$. This suggests reparameterizing $\phi_i = \alpha_i - \bar{\alpha}$; $\nu = \mu + \bar{\alpha}$; and $\delta = \mu - \bar{\alpha}$. This is called *sweeping*, since the mean is swept from the random effects and onto μ. This reparameterization gives the model:

$$
\begin{aligned}
y_{ij} &= \nu + \phi_i + \epsilon_{ij}, \\
\phi_{-m} &\sim \mathrm{N}_{m-1}(0, \sigma_\alpha^2 K_{m-1}), \\
\phi_m &= -\sum_{i=1}^{m-1} \phi_i,
\end{aligned}
\tag{6.5}
$$

where $\phi_{-m} = (\phi_1, \phi_2, \ldots, \phi_{m-1})^{\mathrm{T}}$; N_{m-1} denotes an $(m-1)$-dimensional multivariate normal distribution; and K_{m-1} is an $(m-1) \times (m-1)$ matrix with $1 - \frac{1}{m}$ on the main diagonal and $-\frac{1}{m}$ everywhere else. Clayton (1995: this volume) also discusses this reparameterization.

The random effects $\{\phi_i\}$ are now no longer *a priori* independent, and ν and δ have flat priors. The unidentified degree of freedom δ can be ignored, since the data and other parameters do not depend on it. Note that the original random effects $\{\alpha_i\}$ are interpretable as deviations from the population mean μ, whilst the reparameterized random effects $\{\phi_i\}$ are interpretable as deviations from the sample mean ν.

The *swept* parameterization gives posterior correlations

$$
\rho_{\nu,\phi_i} = 0, \quad \rho_{\phi_i,\phi_j} = -\frac{1}{m},
$$

for $i \neq j$. These correlations are not large for any $m > 1$. They do not depend on n or on variance components σ_α^2 and σ_y^2, unlike those produced by hierarchical centring (6.3).

Vines *et al.* (1995) generalize their approach to models with multiple sets of random effects in generalized linear models, and Vines and Gilks (1994) further extend the approach to accommodate random effects and hierarchical interactions of arbitrary order. The technique is to sweep means from high-order interactions onto lower-order terms. For example, for the nested random-effects model (6.4), the reparameterization proceeds in two stages. First, row means $\bar{\beta}_i = \frac{1}{n} \sum_j \beta_{ij}$ are swept from the $\{\beta_{ij}\}$ parameters onto the $\{\alpha_i\}$; then the mean of $\{\alpha_i + \bar{\beta}_i\}$ is swept onto μ, giving $\psi_{ij} = \beta_{ij} - \bar{\beta}_i$; $\phi_i = \alpha_i - \bar{\alpha} + \bar{\beta}_i - \bar{\beta}$; and $\nu = \mu + \bar{\alpha} + \bar{\beta}$; where $\bar{\beta} = \frac{1}{mn} \sum_i \sum_j \beta_{ij}$. This reparameterization gives the model:

$$
\begin{aligned}
y_{ijk} &= \nu + \phi_i + \psi_{ij} + \epsilon_{ijk}, \\
\psi_{i,-n} &\sim \mathrm{N}_{n-1}\left(0, \sigma_\beta^2 K_{n-1}\right), \\
\psi_{i,n} &= -\sum_{j=1}^{n-1} \psi_{ij}, \\
\phi_{-m} &\sim \mathrm{N}_{m-1}\left(0, \sigma_\alpha^2 + \frac{1}{n}\sigma_\beta^2 K_{m-1}\right),
\end{aligned}
$$

where $\psi_{i,-n} = (\psi_{i1}, \psi_{i2}, \ldots, \psi_{i,n-1})^{\mathrm{T}}$, and ν has a flat prior.

Unknown variance components

The above reparameterizations can reduce posterior correlations between random effects even if variance components are unknown. However, lack of information about variance components can itself be a source of slow mixing. Moreover, when this problem arises it seems difficult to cure. In the simple random effects model (6.1), slow mixing tends to occur if the prior on the random-effects variance σ_α^2 gives non-negligible probability to values near zero. An extreme form of this is the improper prior $P(\sigma_\alpha^2) \propto \sigma_\alpha^{-2}$, which gives an improper posterior (see DuMouchel and Waternaux, 1992). Then the MCMC sampler will get stuck for long periods where σ_α^2 and the sample variance of random effects $\frac{1}{m}\sum(\alpha_i - \bar{\alpha})^2$ are both small. At present, the only remedy we can suggest is to use a different prior, perhaps bounding σ_α^2 away from zero.

6.2.4 Nonlinear models

It is difficult to lay down hard and fast rules for reparameterization with nonlinear models. However, much of the experience gained in dealing with such models in the context of maximum likelihood (see for example Ross, 1990) or numerical quadrature (see Hills and Smith, 1992, and references therein) is still of relevance for MCMC, since the aim is often to reparameterize to produce open, independent-normal-like contours as illustrated in Figure 6.1(b).

The advice of Ross (1990) is to reparameterize so that parameters correspond approximately to contrasting features of the data. These are called *stable* parameters. For example, consider the model

$$
E(y_i) = \alpha + \beta e^{-\gamma x_i},
$$

$i = 1, \ldots, m$. If the data are approximately linear, γ will be small. Then $E(y_i) \approx \alpha + \beta - \beta\gamma x_i$, giving high posterior correlations between the parameters. A stable parameterization would be

$$E(y_i) = \alpha' + \beta'[f_i(\gamma) - \bar{f}(\gamma)],$$

where

$$f_i(\gamma) = -\frac{1}{\gamma x^*}e^{-\gamma(x_i - x^*)},$$

$$\bar{f}(\gamma) = \frac{1}{m}\sum f_i(\gamma),$$

and x^* is some central value of x. Now for small γ, $E(y_i) \approx \alpha' + \beta'(x_i - \bar{x})$, as motivated in Section 6.2.2. The stable parameters α' and β' correspond approximately to the mean height of the curve and the slope at x^*.

In general, divination of stable parameterizations must be done on a model-by-model basis, although similarity with well understood models can help. See Ross (1990) for further suggestions; see also Bennett *et al.* (1995: this volume) for an application of MCMC to nonlinear models.

6.2.5 General comments on reparameterization

We have indicated some strategies for reparameterizing commonly used statistical models to reduce posterior correlations. More ambitious strategies aim to reparameterize on the basis of posterior correlations estimated from the output from a preliminary MCMC run (Müller, 1994; Hills and Smith, 1992; Wakefield, 1992). However, the performance of these methods in high dimensions is unknown. Moreover, such methods may have substantial computational overheads, partly through destroying conditional-independence relationships present in the original parameterization.

Conditional-independence relationships allow high-dimensional problems to be tackled efficiently using the Gibbs sampler or a single-component Metropolis–Hastings algorithm (Spiegelhalter *et al.*, 1995b: this volume). For example, the full conditional distribution for α_i in model (6.1) is algebraically independent of $\{y_{kj}; k \neq i, j = 1, \ldots, n\}$ and can therefore be rapidly calculated. 'Natural' parameterizations, which maintain a healthy conditional-independence structure, are therefore advantageous.

Models often admit several natural parameterizations, as illustrated in Section 6.2.3. In the absence of knowledge about which will produce rapid mixing, all might be tried in a random or cyclic order (Gilks, 1995a). The combined strategy will mix well if at least one parameterization mixes well (Tierney, 1994).

For roughly normal or log-concave posterior distributions, the avoidance of high posterior correlations will often be sufficient to produce good mixing (Hills and Smith, 1992). However, other forms of posterior can yield poor mixing even in the absence of posterior correlations. For example, the Metropolis algorithm can be non-geometrically ergodic with heavy-tailed distributions (see Roberts, 1995: this volume). Hills and Smith (1992) suggest a signed root transformation to help in such situations. Another cause of slow mixing is multimodality in $\pi(.)$. Reparameterization may not help here, and we may need to turn to other strategies.

6.3 Random and adaptive direction sampling

The Gibbs sampler moves in directions parallel to the coordinate axes and, as we have seen, these directions may not be conducive to rapid mixing. Linear reparameterization is equivalent to changing the directions in which the Gibbs sampler moves. For example, Gibbs sampling with the reparameterization described in Section 6.2.1 is equivalent to Gibbs sampling in directions $X_{.2} = X_{.1}$ and $X_{.2} = -X_{.1}$ in the original parameterization. When it is not clear how to choose sampling directions to produce rapid mixing, a solution might be to construct a sampler which can sample in any direction. In this section, we describe some such samplers, the simplest being the *hit-and-run* algorithm.

6.3.1 The hit-and-run algorithm

The hit-and-run algorithm (Schmeiser and Chen, 1991) is a MCMC sampler which chooses sampling directions in $\mathbb{R}^k$ at random. At each iteration t, the hit-and-run algorithm first randomly samples a direction e_t (a unit vector of dimension k). Then X_{t+1} is chosen according to the full conditional distribution along the straight line passing through X_t in direction e_t. Thus each hit-and-run iteration comprises:

Step 1: sample e_t;
Step 2: sample a scalar r_t from density $f(r) \propto \pi(X_t + re_t)$;
Step 3: set $X_{t+1} = X_t + r_t e_t$.

Often, but not necessarily, e_t is chosen uniformly on the unit k-dimensional sphere. This may be done by generating k independent standard normal random variates $\{z_i; \ i = 1, \ldots, k\}$, and then setting

$$e_{t.i} = \frac{z_i}{\sqrt{\sum_j z_j^2}},$$

for $i = 1, \ldots, k$.

Under extremely weak regularity conditions, the hit-and-run algorithm is irreducible. Moreover, it can often mix better than the Gibbs sampler. Consider for example the target density in Figure 6.1(a). Most of the sampling directions generated by the hit-and-run algorithm will be away from the diagonal line $X_{.2} = X_{.1}$, and consequently X_{t+1} will tend to lie near X_t. Occasionally, however, a sampling direction close to the diagonal line will be generated, and then X_{t+1} will have the opportunity to move well away from X_t. This would not happen for the Gibbs sampler without reparameterization. The hit-and-run algorithm can also work well in multimodal problems, allowing mode-hopping when the sampling direction traverses two or more modes of $\pi(.)$. The algorithm can be especially useful as an exploratory tool for discovering effective sampling directions.

If $\pi(.)$ has moderately sharp ridges or spikes, random directions may not pick out regions of high probability content sufficiently often to promote rapid mixing. These problems are exacerbated in high dimensions, where the method may perform poorly.

6.3.2 Adaptive direction sampling (ADS)

ADS (Gilks et al., 1994; Roberts and Gilks, 1994) can be thought of as a generalization of the hit-and-run algorithm. It is an *adaptive* method in the sense that the direction to be sampled, e_t, can be chosen on the basis of other previously sampled points.

In general, care must be taken when using adaptive methods since the stationarity of the target distribution $\pi(.)$ can be compromised by adaptation (see for example Gelfand and Sahu, 1994). ADS avoids this problem by enlarging the Markov chain space to allow points on which adaptation is based to be part of the Markov chain state vector.

Specifically, at each iteration we store m points $\{X_t^{(1)}, X_t^{(2)}, \ldots, X_t^{(m)}\}$. These m points are called the *current set*. Each point of the current set is a k-vector in its own right. At each iteration, a point $X_t^{(c)}$ is randomly selected from the current set, and is updated. Its new value is sampled along a line passing through $X_t^{(c)}$ in a direction e_t (not necessarily a unit vector) which may depend on any or all of the points in the current set. The remaining points in the current set are not changed. Thus each iteration of ADS comprises:

Step 1: select one point $X_t^{(c)}$ at random from current set;

Step 2: choose a k-vector e_t;

Step 3: sample a scalar r_t from a density $f(r)$;

Step 4: set $X_{t+1}^{(c)} = X_t^{(c)} + r_t e_t$;

Step 5: set $X_{t+1}^{(i)} = X_t^{(i)}$ for $i \neq c$.

With the general framework for constructing direction e_t and density $f(.)$ described below, the stationary distribution for ADS is the distribution of m independently sampled points from the target density $\pi(.)$. The simplest special case of ADS is the hit-and-run algorithm, where $m = 1$, and e_t and r_t are sampled as described in Section 6.3.1.

A more interesting special case of ADS is the *snooker algorithm*. For this, the sampling direction is $e_t = X_t^{(i)} - X_t^{(c)}$, where $X_t^{(i)}$ is a second point chosen at random from the current set; and the density

$$f(r) \propto \pi(X_t^{(c)} + re_t)|1 - r|^{k-1}.$$

Thus $X_{t+1}^{(c)}$ lies along the straight line passing through $X_t(c)$ and $X_t(i)$. This is illustrated in Figure 6.2. Heuristically, the idea of the snooker algorithm is that, as the algorithm proceeds, sampling directions are increasingly likely to traverse regions of high probability under $\pi(.)$, since both $X_t^{(c)}$ and $X_t^{(i)}$ will increasingly tend to be in such regions. Note that $f(r)$ is just the full conditional density along the line $X_t^{(c)} + re_t$, multiplied by $|1 - r|^{k-1}$. This multiplier derives from a Jacobian matrix: it arises because the sampling direction for updating $X_t^{(c)}$ depends on $X_t^{(c)}$ itself, unlike the hit-and-run algorithm and Gibbs sampler.

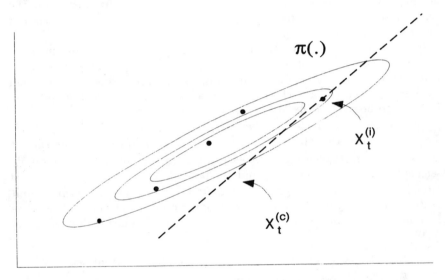

Figure 6.2 *Illustrating the snooker algorithm. Ellipses denote contours of the target density $\pi(.)$; dots denote points in the current set; and the broken line is the sampling direction for updating $X_t^{(c)}$.*

The snooker algorithm has some desirable theoretical properties, and can work well in problems where the Gibbs sampler fails badly. However, the increased computational burden it demands, and its inconsistent performance on some problems that simpler algorithms solve easily, suggest the need for schemes which make more productive use of the adaptive information contained in the current set.

For the general form of ADS, the sampling direction is $e_t = u_t X_t^{(c)} + V_t$, where u_t is a random scalar drawn independently of the current set, and

V_t is a random k-vector depending on the $m-1$ points $\{X_t^{(i)}; i \neq c\}$. Then

$$f(r) \propto \pi(X_t^{(c)} + re_t)|1 + ru_t|^{k-1}.$$

ADS can be generalized still further to Metropolized versions (*Adaptive Metropolis Sampling*, AMS). Like ADS, AMS updates a randomly chosen point $X_t^{(c)}$ from a current set of m points, but does this using a Metropolis–Hastings step, where the proposal for a new current set $\{Y^{(i)}; i = 1, \ldots, m\}$ has density $q(\{X_t^{(i)}\}; \{Y^{(i)}\})$, and depends on any or all of the points in the current set $\{X_t^{(i)}; i = 1, \ldots, m\}$. The acceptance probability is

$$\min\left\{1, \frac{q(\{Y^{(i)}\}; \{X_t^{(i)}\}) \prod_{i=1}^m \pi(Y^{(i)})}{q(\{X_t^{(i)}\}; \{Y^{(i)}\}) \prod_{i=1}^m \pi(X_t^{(i)})}\right\}$$

The stationary distribution for AMS is the distribution of m independently sampled points from the target density $\pi(.)$. As for ADS, successful implementation of AMS requires effective use of the adaptive information contained in the current set.

AMS provides a general framework for adaptation where the information from which the adaptation is based depends on only a finite history (the m points of the current set). Adaptive methods where the whole history of the chain is taken into account cannot be considered in this way. It is also doubtful that *purely* adaptive methods are advisable. Adaptation to the 'wrong' information, obtained early on in the MCMC, might worsen mixing. We might then never discover that the early information was misleading. Mathematically, this problem manifests itself in that many adaptive algorithms are not *geometrically convergent* (see Roberts, 1995: this volume). It is always advisable when using adaptive methods to use the adaptive strategy in hybridization with some fixed non-adaptive strategy.

6.4 Modifying the stationary distribution

This section describes techniques which aim to improve mixing by modifying the stationary distribution $\pi(.)$ of the Markov chain.

6.4.1 Importance sampling

Suppose we can devise a MCMC sampler which mixes well but has a modified stationary distribution $\pi^*(.)$, where $\pi^*(X) \approx \pi(X)$ for all X. We can estimate the expectation under $\pi(.)$ of an arbitrary function $g(X)$ of interest by importance reweighting the output $\{X_t; t = 1, \ldots, n\}$ from the chain with stationary distribution $\pi^*(.)$ (Fosdick, 1963; Hastings, 1970). Thus,

$$E_\pi g(X) \approx \frac{\sum_t w_t g(X_t)}{\sum_t w_t} \tag{6.6}$$

where the importance weight $w_t = \pi(X_t)/\pi^*(X_t)$. See, for example, Ripley (1987) for further discussion of importance sampling.

Jennison (1993) suggests using the form $\pi^*(X) = \pi(X)^{\frac{1}{T}}$, where $T > 1$. This form of modification of the target $\pi(.)$ is used in the optimization technique of simulated annealing, where T is called *temperature*. Heating the target $(T > 1)$ will flatten $\pi(.)$ and may make it easier for the modified MCMC sampler to explore the support of $\pi(.)$. For example, suppose $\pi(.)$ is multimodal with well-separated modes which contain most of the probability content of $\pi(.)$. For this $\pi(.)$, a Metropolis algorithm proposing mainly local moves will mix poorly, since the chain will be trapped for long periods at one mode before escaping to another mode. Heating the target will flatten the modes and place more probability between them, so the modified MCMC sampler will travel more easily between modes. Note that this $\pi^*(.)$ is unnormalized, but this presents no problem for MCMC methods.

Another useful form of modification of the target distribution is to add stepping stones between regions of the state-space which contain substantial amounts of $\pi(.)$ but which do not communicate well (i.e. between which the MCMC sampler seldom moves). This is illustrated in Figure 6.3, in which the support D of $\pi(.)$ is concentrated in two disjoint regions. The Gibbs sampler (or a single-component Metropolis–Hastings algorithm making moves parallel to the axes) applied to this problem would be reducible, since moving between the two regions would entail an intermediate move X_t outside D, where $\pi(X_t) = 0$. This problem can be avoided by placing a stepping stone E, containing a small amount of probability ϵ, such that E communicates with both regions of D as illustrated in Figure 6.3. Then $\pi^*(X) \propto \pi(X) + \epsilon e(X)$, where $e(X)$ is a density having support on E. If D and E are disjoint, importance reweighting as in (6.6) dictates that output samples which belong to E should be ignored. This technique can be useful in problems which place intricate constraints on X, as can occur for example in genetics applications (Sheehan and Thomas, 1993; see also Thomas and Gauderman, 1995: this volume).

In general, if $\pi^*(X)$ differs substantially from $\pi(X)$, importance weights will be unstable and the variance of (6.6) will be inflated, which may necessitate lengthening the MCMC run. Therefore, the success of the method relies on being able to obtain a rapidly mixing MCMC with only slight modification of the stationary distribution $\pi(.)$. This may be unrealistic for problems where mixing is very slow. However, importance reweighting of MCMC output is useful for other reasons, for example for exploring the impact of small changes in the prior or likelihood of a model (Besag *et al.*, 1995).

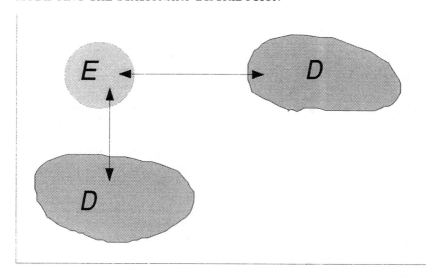

Figure 6.3 *Adding a stepping stone E to allow communication between disjoint regions D of the support of $\pi(.)$, as indicated by the arrows.*

6.4.2 Metropolis-coupled MCMC

Geyer (1991) proposes running in parallel m MCMC chains with different stationary distributions $\pi_i(X)$, $i = 1, \ldots, m$, where $\pi_1(X) = \pi(X)$ and $\{\pi_i(X);\ i > 1\}$ are chosen to improve mixing. For example, incremental heating of the form

$$\pi_i(X) = \pi(X)^{\frac{1}{1+\lambda(i-1)}}, \qquad \lambda > 0,$$

is often convenient, but not always desirable; see Geyer and Thompson (1995). After each iteration an attempt is made to swap the states of two of the chains using a Metropolis–Hastings step. Let $X_t^{(i)}$ denote the state of chain i at iteration t. Suppose that after iteration t a swap between chains i and j is proposed. The probability of accepting the proposed swap is

$$\min\left\{1, \frac{\pi_i(X_t^{(j)})\pi_j(X_t^{(i)})}{\pi_i(X_t^{(i)})\pi_j(X_t^{(j)})}\right\}.$$

At the end of the run, output from the modified chains $\{X_t^{(i)};\ t > 0,\ i > 1\}$ can be discarded. This method is called *Metropolis-coupled* MCMC (or MCMCMC).

Heuristically, swapping states between chains will confer some of the rapid mixing of the modified chains upon the unmodified chain. For example, suppose that $\pi(.)$ has well separated modes and that a modified chain moves freely between modes. Proposing swaps between this modified chain and the unmodified chain will sometimes result in the unmodified chain changing modes, thereby improving mixing. Proposed swaps will seldom be accepted if $\pi_i(X)/\pi_j(X)$ is very unstable; this is the reason for using several chains which differ only gradually with i.

An obvious disadvantage of Metropolis-coupled MCMC is that m chains are run but the output from only one is utilized. However, Metropolis-coupled MCMC is ideally suited to implementation on a parallel processing machine or even on a network of workstations (each processor being assigned one chain), since each chain will in general require about the same amount of computation per iteration, and interactions between chains are simple.

6.4.3 Simulated tempering

An idea closely related to Metropolis-coupled MCMC is *simulated tempering* (Marinari and Parisi, 1992; Geyer and Thompson, 1995). Instead of running in parallel m MCMC samplers with different stationary distributions, they are run in series and randomly interchanged. Thus simulated tempering produces one long chain, within which are embedded variable length runs from each sampler and occasional switches between samplers. The term 'simulated annealing' arises from an analogy with repeated heating and cooling of a metal.

Let $\pi_i(.)$ denote the stationary distribution of the ith sampler, where $i = 1$ denotes the unmodified (cold) sampler, and let I_t indicate which sampler is current at iteration t of the simulated tempering chain. The state of the chain at time t comprises the pair (X_t, I_t). One iteration of simulated tempering involves an update of the current point X_t using sampler I_t, followed by a Metropolis–Hastings update of I_t.

Consider now the Metropolis–Hastings update of I_t. Let $q_{i,j}$ denote the probability that sampler j is proposed given that the current sampler is the ith. Geyer and Thompson (1995) suggest setting $q_{i,i+1} = q_{i,i-1} = 0.5$ if $1 < i < m$ and $q_{1,2} = q_{m,m-1} = 1$, so that only adjacent samplers are proposed. Then the proposal to change to sampler j is accepted with probability

$$\min\left\{ 1, \frac{c_j \pi_j(X_t) q_{j,i}}{c_i \pi_i(X_t) q_{i,j}} \right\}, \tag{6.7}$$

where $i = I_t$ and the $\{c_i; \ i = 1, \ldots, m\}$ are conveniently chosen constants (see below).

The stationary distribution of the simulated tempering chain is $\pi(X, I) \propto c_I \pi_I(X)$. Since we are usually interested only in $\pi_1(.) = \pi(.)$, at the end of the run all samples for which $I_t \neq 1$ are normally discarded. After an initial burn-in, all samples $\{X_t\}$ for which $I_t = 1$ are retained; there is no need for a new burn-in after each change of sampler.

The constants $\{c_i; \ i = 1, \ldots, m\}$ are chosen so that the chain divides its time roughly equally among the m samplers. If the $\pi_i(.)$ were normalized, this would be achieved by setting $c_i = 1$ for all i. However, $\pi_i(.)$ will usually be known only up to a normalizing constant, so it will be necessary to set $c_i \approx 1/\int \pi_i(X)dX$. Thus simulated tempering requires estimation of normalizing constants, unlike Metropolis-coupled MCMC. Precise estimation of the $\{c_i\}$ is not required for valid analysis of $\pi(.)$, but even approximate estimation of the $\{c_i\}$ can be tricky. Geyer (1993) suggests several *ad hoc* techniques, the most formal of which is *reverse logistic regression*.

Reverse logistic regression (Geyer, 1993) requires a preliminary run of Metropolis-coupled MCMC which at least mixes a little. The $\{c_i\}$ are then estimated by maximizing with respect to $\{c_i\}$ a log quasi-likelihood

$$l_n(c) = \sum_t \sum_{i=1}^m \log p_i(X_t^{(i)}, c)$$

in the notation of Section 6.4.2, where

$$p_i(X, c) = \frac{c_i \pi_i(X)}{\sum_{j=1}^m c_j \pi_j(X)}.$$

An arbitrary constraint on the $\{c_i\}$ is required to identify the optimization. Note that reverse logistic regression 'forgets' the sample i from which each $X_t^{(i)}$ belongs.

Geyer and Thomson (1995) suggest setting the number of samplers m and the 'spacing' of their stationary distributions (for example, the temperature spacing λ) so that average acceptance rates in (6.7) are between 20% and 40%. This range of acceptance rates is also recommended for Metropolis algorithms in other settings (Gelman *et al.*, 1995; Roberts, 1995: this volume). See Geyer (1993) and Geyer and Thomson (1995) for further implementational suggestions.

If the mth sampler draws samples independently from $\pi_m(.)$, the simulated tempering chain will 'forget' its past (i.e. *regenerate*) whenever sampler m is used. This is the basis of *regenerative simulation* techniques (Mykland *et al.*, 1995; see also Tierney, 1995: this volume). In particular, sample paths between regenerations are independent so Monte-Carlo standard errors can be estimated easily. Also, there is no need for an initial burn-in if the chain is begun with a draw from sampler m.

6.4.4 Auxiliary variables

Adding variables can often simplify calculations and lead to improved mixing. For example, when some data x_m are missing, it may be difficult to work with the marginal posterior density $\pi(\theta) \propto P(\theta) \int P(x_o, x_m|\theta)dx_m$ of the model parameters given the observed data x_o, particularly if the marginal is not available in closed form. In general, it will be far simpler to run the MCMC on an augmented state vector (θ, x_m) and a full posterior $\pi^*(\theta, x_m) \propto P(\theta)P(x_o, x_m|\theta)$, sampling both missing data and model parameters in the MCMC. An important example occurs in survival analysis, where x_m is the set of true (but unknown) failure times for censored individuals.

Added variables are called *auxiliary variables* by Besag and Green (1993), and the idea of running a MCMC on an augmented state vector is the essence of *data augmentation* (Tanner and Wong, 1987). Auxiliary variables U need not have an immediate interpretation such as missing data. The task is to choose a convenient conditional density $P(U|X)$ and an MCMC sampler which samples both X and U with stationary distribution $\pi^*(X, U) = \pi(X)P(U|X)$, such that the resulting chain is rapidly mixing. Note that the marginal distribution for X is $\pi(X)$, as required. Note also that, in the Bayesian context, $\pi(X)$ implicitly conditions on the observed data; therefore $\pi(U|X)$ may also condition on the observed data. Having run the MCMC, the $\{X_t\}$ samples can be used for inference about $\pi(X)$, and the $\{U_t\}$ samples can be ignored.

The most famous example of the use of auxiliary variables is the *Swendsen–Wang algorithm* (Swendsen and Wang, 1987), which finds application in certain lattice models; this is discussed in Green (1995: this volume). Some of the strategies discussed above can be viewed as auxiliary variables methods, as can adaptive rejection Metropolis sampling (Gilks *et al.*, 1995a; Gilks, 1995b: this volume).

Here we somewhat speculatively suggest another auxiliary variable strategy, for use when the target density $\pi(.)$ can be approximated by a more convenient density $\pi_0(.)$ for most X, but not everywhere. For example, $\pi_0(.)$ might be a multivariate normal approximation to the target density, permitting independent sampling. Another example might be a posterior density having a simple form apart from an awkward multiplicative term arising perhaps through an ascertainment mechanism. Such terms often make little difference for moderately probable parameter values, but typically involve numerical integration. Here $\pi_0(.)$ would be the posterior without the awkward multiplier.

As usual we do not need to assume that $\pi(.)$ and $\pi_0(.)$ are normalized densities. Define a region A such that $\pi(X) \approx \pi_0(X)$ for all $X \in A$; see Figure 6.4. Let c be a constant such that $c \leq \pi(X)/\pi_0(X)$ for all $X \in A$. Region A might be chosen to make the calculation of c easy. Define a scalar auxiliary variable U taking values 0 and 1, with

$$P(U = 1|X) = \begin{cases} c\frac{\pi_0(X)}{\pi(X)} & \text{if } x \in A \\ 0 & \text{otherwise} \end{cases}. \tag{6.8}$$

Note that c ensures $P(U = 1|X) \leq 1$. Two different MCMC samplers S_1 and S_0 are employed for updating X; S_1 having stationary density proportional to $\pi_0(X)I(X \in A)$, and S_0 having stationary density proportional to $\pi(X) - c\pi_0(X)I(X \in A)$. Here $I(.)$ is the indicator function, taking the value 1 when its argument is true, and zero otherwise. Each iteration of the algorithm then proceeds as follows:

Step 1: If $U_t = 1$ sample X_{t+1} using sampler S_1
else sample X_{t+1} using sampler S_0;

Step 2: Sample U_{t+1} directly from (6.8).

This chain has the required stationary distribution $\pi^*(X, U)$.

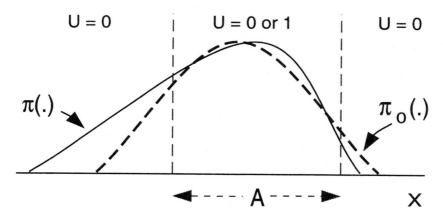

Figure 6.4 *Illustrating the use of an auxiliary variable U; $\pi_0(X)$ approximates $\pi(X)$ in region A. See text for explanation.*

If independent sampling from $\pi_0(.)$ is possible, S_1 might sample independently from $\pi_0(.)$, rejecting any samples falling outside A; and S_0 might be a Metropolis algorithm. If A and c are chosen well, the chain should spend most of its time with $U_t = 1$, where mixing is rapid. When $U_t = 0$, the chain may mix less well, but careful choice of S_0 may help prevent the chain from getting stuck in one place for many iterations. Whenever $U_t = 1$, the chain *regenerates*, and techniques of regenerative simulation can be employed to estimate Monte-Carlo standard errors (Mykland *et al.*, 1995; Tierney, 1995: this volume; see also Section 6.4.3 below).

As noted above, $\pi(X_t)$ may be particularly expensive to compute. While $U_t = 1$, $\pi(X_t)$ need not be evaluated at Step 1 since sampler S_1 needs to evaluate only $\pi_0(X_t)$. However, $\pi(X_t)$ must be evaluated in Step 2. If an upper bound $d \geq \pi(X)/\pi_0(X)$ for all $X \in A$ is available, considerable computational savings can be made in Step 2. According to (6.8), if $X_t \notin A$, we must have $U_{t+1} = U_t = 0$. If $X_t \in A$, Step 2 can be performed by:

Step 2a: Set $U_{t+1} = 1$ with probability $\frac{c}{d}$;

Step 2b: If now $U_{t+1} = 0$, reset $U_{t+1} = 1$ with
probability $c\frac{\pi_o(X_t)}{\pi(X_t)} - \frac{c}{d}$.

If c and d squeeze $\pi(X)/\pi_0(X)$ tightly within region A, U_{t+1} will usually be set to 1 at Step 2a, and $\pi(X_t)/\pi_0(X_t)$ will not need to be evaluated. Thus if the chain spends most of the time with $U_t = 1$, good mixing will be achieved with minimal computation.

The above techniques can be readily generalized to single-component Metropolis–Hastings samplers, so that for each full conditional distribution $\pi(. \mid \{X_{t.j}; j \neq i\})$ an approximation $\pi_0(. \mid \{X_{t.j}; j \neq i\})$ is devised, and an auxiliary variable U_i is constructed analogously to (6.8).

6.5 Methods based on continuous-time processes

There are a number of MCMC methods derived from the properties of continuous-time Markov processes which can be used to simulate from target densities. See Chapter 5 of Neal (1993) for a review of many of these methods and their motivations from a physical perspective.

First, we consider a k-dimensional diffusion process (a continuous-time, continuous-sample-path Markov chain in k dimensions) which has a stationary distribution $\pi(.)$. Diffusion processes are best described by *stochastic differential equations* of the form

$$dX_t = \Sigma^{\frac{1}{2}}dB_t + \mu(X_t)dt \qquad (6.9)$$

where dX_t is the change $X_{t+dt} - X_t$ in state vector X occuring in an infinitessimally small interval of time dt; Σ is a constant positive-definite $k \times k$ matrix; μ is a vector-valued function of length k; and dB_t is an increment of k-dimensional *Brownian motion*. Loosely speaking, (6.9) says that, given X_t,

$$X_{t+dt} \sim N_k(X_t + \mu(X_t)dt, \Sigma dt), \qquad (6.10)$$

where N_k denotes a k-dimensional multivariate normal density.

If we set

$$\mu(X) = \frac{1}{2}\nabla \log \pi(X), \qquad (6.11)$$

where ∇ denotes $(\frac{\partial}{\partial X_{.1}}, \frac{\partial}{\partial X_{.2}}, \ldots, \frac{\partial}{\partial X_{.k}})^{\mathrm{T}}$, and if $\Sigma = I_k$ (the identity matrix) then, under suitable regularity conditions (see for example Roberts and Tweedie, 1995), $\pi(.)$ is the unique stationary and limiting distribution for the process X, and X is called the *Langevin* diffusion for $\pi(.)$. From now on, we shall assume that (6.11) holds.

Thus, if we were able to run this continuous-time process, at each time t after a suitable burn-in, the sample X_t could be regarded as a sample from the target density $\pi(.)$. In practice of course, simulation from diffusions is impossible and a discrete-time approximation is necessary. A natural choice is suggested by (6.10):

$$X_{t+\delta} \sim N_k(X_t + \delta\mu(X_t), \delta I_k), \qquad (6.12)$$

for some small $\delta > 0$. Neal (1993) discusses the interpretation of this algorithm in terms of Hamiltonian dynamics, the use of correlated noise, and issues related to time discretization.

Unfortunately, the time discretization can have dire consequences for the stationary distribution of the algorithm. In particular, it is quite common to find that the algorithm is not even recurrent, no matter how fine the time discretization δ (Roberts and Tweedie, 1995). To counter this, it is advisable to use a correcting Metropolis–Hastings rejection step. Therefore, given X_t, the Langevin–Hastings algorithm proceeds by generating a proposal X' from $N_k(X_t + \delta\mu(X_t), \delta I_k)$, and accepting this proposal (setting $X_{t+\delta} = X'$) with probability the minimum of 1 and

$$\frac{\pi(X')}{\pi(X_t)} \exp\left\{ -\frac{1}{2}[\mu(X') + \mu(X_t)]^T [2(X' - X_t) + \delta\{\mu(X') - \mu(X_t)\}] \right\}.$$

This correction preserves the stationarity of $\pi(.)$, but the algorithm produced is very often non-geometric (Roberts and Tweedie, 1995).

Recent work by Hwang *et al.* (1993) has demonstrated that, for Gaussian target densities, the Langevin scheme described above can be improved upon by a non-reversible diffusion process.

The Langevin–Hastings algorithm can be considered as an alternative to the random-walk Metropolis algorithm, where the proposal distribution is adjusted by considering a local property (the gradient of $\log \pi(.)$). This adjustment tends to propose points in the uphill direction from X_t. Thus the chain is nudged in the direction of modes. If the local properties of $\pi(.)$ are erratic however, this nudge can be misleading, and could for example result in wildly overshooting modes. Therefore, care has to taken when using this method, to check that the target density is sufficiently smooth, and has no zeros in the interior of the state-space. Alternatively, algorithms with truncated proposals of the form

$$N_k(X_t + \min\{b, \delta\mu(X_t)\}, \delta I_k)$$

for large fixed b, retain most of the problem specific advantages of the Langevin–Hastings algorithm, without the potential inherent instability of the algorithm.

Extensions of this approach include the use of jump-diffusions, perhaps allowing the Markov chain to jump between different-dimensional subspaces of the parameter space, whilst following a diffusion sample path within each subspace. See Phillips and Smith (1995: this volume) and Grenander and Miller (1994) for applications.

6.6 Discussion

Our discussion to this point has been focused mainly on improving mixing. We have said little about reducing Monte-Carlo standard errors. Perfect mixing is achieved by independent sampling from $\pi(.)$, but a suitable choice of proposal can reduce variances below those of independent sampling (Peskun, 1973). To reduce Monte-Carlo standard errors in single-component Metropolis–Hastings samplers, Besag and Green (1993) suggest using proposal distributions which incorporate an antithetic effect, to encourage $X_{t,i}$ to flip from one side of its full conditional density to the other. Importance reweighting of samples generated from a distribution with heavier tails than $\pi(.)$ can also reduce Monte-Carlo standard errors (Geweke, 1989). The performance of these strategies may be sensitive to the functionals $g(.)$ of interest, but a well mixing sampler will generally deliver small Monte-Carlo standard errors for all functionals of interest. Small Monte-Carlo standard errors may be achieved most easily at the expense of rapid mixing; zero variances are guaranteed with a proposal which sets $X_{t+1} = X_t$ with probability 1.0! Thus we regard rapid mixing as a higher ideal than variance reduction.

In this chapter we have reviewed a variety of strategies for reducing run times. None are guaranteed to achieve reductions, as all are justified to some extent heuristically. However, an appreciation of the source of the problem in a given application may suggest which strategy might improve the MCMC. We recommend first trying some reparameterizations, or equivalently some carefully tailored Metropolis–Hastings proposal densities. If these do not help, one of the techniques of Section 6.4 might. If still no progess has been made, there is enormous scope for experimentation within the frameworks outlined in Sections 6.3 and 6.5.

The possibilites do not end there. Markov chain mixing can be improved by *mixing strategies* within a single chain (Tierney, 1994), i.e. using different strategies at different iterations of the chain (note two different uses of the word 'mixing' here). For example, different reparameterizations and auxiliary variables might be used at different iterations. As another example, the blocking scheme (which defines the components in a single-component sampler) can be altered as the simulation proceeds; this is the essence of *multigrid methods* (see Goodman and Sokal, 1989, for an extensive review). When mixing strategies, it is important to ensure that the choice of strategy at iteration t does not depend on X_t, to avoid disturbing the stationary distribution of the chain.

MCMC methodology has had a profound effect on liberating statistical modelling, and application of sophisticated models has been made feasible through specially developed software, in particular BUGS (Spiegelhalter *et al.*, 1995a). Having adjusted to the new freedom, many practitioners are now working at the operational limits of existing MCMC methodology, the limits essentially being met in the form of exasperatingly slow mixing. Thus there is an urgent need to develop methods for improving mixing for ever greater levels of model generality. Without very general and robust methods, reliable general-purpose software will not be attainable, and without such software, much that MCMC has to offer will be inaccessible to the greater corpus of applied statisticians.

A remarkable feature of the general framework of MCMC is the scope it affords for ingenuity and creativity in developing rapidly mixing chains for classes of problem. The techniques reviewed in this chapter give some indication of the possibilities. Undoubtedly there will be many exciting new developments in the future.

Acknowledgement

We are grateful to Nicky Best for many helpful comments on an earlier version of this chapter.

References

Bennett, J. E., Racine-Poon, A. and Wakefield, J. C. (1995) MCMC for nonlinear hierarchical models. In *Markov Chain Monte Carlo in Practice* (eds W. R. Gilks, S. Richardson and D. J. Spiegelhalter), pp. 339–357. London: Chapman & Hall.

Besag, J. and Green, P. J. (1993) Spatial statistics and Bayesian computation. *J. R. Statist. Soc.* B, **55**, 25–37.

Besag, J., Green, P. J., Higdon, D. and Mengersen, K. (1995) Bayesian computation and stochastic systems. *Statist. Sci.*, **10**, 3–41.

Clayton, D. G. (1995) Generalized linear mixed models. In *Markov Chain Monte Carlo in Practice* (eds W. R. Gilks, S. Richardson and D. J. Spiegelhalter), pp. 275–301. London: Chapman & Hall.

DuMouchel, W. and Waternaux, C. (1992) Discussion on hierarchical models for combining information and for meta-analyses (by C. N. Morris. and L. Normand). In *Bayesian Statistics 4* (eds J. M. Bernardo, J. Berger, A. P. Dawid and A. F. M. Smith), pp. 338–339. Oxford: Oxford University Press.

Fosdick, L. D. (1963) Monte Carlo calculations on the Ising lattice. *Meth. Comput. Phys.*, **1**, 245–280.

Gelfand, A. E. and Sahu, S. K. (1994) On Markov chain Monte Carlo acceleration. *J. Comp. Graph. Statist.*, **3**, 261–267.

Gelfand, A. E., Sahu, S. K. and Carlin, B. P. (1995a) Efficient parametrizations for normal linear mixed models. *Biometrika*, (in press).

Gelfand, A. E., Sahu, S. K. and Carlin, B. P. (1995b) Efficient parametrizations for generalized linear mixed models. In *Bayesian Statistics 5* (eds J. M. Bernardo, J. Berger, A. P. Dawid and A. F. M. Smith). Oxford: Oxford University Press (in press).

Gelman, A., Roberts, G. O. and Gilks, W. R. (1995) Efficient Metropolis jumping rules. In *Bayesian Statistics 5* (eds J. M. Bernardo, J. Berger, A. P. Dawid and A. F. M. Smith). Oxford: Oxford University Press (in press).

Geweke, J. (1989) Bayesian inference in econometric models using Monte Carlo integration. *Econometrika*, **57**, 1317–1339.

Geyer, C. J. (1991) Markov chain Monte Carlo maximum likelihood. In *Computing Science and Statistics: Proceedings of the 23rd Symposium on the Interface* (ed. E. M. Keramidas), pp. 156–163. Fairfax Station: Interface Foundation.

Geyer, C. J. (1993) Estimating normalising constants and reweighting mixtures in Markov chain Monte Carlo. *Technical Report 589*, School of Statistics, University of Minnesota.

Geyer, C. J. and Thompson, E. A. (1995) Annealing Markov chain Monte Carlo with applications to pedigree analysis. *J. Am. Statist. Ass.*, (in press).

Gilks, W. R. (1995a) Discussion on efficient parametrizations for generalized linear mixed models (by A. E. Gelfand, S. K. Sahu, and B. P. Carlin. In *Bayesian Statistics 5* (eds J. M. Bernardo, J. Berger, A. P. Dawid and A. F. M. Smith). Oxford: Oxford University Press (in press).

Gilks, W. R. (1995b) Full conditional distributions. In *Markov Chain Monte Carlo in Practice* (eds W. R. Gilks, S. Richardson and D. J. Spiegelhalter), pp. 75–88. London: Chapman & Hall.

Gilks, W. R., Best, N. G. and Tan, K. K. C. (1995a) Adaptive rejection Metropolis sampling within Gibbs sampling. *Appl. Statist.*, (in press).

Gilks, W. R., Richardson, S. and Spiegelhalter, D. J. (1995b) Introducing Markov chain Monte Carlo. In *Markov Chain Monte Carlo in Practice* (eds W. R. Gilks, S. Richardson and D. J. Spiegelhalter), pp. 1–19. London: Chapman & Hall.

Gilks, W. R., Roberts, G. O. and George, E. I. (1994) Adaptive direction sampling. *The Statistician*, **43**, 179–189.

Goodman, J. and Sokal, A. D. (1989) Multigrid Monte-Carlo method: conceptual foundations. *Phys. Rev. D*, **40**, 2035–2071.

Green, P. J. (1995) MCMC in image analysis. In *Markov Chain Monte Carlo in Practice* (eds W. R. Gilks, S. Richardson and D. J. Spiegelhalter), pp. 381–399. London: Chapman & Hall.

Grenander, U. and Miller, M. I. (1994) Representation of knowledge in complex systems. *J. R. Statist. Soc. B*, **56**, 549–603.

Hastings, W. K. (1970) Monte Carlo sampling methods using Markov chains and their applications. *Biometrika*, **57**, 97–109.

Hills, S. E. and Smith, A. F. M. (1992) Parameterization issues in Bayesian inference. In Bayesian Statistics 4 (eds J. M. Bernardo, J. Berger, A. P. Dawid and A. F. M. Smith), pp. 227–246. Oxford: Oxford University Press.

Hwang, C.-R., Hwang-Ma S.-Y. and Shen, S.-J. (1993). Accelerating Gaussian diffusions. *Ann. Appl. Prob.*, **3**, 897–913.

Jennison, C. (1993) Discussion on the meeting on the Gibbs sampler and other Markov chain Monte Carlo methods. *J. R. Statist. Soc. B*, **55**, 54–56.

Marinari, E. and Parisi, G. (1992). Simulated tempering: a new Monte Carlo scheme. *Europhys. Lett.*, **19**, 451–458.

Müller, P. (1994) A generic approach to posterior integration and Gibbs sampling. Technical report, Institute of Statistics and Decision Sciences, Duke University.

Mykland, P. Tierney, L. and Yu, B. (1995) Regeneration in Markov chain samplers. *J. Am. Statist. Ass.*, **90**, 233–241.

Neal, R. M. (1993) Probabilistic inference using Markov chain Monte Carlo methods. Technical report, Department of Computer Science, University of Toronto.

Peskun, P. H. (1973) Optimum Monte-Carlo sampling using Markov chains. *Biometrika*, **60**, 607–612.

Phillips, D. B. and Smith, A. F. M. (1995) Bayesian model comparison via jump diffusions. In *Markov Chain Monte Carlo in Practice* (eds W. R. Gilks, S. Richardson and D. J. Spiegelhalter), pp. 215–239. London: Chapman & Hall.

Ripley, B. D. (1987) *Stochastic Simulation*. New York: Wiley.

Roberts, G. O. (1995) Markov chain concepts related to sampling algorithmms. In *Markov Chain Monte Carlo in Practice* (eds W. R. Gilks, S. Richardson and D. J. Spiegelhalter), pp. 45–57. London: Chapman & Hall.

Roberts, G. O. and Gilks, W. R. (1994) Convergence of adaptive direction sampling. *J. Multiv. Anal.*, **49**, 287–298.

Roberts, G. O. and Tweedie, R. L. (1995) Stability of Langevin algorithms. *In preparation*.

Ross, G. S. J. (1990) *Nonlinear Estimation*. New York: Springer-Verlag.

Schmeiser, B. and Chen, M.-H. (1991) General hit-and-run Monte Carlo sampling for evaluating multidimensional integrals. Technical report, School of Industrial Engineering, Purdue University.

Sheehan, N. and Thomas, A. (1993) On the irreducibility of a Markov chain defined on a space of genotype configurations by a sampling scheme. *Biometrics*, **49**, 163–175.

Spiegelhalter, D. J., Thomas, A. and Best, N. G. (1995a). Computation on Bayesian graphical models. In *Bayesian Statistics 5* (eds J. M. Bernardo, J. Berger, A. P. Dawid and A. F. M. Smith). Oxford: Oxford University Press (in press).

Spiegelhalter, D. J., Best, N. G., Gilks, W. R. and Inskip, H. (1995b) Hepatitis B: a case study in MCMC methods. In *Markov Chain Monte Carlo in Practice* (eds W. R. Gilks, S. Richardson and D. J. Spiegelhalter), pp. 21–43. London: Chapman & Hall.

Swendsen, R. H. and Wang, J.-S. (1987) Nonuniversal critical dynamics in Monte carlo simulations. *Phs. Rev. Lett.*, **58**, 86–88.

Tanner, M. A. and Wong, W. H. (1987). The calculation of posterior distributions by data augmentation. *J. Am. Statist. Ass.*, **82**, 528–540.

Thomas, D. C. and Gauderman, W. J. (1995) Gibbs sampling methods in genetics. In *Markov Chain Monte Carlo in Practice* (eds W. R. Gilks, S. Richardson and D. J. Spiegelhalter), pp. 419–440. London: Chapman & Hall.

Tierney, L. (1994) Markov chains for exploring posterior distributions (with discussion). *Ann. Statist.*, **22**, 1701–1762.

Tierney, L. (1995) Introduction to general state-space Markov chain theory. In *Markov Chain Monte Carlo in Practice* (eds W. R. Gilks, S. Richardson and D. J. Spiegelhalter), pp. 59–74. London: Chapman & Hall.

Vines, S. K. and Gilks, W. R. (1994) Reparameterising random interactions for Gibbs sampling. Technical report, MRC Biostatistics Unit, Cambridge.

Vines, S. K., Gilks, W. R. and Wild, P. (1995) Fitting multiple random effects models. Technical report, MRC Biostatistics Unit, Cambridge.

Wakefield, J. C. (1992) Discussion on parameterization issues in Bayesian inference (by S. E. Hills and A. F. M. Smith). In *Bayesian Statistics 4* (eds J. M. Bernardo, J. Berger, A. P. Dawid and A. F. M. Smith), pp. 243–244. Oxford: Oxford University Press.

7

Implementing MCMC

Adrian E Raftery

Steven M Lewis

7.1 Introduction

When implementing MCMC, it is important to determine how long the simulation should be run, and to discard a number of initial 'burn-in' iterations (see Gilks *et al.*, 1995: this volume). Saving all simulations from a MCMC run can consume a large amount of storage, especially when consecutive iterations are highly correlated, necessitating a long run. Therefore it is sometimes convenient to save only every k^{th} iteration ($k > 1$). This is sometimes referred to as *thinning* the chain. While neither burn-in nor thinning are mandatory practices, they both reduce the amount of data saved from a MCMC run.

In this chapter, we outline a way of determining in advance the number of iterations needed for a given level of precision in a MCMC algorithm. The method is introduced in Section 7.2, and in Section 7.3 we describe the **gibbsit** software which implements it and is available free of charge from StatLib. In Section 7.4, we show how the output from this method can also be used to diagnose lack of convergence or slow convergence due to bad starting values, high posterior correlations, or 'stickiness' (slow mixing) of the chain. In Section 7.5, we describe how the methods can be combined with ideas of Müller (1991) and Gelman *et al.* (1995) to yield an automatic generic Metropolis algorithm.

For simplicity, the discussion is in the context of a single long chain. However, as discussed in Section 7.6, the same basic ideas can also be used to determine the number of iterations and diagnose slow convergence when multiple sequences are used, as advocated by Gelman and Rubin (1992b): see also Gelman (1995: this volume).

7.2 Determining the number of iterations

In any practical application of MCMC, there are a number of important decisions to be made. These include the number of iterations, the spacing between iterations retained for the final analysis, and the number of initial burn-in iterations discarded. A simple way of making these decisions was proposed by Raftery and Banfield (1991) and Raftery and Lewis (1992a).

To use MCMC for Bayesian inference, the sample generated during a run of the algorithm should 'adequately' represent the posterior distribution of interest. Often, interest focuses on posterior quantiles of functions of the parameters, such as Bayesian confidence intervals and posterior medians, and then the main requirement is that MCMC estimates of such quantities be approximately correct.

We thus consider probability statements regarding quantiles of the posterior distribution of a function U of the parameter vector θ. A quantile is the same thing as a percentile except that it is expressed in fractions rather than in percentages. Suppose we want to estimate the posterior probability $P(U \leq u \mid \text{Data})$ to within $\pm r$ with probability s, where U is a function of θ. We will find the approximate number of iterations required to do this, when the actual quantile of interest is q. For example, if $q = 0.025$, $r = 0.0125$ and $s = 0.95$, this corresponds to requiring that the cumulative distribution function of the 0.025 quantile be estimated to within ± 0.0125 with probability 0.95. This might be a reasonable requirement if, roughly speaking, we want reported 95% intervals to have actual posterior probability between 0.925 and 0.975. We run the MCMC algorithm for an initial M iterations that we discard, and then for a further N iterations of which we store every k^{th}. Our problem is to determine M, N and k.

We first calculate U_t for each iteration t where U_t is the value of U for the tth iteration, and then form

$$Z_t = \left\{ \begin{array}{ll} 1 & \text{if} \quad U_t \leq u \\ 0 & \text{otherwise} \end{array} \right. .$$

The sequence $\{Z_t\}$ is a binary 0-1 process that is derived from a Markov chain, but is not itself a Markov chain. Nevertheless, it seems reasonable to suppose that the dependence in $\{Z_t\}$ falls off fairly rapidly with increasing lag, and hence that if we form the new process $\{Z_t^{(k)}\}$, where

$$Z_t^{(k)} = Z_{1+(t-1)k},$$

consisting of every k^{th} iteration from the original chain, then $\{Z_t^{(k)}\}$ will be approximately a Markov chain for k sufficiently large.

To determine k, we form the series $\{Z_t^{(k)}\}$ for $k = 1, 2, \ldots$. For each k, we then compare the fit of a first-order Markov chain model with that of a second-order Markov chain model, and choose the smallest value of k for which the first-order model is preferred. We compare the models by calculating G^2, the likelihood-ratio test statistic between the second-order Markov model and the first-order Markov model, and then use the Bayesian Information Criterion (BIC), $G^2 - 2\log n$, where n is the number of iterations in a pilot sample, to see which of the two models better fits the pilot sample. BIC was introduced by Schwarz (1978) in another context and generalized to log-linear models by Raftery (1986); it provides an approximation to twice the logarithm of the Bayes factor for the second-order model: see also Raftery (1995: this volume).

Next, assuming that $\left\{ Z_t^{(k)} \right\}$ is indeed a first-order Markov chain, we now determine $M = mk$, the number of burn-in iterations to be discarded. In what follows, we use standard results for two-state Markov chains (see for example Cox and Miller, 1965). Let

$$Q \; = \; \begin{pmatrix} 1 - \alpha & \alpha \\ \beta & 1 - \beta \end{pmatrix}$$

be the transition matrix for $\left\{ Z_t^{(k)} \right\}$, where α is the probability of changing from the first state to the second state and β is the probability of changing from the second state to the first state. The equilibrium distribution is then $\pi = (\pi_0, \pi_1) = (\alpha + \beta)^{-1}(\beta, \alpha)$, where $\pi_0 = P(U \leq u \mid \text{Data})$ and $\pi_1 = 1 - \pi_0$. The ℓ-step transition matrix is

$$Q^\ell \; = \; \begin{pmatrix} \pi_0 & \pi_1 \\ \pi_0 & \pi_1 \end{pmatrix} + \frac{\lambda^\ell}{\alpha + \beta} \begin{pmatrix} \alpha & -\alpha \\ -\beta & \beta \end{pmatrix},$$

where $\lambda = (1 - \alpha - \beta)$. Suppose we require that $P(Z_m^{(k)} = i \mid Z_0^{(k)} = j)$ be within ε of π_i for $i, j = 0, 1$. If $e_0 = (1, 0)$ and $e_1 = (0, 1)$, then

$$P(Z_m^{(k)} = i \mid Z_0^{(k)} = j) = e_j Q^m e_i^T,$$

and so the requirement becomes

$$\lambda^m \; \leq \; \frac{(\alpha + \beta)\,\varepsilon}{\max(\alpha, \beta)},$$

which holds when

$$m = m^\star \; = \; \frac{\log\left\{ \frac{(\alpha+\beta)\varepsilon}{\max(\alpha,\beta)} \right\}}{\log \lambda},$$

assuming that $\lambda > 0$, which is usually the case in practice. Thus

$$M \; = \; m^\star k.$$

To determine N, we note that the estimate of $P(U \leq u \mid \text{Data})$ is $\overline{Z}_n^{(k)} = \frac{1}{n} \sum_{t=1}^n Z_t^{(k)}$. For n large, $\overline{Z}_n^{(k)}$ is approximately normally distributed with mean q and variance

$$\frac{1}{n}\frac{(2-\alpha-\beta)\,\alpha\beta}{(\alpha+\beta)^3}.$$

Thus the requirement that $P(q - r \le \overline{Z}_n^{(k)} \le q + r) = s$ will be satisfied if

$$n = n^{\star} \;\; = \;\; \frac{(2-\alpha-\beta)\,\alpha\beta}{(\alpha+\beta)^3}\left\{\frac{\Phi^{-1}\left(\frac{1}{2}(s+1)\right)}{r}\right\}^2,$$

where $\Phi(\cdot)$ is the standard normal cumulative distribution function. Thus we have

$$N \;\; = \;\; n^{\star}k.$$

Use of a thinned chain with only N simulations to perform the final inference results in less precision than if no thinning had been done. Thus, if our value of N is used without thinning (i.e. if all the iterations are used), then the probability of satisfying our accuracy criterion is greater than required, and so our criterion is conservative in this case.

The method requires that the MCMC algorithm be run for an initial number of iterations in order to obtain a pilot sample of parameter values. As a rough guide to the size of this pilot sample, we note that the required N will be minimized if successive values of $\{Z_t\}$ are independent, implying that $\alpha = 1 - \beta = \pi_1 = 1 - q$, in which case $M = 0$, $k = 1$ and

$$N = N_{\min} \;\; = \;\; \left\{\Phi^{-1}\left(\frac{s+1}{2}\right)\right\}^2\frac{q(1-q)}{r^2}.$$

The user needs to give only the required precision, as specified by the four quantities q, r, s and ε. Of these, the result is by far the most sensitive to r, since $N \propto r^{-2}$. For example, when $q = 0.025$, $r = 0.0125$ and $s = 0.95$, we have $N_{\min} = 600$. $N_{\min}$ is, in general, a reasonable choice for the size of the pilot sample.

Instead of specifying the required precision in terms of the error in the cumulative distribution function at the quantile, which is what r refers to, it may be more natural to specify the required precision in terms of the error in an estimate of the quantile itself; see Raftery and Lewis (1992a) for further details.

7.3 Software and implementation

The method is implemented in the Fortran program `gibbsit`, which can be obtained free of charge by sending the e-mail message 'send gibbsit from general' to *statlib@stat.cmu.edu*. An S version, written by the authors with Karen Vines, may be obtained by sending the message 'send gibbsit from S' to the same address. An XLISP-STAT version has been written by Jan de Leeuw and is available at the URL http://ftp.stat.ucla.edu. Despite its name, gibbsit can be used for any MCMC, not just the Gibbs sampler. The program takes as input the pilot MCMC sample for the quantity of interest and the values of q, r and s, and it returns as output the estimated values of M, N and k.

We recommend that gibbsit be run a second time after the $(M + N)$ iterations have been produced, to check that the N iterations recommended on the basis of the pilot sample were indeed adequate. If not, i.e. if the value of $(M + N)$ from the second call to gibbsit is appreciably more than that from the first call, then the MCMC algorithm should be continued until the total number of iterations is adequate.

In any practical setting, several quantities will be of interest, and perhaps several quantiles of each of these. We recommend that gibbsit be called for each quantile of primary interest, and that the maximum values of M and N be used. Typically, tail quantiles are harder to estimate than central quantiles such as medians, so one reasonable routine practice would be to apply gibbsit to each quantity of primary interest twice, with $q = 0.025$ and $q = 0.975$.

7.4 Output analysis

In this section, we assume that the gibbsit program has been run. As output from this program, we will have values for M, N and k. It would be possible to just take these parameters for a final run of the MCMC, but this would not be recommended. Performing a little output analysis on the initial $N_{\min}$ iterations may indicate ways in which the MCMC algorithm could be improved, as we now discuss.

The gibbsit program outputs, M, N and k, can be used for diagnostic purposes. These outputs can be combined to calculate

$$I \;=\; \frac{M + N}{N_{\min}}$$

(Raftery and Lewis, 1992b). This statistic measures the increase in the number of iterations due to dependence in the sequence. Values of I much greater than 1 indicate a high level of dependence. We have found that values of I greater than about 5 often indicate problems that might be alleviated by changing the implementation. Such dependence can be due to a bad starting value (in which case, other starting values should be tried), to high posterior correlations (which can be remedied by crude correlation-removing transformations), or to 'stickiness' in the Markov chain (sometimes removable by changing the MCMC algorithm); see Gilks and Roberts (1995: this volume). It may seem surprising that a bad starting value can lead to high values of N as well as M. This happens because progress away from a bad starting value tends to be slow and gradual, leading to a highly autocorrelated sequence and high values of N, since the entire pilot sequence (including the initial values) is used to estimate N, as well as M and k.

It is also important to examine plots of successive iterations of the initial $N_{\min}$ iterations for at least the key parameters of the model. For example, in hierarchical models, it is important to look at a plot of the random-effects variance or an equivalent parameter. If the starting value for this parameter is too close to zero, componentwise MCMC (such as the Gibbs sampler) can get stuck for a long time close to the starting value. A plot of the simulated values of this parameter will show whether or not the algorithm remained stuck near the starting value. This problem arises in the following example.

7.4.1 An example

We illustrate these ideas with an example from the analysis of longitudinal World Fertility Survey data (Raftery *et al.*, 1995; Lewis, 1993). The data are complete birth histories for about 5 000 Iranian women, and here we focus on the estimation of unobserved heterogeneity. Let π_{it} be the probability that woman i had a child in calendar year t. Then a simplified version of the model used is

$$\log \left(\frac{\pi_{it}}{1 - \pi_{it}} \right) \;=\; \eta + \delta_i$$

$$\delta_i \;\sim\; N\left(0, \sigma^2\right), \tag{7.1}$$

independently. The prior on η is Gaussian and the prior on σ^2 is inverted gamma (i.e. σ^{-2} has a gamma distribution) with a shape parameter of 0.5 and a scale parameter of 0.02 (Lewis, 1994). The δ_i are random effects representing unobserved sources of heterogeneity in fertility such as fecundability and coital frequency. There are also measured covariates in the model, but these are omitted here for ease of exposition.

Figure 7.1 shows a run of a MCMC algorithm starting with a value of σ^2 close to zero, namely $\sigma^2 = 10^{-4}$, and with values of the δ_i randomly generated from $N(0, 10^{-4})$. (In Figures 7.1 and 7.2, the starting value has been omitted for reasons of scaling.) The σ^2 series seems highly autocorrelated and a run of the gibbsit program confirms this. With $q = 0.025$, $r = 0.0125$ and $s = 0.95$, we obtain $N = 4258$, $k = 2$ and $M = 42$. Here $N_{\min} = 600$, so that $I = 7.2$. The high value of I and the trend-like appearance of Figure 7.1(a) suggest that there is a starting value problem. By contrast, the values of δ_9 *in the same run* are much less correlated (Figure 7.1(b)) with $I = 2.8$, so that diagnostics based on that series alone would mislead.

Figure 7.2 shows three other series of σ^2 from different starting values, illustrating a simple trial-and-error approach to the choice of an ade-

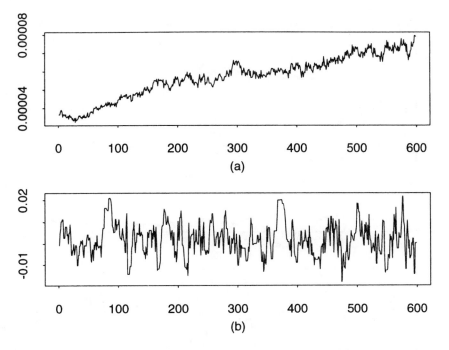

Figure 7.1 *MCMC output for the model in (7.1) for the Iranian World Fertility Survey data starting with $\sigma^2 = 10^{-4}$: (a) series of σ^2 values; (b) series of values of δ_9, the random effect for woman 9 in the survey.*

quate starting value. Figure 7.2(a) starts with $\sigma^2 = 0.1$, and the method of Raftery and Lewis (1992b) yields $I = 3.3$. Figure 7.2(b) starts with $\sigma^2 = 1.0$ and has $I = 2.8$. Figure 7.2(c) starts with $\sigma^2 = 0.25$ and has $I = 2.4$. The results of these trajectories all seem satisfactory, suggesting that the results are relatively insensitive to the starting value when our diagnostics do not indicate there to be a problem.

This example bears out our main points. It is important to monitor the MCMC run for all the key parameters, and to start again with different starting values when the diagnostics suggest doing this.

7.5 Generic Metropolis algorithms

In this section, we develop a generic Metropolis algorithm (Metropolis *et al.*, 1953) which could be fully automated, although we have not yet done so. This combines the methods of Sections 7.2 and 7.4 with ideas of Müller (1991) and Gelman *et al.* (1995). In the basic algorithm, one parameter is updated at a time, and the proposal distribution is normal centred at

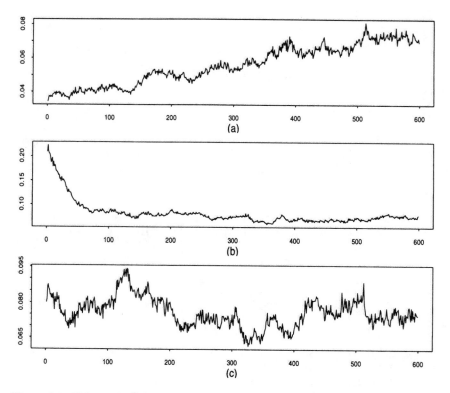

Figure 7.2 *Values of σ^2 for three runs of the same MCMC algorithm as in Figure 7.1, with different starting values: (a) $\sigma^2 = 0.1$; (b) $\sigma^2 = 1$; (c) $\sigma^2 = 0.25$.*

the current value. The user has only to specify the variance of the proposal distribution for each parameter, and here we outline a strategy for doing this. This version of the Metropolis algorithm is sometimes referred to as 'Metropolis within Gibbs' because one parameter at a time is being updated. We prefer not using this name since this form of Metropolis algorithm is not a Gibbs sampler. The latter name is used when sampling from the full conditional distributions, which is not required for this generic algorithm. For a related discussion, see Gilks *et al.* (1995: this volume).

The strategy consists of three applications of MCMC. We use θ to denote the vector of parameters and a_j^2 to denote the variance of the proposal function for θ_j (the jth component of θ). The performance of a Metropolis algorithm can be very sensitive to the values of a_j^2, which must be set by the user: if they are too large, the chain will almost never move from the current state, while if they are too small, the chain will move frequently but slowly. In either situation, it will take a long time to obtain an adequate sample from the posterior distribution.

What values should be used for a_j? A number of values have been suggested in the literature. Tierney (1991) suggested setting the standard deviation of the proposal distribution at a fixed multiple, such as 0.5 or 1, of the current estimated variance matrix. Others have suggested using a variable schedule, based in some way on how well the chain is behaving, to determine a good value for a_j (Müller, 1991; Clifford, 1994).

Gelman *et al.* (1995) have studied what the 'optimal' variance should be for the case where one is simulating a univariate standard normal distribution using a Metropolis algorithm with a normal distribution centred at the current value as the proposal distribution. Two different optimality criteria were considered: the asymptotic efficiency of the Markov chain and the geometric convergence rate of the Markov chain. In either case, Gelman *et al.* found that, for the univariate standard normal distribution, the optimal proposal standard deviation ($s.d.$) is about 2.3. We might then expect that, for an arbitrary univariate normal target distribution, the optimal proposal standard deviation should be roughly 2.3 times the standard deviation of the target distribution. See Roberts (1995: this volume) for further discussion of these results.

This result is readily incorporated in the generic three-simulation strategy outlined here. The generic three-simulation strategy consists of the following steps:

Simulation 1

- Assign a large value to a_j, such as 2.3 times the approximate marginal standard deviation of θ_j.

- Run MCMC for $N_{\min}$ scans to obtain a pilot sample.

Simulation 2

- Use the pilot sample to calculate conditional standard deviations of each component of θ given the sample estimates for the other components. This is done using linear regression (regressing θ_j on θ_{-j}, where θ_{-j} denotes all of θ except the jth component).

- Assign $a_j = 2.3\,s.d.\,(\theta_j \mid \theta_{-j})$.

- Run another MCMC for $N_{\min}$ scans to obtain a second sample. The second sample is necessary to obtain reasonable estimates of the conditional standard deviations.

Simulation 3

- Calculate conditional standard deviations of the parameters from the second sample.

- Reassign $a_j = 2.3\,s.d.\,(\theta_j \mid \theta_{-j})$.

- Run the `gibbsit` program to obtain k, M and N.

- Run MCMC for $(M + N)$ scans.

- Rerun the `gibbsit` program to check that N is big enough.

- If N is big enough, i.e. if the `gibbsit` program says that the number of iterations required is no greater than the number of iterations actually run, make inference using the final sample.

- If not, return to the beginning of Simulation 3.

7.5.1 An example

To illustrate the three-simulation strategy, we use it to simulate from the following trivariate normal distribution for $\theta = (\theta_0, \theta_1, \theta_2)^{\mathrm{T}}$:

$$
\theta \;\sim\; \mathrm{N}\left(\begin{bmatrix} 0 \\ 0 \\ 0 \end{bmatrix}, V = \begin{bmatrix} 99 & -7 & -7 \\ -7 & 1 & 0 \\ -7 & 0 & 1 \end{bmatrix} \right).
$$

The conditional variances are

$$
\begin{aligned}
V\left(\theta_0 \mid \theta_1, \theta_2\right) &= 1 \\
V\left(\theta_1 \mid \theta_0, \theta_2\right) &= 1/50 \\
V\left(\theta_2 \mid \theta_0, \theta_1\right) &= 1/50.
\end{aligned}
$$

This kind of posterior covariance matrix is common for regression parameters, where θ_0 is the intercept and the independent variables corresponding to θ_1 and θ_2 have non-zero means. It is a challenging example because the conditional variances are many times smaller than the marginal variances (even though there are no very high pairwise correlations).

We start by generating a small pilot sample of $N_{\min}$ draws. To determine the size of this pilot sample, we must first select values for q, r and s. We take $q = 0.025$, $r = 0.0125$ and $s = 0.95$, which corresponds to requiring that the estimated cumulative distribution function of the 0.025 quantile of θ_0 falls within the interval $(0.0125, 0.0375)$ with probability 0.95. This leads to $N_{\min} = 600$. The proposal standard deviations for obtaining the pilot sample are set to $a_0 = 2.3\sqrt{99} = 22.9$ and $a_1 = a_2 = 2.3\sqrt{1} = 2.3$. Since we know that the actual marginal distribution of θ_0 is $\mathrm{N}(0, 99)$, we can find the true 0.025 quantile for θ_0, $F^{-1}(0.025)$, which is -19.5 (where

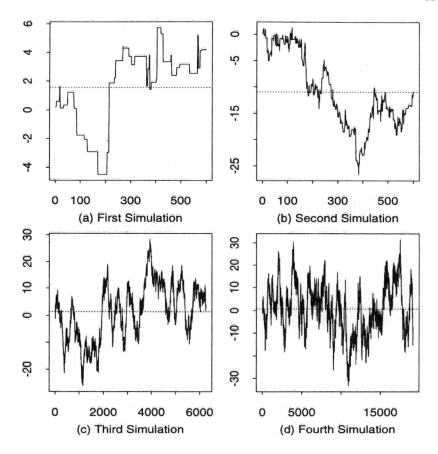

Figure 7.3 *Sequence plots of intercept estimate for trivariate normal example.*

F denotes the cumulative distribution function of the marginal distribution of θ_0).

Figure 7.3 plots successive iterations of the intecept parameter, θ_0. Figure 7.3(a) shows the 600-iteration pilot sample; this is clearly too few to obtain a good sample from the posterior. Next we calculated the sample covariance matrix for the pilot sample and from this the conditional variance of each parameter given the other two. These are reported in Table 7.1, which summarizes the results. It took four simulations before the number of iterations indicated by the `gibbsit` program was less than the actual number of iterations run during the current simulation.

The sample covariance matrix for the pilot sample is far from the truth. When we input the pilot sample into the `gibbsit` program, it says that M

variable	true value	simulation			
		first	second	third	fourth
$M + N$		600	600	6 313	13 353
a_0		22.9	2.2	2.2	2.3
a_1		2.3	0.33	0.32	0.33
a_2		2.3	0.31	0.31	0.33
$E(\theta_0)$	0	1.6	−11.0	1.4	0.5
$E(\theta_1)$	0	−0.8	0.9	−0.2	−0.1
$E(\theta_2)$	0	0.6	0.7	0	0
V_{11}	99	7.4	48	106	106
V_{12}	−7	−0.6	−4.7	−6.4	−7.4
V_{13}	−7	−0.4	−2.0	−8.6	−7.6
V_{22}	1	0.2	1.0	1.0	1.1
V_{23}	0	−0.1	−0.4	−0.1	−0.1
V_{33}	1	0.1	0.6	1.3	1.2
$V(\theta_0 \mid \theta_1, \theta_2)$	1	0.9	0.9	1.0	1.0
$V(\theta_1 \mid \theta_0, \theta_2)$	0.02	0.02	0.02	0.02	0.02
$V(\theta_2 \mid \theta_0, \theta_1)$	0.02	0.02	0.02	0.02	0.02
k		1	1	1	1
M		229	57	148	138
N		46 992	6 256	13 205	12 276
$\hat{F}\left(F^{-1}(0.025)\right)$	0.025		0.082	0.021	0.020

Table 7.1 *Three-simulation strategy results for the trivariate normal example: the input control parameters for each simulation are reported in the upper part of the table; the resulting estimates from each simulation are shown in the lower part*

should equal 229 and N should be set to 46 992. Hence

$$I = \frac{M + N}{N_{\min}} = 78.7,$$

so that the pilot sample is clearly not large enough. We recalculated the proposal standard deviations as $a_0 = 2.3\sqrt{0.92} = 2.21$, $a_1 = 2.3\sqrt{0.020} = 0.33$ and $a_2 = 2.3\sqrt{0.018} = 0.31$. These were used as inputs to the second simulation. We then obtained another sample of $N_{\min} = 600$ draws from the posterior. The results are shown in the second simulation column of Table 7.1. The second sample covariance matrix was much closer to the true covariance matrix than was the pilot sample covariance matrix. It was not correct, but we were heading in the right direction.

The conditional variances from the second sample were then used to recalculate proposal standard deviations one more time. These are shown as inputs to the third simulation. Also, using the second Metropolis sample as input to the `gibbsit` program produced values for k, N and M to use for a third run of the Metropolis algorithm. In our case, `gibbsit` indicated that k was 1, M should be set to 57 and N should be set at 6 256.

We then used the recalculated proposal standard deviations to perform a third Metropolis run for a total of $(M + N) = 6\,313$ iterations. The marginal and conditional sample variances for this third sample were reasonably close to the correct values. However, when we ran the `gibbsit` program it said that $k = 1$, M should be 148 and N should be 13 205. The new N was considerably greater than the number of iterations used for the third sample, so we repeated the third simulation.

Accordingly, we obtained a fourth sample containing $(M + N) = 13\,353$ iterations. The marginal and conditional sample variances for this fourth sample are even closer to the correct values, and the sequence looks reasonably stationary (Figure 7.3(d)). As an additional check, we calculated the sample estimate of the cumulative distribution function at the true 0.025 quantile by finding the proportion of simulations of θ_0 less than or equal to -19.5 to be

$$\hat{F}\left(F^{-1}\left(0.025\right)\right) \;=\; 0.020$$

using the fourth sample. This estimate is between 0.0125 and 0.0375, satisfying our original criterion. Finally, when we use `gibbsit` on the fourth sample it says that N should be 12 276, which is less than the number of iterations used in the fourth sample. We conclude that the fourth sample is good enough to perform inference with.

7.6 Discussion

We have described a way of determining the number of MCMC iterations needed, together with the `gibbsit` software that implements it. We have shown how this can also be used to diagnose slow convergence, and to help design a fully automatic generic Metropolis algorithm. The method is designed for the common situation where interest focuses on posterior quantiles of parameters of interest, but it can also be applied to the estimation of probabilities rather than parameters, which arises in image processing, the analysis of pedigrees and expert systems, for example (Raftery and Lewis, 1992a).

For simplicity, we have described our methods in the context of a single long run of MCMC iterations. However, Gelman and Rubin (1992b) have argued that one should use several independent sequences, with starting points sampled from an overdispersed distribution; see also Gelman (1995: this volume). Our method can be used in this situation also, with `gibbsit` applied to pilot sequences from each starting point, and the N iterations distributed among the different sequences. Various implementations are possible, and more research is needed on this topic.

Is it really necessary to use multiple sequences? Gelman and Rubin (1992b) argue that, in their absence, MCMC algorithms can give misleading answers, and this is certainly true. However, the creation of an overdispersed starting distribution can be a major chore and can add substantially to the complexity of an MCMC algorithm (Gelman and Rubin, 1992b: Section 2.1). There is clearly a trade-off between the extra work and cost required to produce and use an overdispersed starting distribution and whatever penalty or cost might be experienced on those rare occasions where MCMC without the overdispersed starting distribution arrives at misleading parameter estimates; the answer ultimately depends on the application.

In our experience, MCMC algorithms often do converge rapidly even from poorly chosen starting values, and when they do not, simple diagnostics such as those of Section 7.4 usually reveal the fact. Then simple trial and error with new starting values often leads rapidly to a satisfactory starting value, as illustrated in Section 7.4.1. However, there are cases where diagnostics will not reveal a lack of convergence (e.g. Gelman and Rubin, 1992a), and so multiple sequences should certainly be used for final inferences when much is at stake.

Diagnostics such as those of Section 7.4 should be used even with multiple sequences. This is because a bad starting value close to a local mode (which is what the Gelman–Rubin multiple sequence methods are designed to protect against) is only one of the possible causes of slow convergence in MCMC. Others include high posterior correlations (which can be removed by approximate orthogonalization; see Hills and Smith, 1992), and 'stickiness' of the chain. The latter often arises in hierarchical models when the algorithm enters parts of the parameter space where the random effects variance is small, and can require redesigning the MCMC algorithm itself (e.g. Besag and Green, 1993).

Acknowledgments

This research was supported by ONR contract N00014-91-J-1074 and by NIH grant 5R01HD26330. We are grateful to Andrew Gelman for stimulating discussions and to Wally Gilks for helpful comments.

References

Besag, J. E. and Green, P. J. (1993) Spatial statistics and Bayesian computation. *J. R. Statist. Soc.* B, **55**, 25–37.

Clifford, P. (1994) Discussion on approximate Bayesian inference with the weighted likelihood bootstrap (by M. A. Newton and A. E. Raftery). *J. R. Statist. Soc.* B, **56**, 34–35.

Cox, D. R. and Miller, H. D. (1965) *The Theory of Stochastic Processes.* London: Chapman & Hall.

Gelman, A. (1995) Inference and monitoring convergence. In *Markov Chain Monte Carlo in Practice* (eds W. R. Gilks, S. Richardson and D. J. Spiegelhalter), pp. 131–143. London: Chapman & Hall.

Gelman, A. and Rubin, D. B. (1992a) A single series from the Gibbs sampler provides a false sense of security. In *Bayesian Statistics 4* (eds J. M. Bernardo, J. O. Berger, A. P. Dawid and A. F. M. Smith), pp. 625–632. Oxford: Oxford University Press.

Gelman, A. and Rubin, D. B. (1992b) Inference from iterative simulation using multiple sequences (with discussion). *Statist. Sci.*, **7**, 457–511.

Gelman, A. Roberts, G. O. and Gilks, W. R. (1995) Efficient Metropolis jumping rules. In *Bayesian Statistics 5* (eds J. M. Bernardo, J. O. Berger, A. P. Dawid and A. F. M. Smith). Oxford: Oxford University Press (in press).

Gilks, W. R. and Roberts, G. O. (1995) Strategies for improving MCMC. In *Markov Chain Monte Carlo in Practice* (eds W. R. Gilks, S. Richardson and D. J. Spiegelhalter), pp. 89–114. London: Chapman & Hall.

Gilks, W. R., Richardson, S. and Spiegelhalter, D. J. (1995) Introducing Markov chain Monte Carlo. In *Markov Chain Monte Carlo in Practice* (eds W. R. Gilks, S. Richardson and D. J. Spiegelhalter), pp. 1–19. London: Chapman & Hall.

Hills, S. E. and Smith, A. F. M. (1992) Parametrization issues in Bayesian inference (with discussion). In *Bayesian Statistics 4* (eds J. M. Bernardo, J. O. Berger, A. P. Dawid and A. F. M. Smith), pp. 227–246. Oxford: Oxford University Press.

Lewis, S. M. (1993) Discussion on the meeting on the Gibbs sampler and other Markov chain Monte Carlo methods. *J. R. Statist. Soc.* B, **55**, 79–81.

Lewis, S. M. (1994) Multilevel modeling of discrete event history data using Markov chain Monte Carlo methods. Ph.D. thesis, Department of Statistics, University of Washington.

Metropolis, N., Rosenbluth, A. W., Rosenbluth, M. N., Teller, A. H. and Teller, E. (1953) Equations of state calculations by fast computing machines. *J. Chem. Phys.*, **21**, 1087–1092.

Müller, P. (1991) A generic approach to posterior integration and Gibbs sampling. Technical Report, Institute of Statistics and Decision Sciences, Duke University.

Raftery, A. E. (1986) A note on Bayes factors for log-linear contingency table models with vague prior information. *J. R. Statist. Soc.* B, **48**, 249–250.

Raftery, A. E. (1995) Hypothesis testing and model selection. In *Markov Chain Monte Carlo in Practice* (eds W. R. Gilks, S. Richardson and D. J. Spiegelhalter), pp. 163–187. London: Chapman & Hall.

Raftery, A. E. and Banfield, J. D. (1991) Stopping the Gibbs sampler, the use of morphology, and other issues in spatial statistics. *Ann. Inst. Statist. Math.*, **43**, 32–43.

Raftery, A. E. and Lewis, S. M. (1992a) How many iterations in the Gibbs sampler? In *Bayesian Statistics 4* (eds J. M. Bernardo, J. O. Berger, A. P. Dawid and A. F. M. Smith), pp. 765–776. Oxford: Oxford University Press.

Raftery, A. E. and Lewis, S. M. (1992b) One long run with diagnostics: implementation strategies for Markov chain Monte Carlo. *Statist. Sci.*, **7**, 493–497.

Raftery, A. E., Lewis, S. M., Aghajanian, A. and Kahn, M. J. (1995) Event history modeling of World Fertility Survey data. *Math. Pop. Stud.*, (in press).

Roberts, G. O. (1995) Markov chain concepts related to sampling algorithms. In *Markov Chain Monte Carlo in Practice* (eds W. R. Gilks, S. Richardson and D. J. Spiegelhalter), pp. 45–57. London: Chapman & Hall.

Schwarz, G. (1978) Estimating the dimension of a model. *Ann. Statist.*, **6**, 461–464.

Tierney, L. (1991) Exploring posterior distributions using Markov chains. In *Computer Science and Statistics: Proc. 23rd Symp. Interface* (ed. E. Keramidas), pp. 563–570. Fairfax Station: Interface Foundation.

8

Inference and monitoring convergence

Andrew Gelman

8.1 Difficulties in inference from Markov chain simulation

Markov chain simulation is a powerful tool, so easy to apply that there are risks of serious errors:

1. Inappropriate modeling: the assumed model may not be realistic from a substantive standpoint or may not fit the data.

2. Errors in calculation or programming: the stationary distribution of the simulation process may not be the same as the desired target distribution; or the algorithm, as programmed, may not converge to any proper distribution.

3. Slow convergence: the simulation can remain for many iterations in a region heavily influenced by the starting distribution. If the iterations are used to summarize the target distribution, they can yield falsely precise inference.

The first two errors can occur with other statistical methods (such as maximum likelihood), but the complexity of Markov chain simulation makes mistakes more common. In particular, it is possible to program a method of computation such as the Gibbs sampler or the Metropolis algorithm that only depends on local properties of the model (e.g., conditional distributions for the Gibbs sampler or density ratios for the Metropolis algorithm) without ever understanding the large-scale features of the joint distribution. For a discussion of this issue in the context of probability models for images, see Besag (1986).

Slow convergence is a problem with deterministic algorithms as well. Consider, for example, the literature about convergence of EM and related algorithms (e.g., Meng and Pedlow, 1992). In deterministic algorithms, the two most useful ways of measuring convergence are (a) monitoring individual summaries, such as the increasing likelihood in the EM and ECM algorithms or the symmetry of the covariance matrix in the SEM algorithm (Dempster *et al.*, 1977; Meng and Rubin, 1991, 1993), and (b) replicating the algorithm with different starting points and checking that they converge to the same point (or, if not, noting multiple solutions). We apply both general approaches to Markov chain simulation, but we must overcome the difficulties that (a) the algorithm is stochastic, so we cannot expect any summary statistic to increase or decrease monotonically, and (b) convergence is to a distribution, rather than a point.

This chapter presents an overview of methods for addressing two practical tasks: monitoring convergence of the simulation and summarizing inference about the target distribution using the output from the simulations. The material in Sections 8.3–8.5 is presented in more detail, with an example, in Gelman and Rubin (1992b). Section 8.6 introduces and provides references to various methods in the recent statistical literature for using inference from the simulation to improve the efficiency of the Markov chain algorithm.

The practical task in monitoring convergence is to estimate how much the inference based on Markov chain simulations differs from the desired target distribution. Our basic method, inspired by the analysis of variance, is to form an overestimate and an underestimate of the variance of the target distribution, with the property that the estimates will be roughly equal at convergence but not before.

8.2 The risk of undiagnosed slow convergence

The fundamental problem of inference from Markov chain simulation is that there will always be areas of the target distribution that have not been covered by the finite chain. As the simulation progresses, the ergodic property of the Markov chain causes it eventually to cover all the target distribution but, in the short term, the simulations cannot, in general, tell us about areas where they have not been. Incidentally, this is a general problem whenever convergence is slow, even in a distribution that has a single mode. It has happened several times in our experience that a single sequence of Markov chain simulation appeared to have 'converged', but independent replications of the sequence revealed that within-sequence movement was just too slow to detect.

In our own experience of applying Markov chain simulation to probability models and Bayesian posterior distributions, we have commonly noticed poor convergence by examining multiple independent simulations. In many of these settings, any single simulated sequence would have appeared to have converged perfectly if examined alone. Some of these examples have been published as Figures 1–3 of Gelman and Rubin (1992a) (note the title of that article) and Figure 4 of Gelman and Rubin (1992b). In these examples, methods for diagnosing lack of convergence from a single sequence (e.g., Hastings, 1970; Geyer, 1992; Raftery and Lewis, 1992, 1995: this volume) all fail, because the simulations are moving so slowly, or are stuck in separate places in the target distribution. Here we present yet another example, from our current applied research.

Figure 8.1 displays an example of slow convergence from a Markov chain simulation for a hierarchical Bayesian model for a pharmacokinetics problem (see Gelman *et al.*, 1995a, for details). The simulations were done using a Metropolis-approximate Gibbs sampler (as in Section 4.4 of Gelman, 1992). Due to the complexity of the model, each iteration was expensive in computer time, and it was desirable to keep the simulation runs as short as possible. Figure 8.1 displays time-series plots for a single parameter in the posterior distribution in two independent simulations, each of length 500. The simulations were run in parallel simultaneously on two workstations in a network. It is clear from the separation of the two sequences that, after 500 iterations, the simulations are still far from convergence. However, either sequence alone looks perfectly well behaved.

Interestingly, we do not yet know whether the slow convergence exhibited in Figure 8.1 is due to an inappropriate model, programming mistakes, or just slow movement of the Markov chain. As is common in applied statistics, we have repeatedly altered our model in the last several months of research on this problem, as we have gained understanding about the relation between the model, the data, and our prior information. The simulations from a previous version of the model had reached approximate convergence before 500 iterations, which leads us to suspect an error of some sort in the simulation leading to Figure 8.1. (One of the strengths of Markov chain simulation is that it allows us to change our model with only minor alterations in the computer program.) Several times in this research, we have noted poor convergence by comparing parallel sequences, and each time we have investigated and found a substantive flaw in the model or a programming error, or else we have had to alter our Markov chain simulation to run more efficiently. Another approach would be to run the simulation indefinitely and wait until the lines begun in Figure 8.1 overlap. Due to the slowness in computing the simulation draws at each step, we would prefer to avoid this approach. Also, our model is still in a preliminary stage, and so any investment made now in computational efficiency (or in debugging!) can be expected to pay off repeatedly in computations with future versions of the model.

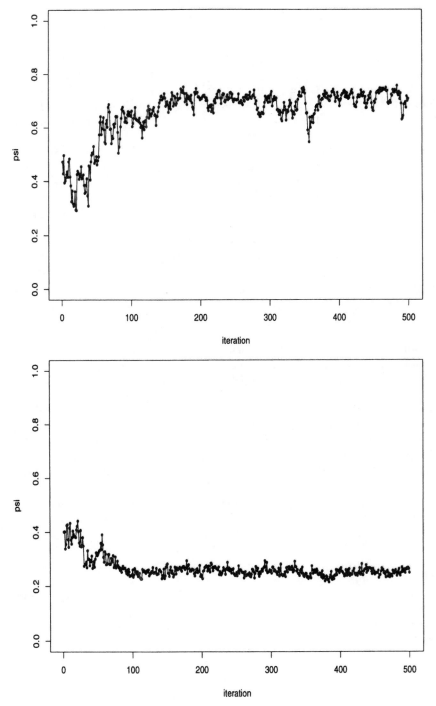

Figure 8.1 *Time series of the value of a single parameter from two parallel simulations of length 500 from iterative simulation of a complicated multiparameter model; lack of convergence is evident by comparing the simulations but cannot be detected from either sequence alone.*

8.3 Multiple sequences and overdispersed starting points

Our general approach to monitoring convergence of Markov chain simulations is based on plots such as Figure 8.1. We have always found it useful to simulate at least two parallel sequences, typically four or more. If the computations are implemented on a network of workstations or a parallel machine, it makes sense to run as many parallel simulations as there are free workstations or machine processors. The recommendation to always simulate multiple sequences is not new in the iterative simulation literature (e.g., Fosdick, 1959) but is somewhat controversial (see the discussion of Gelman and Rubin, 1992b, and Geyer, 1992). In our experience with Bayesian posterior simulation, however, we have found that the added information obtained from replication (as in Figure 8.1) outweighs any additional costs required in multiple simulations.

It is desirable to choose starting points that are widely dispersed in the target distribution. Overdispersed starting points are an important design feature because starting far apart can make lack of convergence apparent and, for the purposes of inference, ensure that all major regions of the target distribution are represented in the simulations. For many problems, especially with discrete or bounded parameter spaces, it is possible to pick several starting points that are far apart by inspecting the parameter space and the form of the distribution. For example, the proportion in a two-component mixture model can be started at values of 0.1 and 0.9 in two parallel sequences.

In more complicated situations, more work may be needed to find a range of dispersed starting values. In practice, we have found that any additional effort spent on approximating the target density is useful for understanding the problem and for debugging; after the Markov chain simulations have been completed, the final estimates can be compared to the earlier approximations. In complicated applied statistical problems, it is standard practice to improve models gradually as more information becomes available, and the estimates from each model can be used to obtain starting points for the computation in the next stage.

Before running the Markov chain simulations, it is important to have a rough idea of the extent of the target distribution. In many problems, initial estimates can be obtained using the data and a simpler model: for example, approximating a hierarchical generalized linear model by a linear regression or nonhierarchical generalized linear model computation. In other problems, including the example used for Figure 8.1, the prior distribution is informative and can be used to construct rough bounds on the parameters of the model. In problems without strong prior distributions, it is often useful to locate the mode or modes of the target distribution using some deterministic algorithm such as stepwise ascent, EM, or ECM (Dempster *et al.*, 1977; Meng and Rubin, 1993). (Once the Markov chain simulation algorithm has been programmed, it is often easily altered to find modes, by replacing random jumps with deterministic steps to move to higher points in the target density.) It is also useful to estimate roughly the scale of the target distribution near the modes, which can often be done by computing the second-derivative matrix of the log-posterior density at each mode. For continuous-parameter problems, starting points for parallel Markov chain simulations can be drawn from an approximate student-t mixture distribution based on the posterior modes, possibly corrected by importance resampling; see Gelman and Rubin (1992b) for details. If the target distribution is multimodal, or suspected to be multimodal, it is a good idea to start at least one sequence at each mode. If the number of modes is large, the simulation algorithm should be designed to jump between modes frequently. As we have seen, preliminary estimation is not always easy, but the effort generally pays off in greater understanding of the model and confidence in the results.

8.4 Monitoring convergence using simulation output

Our recommended general approach to monitoring convergence is based on detecting when the Markov chains have 'forgotten' their starting points, by comparing several sequences drawn from different starting points and checking that they are indistinguishable. There are many possible ways to compare parallel sequences, the most obvious approach being to look at time-series plots such as in Figure 8.1 overlaid, and see if the two sequences can be distinguished. Here, we outline a more quantitative approach based on the analysis of variance; approximate convergence is diagnosed when the variance between the different sequences is no larger than the variance within each individual sequence.

A more general formulation of the method presented here is to identify 'convergence' with the condition that the empirical distribution of simulations obtained separately from each sequence is approximately the same as the distribution obtained by mixing all the sequences together. Before the parallel sequences have converged, the simulations collected within each single sequence will be much less variable than the simulations collected from all the sequences combined; consider Figure 8.1, for example.

The approach we have found most convenient is based on separately monitoring the convergence of all scalar summaries of interest from the target distribution. For example, we may be interested in all the parameters in the distribution and various predictive quantities; each of these parameters can be considered separately as a scalar summary. We will defer until the end of this section a discussion of what scalar summaries should be monitored. For the purpose of defining the method, we shall consider a single summary at a time, and label it ψ. We shall assume m parallel simulations, each of length n.

For each scalar summary of interest, we would like a numerical equivalent of the comparison in Figure 8.1 that states: 'The two sequences are much farther apart than we could expect, based on their internal variability.' For each scalar summary ψ, we label the m parallel sequences of length n as (ψ_{ij}), $j = 1, \ldots, n; i = 1, \ldots, m$, and we compute two quantities, the between-sequence variance B and the within-sequence variances W:

$$B = \frac{n}{m-1} \sum_{i=1}^{m} (\overline{\psi}_{i.} - \overline{\psi}_{..})^2, \quad \text{where} \quad \overline{\psi}_{i.} = \frac{1}{n} \sum_{j=1}^{n} \psi_{ij}, \quad \overline{\psi}_{..} = \frac{1}{m} \sum_{i=1}^{m} \overline{\psi}_{i.}$$

$$W = \frac{1}{m} \sum_{i=1}^{m} s_i^2, \quad \text{where} \quad s_i^2 = \frac{1}{n-1} \sum_{j=1}^{n} (\psi_{ij} - \overline{\psi}_{i.})^2.$$

The between-sequence variance B contains a factor of n because it is based on the variance of the within-sequence means, $\overline{\psi}_{i.}$, each of which is an average of n values ψ_{ij}.

From the two variance components, we construct two estimates of the variance of ψ in the target distribution. First,

$$\widehat{\text{var}}(\psi) = \frac{n-1}{n} W + \frac{1}{n} B$$

is an estimate of the variance that is *unbiased* under stationarity (that is, if the starting points of the simulations were actually drawn from the target distribution), but is an *overestimate* under the more realistic assumption that the starting points are overdispersed. We call $\widehat{\text{var}}(\psi)$ a 'conservative' estimate of the variance of ψ under overdispersion.

Meanwhile, for any finite n, the within-sequence variance W should *underestimate* the variance of ψ because the individual sequences have not had time to range over all of the target distribution and, as a result, will have less variability. In the limit as $n \to \infty$, both $\widehat{\text{var}}(\psi)$ and W approach $\text{var}(\psi)$, but from opposite directions.

We can now monitor the convergence of the Markov chain by estimating the factor by which the conservative estimate of the distribution of ψ might be reduced: that is, the ratio between the estimated upper and lower bounds for the standard deviation of ψ, which we call the 'estimated potential scale reduction',

$$\sqrt{\widehat{R}} = \sqrt{\frac{\widehat{\text{var}}(\psi)}{W}}.$$

(This is $\hat{R}$ rather than R because the numerator and denominator are merely *estimated* upper and lower bounds on the variance.) As the simulation converges, the potential scale reduction declines to 1, meaning that the parallel Markov chains are essentially overlapping. If the potential scale reduction is high, then we have reason to believe that proceeding with further simulations may improve our inference about the target distribution.

For example, the estimate $\widehat{\text{var}}(\psi)$ derived from the two simulations of Figure 8.1 would just about cover the range of both sequences, and is about 2.5^2, while the average within variance W measures just the variance within each sequence and is about 0.5^2. The estimated potential scale reduction $\sqrt{\widehat{R}}$ is about 5 for this example, indicating poor convergence and a potential for Bayesian credible intervals for ψ to shrink by as much as a factor of 5 once convergence is eventually reached.

In general, if $\widehat{R}$ is not near 1 for all scalar summaries of interest, it is probably a good idea to continue the simulation runs (and perhaps alter the simulation algorithm itself to make the simulations more efficient, as we discuss in Section 8.6). In practice, we generally run the simulations until the values of $\widehat{R}$ are all less than 1.1 or 1.2. Using this method, we never actually have to look at graphs such as Figure 8.1; the potential scale reductions are all computed automatically.

There is still the question of which scalar summaries to monitor, although the above approach simplifies the problem in practice by making monitoring so easy that we can, and have, monitored over a hundred summaries for a single problem and just scanned for values of $\widehat{R}$ greater than 1.2 as indicating poor convergence. We have no problem monitoring all parameters and hyperparameters of a model and also examining predictive simulations of interest and other summaries of interest such as the ratio between two variance components. Tables 2 and 3 of Gelman and Rubin (1992b) provide an example of monitoring several summaries at once. In addition, the method could be generalized to monitor convergence of vector summaries, in which case B, W, and $\widehat{\text{var}}(\psi)$ become matrices whose eigenvalues can be compared to estimate the potential reduction in the scale of vector inferences.

Another issue to consider is sampling variability in the quantities W and B; we do not want to falsely declare convergence when $\widehat{R}$ just happens to be near 1 in a short simulation run. In practice, sampling variability in convergence monitoring statistics is not a serious concern because, regardless of convergence, one will almost always run the simulations long enough to obtain a fairly good estimate of the variance in the target distribution. In addition, if several scalar summaries are being monitored, it is extremely unlikely that they will all appear to have converged by 'luck', especially if the number of parallel simulations m is fairly large (at least 10, say). For theoretical completeness, however, it is possible to correct the above estimates for sampling variability, leading to a slightly different estimate of R; details appear in Gelman and Rubin (1992b), and a computer program 'itsim' in the S language is available on Statlib or from the author.

A potentially useful improvement for monitoring convergence is to create an underestimate of $\text{var}(\psi)$ that is more efficient than W, by making use of the autocorrelated time-series structure of the iterations within each series. Hastings (1970) discusses this approach, and Geyer (1992) reviews some more recent theoretical results in this area. Both these references attempt to estimate var(ψ) from a single Markov chain sequence, which is a hopeless task in many practical applications (see Section 8.2), but can be useful as improved *under*estimates for use in place of W in the formula for $\hat{R}$.

In addition, several methods have been proposed in recent years to use the Markov chain transition probabilities, which are known in most applications of Markov chain simulation, to diagnose lack of convergence more efficiently. At convergence, the simulations in any sequence should look just as 'likely' backward as forward, and the joint distribution of successive simulations in a sequence should be symmetric. Cui *et al.* (1992) and Zellner and Min (1994) construct scalar summaries based on these principles that can diagnose poor convergence in cases where summaries based only on the simulation output (and not the transition probabilities) fail. Liu *et al.* (1992) and Roberts (1995a) construct somewhat similar scalar summaries using information from multiple sequences. All these methods are most effective when used in addition to the more basic analysis-of-variance approach for monitoring scalar summaries of interest.

8.5 Output analysis for inference

Our main practical concern in Bayesian inference is to make reliable inferences about the target distribution: for example, claimed 95% regions that include *at least* 95% of the mass of the target distribution, with exact coverage as the length of the Markov chain simulations approach infinity. In this section, we elaborate on the basic methods of output analysis described in Gilks *et al.* (1995: this volume).

The simplest and most generally useful idea in inference is to use the empirical distribution of the simulated draws, as in multiple imputation (Rubin, 1987), with the iterations from all the parallel simulations mixed in together. If θ is the vector variable from which N values have been simulated, this means computing any moments of the posterior distribution using sample moments of the N draws of θ, estimating 95% posterior intervals of any scalar summary ψ by the 2.5% and 97.5% order statistics of the N simulated values ψ, and so forth. This approach is generally reliable if based on multiple sequences with overdispersed starting points. Intervals obtained before convergence should be overdispersed and conservative. Once approximate convergence has been reached, the intervals and other summaries of the target distribution should be accurate, up to the granularity of the finite number of simulation draws. If the early parts of the simulated sequences have been discarded in monitoring convergence, they should also be discarded for the purposes of inference.

It has sometimes been suggested that inferences should be based on every

k^{th} iteration of each sequence, with k set to some value high enough that successive draws of θ are approximately independent. This strategy, known as *thinning*, can be useful when the set of simulated values is so large that reducing the number of simulations by a factor of k gives important savings in storage and computation time. Except for storage and the cost of handling the simulations, however, there is no advantage in discarding intermediate simulation draws, even if highly correlated. The step of mixing the simulations from all m sequences and then choosing at random destroys any serial dependence in the simulated sequences, and even correlated draws add some information. A quantitative treatment of these issues is given by Geyer (1992), for the case of estimating the mean of a scalar summary using simulation draws.

Suppose we are interested in the distribution of a scalar summary, ψ, for a multivariate target distribution, $P(\theta)$. If we know the mathematical form of the conditional density of ψ given the other components of θ, then we can obtain a better estimate of the density of ψ by averaging the conditional densities over the simulated values of θ:

$$\hat{P}(\psi) = \frac{1}{N} \sum_{\ell=1}^{N} P(\psi | \theta^{(\ell)}(-\psi)),$$

where $\{\theta^{(\ell)}, \ell = 1, \ldots, N\}$ represents the simulation draws of θ (from, say, the last halves of a set of simulated sequences that have reached approximate convergence), and $\theta^{(\ell)}(-\psi)$ represents all the components of $\theta^{(\ell)}$ except for ψ. The application of this method to Markov chain simulation, specifically the Gibbs sampler, was suggested by Tanner and Wong (1987) and Gelfand and Smith (1990) who termed it 'Rao–Blackwellization'. A theoretical discussion of its effectiveness is given by Liu *et al.* (1994).

8.6 Output analysis for improving efficiency

An interesting area of current research combines the ideas of inference and efficiency. It is possible to improve monitoring and inference through more effective use of the information in the Markov chain simulation. Most obviously, we note that the early part of a simulation is often far from convergence, and we can crudely create simulations that are closer to convergence by simply discarding the early parts of the simulated sequences. In our applications, we have followed the simple but effective approach of discarding the first half of each simulated sequence and applying the procedures of Sections 8.4 and 8.5 to the remainder.

Conversely it is possible, both in theory and practice, to use inference about the target distribution to improve the efficiency of the simulation algorithm. Many different ideas apply to this problem, including the adaptive methods discussed by Gilks and Roberts (1995: this volume). A Gibbs sampler is generally most efficient when the jumps are along the principal components of the target distribution; inference from early simulations can be used to reparameterize (Hills and Smith, 1992). For the Metropolis algorithm, normal distribution theory suggests that the most efficient proposal density is shaped like the target distribution scaled by a factor of about $2.4/\sqrt{d}$, where d is the dimension of the target distribution (Gelman et al., 1995b; Muller, 1993; Roberts, 1995b: this volume). The scale and shape of the target distribution can again be estimated from early simulation draws, and the simulations can be adaptively altered as additional information arrives. Inference from multiple simulated sequences is useful here, so that the early estimates of the target distribution are conservatively spread. Other related approaches suggested by normal distribution theory for the Metropolis algorithm involve adaptively altering Metropolis jumps so that the frequency of acceptance is in the range of 0.25 to 0.5, or optimizing the average distance jumped (Gelman et al., 1995b). These approaches have not yet reached the stage of automatic implementation; as Gelfand and Sahu (1994) demonstrate, transition rules that are continually adaptively altered have the potential for converging to the wrong distribution. We anticipate that the interaction between methods of inference, monitoring convergence, and improvements in efficiency will ultimately lead to more automatic, reliable, and efficient iterative simulation algorithms.

Acknowledgements

We thank Jiming Jiang for the figures, Xiao-Li Meng for helpful comments, the American Lung Association for financial support, and the National Science Foundation for grants DMS-9404305 and DMS-9457824.

References

Besag, J. (1986) On the statistical analysis of dirty pictures (with discussion). *J. R. Statist. Soc.* B, **48**, 259–302.

Cui, L., Tanner, M. A., Sinha, D. and Hall, W. J. (1992) Monitoring convergence of the Gibbs sampler: further experience with the Gibbs stopper. *Statist. Sci.*, **7**, 483–486.

Dempster, A. P., Laird, N. M. and Rubin, D. B. (1977) Maximum likelihood from incomplete data via the EM algorithm (with discussion). *J. R. Statist. Soc.* B, **39**, 1–38.

Fosdick, L. D. (1959) Calculation of order parameters in a binary alloy by the Monte Carlo method. *Phys. Rev.*, **116**, 565–573.

Gelfand, A. and Sahu, S. K. (1994) On Markov chain Monte Carlo acceleration. *J. Comp. Graph. Statist.*, **3**, 261–267.

Gelfand, A. E. and Smith, A. F. M. (1990) Sampling-based approaches to calculating marginal densities. *J. Am. Statist. Ass.*, **85**, 398–409.

Gelman, A. (1992) Iterative and non-iterative simulation algorithms. *Comput. Sci. Statist.*, **24**, 433–438.

Gelman, A. and Rubin, D. B. (1992a) A single sequence from the Gibbs sampler gives a false sense of security. In *Bayesian Statistics 4* (eds J. M. Bernardo, J. O. Berger, A. P. Dawid and A. F. M. Smith), pp. 625–631. Oxford: Oxford University Press.

Gelman, A. and Rubin, D. B. (1992b) Inference from iterative simulation using multiple sequences (with discussion). *Statist. Sci.*, **7**, 457–511.

Gelman, A., Bois, F. Y. and Jiang, J. (1995a) Physiological pharmacokinetic analysis using population modeling and informative prior distributions. Technical report, Department of Statistics, University of California, Berkeley.

Gelman, A., Roberts, G. O. and Gilks, W. R. (1995b) Efficient Metropolis jumping rules. In *Bayesian Statistics 5* (eds J. M. Bernardo, J. O. Berger, A. P. Dawid and A. F. M. Smith). Oxford: Oxford University Press (in press).

Geyer, C. J. (1992) Practical Markov chain Monte Carlo (with discussion). *Statist. Sci.*, **7**, 473–511.

Gilks, W. R. and Roberts, G. O. (1995) Strategies for improving MCMC. In *Markov Chain Monte Carlo in Practice* (eds W. R. Gilks, S. Richardson and D. J. Spiegelhalter), pp. 89–114. London: Chapman & Hall.

Gilks, W. R., Richardson, S. and Spiegelhalter, D. J. (1995) Introducing Markov chain Monte Carlo. In *Markov Chain Monte Carlo in Practice* (eds W. R. Gilks, S. Richardson and D. J. Spiegelhalter), pp. 1–19. London: Chapman & Hall.

Hastings, W. K. (1970) Monte Carlo sampling methods using Markov chains and their applications. *Biometrika*, **57**, 97–109.

Hills, S. E. and Smith, A. F. M. (1992) Parameterization issues in Bayesian inference (with discussion). In *Bayesian Statistics 4* (eds J. M. Bernardo, J. O. Berger, A. P. Dawid and A. F. M. Smith), pp. 227–246. Oxford: Oxford University Press.

Liu, C., Liu, J. and Rubin, D. B. (1992) A variational control variable for assessing the convergence of the Gibbs sampler. *Proc. Statist. Comput. Sect. Am. Statist. Ass.*, 74–78.

Liu, J., Wong, W. H. and Kong, A. (1994) A covariance structure of the Gibbs sampler with applications to the comparisons of estimators and augmentation schemes. *Biometrika*, **81**, 27–40.

Meng, X. L. and Pedlow, S. (1992) EM: a bibliographic review with missing articles. *Proc. Statist. Comput. Sect. Am. Statist. Ass.*, 24–27.

Meng, X. L. and Rubin, D. B. (1991) Using EM to obtain asymptotic variance-covariance matrices: the SEM algorithm. *J. Am. Statist. Ass.*, **86**, 899–909.

Meng, X. L. and Rubin, D. B. (1993) Maximum likelihood estimation via the ECM algorithm: a general framework. *Biometrika*, **80**, 267–278.

Müller, P. (1993) A generic approach to posterior integration and Gibbs sampling. Technical report, Institute of Statistics and Decision Sciences, Duke University.

Raftery, A. E. and Lewis, S. M. (1992) How many iterations in the Gibbs sampler? In *Bayesian Statistics 4* (eds J. M. Bernardo, J. O. Berger, A. P. Dawid and A. F. M. Smith), pp. 763–773. Oxford: Oxford University Press.

Raftery, A. E. and Lewis, S. M. (1995) Implementing MCMC. In *Markov Chain Monte Carlo in Practice* (eds W. R. Gilks, S. Richardson and D. J. Spiegelhalter), pp. 115–130. London: Chapman & Hall.

Roberts, G. O. (1995a) Methods for estimating L^2 convergence of Markov chain Monte Carlo. In *Bayesian Statistics and Econometrics: Essays in honor of Arnold Zellner* (eds D. Berry, K. Chaloner and J. Geweke). New York: Wiley (in press).

Roberts, G. O. (1995b) Markov chain concepts related to sampling algorithms. In *Markov Chain Monte Carlo in Practice* (eds W. R. Gilks, S. Richardson and D. J. Spiegelhalter), pp. 45–57. London: Chapman & Hall.

Rubin, D. B. (1987) *Multiple Imputation for Nonresponse in Surveys*. New York: Wiley.

Tanner, M. A. and Wong, W. H. (1987) The calculation of posterior distributions by data augmentation (with discussion). *J. Am. Statist. Ass.*, **82**, 528–550.

Zellner, A. and Min, C. (1994) Gibbs sampler convergence criteria. *Technical report*, Graduate School of Business, University of Chicago.

9

Model determination using sampling-based methods

Alan E Gelfand

9.1 Introduction

Responsible data analysis must address the issue of model determination, which we take as consisting of two components: model assessment or checking and model choice or selection. Since in practice, apart from rare situations, a model specification is never 'correct' we must ask (i) is a given model adequate? and (ii) within a collection of models under consideration, which is the best?

All the complex models featured in this volume may be viewed as the specification of a joint distribution of observables (data) and unobservables (model parameters, missing data or latent variables). Often the model can be represented as a directed acyclic graph (Lauritzen and Spiegelhalter, 1988; Whittaker, 1990) in which case this joint density can be immediately written down (see Gilks, 1995: this volume). In some cases, such as in spatial modelling, this joint density is uniquely determined through suitable local specification (Besag and Green, 1993). Regardless, it is this joint density whose performance must be examined with respect to model determination.

Continuing in this generality, inference is based upon features of the conditional distribution of the unobservables in the model, given the observables. On the other hand, model performance is addressed through features of the marginal distribution of the observables, in particular through suitable comparison of what the model predicts with what has been observed. Fortunately, sampling-based methods using MCMC techniques not only enable desired inferences to be performed, but also facilitate model determination.

This chapter describes how to perform model determination in practice. Our approach is informal, adopting an exploratory data analysis (EDA) viewpoint and creating a variety of diagnostic tools for this purpose. We illustrate our approach with an example which, though simpler than usual applications, conveniently enables us to reveal the key ideas.

9.2 Classical approaches

We note that classical approaches to model adequacy also involve comparison of observed with predicted. For instance, theoretical Q-Q plots (usually applied to residuals) compare an empirical cumulative distribution function with a theoretical one (Chambers *et al.*, 1983). Goodness-of-fit tests aggregate these comparisons in some fashion to obtain a global discrepancy measure (D'Agostino and Stephens, 1986). Similarly, residual analysis compares an individual observation with an estimate of its theoretical mean (customarily 'observed minus expected' or more generally perhaps a deviance residual). Such residuals are aggregated to form a global measure of adequacy, for example: R^2; error sum of squares; PRESS statistic; Mallows C_p; deviance; and so on (Cook and Weisberg, 1982; McCullagh and Nelder, 1989). The emergent suggestion, which we pursue here, is that model checking should be administered both locally and globally.

Classical approaches to model choice are unsatisfactory. Standard Neyman–Pearson theory provides an 'optimal' test through the likelihood ratio for the rather useless comparison of two distinct completely specified (no unobservables) models . More generally, likelihood ratio tests enable choice between models only in the nested case where there is an unambiguous null hypothesis. Selection is based upon an asymptotic χ^2 approximation (deviance difference) which may be poor for small sample sizes. Also, such asymptotics will often be inapplicable to models which are nonregular (in the sense that the number of parameters tends to infinity as the number of observations increases): a problem recognized as early as 1948 by Neyman and Scott. Perhaps most distressingly, classical theory offers little for comparison of non-nested models (though see Cox, 1961).

As a side remark, the notion of nesting is really only well defined in special cases. For instance in linear models or, more broadly, generalized linear models where the mean is linear on a transformed scale, nesting arises naturally through the columns of the design matrix. However, consider a reduced model with observations $\{y_1, y_2, \ldots, y_n\}$ independently and identically distributed (*i.i.d.*) from a normal distribution $N(\mu_1, \sigma^2)$; versus a full model which is a mixture model with $\{y_1, y_2, \ldots, y_n\}$ being *i.i.d.* from

$$\alpha N(\mu_1, \sigma^2) + (1 - \alpha)N(\mu_2, \sigma^2), \qquad 0 \le \alpha \le 1.$$

The reduced model can arise from either $\alpha = 1$ or $\mu_1 = \mu_2$, giving an identifiability problem with regard to the parameters. Starting in the four dimensional space $(\mu_1, \mu_2, \sigma^2, \alpha)$ we approach the reduced model as $\alpha \to 1$ or as $\mu_1 - \mu_2 \to 0$. Is the reduced model acceptable because there is little contamination or because the two component densities have very similar means? We cannot identify which of these quite different explanations is appropriate. Moreover, starting with a four-dimensional full model, we would expect to impose two constraints rather than one to obtain a two-dimensional reduced model.

In a similar vein, consider two Gaussian nonlinear regression models:

$$M_1: E(y_i) = \beta_0(1 + \beta_1\beta_2^{x_i}), \qquad (0 \le \beta_2 \le 1);$$

$$M_2: E(y_i) = \gamma_0 e^{-\gamma_1\gamma_2^{x_i}}, \qquad (0 \le \gamma_2 \le 1).$$

A strategy for comparing M_1 and M_2 within classical likelihood ratio testing requires the creation of an encompassing or full model having these as reduced models. Then we could investigate the difference in fit between the full model and each reduced model. However, the additive form

$$E(y_i) = \beta_0(1 + \beta_1\beta_2^{x_i}) + \gamma_0 e^{-\gamma_1\gamma_2^{x_i}}$$

has an unidentifiable asymptote $\beta_0 + \gamma_0$ as $x_i \to \infty$, and if we set $\beta_0 = \gamma_0$ we can retrieve neither M_1 nor M_2 as reduced models. Similarly, the multiplicative form

$$E(y_i) = \beta_0(1 + \beta_1\beta_2^{x_i})(\gamma_0 e^{-\gamma_1\gamma_2^{x_i}})$$

yields M_1 if either $\gamma_1 = 0$ or $\gamma_2 = 0$ and yields M_2 if either $\beta_1 = 0$ or $\beta_2 = 0$, again a lack of identifiability.

A related issue is worthy of comment. Even when the likelihood ratio test is applicable and the usual asymptotics are valid, the test is inconsistent. As the sample size tends to infinity, the probability that the full model will be selected, given the reduced one is true, does not approach zero. The likelihood ratio gives too much weight to the higher-dimensional model suggesting the need to impose a reduction or penalty on the likelihood for dimension. There is, by now, a considerable literature on penalized likelihoods but no resolution emerges. A general penalty function would depend upon both dimension p and sample size n and would be increasing and unbounded in both. Penalty functions which do not depend upon n will not eliminate inconsistency. Those that do may over-penalize, placing too much weight on parsimony. Whether the asymptotic behaviour of a modified likelihood ratio is of practical concern is likely to depend upon the particular setting but, in any event, penalized likelihood reduces model performance to a single number. Depending upon the intended use for the model, performance at individual observations may also be of value.

9.3 The Bayesian perspective and the Bayes factor

As noted in the introduction, a model is defined by the specification of a joint distribution for observables which we denote by Y, and unobservables which we denote by θ. We use $f()$ as a generic notation for densities involving Y and θ under this model. We shall adopt a Bayesian perspective. It seems clear that, while inference proceeds from $f(\theta|Y)$, model determination does not. The posterior $f(\theta|Y)$ does not allow us to criticize a model in light of the observed data nor does it permit comparison amongst models since different models will have different sets of unobservables. Rather, it is $f(Y)$ which can assess model performance. Regardless of the model, $f(Y)$ is a density over the space of observables which can be compared with what was actually observed (Box, 1980). If Y_{obs} denotes the actual observations and $f(Y|M_i)$ denotes the marginal density under model M_i, $i = 1, 2$, the *Bayes factor*

$$\mathrm{BF} = \frac{f(Y_{obs}|M_1)}{f(Y_{obs}|M_2)}$$

provides the relative weight of evidence for model M_1 compared to model M_2. The Bayes factor arises formally as the ratio of the posterior odds for M_1 versus M_2 to the prior odds for M_1 versus M_2, and rough calibration of the Bayes factor has been proposed, (e.g. Jeffreys, 1961, or Pettit and Young, 1990). A table elaborating this calibration appears in Raftery (1995: this volume).

Hence, for many purists, model adequacy requires no more than calculation of $f(Y_{obs}|M_i)$ and model selection requires no more than the Bayes factor (Kass and Raftery, 1995). No further discussion is needed. In principle, we do not disagree with this formalism, though different utility functions for model performance can lead to different and arguably comparably good formal criteria (Kadane and Dickey, 1978). Rather, we are troubled by interpretation and computation of the Bayes factor for the sorts of complex models we consider in this volume. We clarify this concern in what follows, again noting that examination of model performance at the level of the individual observation can provide added insight.

A serious problem with the Bayes factor is that, for many of the models we discuss in this book, at least some part of the prior specification is vague so that $f(\theta)$ is improper. But then, even if the posterior $f(\theta|Y)$ is proper, we must necessarily have $f(Y)$ improper. How then can we assess whether $f(Y_{obs})$ is large? How can we calibrate the Bayes factor? The answer is that we cannot and hence we cannot interpret these quantities. The reader might appreciate this most easily by realizing that $c \cdot f(\theta)$ carries the same prior information as $f(\theta)$ when $f(\theta)$ is improper, and hence any multiple of a calculated Bayes factor can be arbitrarily obtained. Purists might argue that this problem should never arise because $f(\theta)$ should never be improper; we always know enough about an unknown quantity to specify a range for it, so we can employ a proper prior. However, in practice this introduces awkward normalizations, and renders implementation of sampling-based fitting methods much more difficult.

Another problem with the Bayes factor is its computation. Apart from special cases where substantial data reduction occurs due to sufficiency, even a proper $f(Y)$ is an n-dimensional density which will not generally be available explicitly, and must be computed as a marginalization by integrating over θ. That is

$$f(Y) = \int f(Y|\theta)f(\theta)d\theta. \tag{9.1}$$

The high-dimensional integration in (9.1) is not trivial to carry out; a good importance-sampling density is needed. MCMC methods can be helpful in its calculation (see Section 9.6 and Raftery, 1995: this volume), but in general Monte Carlo integration for (9.1) will tend to be very unstable. If $f(Y_{obs})$ cannot be calculated confidently, then surely this is so for the Bayes factor, a ratio of such integrations.

Also, in the case of nested models, the Bayes factor may exhibit the so-called 'Lindley paradox'. Suppose the data points to rejection of the reduced model. If the prior is proper but is sufficiently diffuse, it can place little mass on alternatives so that the denominator of the Bayes factor is much smaller than the numerator, resulting in support for the reduced model in spite of the data. The Bayes factor tends to place too much weight on parsimonious selection.

We conclude this section by noting that use of the Bayes factor often seems inappropriate in real applications. If, as is usually the case, none of the models under consideration are correct, what sense can we make of prior or posterior odds? For nested models, since one is contained within the other, again what sense can we make of odds? The reduced model usually has no probability under the full model unless an artificial point mass is introduced at the reduced model. More generally, if the models under consideration are proxies for an unknown underlying model, why is the Bayes factor a suitable summary?

9.4 Alternative predictive distributions

The density $f(Y)$ might be called the *prior predictive* density. That is, from (9.1), the conditional density of the observables is averaged against prior knowledge about the unobservables. This suggests that many other predictive densities might be examined.

9.4.1 Cross-validation predictive densities

Suppose, as before, that Y is the set of obervations $\{y_r \, ; \, r = 1, 2, \ldots, n\}$. Then the *cross-validation predictive* densities are the set $\{f(y_r|Y_{(r)}) \, ; \, r = 1, 2, \ldots, n\}$ where $Y_{(r)}$ denotes all elements of Y except y_r. The density $f(y_r|Y_{(r)})$ is attractive in that it suggests what values of y_r are likely when the model is fitted to all of the observations except y_r. The actual $y_{r,obs}$ can then be compared with this density in a variety of ways to see whether it is likely under the model, that is, to see whether the observation supports the model (see Gelfand *et al.*, 1992).

Cross-validation predictive densities are usually calculated by the equation:

$$f(y_r|Y_{(r)}) = \int f(y_r|\theta, Y_{(r)}) \cdot f(\theta|Y_{(r)})d\theta, \tag{9.2}$$

rather than through the equivalent equation, $f(y_r|Y_{(r)}) = f(Y)/f(Y_{(r)})$. In the case of conditionally independent observations given θ, $f(y_r|\theta, Y_{(r)}) = f(y_r|\theta)$.

Usually, $f(y_r|Y_{(r)})$ will be a proper density even if $f(Y)$ is not, and hence we need not worry about its interpretation for a given model or its comparison between models. The propriety of $f(y_r|Y_{(r)})$ is revealed in (9.2) since $f(y_r|\theta, Y_{(r)})$ is derived from the proper density $f(Y|\theta)$, and $f(\theta|Y_{(r)})$ is usually proper if $f(\theta|Y)$ is. In fact, $f(\theta|Y_{(r)})$ is often very similar to $f(\theta|Y)$ which is useful in the computation of (9.2). We note that when $f(Y)$ is proper, under mild conditions the set $\{f(y_r|Y_{(r)})\}$ is equivalent to $f(Y)$ in that each uniquely determines the other. Hence $\{f(y_r|Y_{(r)})\}$ contains the same information about model performance as $f(Y)$.

The distributions $f(y_r|Y_{(r)})$ are attractive in permitting us to work with univariate distributions. Indeed, single element deletion is standard in classical regression diagnostics, although we can also work with small set deletion (say two or three elements) rather than a single element (Peña and Tiao, 1992). When n is large, we may randomly or systematically (depending upon the context) select a subset of elements of Y within which we study model performance using $f(y_r|Y_{(r)})$.

The product of cross-validation predictive densities $\prod_{r=1}^{n} f(y_r|Y_{(r),obs})$ has been proposed as a surrogate for $f(Y)$, giving rise to the *pseudo-Bayes factor*:

$$\prod_{r=1}^{n} \frac{f(y_{r,obs}|Y_{(r),obs}, M_1)}{f(y_{r,obs}|Y_{(r),obs}, M_2)},$$

as a surrogate for the Bayes factor (Geisser and Eddy, 1979; Gelfand *et al.*, 1992).

9.4.2 Posterior predictive densities

The *posterior predictive* density, $f(Y|Y_{obs})$, is the predictive density of a new independent set of observables under the model, given the actual observables (Rubin, 1984; Aitkin, 1991). Similar to (9.1) it would be computed as

$$f(Y|Y_{obs}) = \int f(Y|\theta) \cdot f(\theta|Y_{obs})d\theta \qquad (9.3)$$

since $f(Y|\theta) = f(Y|\theta, Y_{obs})$. Comparision with (9.1) shows that the posterior predictive density averages the conditional density of the observables, against the posterior knowledge about the unobservables. The posterior predictive density offers a conservative view of model performance. If the observations $Y = Y_{obs}$ are not in agreement with $f(Y|Y_{obs})$, which is derived under the model using these observations, then the model is unlikely to receive support under any other criterion.

By marginalizing (9.3), we obtain

$$f(y_r|Y_{obs}) = \int f(y_r|\theta) \cdot f(\theta|Y_{obs})d\theta. \qquad (9.4)$$

Note that, for conditionally independent observations, (9.2) and (9.4) should be quite similar unless $y_{r,obs}$ is very influential in the fitting. A comparison of $f(y_{r,obs}|Y_{obs})$ and $f(y_{r,obs}|Y_{(r),obs})$ provides information on the 'influence' of $y_{r,obs}$.

9.4.3 Other predictive densities

We mention two other predictive densities. For instance, in time-series settings, the *forecasting predictive* density may be natural (Nandram and Petrucelli, 1995). If Y is an observed time series, $y_1, y_2, \ldots, y_n$, the forecasting predictive density at time t is $f(y_t|y_1, y_2, \ldots, y_{t-1})$.

The *intrinsic predictive* density is defined as $f(Y_{(s)}|Y_s)$ where Y_s is the smallest set of elements of Y such that $f(\theta|Y_s)$ is proper. It can be used to create a so-called intrinsic Bayes factor, to replace the Bayes factor when improper priors are used (Berger and Pericchi, 1994).

9.5 How to use predictive distributions

Having described a variety of predictive distributions, how should we use them in conjunction with the observations to investigate model determination? Rather than rigid insistence upon the use of $f(Y)$, we adopt an informal, exploratory stance, incorporating the spirit of modern regression diagnostics. A global summary measure will tell us whether or not a model is working well, but more detailed inspection is required to discover the reasons for poor global performance. It is easiest, both conceptually and technically, to do this for each individual observation. Hence, we assume that a univariate predictive density (such as in Section 9.4) has been selected. For the observable y_r, we compare this density with the actual $y_{r,obs}$ through a checking function $g(y_r; y_{r,obs})$, and take the expectation of g with respect to the predictive density to quantify the comparison. A few examples should prove helpful. For concreteness we use the cross-validation predictive density, $f(y_r|Y_{(r),obs})$, in these illustrations.

(i) Suppose $g(y_r; y_{r,obs}) = y_{r,obs} - y_r$, a 'residual' form. Then the comparison becomes

$$d_{1r} = E(y_{r,obs} - y_r \mid Y_{(r),obs}) = y_{r,obs} - E(y_r|Y_{(r),obs}),$$

the difference between the actual value of the observable and the expectation of the observable under the predictive distribution. From a different perspective, in many applications, y_r is modelled as $y_r = h_r(\theta) + \epsilon_r$ for some specified function $h()$, where ϵ_r is some sort of random noise with zero mean. But then, if we consider the posterior mean of the observed residual $E(y_{r,obs} - h_r(\theta)|Y_{(r),obs})$, as in Chaloner and Brant (1988), we may easily show that this expectation is d_{1r}. The absolute value $|d_{1r}|$ may be considered, though the pattern of signs may be of interest. A standarized form for d_{1r} is

$$d_{1r}^* = \frac{d_{1r}}{\sqrt{\mathrm{var}(y_r|Y_{(r),obs})}}.$$

(ii) Suppose $g_2(y_r; y_{r,obs}) = I(-\infty < y_r \leq y_{r,obs}]$ where I denotes the indicator function, taking the value 1 if its argument is true and zero otherwise. Taking expectations yields

$$d_{2r} = P(y_r \leq y_{r,obs}|Y_{(r),obs}),$$

where P denotes probability. Hence d_{2r} measures how extreme (unlikely) $y_{r,obs}$ is under $f(y_r|Y_{(r),obs})$. If y_r is discrete, we might instead calculate $P(y_r = y_{r,obs}|Y_{(r),obs})$.

(iii) Suppose $g_3(y_r; y_{r,obs}) = I(y_r \in B_r)$ where

$$B_r = \{y_r : f(y_r|Y_{(r),obs}) \leq f(y_{r,obs}|Y_{(r),obs})\}.$$

Taking expectations, we obtain

$$d_{3r} = P(B_r|Y_{(r),obs}).$$

B_r consists of the set of y_r values which are less likely than $y_{r,obs}$ under the predictive density $f(y_r|Y_{(r),obs})$ so that d_{3r} indicates how likely it is to see a predictive density ordinate smaller than that corresponding to $y_{r,obs}$. Small d_{3r} indicates that $y_{r,obs}$ does not support the model.

(iv) Suppose

$$g_4(y_r; y_{r,obs}) = \frac{1}{2\epsilon}I[y_{r,obs} - \epsilon \leq y_r \leq y_{r,obs} + \epsilon],$$

the indicator function of a small interval around $y_{r,obs}$. Taking expectations and letting $\epsilon \to 0$ we obtain

$$d_{4r} = f(y_{r,obs}|Y_{(r),obs}),$$

which is known as the *conditional predictive ordinate* (CPO). See Pettit and Young (1990) and references therein. Small d_{4r} indicates that $y_{r,obs}$ does not support the model.

One of the principal attractions of d_{4r} arises from the obvious relationship

$$\log d_{4r} = \log f(Y_{obs}) - \log f(Y_{(r),obs}).$$

Thus the log(CPO) for the rth observation contrasts the prior predictive density of all the observations with that of all but the rth observation, and may be interpreted as the incremental contribution to the adequacy of the model attributable to the rth observation.

For model choice, consider the CPO ratio:

$$C_r = \frac{f(y_{r,obs}|Y_{(r),obs}, M_1)}{f(y_{r,obs}|Y_{(r),obs}, M_2)}.$$

Note that $\prod_{r=1}^{n} C_r$ is the previously defined *pseudo-Bayes factor* and

$$\log C_r = \log \text{BF} - \log \text{BF}_{(r)},$$

where $\text{BF}_{(r)}$ is the Bayes factor for Model 1 relative to Model 2 using all but the rth observation. If $C_r > 1$, i.e. $\log C_r > 0$, $y_{r,obs}$ supports Model 1; if $C_r < 1$, i.e. $\log C_r < 0$ it supports Model 2. If we plot $\log C_r$ versus r, we can see how many observations support each model, which observations support which model and the strength of that support.

In choosing amongst more than two models, we can no longer work with CPO ratios. However, we can still plot d_{4r} versus r for each model to compare the models visually. With such a *CPO plot* we can see which models do better than others, which points are poorly fit under all models, which models are essentially indistinguishable, and so on. See the illustrative example of Section 9.7.

We need not necessarily adopt checking functions to assess model adequacy. For instance, continuing to work with $f(y_r|Y_{(r),obs})$, suppose we generate a sample $y_{r1}^*, \ldots, y_{rm}^*$ from this predictive distribution. (Performing such sampling is discussed in Section 9.6.) In practice m need not be too large, say 100 to 200. We could note how consonant $y_{r,obs}$ is with these y_{rj}^*, for instance by using these samples to create $(1 - \alpha)$ equal tail predictive intervals for each r. We would then expect roughly $(1 - \alpha)$ of the $y_{r,obs}$ to fall in their respective intervals. See the example of Section 9.7 and Gelman and Meng (1995: this volume).

9.6 Computational issues

In general, output from a MCMC simulation (Gelfand and Smith, 1990; Tierney, 1994) can be utilized, with some additional sampling, to implement the diagnostics described in the previous section. Assume that the model has been fitted by a sampling-based method yielding a collection $\{\theta_j^* ; j = 1, 2, \ldots, m\}$, which we treat as a sample from $f(\theta|Y_{obs})$. We note that, while analytic approximations might be considered in obtaining predictive densities, the accuracy of such approximations is unknown in practice (Gelfand and Dey, 1994). More troubling, such an approach will typically not yield an approximating function but, rather, its value at some value of θ, impeding study of the behaviour of these densities. Moreover, calculating expectations and other features of these distributions is not feasible through analytic approximation but, as we shall see, is routine using sampling-based approaches.

Using the sample $\{\theta_j^* ; j = 1, \ldots, m\}$, the computational problems are the following:

(i)   how to estimate, as a function of the Y, a predictive density of interest;

(ii) how to compute an expectation (such as of a checking function) under this density;

(iii) how to sample from this density, in the spirit of the last paragraph of Section 9.5.

We consider these problems in turn. See Gelfand and Dey (1994) for a fuller discussion of the material in this section.

9.6.1 Estimating predictive densities

For simplicity, we consider the density estimation question through three illustrative cases.

Posterior predictive density estimation

The simplest case is the posterior predictive density, $f(Y|Y_{obs})$. An immediate Monte Carlo integration from (9.3) is:

$$\hat{f}(Y|Y_{obs}) = \frac{1}{m} \sum_{j=1}^{m} f(Y|\theta_j^*).$$

Cross-validation predictive density estimation

For the cross-validation predictive density, in general we may write:

$$f(y_r|Y_{(r)}) = \frac{f(Y)}{f(Y_{(r)})}$$

$$= \frac{1}{\int \frac{f(Y_{(r)},\theta)}{f(Y,\theta)} \cdot f(\theta|Y)d\theta}$$

$$= \frac{1}{\int \frac{1}{f(y_r|Y_{(r)},\theta)} \cdot f(\theta|Y)d\theta}. \tag{9.5}$$

An immediate Monte Carlo integration of (9.5) yields

$$\hat{f}(y_r|Y_{(r),obs}) = \frac{1}{\frac{1}{m}\sum_{j=1}^{m} \frac{1}{f(y_r|Y_{(r),obs},\theta_j^*)}}, \tag{9.6}$$

the harmonic mean of the $\{f(y_r|Y_{(r),obs},\theta_j^*)\,; \, j = 1,\ldots,m\}$. If the $\{y_r\}$ are conditionally independent given θ, $f(y_r|Y_{(r),obs},\theta_j^*)$ simplifies to $f(y_r|\theta_j^*)$. We anticipate that $f(\theta|Y)$ will serve as a good importance sampling density for $f(\theta|Y_{(r)})$ and thus that the Monte Carlo integration in (9.6) will be fairly stable. Importance sampling is discussed in more detail by Gilks and Roberts (1995) and Raftery (1995) in this volume.

Prior predictive density estimation

Consider now the calculation of the prior predictive density $f(Y)$. For any proper density $h(\theta)$ note that

$$(f(Y))^{-1} = \int \frac{h(\theta)}{f(Y|\theta)f(\theta)} \cdot f(\theta|Y)d\theta. \tag{9.7}$$

From (9.7), a Monte Carlo integration yields

$$\hat{f}(Y) = \frac{1}{\frac{1}{m}\sum_{j=1}^{m} \frac{h(\theta_j^*)}{f(Y|\theta_j^*)f(\theta_j^*)}}. \tag{9.8}$$

In (9.7) and (9.8), $h(\theta)$ plays the role of an importance sampling density for the posterior, so the better h approximates the posterior the better $\hat{f}(Y)$ will be. After transforming the components of θ to the real line, a convenient choice for $h()$ might be a multivariate normal or t density with mean and covariance computed from the θ_j^*. Alternatively, if $f(\theta)$ is proper, we might take $h(\theta) = f(\theta)$, whence (9.8) becomes the harmonic mean of the $\{f(Y|\theta_j^*)\}$. However, since $f(\theta)$ and $f(\theta|Y)$ will usually be quite different, typically this estimate is very unstable.

9.6.2 Computing expectations over predictive densities

With regard to expectations over predictive densities, again it is easiest to use an illustration. Suppose we seek

$$E(g(y_r)|Y_{(r),obs}) = E_1 E_2(g(y_r)) = E_1(a_r(\theta, Y_{(r),obs})), \tag{9.9}$$

where E_1 denotes a conditional expectation over θ given $Y_{(r),obs}$; E_2 denotes a conditional expectation over y_r given $Y_{(r),obs}$ and θ; and $a_r(\theta, Y_{(r),obs}) =$

$E_2(g(y_r))$. For conditionally independent $\{y_r\}$ (usually the case when using cross-validation predictive distributions), a_r is free of $Y_{(r),obs}$. Then the right-most expression in (9.9) can be written as

$$\int a_r(\theta)f(\theta|Y_{(r),obs})d\theta = f(y_{r,obs}|Y_{(r),obs}) \int \frac{a_r(\theta)}{f(y_{r,obs}|\theta)} f(\theta|Y_{obs})d\theta,$$

whence a Monte Carlo integration for (9.9) becomes

$$\hat{E}(g(y_r)|Y_{(r),obs}) = \hat{f}(y_{r,obs}|Y_{(r),obs}) \cdot \frac{1}{m} \sum_{j=1}^{m} \frac{a_r(\theta_j^*)}{f(y_{r,obs}|\theta_j^*)}. \qquad (9.10)$$

In (9.10), $\hat{f}(y_{r,obs}|Y_{(r),obs})$ is available from (9.6).

Suppose $a_r(\theta)$ is unavailable explicitly but we draw samples $\{y_{rj}^* \; ; \; j = 1, \ldots, m\}$ from $f(y_r|Y_{(r),obs})$, as described in Section 9.6.3. Then the left-most expression in (9.9) permits the trivial Monte Carlo integration

$$\hat{E}(g(y_r)|Y_{(r),obs}) = \frac{1}{m} \sum g(y_{rj}^*).$$

Such an integration may be preferable to (9.10).

9.6.3 Sampling from predictive densities

Suppose we wish to sample from $f(Y|Y_{obs})$. If for each θ_j^*, we sample a data set Y_j^* from $f(Y|\theta_j^*)$, then marginally Y_j^* is a sample from $f(Y|Y_{obs})$. Consequently $y_{r,j}^*$, the rth element of Y_j^*, is a sample from $f(y_r|Y_{obs})$. Usually, this straightforward sampling suffices to investigate model adequacy such as proposed at the end of Section 9.5.

However, suppose we wish to make draws from $f(y_r|Y_{(r),obs})$. If θ_j^{**} is a sample from $f(\theta|Y_{(r),obs})$ and we draw Y_j^{**} from $f(Y|\theta_j^{**})$, then Y_j^{**} is a sample from $f(Y|Y_{(r),obs})$ and $y_{r,j}^{**}$ is a sample from $f(y_r|Y_{(r),obs})$, where $y_{r,j}^{**}$ is the rth element of Y_j^{**}. How can we obtain the θ_j^{**}? Fortunately, to do this we need not rerun the MCMC sampler using only $Y_{(r),obs}$. Rather, suppose for each θ_j^*, we compute the importance ratio

$$\frac{f(\theta_j^*|Y_{(r),obs})}{f(\theta_j^*|Y_{obs})} \quad \propto \quad \frac{1}{f(y_{r,obs}|Y_{(r),obs},\theta_j^*)}$$
$$= \quad w_j.$$

We can now resample with replacement from the collection $\{\theta_j^*\}$ with probabilities proportional to $\{w_j\}$. The resulting sample is approximately from $f(\theta|Y_{(r),obs})$. In practice, we would often not need to bother to do this resampling.

Lastly, suppose we wish to sample from $f(Y)$, assuming it is proper (i.e. $f(\theta)$ is proper). If we sample first $\tilde{\theta}_j$ from $f(\theta)$ and then $\tilde{Y}_j$ from $f(Y|\tilde{\theta}_j)$, then $\tilde{Y}_j$ will be a sample from $f(Y)$.

9.7 An example

Our example involves a small data set, but permits interesting modelling and enables illustration of the various model determination diagnostics we have discussed. Taken from Draper and Smith (1981, p. 524), it studies the growth of orange trees over time. Seven measurements of trunk circumference (in millimetres) for each of five trees are presented in Table 9.1. Measurements were taken on the same days for each tree, with days counted from January 1, 1969.

			tree number		
time	1	2	3	4	5
118	30	33	30	32	30
484	58	69	51	62	49
664	87	111	75	112	81
1004	115	156	108	167	125
1231	120	172	115	179	142
1372	142	203	139	209	174
1582	145	203	140	214	177

Table 9.1 *Orange tree circumferences (mm)*

We investigate several nonlinear mixed-effects models. In all cases, errors are assumed *i.i.d.* $N(0, \sigma^2)$. Let y_{ij} denote the trunk circumference for the ith tree on the jth measurement day, and t_j denote the jth measurement day. We consider four models:

$$\text{Model 1: } E(y_{ij}) = \beta_0 + b_i + \epsilon_{ij}$$

$$\text{Model 2: } E(y_{ij}) = \beta_0 + \beta_1 t_j + b_i + \epsilon_{ij}$$

$$\text{Model 3: } E(y_{ij}) = \frac{\beta_0}{1 + \beta_1 \exp(\beta_2 t_j)} + \epsilon_{ij}$$

$$\text{Model 4: } E(y_{ij}) = \frac{\beta_0 + b_i}{1 + \beta_1 \exp(\beta_2 t_j)} + \epsilon_{ij}.$$

Model 1 is a simple random-effects model ignoring time (hence rather silly). Model 2 is a linear growth curve. Models 3 and 4 are logistic growth curves, but Model 3 differs from Model 4 in presuming a common growth curve for all trees (no random effects). The only nesting is Model 1 within Model 2, Model 1 within Model 4 ($\beta_1 = 0$) and Model 3 within Model 4. When present, the b_i are assumed *i.i.d.* $N(0, \tau^2)$. In our notation, Y is the set of y_{ij} and θ is the collection of β_0, β_1, β_2, $\{b_i\}$, σ^2 and τ^2. The joint prior distribution for β_0, β_1, β_2, σ^2 and τ^2 is assumed proportional to

$$\sigma^{-2} \tau^{-3} \exp\{-2(\sigma^{-2} + \tau^{-2})\}.$$

Hence $f(\theta)$ is improper, but a proper posterior results.

We first consider model adequacy, drawing samples from the posterior predictive distributions $f(y_r|Y_{obs})$. We summarize our findings in Table 9.2 by recording, for each model, the percentage of the 35 observations which fell in their respective 50% and 95% equal-tailed predictive intervals.

Model	50% predictive interval	95% predictive interval
1	89	100
2	29	51
3	46	100
4	60	86

Table 9.2 *Model adequacy diagnostics for the orange tree data, giving the percentage of the 35 observations $\{y_{ij,obs}\}$ which were contained in their predictive interval*

An adequate model should support these percentages. In this sense, Models 3 and 4 seem adequate. The intervals under Model 1 are too long; the predictive distributions are too dispersed. Conversely, under Model 2, the intervals are too short.

Turning to model choice, we present a CPO plot for the four models in Figure 9.1. Clearly Model 4 is best and Model 1 is worst. The pseudo-Bayes factors strongly confirm this. The $\log_{10}$(pseudo-Bayes factor) for Model 4 versus Model 1 is 29.2; for Model 4 versus Model 2 it is 9.3; and for Model 4 versus Model 3 it is 15.3. Unfortunately, unlike the Bayes factor, the pseudo-Bayes factor does not arise as a relative odds, so it is not obvious how to calibrate its magnitude.

9.8 Discussion

At the outset, we note that model construction is an art where considerations regarding the intended use for the model play an important part. In practice, except in rare situations, it is an evolutionary process with subjective decisions being made along the way and no choice being objectively 'best'. This subjectivity manifests itself in our proposed diagonostic approach and should not be viewed negatively.

We advocate a more detailed investigation of model performance than may be captured through a single summary number. However, we recognize that in challenging applications the number of models to explore may be enormous and that such refined examination may not be feasible for all models. In such cases, summary numbers might be used to reduce the set of models to a manageable size, allowing the remaining models to be individually studied. It is likely that such refined examination will

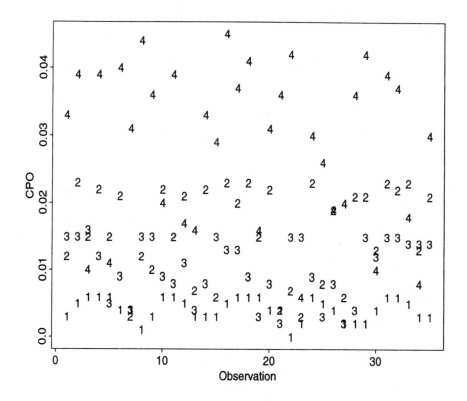

Figure 9.1 *CPO plot for orange tree data: model indicated by number.*

involve comparison of non-nested models so that classical theory will not offer much. Fortunately, the Bayesian viewpoint successfully handles such comparisons.

Lastly, for the complicated modelling scenarios we envision, we will probably know little about model unknowns, so prior specifications will be rather vague (though when we do have prior information we would certainly use it). Hence our inference will be close to that from likelihood methods but, in avoiding asymptotics, we may obtain better estimates of variability. The sampling-based technology implemented through MCMC methods enables not only the fitting of such models but also, as we have seen, rather straightforward investigation of model adequacy and choice.

References

Aitkin, M. (1991) Posterior Bayes factors. *J. R. Statist. Soc* B, 111–142.

Berger, J. O. and Pericchi, L. R. (1994) The intrinsic Bayes factor. Technical report, Department of Statistics, Purdue University.

Besag, J. and Green, P. (1993) Spatial statistics and Bayesian computation. *J. R. Statist. Soc* B, **56**, 25–38.

Box, G. E. P. (1980) Sampling and Bayes' inference in scientific modeling and robustness. *J. R. Statist. Soc.* A, **143**, 383–430.

Chaloner, K and Brant, R. (1988) A Bayesian approach to outlier detection and residual analysis. *Biometrika*, **75**, 651–659.

Chambers, J. M., Cleveland, W. S., Kleiner, B. and Tukey, P. (1983) *Graphical Methods for Data Analysis.* California: Wadsworth and Brooks/Cole.

Cook, R. D. and Weisberg, S. (1982) *Residuals and Influence in Regression.* London: Chapman & Hall.

Cox, D. R. (1961) Tests of separate families of hypotheses. In *Proceedings of the 4th Berkeley Symposium, Vol 1*, pp. 105–123. Berkeley: University of California Press.

D'Agostino, R. B. and Stephens, M.A (1986) *Goodness of Fit Techniques.* New York: Marcel Dekker.

Draper, N. and Smith, H. (1981) *Applied Regression Analysis.* New York: Wiley.

Geisser, S. and Eddy, W. (1979) A predictive approach to model selection. *J. Am. Statist. Ass.*, **74**, 153–160.

Gelfand, A. E. and Dey, D. K. (1994) Bayesian model choice: asymptotics and exact calculations. *J. R. Statist. Soc.* B, **56**, 501–514.

Gelfand, A. E. and Smith, A. F. M. (1990) Sampling based approaches to calculating marginal densities. *J. Am. Statist. Ass.*, **85**, 398–409.

Gelfand, A. E., Dey, D. K. and Chang, H. (1992) Model determination using predictive distributions with implementation via sampling-based methods. In *Bayesian Statistics 4* (eds J. M. Bernardo, J. O. Berger, A. P. Dawid and A. F. M. Smith), pp. 147–167. Oxford: Oxford University Press.

Gelman, A. and Meng, X.-L. (1995) Model checking and model improvement. In *Markov Chain Monte Carlo in Practice* (eds W. R. Gilks, S. Richardson and D. J. Spiegelhalter), pp. 189–201. London: Chapman & Hall.

Gilks, W. R. (1995) Full conditional distributions. In *Markov Chain Monte Carlo in Practice* (eds W. R. Gilks, S. Richardson and D. J. Spiegelhalter), pp. 75–88. London: Chapman & Hall.

Gilks, W. R. and Roberts, G. O. (1995) Strategies for improving MCMC. In *Markov Chain Monte Carlo in Practice* (eds W. R. Gilks, S. Richardson and D. J. Spiegelhalter), pp. 89–114. London: Chapman & Hall.

Jeffreys, H. (1961) *Theory of Probability.* London: Oxford University Press.

Kadane, J. B. and Dickey, J. (1978) Decision theory and the simplification of models. In *Evaluation of Econometric models* (eds J. Kmenta and J. B. Ramsey), pp. 245–268. New York: Academic Press.

Kass, R. E. and Raftery, A. (1995) Bayes factors. *J. Am. Statist. Soc*, **90**, 773–795.

Lauritzen, S. and Spiegelhalter, D. (1988) Local computations with probabilities on graphical structures and their application to expert systems. *J. R. Statist. Soc.* B, **50**, 157–194.

McCullagh, P. and Nelder, J. A. (1989) *Generalized Linear Models.* London: Chapman & Hall.

Nandram, B. and Petrucelli, J. (1995) Bayesian analysis of autoregressive time series panel data: a Gibbs sampler approach. *J. Bus. Econ. Statist.*, (in press).

Neyman, J. and Scott, E. L. (1948) Consistent estimates based on partially consistent observations. *Econometrica*, **16**, 1–32.

Peña, D. and Tiao, G. C. (1992) Bayesian robustness functions for linear models. In *Bayesian Statistics 4* (eds J. M. Bernardo, J. O. Berger, A. P. Dawid and A. F. M. Smith), pp. 365–388. Oxford: Oxford University Press.

Pettit, L. I. and Young K. D. S. (1990) Measuring the effect of observations on Bayes factors. *Biometrika*, **77**, 455–466.

Raftery, A. E. (1995) Hypothesis testing and model selection. In *Markov Chain Monte Carlo in Practice* (eds W. R. Gilks, S. Richardson and D. J. Spiegelhalter), pp. 163–187. London: Chapman & Hall.

Rubin, D. B. (1984) Bayesianly justifiable and relevant frequency calculations for the applied statistician. *Ann. Statist.*, **12**, 1151–1172.

Tierney, L. (1994) Markov chains for exploring posterior distributions (with discussion). *Ann. Statist.*, **22**, 1701–1762.

Whittaker, J. (1990) *Graphical Models in Applied Multivariate Statistics.* Chichester: Wiley.

10

Hypothesis testing and model selection

Adrian E Raftery

10.1 Introduction

To motivate the methods described in this chapter, consider the following inference problem in astronomy (Soubiran, 1993). Until fairly recently, it had been believed that the Galaxy consists of two stellar populations, the disk and the halo. More recently, it has been hypothesized that there are in fact three stellar populations, the old (or thin) disk, the thick disk, and the halo, distinguished by their spatial distributions, their velocities, and their metallicities. These hypotheses have different implications for theories of the formation of the Galaxy. Some of the evidence for deciding whether there are two or three populations is shown in Figure 10.1, which shows radial and rotational velocities for $n = 2\,370$ stars.

A natural model for this situation is a mixture model with J components, namely

$$y_i = \sum_{j=1}^{J} \rho_j f(y_i|\theta_j), \qquad i = 1, \ldots, n, \tag{10.1}$$

where $y = (y_1, \ldots, y_n)$ are the observations, $\rho = (\rho_1, \ldots, \rho_J)$ is a probability vector, $f(\cdot|\theta)$ is a probability density for each θ, and θ_j is the parameter vector for the jth component. The inference problem can then be cast as one of choosing between the two-component and the three-component models, i.e. between $J = 2$ and $J = 3$ in (10.1). This problem is also discussed by Robert (1995: this volume).

Standard frequentist hypothesis testing theory does not apply to this problem because the regularity conditions that it requires do not hold, and because the two- and three-component models are not nested (Titterington *et al.*, 1985). Various *ad-hoc* adjustments have been proposed, however.

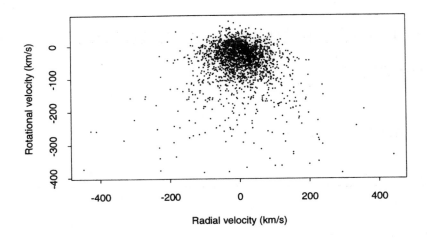

Figure 10.1 *Radial and rotational velocities for 2 370 stars in the Galaxy. Source: Soubiran (1993).*

The standard Bayesian solution consists of calculating the Bayes factor for the two-component model against the three-component model, but as far as I know this has not been worked out analytically, and it is a difficult calculation. A heuristic approximation based on cluster analysis has been proposed (Banfield and Raftery, 1993). As many of the chapters in the volume show, posterior simulation via MCMC is a good way to estimate the parameters in a model such as (10.1). In this chapter, I will describe ways of utilizing MCMC output for hypothesis testing using Bayes factors.

In Section 10.2, the Bayesian approach to hypothesis testing, model selection and accounting for model uncertainty is reviewed. These are all based on Bayes factors. The Bayes factor comparing two competing models is the ratio of the marginal likelihoods under the two models. The marginal likelihood of a model is the probability of the data with all the model parameters integrated out. In Sections 10.3 and 10.4, I review ways of estimating the marginal likelihood of a model from posterior simulation output. In Section 10.3, the harmonic mean estimator and other importance-sampling and related estimators are reviewed. In Section 10.4, two new estimators, the Laplace–Metropolis estimator and the Candidate's estimator, are introduced. These both consist of adjusting the maximized likelihood, or another likelihood value. In Section 10.5, I return to the mixture problem and apply the methods described to the motivating problem from astronomy. In Section 10.6, several other solutions to the problem are briefly reviewed, and outstanding issues are discussed.

10.2 Uses of Bayes factors

The standard Bayesian solution to the hypothesis testing problem is to represent both the null and alternative hypotheses as parametric probability models, and to compute the Bayes factor for one against the other. The Bayes factor B_{10} for a model M_1 against another model M_0 given data D is the ratio of posterior to prior odds, namely

$$B_{10} = \frac{P(D|M_1)}{P(D|M_0)}, \qquad (10.2)$$

the ratio of the marginal likelihoods. In (10.2),

$$P(D|M_k) = \int P(D|\theta_k, M_k) P(\theta_k|M_k) d\theta_k, \qquad (10.3)$$

where θ_k is the vector of parameters of model M_k, and $P(\theta_k|M_k)$ is its prior density ($k = 0, 1$).

One important use of the Bayes factor is as a summary of the evidence for M_1 against M_0 provided by the data. It can be useful to consider twice the logarithm of the Bayes factor, which is on the same scale as the familiar deviance and likelihood-ratio test statistics. Table 10.1 gives a rounded scale for interpreting B_{10}, which is based on that of Jeffreys (1961), but is more granular and slightly more conservative than his.

B_{10}	$2 \log B_{10}$	evidence for M_1
< 1	< 0	negative (supports M_0)
1 to 3	0 to 2	barely worth mentioning
3 to 12	2 to 5	positive
12 to 150	5 to 10	strong
> 150	> 10	very strong

Table 10.1 *Calibration of the Bayes factor B_{10}*

The model selection problem arises when one initially considers several models, perhaps a very large number, and wishes to select one of them. A Bayesian solution to this problem is to choose the model with the highest posterior probability. However, the purpose of choosing a model needs to be kept in mind. If the principal decision-making problem is truly one of choosing a model, then utilities should be introduced (Kadane and Dickey, 1980).

The marginal likelihoods yield posterior probabilities of all the models, as follows. Suppose that K models, $M_1, \ldots, M_K$, are being considered. Then, by Bayes theorem, the posterior probability of M_k is

$$P(M_k|D) = \frac{P(D|M_k)\, P(M_k)}{\sum_{r=1}^{K} P(D|M_r)\, P(M_r)}. \tag{10.4}$$

In the examples which follow, I will take all the models to have equal prior probabilities, corresponding to prior information that is 'objective' or 'neutral' between competing models (e.g. Berger, 1985), but other prior information about the relative plausibility of competing models can easily be taken into account.

The ultimate goal of an investigation is often estimation, prediction or decision-making, rather than model selection *per se*. In that case, selecting a single model ignores model uncertainty and so will underestimate uncertainty about quantities of interest, thus, for example, biasing policy choices in favour of policies that are riskier than the analysis appears to indicate (Hodges, 1987). It is better to take account explicitly of model uncertainty.

The posterior model probabilities given by (10.4) lead directly to solutions of the prediction, decision-making and inference problems that take account of model uncertainty. The posterior distribution of a quantity of interest Δ, such as a structural parameter to be estimated, a future observation to be predicted, or the utility of a course of action, is

$$P(\Delta|D) = \sum_{k=1}^{K} P(\Delta|D, M_k)\, P(M_k|D), \tag{10.5}$$

where $P(\Delta|D, M_k) = \int P(\Delta|D, \theta_k, M_k) P(\theta_k|D, M_k) d\theta_k$. The overall posterior mean and variance are

$$E[\Delta|D] = \sum_{k=1}^{K} \hat{\Delta}_k P(M_k|D), \tag{10.6}$$

$$\mathrm{var}[\Delta|D] = \sum_{k=1}^{K} \left(\mathrm{var}[\Delta|D, M_k] + \hat{\Delta}_k^2 \right) P(M_k|D) - E[\Delta|D]^2, \tag{10.7}$$

where $\hat{\Delta}_k = E[\Delta|D, M_k]$, the posterior mean under model M_k (Raftery, 1993a).

The number of models considered can be very large, and then direct evaluation of (10.5) will often be impractical. For example, in variable selection for regression with p candidate independent variables, the number of possible models is $K = 2^p$, so that if $p = 20$, K is about one million. In such cases, two strategies for the evaluation of (10.5) have been proposed. One of these, Occam's window, consists of selecting and averaging over a much smaller set of models by excluding models that are much less likely than the most likely one *a posteriori*, and also optionally excluding models that have more likely models nested within them (Madigan and Raftery, 1994). The other, MCMC model composition (MC^3), is a MCMC algorithm that moves through model space and generates a sample from the posterior distribution of the 'true' model, which is used to approximate (10.5) (Madigan and York, 1995; see also George and McCulloch, 1995: this volume). Bayes factors and posterior model probabilities calculated from posterior simulation of the model parameters can be used as part of either Occam's window or MC^3 algorithms.

For reviews of Bayes factors and their use in accounting for model uncertainty, see Kass and Raftery (1995) and Draper (1995). See also Gelfand (1995: this volume).

10.3 Marginal likelihood estimation by importance sampling

The marginal likelihood, $P(D|M_k)$, is the key quantity needed for Bayesian hypothesis testing, model selection and accounting for model uncertainty. In this section and the next, I deal only with estimating the marginal likelihood for a single model, and so I will drop the notational dependence on M_k. Here I outline several importance-sampling and related estimators of the marginal likelihood based on a sample $\{\theta^{(t)} : t = 1, \ldots, T\}$ from the posterior distribution of the parameter θ of the model. Such a sample can be generated by direct analytic simulation, MCMC, the weighted likelihood bootstrap (Newton and Raftery, 1994a), or other methods.

Unlike other posterior simulation methods, MCMC algorithms yield samples that are not independent and are only approximately drawn from the posterior distribution. Here I will assume that enough initial 'burn-in' values have been discarded for the approximation to be good; see Gelman (1995) and Raftery and Lewis (1995) in this volume for ways of ensuring this. I will also ignore the dependency between samples for the following reason. Since $\{\theta^{(t)}\}$ defines a Markov chain, it is typically ϕ-mixing at a geometric rate (see Billingsley, 1968; Tierney, 1995: this volume). Thus estimators that are simulation-consistent (as $T \to \infty$) for independent samples from the posterior will usually also be so for MCMC samples, by the laws of large numbers for ϕ-mixing processes in Billingsley (1968). Moreover, MCMC algorithms can be made to yield approximately independent posterior samples by subsampling; see Raftery and Lewis (1995: this volume).

Let $L(\theta) = P(D|\theta)$ be the likelihood and $\pi(\theta) = P(\theta)$ be the prior. Then the marginal likelihood is $P(D) = \int L(\theta)\pi(\theta)d\theta$. Let

$$\|X\|_h = \frac{1}{T} \sum_{t=1}^{T} X(\theta^{(t)}),$$

where $X(\cdot)$ is a function and $\{\theta^{(t)}\}$ is a sample of size T from the probability density $h(\theta)/\int h(\phi)d\phi$, h being a positive function.

Importance sampling can be used to evaluate $P(D)$, as follows. Suppose that we can sample from a density proportional to the positive function $g(\theta)$, say the density $cg(\theta)$, where $c^{-1} = \int g(\theta)d\theta$. Then

$$
\begin{aligned}
P(D) &= \int L(\theta)\pi(\theta)d\theta \\
&= \int L(\theta)\left[\frac{\pi(\theta)}{cg(\theta)}\right] cg(\theta)d\theta.
\end{aligned}
\tag{10.8}
$$

Given a sample $\{\theta^{(t)} : t = 1,\ldots,T\}$ from the density $cg(\theta)$, then, as suggested by (10.8), a simulation-consistent estimator of $P(D)$ is

$$
\begin{aligned}
\hat{P}(D) &= \frac{1}{T}\sum_{t=1}^{T}\frac{L\left(\theta^{(t)}\right)\pi\left(\theta^{(t)}\right)}{cg\left(\theta^{(t)}\right)} \\
&= \left\|\frac{L\pi}{cg}\right\|_g.
\end{aligned}
$$

If c cannot be found analytically, it remains only to estimate it from the MCMC output. A simulation-consistent estimator of c is $\hat{c} = \|\pi/g\|_g$. This yields the general importance-sampling estimator of $P(D)$ with importance-sampling function $g(\cdot)$, namely

$$
\hat{P}_{IS} = \frac{\left\|\frac{L\pi}{g}\right\|_g}{\left\|\frac{\pi}{g}\right\|_g},
\tag{10.9}
$$

(Newton and Raftery, 1994b). Here, g is a positive function; if it is normalized to be a probability density, then (10.9) becomes $\hat{P}_{IS} = \|L\pi/g\|_g$ (Neal, 1994).

The simplest such estimator results from taking the prior as importance sampling function, so that $g = \pi$. This yields

$$
\hat{P}_1(D) = \|L\|_\pi,
$$

a simple average of the likelihoods of a sample from the prior. This is just the simple Monte-Carlo estimator of the integral $P(D) = \int L(\theta)\pi(\theta)d\theta$ (Hammersley and Handscomb, 1964). This was mentioned by Raftery and Banfield (1991), and was investigated in particular cases by McCulloch and Rossi (1991). A difficulty with $\hat{P}_1(D)$ is that most of the $\theta^{(t)}$ have small likelihood values if the posterior is much more concentrated than the prior, so that the simulation process will be quite inefficient. Thus the estimate is dominated by a few large values of the likelihood, and so the variance of $\hat{P}_1(D)$ is large and its convergence to a Gaussian distribution is slow. These problems were apparent in the examples studied in detail by McCulloch and Rossi (1991). The variance of $\hat{P}_1(D)$ is roughly $O(T^{-1}n^{d/2}|W|^{\frac{1}{2}})$ where n

is the data sample size, d is the number of parameters, and W is the prior variance matrix. Therefore, a large simulation sample is required to obtain adequate precision in $\hat{P}_1(D)$ if the data sample is large, the number of parameters is large, or the prior is diffuse.

Posterior simulation produces a sample from the posterior distribution, and so it is natural to take the posterior distribution as the importance-sampling function. This yields the importance-sampling estimator with importance-sampling function $g = L\pi$, namely

$$\hat{P}_2(D) = \frac{1}{\|1/L\|_{\text{post}}}, \tag{10.10}$$

where 'post' denotes the posterior distribution; this is the harmonic mean of the likelihood values (Newton and Raftery, 1994a). It converges almost surely to the correct value, $P(D)$, as $T \to \infty$, but it does not, in general, satisfy a Gaussian central limit theorem, and the variance of $\hat{P}_2(D)^{-1}$ is usually infinite. This manifests itself by the occasional occurrence of a value of $\theta^{(t)}$ with small likelihood and hence large effect, so that the estimator $\hat{P}_2(D)$ can be somewhat unstable.

To avoid this problem, Newton and Raftery (1994a) suggested using as an importance-sampling function in (10.9) a mixture of the prior and posterior densities, namely $g(\theta) = \delta\pi(\theta) + (1-\delta)P(\theta|D)$, where $0 < \delta < 1$. The resulting estimator, $\hat{P}_3(D)$, has the efficiency of $\hat{P}_2(D)$ due to being based on many values of θ with high likelihood, but avoids its instability and does satisfy a Gaussian central limit theorem. However, it has the irksome aspect that one must simulate from the prior as well as the posterior. This may be avoided by simulating all T values from the posterior distribution and imagining that a further $\delta T/(1-\delta)$ values of θ are drawn from the prior, all of them with likelihoods $L(\theta^{(t)})$ equal to their expected value, $P(D)$. The resulting estimator, $\hat{P}_4(D)$, may be evaluated using a simple iterative scheme, defined as the solution x of the equation

$$x = \frac{\delta T/(1-\delta) + \sum[L_t/\{\delta x + (1-\delta)L_t\}]}{\delta T/\{(1-\delta)x\} + \sum\{\delta x + (1-\delta)L_t\}^{-1}},$$

where $L_t = L(\theta^{(t)})$ and the summations are over $t = 1, \ldots, T$.

Another modification of the harmonic mean estimator, $\hat{P}_1(D)$, is

$$\hat{P}_5(D) = \left\| \frac{f}{L\pi} \right\|_{\text{post}}^{-1}, \tag{10.11}$$

where f is a function of θ and is any probability density; this was mentioned by Gelfand and Dey (1994). It is unbiased and simulation-consistent, and satisfies a Gaussian central limit theorem if the tails of f are thin enough, specifically if

$$\int \frac{f^2}{L\pi} \, d\theta < \infty. \tag{10.12}$$

If θ is one-dimensional, if the posterior distribution is normal, and if f is normal with mean equal to the posterior mean and variance equal to κ times the posterior variance, then the mean squared error of $\hat{P}_5(D)$ is minimized when $\kappa = 1$. This suggests that high efficiency is most likely to result if f is roughly proportional to the posterior density.

Meng and Wong (1993) proposed the alternative

$$\hat{P}_6(D) = \frac{\|L\pi g\|_\pi}{\|\pi g\|_{\text{post}}},$$

where g is a positive function. Like $\hat{P}_3(D)$, this has the disadvantage of needing simulation from the prior as well as the posterior. They considered an optimal choice of g and showed how it can be computed from an initial guess. This appears promising but has yet to be extensively tested.

Rosenkranz (1992) evaluated the estimators of the marginal likelihood, $\hat{P}_1(D)$, $\hat{P}_2(D)$, $\hat{P}_3(D)$ and $\hat{P}_4(D)$, in the contexts of normal models, hierarchical Poisson-gamma models for counts with covariates, unobserved heterogeneity and outliers, and a multinomial model with latent variables. She found that analytic approximations via the Laplace method gave greater accuracy for much less computation *and* human time than the posterior simulation estimators; the problem is that the Laplace method is not always applicable. Among the posterior simulation estimators, she found $\hat{P}_3(D)$ with a large value of δ (close to 1) to have the best performance. The harmonic mean estimator, $\hat{P}_2(D)$, is easy to compute and her experience, as well as that of Carlin and Chib (1995), is that with substantial numbers of iterations (at least 5 000), it gave results accurate enough for the granular scale of interpretation in Section 10.2; see also Rosenkranz and Raftery (1994).

There has been less systemmatic evaluation of $\hat{P}_5(D)$ or $\hat{P}_6(D)$. In a very simple one-dimensional example (testing $\mu = 0$ in a $N(\mu, 1)$ model) with a well chosen f function, $\hat{P}_5(D)$ performed very well in a small numerical experiment that I carried out. In more complex and high-dimensional examples, such as the mixture model described in Section 10.1, experience to date is less encouraging. The estimator $\hat{P}_5(D)$ appears to be sensitive to the choice of f function, and it seems hard in practice to ensure that the condition (10.12) holds. If f is not well chosen, $\hat{P}_5(D)$ can give highly inaccurate answers, as illustrated in Section 10.5. More research is required on this issue.

10.4 Marginal likelihood estimation using maximum likelihood

10.4.1 The Laplace–Metropolis estimator

Rosenkranz (1992) found that the Laplace method produces much more accurate estimates of the marginal likelihood than posterior simulation for several very different models and for large amounts of simulation. However, the Laplace method is often not applicable because the derivatives that it requires are not easily available. This is particularly true for complex models of the kind for which posterior simulation, especially MCMC, is often used.

The idea of the Laplace–Metropolis estimator is to avoid the limitations of the Laplace method by using posterior simulation to *estimate* the quantities it needs. The Laplace method for integrals (e.g. de Bruijn, 1970, Section 4.4) is based on a Taylor series expansion of the real-valued function $f(u)$ of the d-dimensional vector u, and yields the approximation

$$\int e^{f(u)}du \approx (2\pi)^{d/2}|A|^{\frac{1}{2}}\exp\{f(u^*)\}, \tag{10.13}$$

where u^* is the value of u at which f attains its maximum, and A is minus the inverse Hessian (second-derivative matrix) of f evaluated at u^*. When applied to (10.3) it yields

$$p(D) \approx (2\pi)^{d/2}|\Psi|^{\frac{1}{2}}P(D|\tilde{\theta})P(\tilde{\theta}), \tag{10.14}$$

where $\tilde{\theta}$ is the posterior mode of $h(\theta) = \log\{P(D|\theta)P(\theta)\}$, Ψ is minus the inverse Hessian of $h(\theta)$ evaluated at $\theta = \tilde{\theta}$, and d is the dimension of θ. Arguments similar to those in the Appendix of Tierney and Kadane (1986) show that in regular statistical models the relative error in (10.14), and hence in the resulting approximation to B_{10}, is $O(n^{-1})$, where n is the data sample size.

Thus the Laplace method requires the posterior mode $\tilde{\theta}$ and $|\Psi|$. The simplest way to estimate $\tilde{\theta}$ from posterior simulation output, and probably the most accurate, is to compute $h(\theta^{(t)})$ for each $t = 1, \ldots, T$ and take the value for which it is largest. If the likelihood is hard to calculate, however, this may take too much computer time. A simple alternative is to use the multivariate median, or L_1 centre, defined as the value of $\theta^{(t)}$ that minimizes $\sum_s |\theta^{(s)} - \theta^{(t)}|$, where $|\cdot|$ denotes L_1 distance; see, e.g., Small (1990). This is suboptimal but often yields values of $h(\theta^{(t)})$ close to those at the posterior mode, and can be much cheaper computationally. Even cheaper is the componentwise posterior median, computed as the estimated posterior median of each parameter individually. This performs well in the majority of cases, but can sometimes give poor results. A fourth possibility is to estimate the posterior mode directly from the posterior sample using nonparametric density estimation.

The matrix Ψ is asymptotically equal to the posterior variance matrix, as sample size tends to infinity, and so it would seem natural to approximate Ψ by the estimated posterior variance matrix from the posterior simulation output. The main problem with this is that it is sensitive to the occasional distant excursions to which MCMC trajectories can be prone, and so I prefer to estimate the posterior variance matrix using a robust but consistent variance matrix estimator with a high breakdown point. This means that the estimator continues to perform well even if the proportion of outliers is high. I use the weighted variance matrix estimate with weights based on the minimum volume ellipsoid estimate of Rousseeuw and van Zomeren (1990); this is implemented in the S-PLUS function `cov.mve` using the genetic algorithm of Burns (1992).

The resulting estimator of the marginal likelihood typically involves less computation than the estimators in Section 10.3. It remains to be systematically evaluated, but initial indications are promising. In a small numerical experiment with a $N(\mu, 1)$ model, it yielded values of the marginal likelihood that were accurate to within about 5% based on 600 iterations. It seems to be more stable than the estimators in Section 10.3. I conjecture that under quite mild conditions (finite posterior fourth moments are probably enough), its relative error is $O(n^{-\frac{1}{2}}) + O(T^{-\frac{1}{2}})$ or less. This has not been proved (but see Kass and Vaidyanathan, 1992, for relevant results and references). If true, this would usually be accurate enough for interpretation on the granular scale of Section 10.2. A short generic S-PLUS function to calculate it is given in the Appendix to this chapter.

The performance of the Laplace–Metropolis estimator has been more thoroughly assessed in the context of a more complex hierarchical model by Lewis and Raftery (1994), and shown to be very accurate in that context.

10.4.2 Candidate's estimator

Bayes theorem says that

$$P(\theta|D) = \frac{P(D|\theta)P(\theta)}{P(D)},$$

from which it follows that

$$P(D) = P(D|\theta) \cdot \left(\frac{P(\theta)}{P(\theta|D)} \right). \tag{10.15}$$

This was pointed out to me by Julian Besag, who noted that it is also related to his 'Candidate's formula' for Bayesian prediction (Besag, 1989), whence the name of the estimator I will describe.

Typically, $P(D|\theta)$ and $P(\theta)$ can be calculated directly. If $P(\theta|D)$ is also available in closed form, then (10.15) allows $P(D)$ to be calculated analytically without integration (Julian Besag, personal communication). This will often not be the case, but one can still use (10.15) if one has a sample from the posterior. One can then simply *estimate* $P(\theta|D)$ by nonparametric density estimation from the posterior sample. This is inexpensive computationally because the density is required at only one point. Practical multivariate density estimators have been proposed by Terrell (1990) and many others.

What value of θ should be used? Equation (10.15) holds for all values of θ, so in principle one could use any value. However, the most precise estimate of $P(\theta|D)$ would usually result from using the posterior mode or a value very close to it. Any value from the central part of the posterior sample should do almost as well.

10.4.3 The data-augmentation estimator

MCMC methods often involve introducing *latent data* z such that, when z is known, the 'complete data likelihood' $P(D, z|\theta)$ has a simple form. This is the idea underlying the EM algorithm (Dempster *et al.*, 1977), its stochastic generalizations (see Diebolt and Ip, 1995: this volume), and its Bayesian analogue, the IP algorithm (Tanner and Wong, 1987). In MCMC methods, values of both z and θ are generated from their joint posterior distribution. The latent data can consist, for example, of individual random effects in a random-effects or hierarchical model, of group memberships in a mixture model (see Section 10.5), or of missing data (see Richardson, 1995: this volume).

Typically z includes quantities that cannot be consistently estimated, in which case the conditions for the Laplace approximation to be good may not hold. Further, the latent data are often very numerous so that the matrix Ψ in (10.14) is of very high dimension, in which case there may also be numerical problems with the Laplace approximation. Rosenkranz (1992) has shown that for calculating marginal likelihoods there can be considerable advantages to integrating over the latent data directly, especially when there are conditional independence properties that can be exploited to reduce the dimensionality of the integrals involved. Here I outline some strategies for doing this.

Latent data with conditional independence

Suppose that the data can be partitioned as $D = (D_1, \ldots, D_n)$ and the latent data as $z = (z_1, \ldots, z_n)$, and that

$$(z_i \perp z_j \mid \theta), \tag{10.16}$$

$$(D_i \perp D_j \mid z_i, z_j, \theta), \qquad i \neq j, \tag{10.17}$$

where $\perp$ denotes independence. This covers many situations, including most random-effects models. A graphical model representation of such a model is shown in Figure 10.2.

Now suppose that the MCMC algorithm has been run, yielding a sample from the joint posterior distribution of θ and z, $P(\theta, z|D)$. We discard the values of z, and are left with a sample from the (marginal) posterior distribution of θ, $P(\theta|D)$. We can now apply the Laplace–Metropolis or the Candidate's estimator as before. Both methods involve evaluation of $\tilde{\theta}$ and

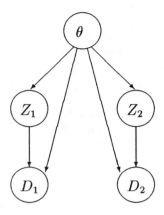

Figure 10.2 *Graphical model representation of the conditional independence model of (10.16) and (10.17), when* $n = 2$.

$\log P(D|\tilde{\theta})$. We can use the fact that

$$\log P(D|\tilde{\theta}) = \sum_{i=1}^{n} \log L_i(\tilde{\theta}),$$

where

$$L_i(\tilde{\theta}) = \int P(D_i|z_i, \tilde{\theta})P(z_i|\tilde{\theta})dz_i. \qquad (10.18)$$

$L_i(\tilde{\theta})$ is typically a low-dimensional integral (often of one dimension), and so it is readily amenable to various integration strategies, including the following:

1. *Analytic evaluation:* If the distribution of z_i is conjugate to that of D_i, the integral in (10.18) can often be evaluated analytically. This was illustrated by Rosenkranz (1992) in the case of hierarchical Poisson-gamma models.

2. *Summation:* If the z_i are discrete (as in Section 10.5), then the integral in (10.18) can be directly and simply evaluated as

$$L_i(\tilde{\theta}) = \sum_{j=1}^{J} P(D_i|z_i = j, \tilde{\theta}) \Pr[z_i = j|\tilde{\theta}]. \qquad (10.19)$$

3. *Laplace method:* The Laplace method can itself be applied to each $L_i(\tilde{\theta})$ individually. If this idea is used with the Laplace–Metropolis

estimator, the result is a kind of 'iterated Laplace method', in which the Laplace method is applied at each of two stages. Analytic expressions for the maximum of the integrand (with respect to z_i) in (10.18) and/or the required second derivatives may be available, making things easier. Otherwise, numerical approximations can be used, especially if the dimension of z_i is low. This is not guaranteed to perform well but the Laplace method has been found to provide remarkable accuracy even for very small 'sample sizes': see Grünwald et al. (1993) for one example of this.

4. *Quadrature:* If the dimension of z_i is low, then numerical quadrature may well work well for evaluating $L_i(\tilde{\theta})$.

Latent data without conditional independence

If conditions such as (10.16) and (10.17) do not hold, it is harder to evaluate $L(\tilde{\theta}) = P(D|\tilde{\theta}) = \int P(D|z,\tilde{\theta})P(z|\tilde{\theta})dz$. One way of doing it is via an importance-sampling estimator using the posterior simulation output $\{(z^{(t)}, \theta^{(t)})\, ;\, t = 1, \ldots, T\}$, as follows:

$$L(\tilde{\theta}) \approx \frac{\sum_{t=1}^{T} w\left(z^{(t)}\right) P(D|z^{(t)}, \tilde{\theta})}{\sum_{t=1}^{T} w\left(z^{(t)}\right)}, \tag{10.20}$$

where

$$w\left(z^{(t)}\right) = \frac{P(z^{(t)}|\tilde{\theta})}{P(z^{(t)}|D)}. \tag{10.21}$$

In (10.21),

$$P(z^{(t)}|D) = \int P(z^{(t)}|\theta, D)P(\theta|D)d\theta$$

$$\approx \frac{1}{T} \sum_{s=1}^{T} P(z^{(t)}|\theta^{(s)}, D). \tag{10.22}$$

Note that in (10.22), superscripts s and t denote the sth and t^{th} samples from the same posterior simulation output: $\{(z^{(1)}, \theta^{(1)}), (z^{(2)}, \theta^{(2)}), \ldots\}$. Then $L(\tilde{\theta})$ is evaluated by substituting (10.22) into (10.21) and then (10.21) into (10.20). This estimator has the advantage that it involves only prior and posterior ordinates, and no integrals. However, it does require the availability of a way of calculating $P(z^{(t)}|\theta^{(s)}, D)$.

If θ is absent from the model, or is assumed known, then (10.20) reduces to the harmonic mean estimator of the marginal likelihood. However, preliminary inspection suggests that if θ is present, the estimator (10.20) will tend not to suffer from the instability of the harmonic mean estimator. This is because the smoothing effect of averaging over the values of θ may help to avoid the overwhelming importance of individual samples that can plague the harmonic mean estimator.

The estimator (10.20) is related to, but not the same as, the estimator of Geyer and Thompson (1992). In their formulation, the $z^{(t)}$ were simulated from $P(z|\theta_0)$ for a fixed θ_0, rather than from the posterior distribution, $P(z|D)$. More recently, Geyer (1993) has proposed an estimator in which the $z^{(t)}$ are simulated from a mixture, $\sum_{j=1}^{J} \gamma_j \, P(z|\theta_j)$. See also Geyer (1995: this volume).

10.5 Application: how many components in a mixture?

I now return to the motivating application of Section 10.1. I first briefly recall the basic ideas of Gibbs sampling for Gaussian mixtures, I then give a simulated example, and finally I return to the problem of the number of disks in the Galaxy.

10.5.1 Gibbs sampling for Gaussian mixtures

I consider the one-dimensional Gaussian mixture of (10.1), where $\theta_j = (\mu_j, v_j)$ and $f(\cdot|\theta_j)$ is a normal density with mean μ_j and variance v_j. I use the prior densities

$$
\begin{aligned}
v_j^{-1} &\sim \text{Ga}(\omega_j/2, \lambda_j/2), \\
\mu_j &\sim \text{N}(\xi_j, v_j/\tau_j), \\
\rho &\sim \text{Dirichlet}(\alpha_1, \ldots, \alpha_J),
\end{aligned}
\tag{10.23}
$$

where $\text{Ga}(a, b)$ denotes a gamma density with mean a/b and variance a/b^2; $\text{N}(a, b)$ denotes a normal distribution with mean a and variance b; and $\rho = (\rho_1, \ldots, \rho_J)$. The prior hyperparameters are chosen so that the prior distribution is relatively flat over the range of values that could be expected given the range of the data. The values used were $\omega_j = 2.56$, $\lambda_j = 0.72 \times \hat{\text{var}}(y)$, $\xi = \bar{y}$, $\tau_j = \left(2.6/(y_{\max} - y_{\min})^2\right)$, and $\alpha_j = 1$, $(j = 1, \ldots, J)$. For derivations and discussion of such 'reference proper priors' for Bayes factors, see Raftery (1993c) and Raftery et al. (1993).

To use the Gibbs sampler, I introduce the latent data $z = (z_1, \ldots, z_n)$, where $z_i = j$ if $y_i \sim \text{N}(\mu_j, v_j)$, i.e. if y_i belongs to the jth component of the mixture. The required conditional posterior distributions are then given by Diebolt and Robert (1994); see also Robert (1995: this volume). The Gibbs sampler is initialized by dividing the data into J equal-sized chunks of contiguous data points, using the resulting sample means and variances for the μ_j and v_j, and setting $\rho_j = 1/J$, $(j = 1, \ldots, J)$. It proceeds by drawing first z, and then ρ, v and μ in turn from their full conditional distributions, and iterating. There is thus no need to initialize z.

10.5.2 A simulated example

To simulate a data set, I generated a sample of size $n = 100$ data points from the model described in Section 10.5.1, with $J = 2$ components, $\mu = (0, 6)$, $v = (1, 4)$, and $\rho = (\frac{1}{2}, \frac{1}{2})$. The Gibbs sampler was run for 600 iterations, of which the first 20 were discarded. The data, together with the true and estimated densities, are shown in Figure 10.3. The estimated density is the mixture of two normal densities corresponding to the estimated posterior means of the parameters. The sequences of μ values are shown in Figure 10.4 together with the corresponding likelihoods.

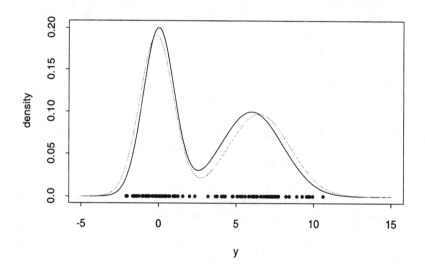

Figure 10.3 *100 points simulated from the $\frac{1}{2}N(0, 1) + \frac{1}{2}N(6, 2^2)$ density, with the true (solid) and estimated (dotted) densities overlaid.*

Corresponding quantities for the three-component model are shown in Figure 10.5. The Gibbs sampler makes some very distant excursions, which occur when one of the groups becomes empty. The algorithm has no problem returning after these excursions, and the corresponding likelihoods are not very low, but these episodes nevertheless require careful handling when estimating Bayes factors.

The various estimators of the log-marginal likelihood are shown in Table 10.2. The Laplace–Metropolis estimator was computed using (10.14),

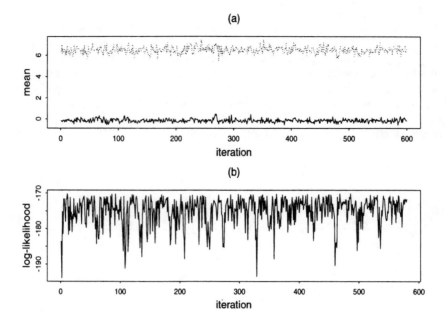

Figure 10.4 *Results of the Gibbs sampler for a two-component Gaussian mixture run on the data of Figure 10.3 with 600 iterations: (a) component means and (b) log-likelihood for each iteration.*

(10.18) and (10.19). The Gelfand–Dey estimator (10.11), $\hat{P}_5(D)$, is computed using for $f(\cdot)$ a product of independent densities. For the μ, v and ρ parameters these are normal with mean and variance equal to those of the simulated values, while for each z_i they are multinomial with probabilities again estimated from the Gibbs sampler output. Maximized log-likelihoods are shown, with maximization over both θ and (θ, z). The Schwarz (1978) approximation to the log-marginal likelihood is given as a general indication of a ballpark value, even though it is not known to be valid in this case. This is defined by

$$\log \hat{P}_{\text{Schwarz}}(D) = \max_t \log P(D|\theta^{(t)}) - \tfrac{1}{2}d\log(rn),$$

where d is the number of parameters and r is the dimension of y_i, (here $r = 1$). Finally, the crude AWE (approximate weight of evidence) approximation of Banfield and Raftery (1993) is also shown, defined as

$$\log \hat{P}_{\text{AWE}}(D) = \max_t \log P(D|\theta^{(t)}, z^{(t)}) - (d + \tfrac{3}{2})\log(rn).$$

The Laplace–Metropolis estimator is the only one that seems to be in the

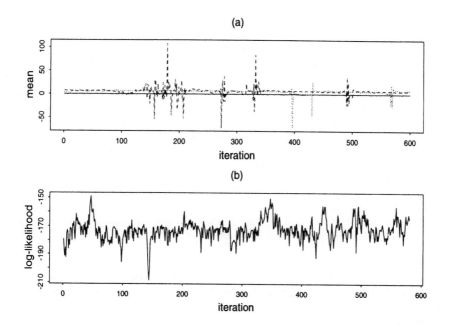

Figure 10.5 *Results of the Gibbs sampler based on a three-component Gaussian mixture model run on the data of Figure 10.3 with 600 iterations: (a) component means and (b) log-likelihood for each iteration.*

right ballpark, and this indicates (using Table 10.1) that there are probably either two or three groups but that the data do not distinguish clearly between these two possibilities. Even though the data were generated from a two-component model, Figure 10.3 shows that there is a second gap in the data around $y = 8$, so that someone analysing the data by eye might well come to the same somewhat uncertain conclusion. For this example, the three ways of calculating the Laplace–Metropolis estimator gave answers that were very close.

The Gelfand–Dey estimator implies that there is decisive evidence for four components against one, two or three. Even though the true values of the marginal likelihoods are not known, this seems somewhat implausible, and suggests the estimator to be rather inaccurate. A more carefully crafted f function seems to be needed. The harmonic mean estimator favours the two-component model, but it seems somewhat inaccurate with this fairly small number of iterations. The AWE estimator, used with the qualitative guidelines of Banfield and Raftery (1993), would, like the Laplace–Metropolis estimator, lead one to consider the two- and three-component

	number of components		
estimator	2	3	4
Laplace–Metropolis	−249.8	−248.6	−251.7
harmonic mean	−188.1	−202.1	−203.7
$\log \hat{P}_5(D)$	37.6	127.0	181.6
Schwarz	−249.7	−256.7	−263.6
AWE	−200.6	−197.7	−211.7
$\max_t \ell(\theta^{(t)})$	−238.2	−238.3	−238.3
$\max_t \ell(\theta^{(t)}, z^{(t)})$	−170.1	−148.9	−144.5
number of parameters	5	8	11

Table 10.2 *Log marginal likelihood estimates for the simulated data based on 600 Gibbs sampler iterations. The log marginal likelihood for the one-component model is −271.8 (calculated analytically).*

models, and so it is as accurate here as its creators claimed! Except for $\hat{P}_5(D)$, these estimators estimate the Bayes factor between the two- and three-component models better than they estimate the marginal likelihoods themselves.

10.5.3 How many disks in the Galaxy?

The Gibbs sampler was run using the rotational velocities for the 2 370 stars shown in Figure 10.1 for the two- and three-component models. Only 100 iterations were used because of computational constraints, and the first 50 of these were discarded. The trajectories were quite stable in this case, perhaps because of the considerable amount of data.

The log-marginal likelihoods, computed using the Laplace–Metropolis estimator, were: one component: −13141.0; two components: −12615.5; three components: −12609.6. Thus (using Table 10.1) the rotational velocities alone appear to provide strong evidence for a three-component model over a two-component model. This is in agreement with recent astronomical opinion (Soubiran, 1993, and references therein).

Figure 10.6(a) shows the fitted marginal densities of the data under the two competing models. The differences are subtle, but close inspection does support the better fit of the three-component model. The estimated densities under both models are strongly unimodal, so the ability of the method to distinguish between them is impressive. Both models fit quite well, indicating that the choice of a normal distribution for the individual components is satisfactory. The estimated components of the mixture are shown in Figure 10.6(b).

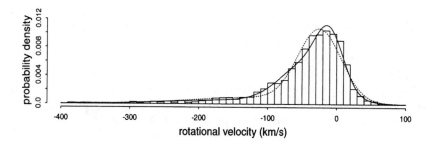

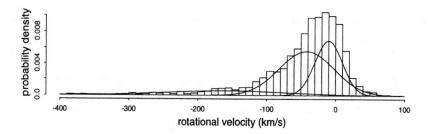

Figure 10.6 *Gibbs sampler results for the astronomy data of Figure 10.1, with histogram of the data: (a) estimated densities for the two-component (dashed) and three-component (solid) models; (b) estimated components of the three-component mixture model.*

Robert (1995: this volume) analyses a related but much smaller data set, choosing between the one- and two-component models using a different approach (Kullback–Leibler divergence).

10.6 Discussion

I have described various methods for computing marginal likelihoods using posterior simulation, facilitating Bayesian hypothesis testing, model selection and accounting for model uncertainty. Research on this topic is at an early stage and much remains to be done. Preliminary experience with the Laplace–Metropolis estimator introduced here is encouraging. The importance-sampling estimators described in Section 10.3 seem to have drawbacks, although future research may overcome these. One way of doing this would be to develop a good automatic way of choosing the f function for the modified harmonic mean estimator, $\hat{P}_5(D)$ of (10.11), suggested by Gelfand and Dey (1994). The ideas of Meng and Wong (1993) seem promising here.

Some general methods of calculating the marginal likelihood have been considered in the statistical physics literature under the name 'free energy estimation'. These approaches are not automatic and require analytical effort to tailor them to statistical applications. However, they may be of use in certain problems. See Neal (1993) for references.

In this chapter, I have concentrated on methods for estimating marginal likelihoods. In some situations, there are methods for directly estimating Bayes factors with posterior simulation. For example, suppose one wishes to compare M_0 with M_1, where the parameters of M_1 are $\theta_1 = (\omega, \psi)$, and θ_0 is specified by setting $\omega = \omega_0$ (so the free parameter of M_0 is $\theta_0 = \psi$). Then Verdinelli and Wasserman (1993) have shown that if $P(\omega|\psi, D, M_1)$ is available in closed form, the Savage density ratio can be used to provide an estimator of the Bayes factor. This is in the form of an unweighted average, and so it may well have good performance.

Unfortunately, this result is often not directly applicable. For example, the comparison of mixture models in Section 10.5 cannot readily be cast in this form. If $P(\omega|\psi, D, M_1)$ is not in closed form, Verdinelli and Wasserman (1993) resort to using an importance-sampling estimator proposed by Chen (1992). This is closely related to the estimator $\hat{P}_5(D)$ (whose performance to date has been mixed), so it needs further evaluation.

Carlin and Chib (1995) suggest including a model indicator variable in the MCMC scheme and defining 'pseudo-priors' for $P(\theta_1|M_2)$ and $P(\theta_2|M_1)$. This involves designing and running a special MCMC algorithm to calculate Bayes factors. Similar suggestions have been made by Carlin and Polson (1991) and George and McCulloch (1993).

In the data-augmentation case, an alternative approach to estimating a Bayes factor is available when the latent data z is present in both models and has the same meaning in each. Then the Bayes factor can be simulation-consistently estimated as the average of the 'complete-data' Bayes factors, namely

$$
\begin{aligned}
B_{10} &\approx T^{-1} \sum_{t=1}^{T} B_{10}(z^{(t)}) \\
&= T^{-1} \sum_{t=1}^{T} \frac{P(D, z^{(t)}|M_1)}{P(D, z^{(t)}|M_0)},
\end{aligned}
$$

where the $z^{(t)}$ are simulated from their posterior distribution under M_0. The $B_{10}(z^{(i)})$ are then often easy to calculate, or at least to approximate fairly well, for example using the Laplace method. When z is present in M_1 but not in M_0, we again recover the harmonic mean estimator of $P(D|M_1)$ (Raftery, 1993b). This is related to previous work of Thompson and Guo (1991) on the calculation of likelihood ratios.

Acknowledgements

This research was supported by ONR contract N00014-91-J-1074. I am grateful to Gilles Celeux for helpful discussions, to Caroline Soubiran for sharing her data, to Julian Besag for pointing out equation (10.15), to Jon Wellner for pointing out a useful reference, and to Wally Gilks, Andrew Gelman, Steven Lewis and David Madigan for helpful comments.

References

Banfield, J. D. and Raftery, A. E. (1993) Model-based Gaussian and non-Gaussian clustering. *Biometrics*, **49**, 803–821.

Berger, J. O. (1985) *Statistical Decision Theory and Bayesian Analysis.* New York: Springer-Verlag.

Besag, J. E. (1989) A candidate's formula: a curious result in Bayesian prediction. *Biometrika*, **76**, 183.

Billingsley, P. (1968) *Convergence of Probability Measures.* New York: Wiley.

Burns, P. J. (1992) A genetic algorithm for robust regression estimation. Technical report, Statistical Sciences, Inc., Seattle, Washington.

Carlin, B. P. and Chib, S. (1995) Bayesian model choice via Markov chain Monte Carlo methods. *J. R. Statist. Soc.* B, **57**, 473–484.

Carlin, B. P. and Polson, N. G. (1991) Inference for nonconjugate Bayesian models using the Gibbs sampler. *Canad. J. Statist.*, **19**, 399–405.

Chen, M-H. (1992) Importance weighted marginal Bayesian posterior density estimation. Technical report, Department of Statistics, Purdue University.

de Bruijn, N. G. (1970) *Asymptotic Methods in Analysis.* Amsterdam: North-Holland.

Diebolt, J. and Ip, E. H. S. (1995) Stochastic EM: methods and application. In *Markov Chain Monte Carlo in Practice* (eds W. R. Gilks, S. Richardson and D. J. Spiegelhalter), pp. 259–273. London: Chapman & Hall.

Diebolt, J. and Robert, C. (1994) Estimation of finite mixture distributions through Bayesian sampling. *J. R. Statist. Soc.* B, **56**, 363–376.

Dempster, A. P., Laird, N. M. and Rubin, D. B. (1977) Maximum likelihood from incomplete data via the EM algorithm (with discussion). *J. R. Statist. Soc.* B, **39**, 1–37.

Draper, D. (1995) Assessment and propagation of model uncertainty (with discussion). *J. R. Statist. Soc.* B, **57**, 45–97.

Gelfand, A. E. (1995) Model determination using sampling-based methods. In *Markov Chain Monte Carlo in Practice* (eds W. R. Gilks, S. Richardson and D. J. Spiegelhalter), pp. 145–161. London: Chapman & Hall.

Gelfand, A. E. and Dey, D. K. (1994) Bayesian model choice: asymptotics and exact calculations. *J. R. Statist. Soc.* B, **56**, 501–514.

Gelman, A. (1995) Inference and monitoring convergence. In *Markov Chain Monte Carlo in Practice* (eds W. R. Gilks, S. Richardson and D. J. Spiegelhalter), pp. 131–143. London: Chapman & Hall.

George, E. I. and McCulloch, R. E. (1993) Variable selection via Gibbs sampling. *J. Am. Statist. Ass.*, **88**, 881–889.

George, E. I. and McCulloch, R. E. (1995) Stochastic search variable selection. In *Markov Chain Monte Carlo in Practice* (eds W. R. Gilks, S. Richardson and D. J. Spiegelhalter), pp. 203–214. London: Chapman & Hall.

Geyer, C. J. (1993) Estimating normalizing constants and reweighting mixtures in Markov chain Monte Carlo. *Technical Report 568*, School of Statistics, University of Minnesota.

Geyer, C. J. (1995) Estimation and optimization of functions. In *Markov Chain Monte Carlo in Practice* (eds W. R. Gilks, S. Richardson and D. J. Spiegelhalter), pp. 241–258. London: Chapman & Hall.

Geyer, C. J. and Thompson, E. A. (1992) Constrained Monte Carlo maximum likelihood for dependent data (with discussion). *J. R. Statist. Soc. B*, **54**, 657–699.

Grünwald, G. K., Raftery, A. E. and Guttorp, P. (1993) Time series of continuous proportions. *J. R. Statist. Soc. B*, **55**, 103–116.

Hammersley, J. M. and Handscomb, D. C. (1964) *Monte Carlo Methods*. London: Chapman & Hall.

Hodges, J. S. (1987) Uncertainty, policy analysis and statistics. *Statist. Sci.*, **2**, 259–291.

Jeffreys, H. (1961) *Theory of Probability*, 3rd edn. Oxford: Oxford University Press.

Kadane, J. B. and Dickey, J. M. (1980) Bayesian decision theory and the simplification of models. In *Evaluation of Econometric Models* (eds J. Kmenta and J. Ramsey), pp. 245–68. New York: Academic Press.

Kass, R. E. and Raftery, A. E. (1995) Bayes factors. *J. Am. Statist. Ass.*, **90**, 773–795.

Kass, R. E. and Vaidyanathan, S. (1992) Approximate Bayes factors and orthogonal parameters, with application to testing equality of two Binomial proportions. *J. R. Statist. Soc. B*, **54**, 129–144.

Lewis, S. M. and Raftery, A. E. (1994) Estimating Bayes factors via posterior simulation with the Laplace–Metropolis estimator. *Technical Report 279*, Department of Statistics, University of Washington.

Madigan, D. and Raftery, A. E. (1994) Model selection and accounting for model uncertainty in graphical models using Occam's window. *J. Am. Statist. Ass.*, **89**, 1335–1346.

Madigan, D. and York, J. (1995) Bayesian graphical models for discrete data. *Int. Statist. Rev.*, (in press).

McCulloch, R. E. and Rossi, P. E. (1991) A Bayesian approach to testing the arbitrage pricing theory. *J. Economet.*, **49**, 141–168.

Meng, X. L. and Wong, W. H. (1993) Simulating ratios of normalizing constants via a simple identity. *Technical Report 365*, Department of Statistics, University of Chicago.

Neal, R. M. (1993) Probabilistic inference using Markov chain Monte Carlo methods based on Markov chains. *Technical Report CRG-TR-93-1*, Department of Computer Science, University of Toronto.

Neal, R. M. (1994) Discussion on approximate Bayesian inference by the weighted likelihood bootstrap (by M. A. Newton and A. E. Raftery). *J. R. Statist. Soc.* B, **56**, 41–42.

Newton, M. A. and Raftery, A. E. (1994a) Approximate Bayesian inference by the weighted likelihood bootstrap (with discussion). *J. R. Statist. Soc.* B, **56**, 3–48.

Newton, M. A. and Raftery, A. E. (1994b) Reply to the discussion on approximate Bayesian inference by the weighted likelihood bootstrap. *J. R. Statist. Soc.* B, **56**, 43–48.

Raftery, A. E. (1993a) Bayesian model selection in structural equation models. In *Testing Structural Equation Models* (eds K. A. Bollen and J. S. Long), pp. 163–180. Beverly Hills: Sage.

Raftery, A. E. (1993b) Discussion on the meeting on the Gibbs sampler and other Markov chain Monte Carlo methods. *J. R. Statist. Soc.* B, **55**, 85.

Raftery, A. E. (1993c) Approximate Bayes factors and accounting for model uncertainty in generalized linear models. *Technical Report 255*, Department of Statistics, University of Washington.

Raftery, A. E. and Banfield, J. D. (1991) Stopping the Gibbs sampler, the use of morphology and other issues in spatial statistics. *Ann. Inst. Statist. Math.* **43**, 32–43.

Raftery, A. E. and Lewis, S. M. (1995) Implementing MCMC. In *Markov Chain Monte Carlo in Practice* (eds W. R. Gilks, S. Richardson and D. J. Spiegelhalter), pp. 115–130. London: Chapman & Hall.

Raftery, A. E., Madigan, D. M. and Hoeting, J. (1993) Accounting for model uncertainty in linear regression models. *Technical Report 262*, Department of Statistics, University of Washington.

Richardson, S. (1995) Measurement error. In *Markov Chain Monte Carlo in Practice* (eds W. R. Gilks, S. Richardson and D. J. Spiegelhalter), pp. 401–417. London: Chapman & Hall.

Robert, C. P. (1995) Mixtures of distributions: inference and estimation. In *Markov Chain Monte Carlo in Practice* (eds W. R. Gilks, S. Richardson and D. J. Spiegelhalter), pp. 441–464. London: Chapman & Hall.

Rosenkranz, S. (1992) The Bayes factor for model evaluation in a hierarchical Poisson model for area counts. Ph.D. dissertation, Department of Biostatistics, University of Washington.

Rosenkranz, S. and Raftery, A. E. (1994) Covariate selection in hierarchical models of hospital admission counts: a Bayes factor approach. *Technical Report 268*, Department of Statistics, University of Washington.

Rousseeuw, P. J. and van Zomeren, B. C. (1990) Unmasking multivariate outliers and leverage points (with discussion). *J. Am. Statist. Ass.*, **85**, 633–651.

Schwarz, G. (1978) Estimating the dimension of a model. *Ann. Statist.*, **6**, 461–464.

Small, C. G. (1990) A survey of multivariate medians. *Int. Statist. Rev.*, **58**, 263–277.

Soubiran, C. (1993) Kinematics of the Galaxy stellar populations from a proper motion survey. *Astron. Astrophys.*, **274**, 181-188.

Tanner, M. and Wong, W. (1987) The calculation of posterior distributions by data augmentation (with discussion). *J. Am. Statist. Ass.*, **82**, 528–550.

Terrell, G. R. (1990) The maximal smoothing principle in density estimation. *J. Am. Statist. Ass.*, **85**, 470–477.

Thompson, E. A. and Guo, S. W. (1991) Evaluation of likelihood ratios for complex genetic models. *IMA J. Math. Appl. Med. Biol.*, **8**, 149–169.

Tierney, L. (1995) Introduction to general state-space Markov chain theory. In *Markov Chain Monte Carlo in Practice* (eds W. R. Gilks, S. Richardson and D. J. Spiegelhalter), pp. 59–74. London: Chapman & Hall.

Tierney, L. and Kadane, J. B. (1986) Accurate approximations for posterior moments and marginal densities. *J. Am. Statist. Ass.*, **81**, 82-86.

Titterington, D. M., Smith, A. F. M. and Makov, U. D. (1985) *Statistical Analysis of Finite Mixture Distributions.* New York: Wiley.

Verdinelli, I. and Wasserman, L. (1993) A note on generalizing the Savage-Dickey density ratio. *Technical Report 573*, Department of Statistics, Carnegie-Mellon University.

Appendix: S-PLUS code for the Laplace–Metropolis estimator

```
mcmcbf <- function (theta, data=NULL, method="likelihood") {

# Inputs:

# theta:   A sequence of samples from the posterior
#          distribution. This is a (niter x p) matrix, where
#          the parameter is of dimension p.

# data:    Input data for calculating likelihoods

# method:  if "likelihood", the empirical posterior mode is
#                           used
#          if "L1center",   the multivariate median is used
#          if "median",     the componentwise posterior
#                           median is used.

# Value:   The Laplace-Metropolis estimator of the marginal
#          likelihood.

# NOTE:    The user must supply functions llik and lprior
#          to calculate the log-likelihood and log-prior for
#          input values of the parameter.

theta <- as.matrix (theta);
```

```
niter <- length (theta[,1]);
p <- length (theta[1,]);

if (method == "likelihood") {
    h <- NULL;
    for (t in (1:niter)) {
        h <- c(h, llik(theta[t,],data) +
            lprior (theta[t,]) );
        hmax <- max(h);
    }
}

if (method == "L1center") {
    L1sum <- NULL;
    oneniter<- as.matrix (rep(1,niter));
    onep <- as.matrix (rep(1,p));
    for (t in (1:niter)) {
        thetat <- theta[t,];
        thetatmat <- oneniter %*% thetat;
        L1sum <- c(L1sum,
            sum(abs((theta-oneniter%*%thetat)%*%onep)));
    }
    argL1center <- min ((1:niter)[L1sum==min(L1sum)]);
    thetaL1center <- theta[argL1center,];
    hmax <- llik(thetaL1center,data) +
        lprior(thetaL1center);
}

if (method=="median") {
    thetamed <- apply (theta,2,median);
    hmax <- llik(thetamed,data) + lprior(thetamed);
}

if (p==1) {
    logdetV <- 2*log(mad(theta[,1]));
} else {
    logdetV <- sum(log(eigen(
                cov.mve(theta,print=F)$cov)$values));
}

hmax + 0.5 * p * log(2*3.14159) + 0.5 * logdetV;
}
```

11

Model checking and model improvement

Andrew Gelman

Xiao-Li Meng

11.1 Introduction

Markov chain simulation, and Bayesian ideas in general, allow a wonderfully flexible treatment of probability models. In this chapter we discuss two related ideas: (1) checking the fit of a model to data, and (2) improving a model by adding substantively meaningful parameters. We illustrate both methods with an example of a hierarchical mixture model, fitted to experimental data from psychology using the Gibbs sampler.

Any Markov chain simulation is conditional on an assumed probability model. As the applied chapters of this book illustrate, these models can be complicated and generally rely on inherently unverifiable assumptions. From a practical standpoint, it is important both to explore how inferences of substantive interest depend on such assumptions, and to check these aspects of the model whenever feasible.

11.2 Model checking using posterior predictive simulation

Bayesian prior-to-posterior analysis conditions on the whole structure (likelihood and prior distribution) of a probability model and can yield very misleading inferences when the model is far from reality. A good Bayesian analysis, therefore, should at least check to see if the fit of a posited model to the data is so poor that the model should be rejected without other evidence. In the classical setting, this checking is often facilitated by a goodness-of-fit test, which quantifies the extremeness of the observed value of a selected measure of discrepancy (e.g., differences between observations and fitted values) by calculating a tail-area probability given that the model under consideration is true. In this section, we discuss how to employ a Bayesian test of model fit using the posterior predictive distribution.

In the classical approach, test statistics cannot depend on unknown quantities and so comparisons are actually made between the data and the best-fitting model from within a family of models (typically obtained via maximum likelihood). The p-value for the test is determined based on the sampling distribution of the data under the assumed model. The main technical problem with the classical method is that, in general, the p-value depends on unknown parameters unless one restricts attention to pivotal test statistics. A pivotal quantity is one whose distribution does not depend on any unknown parameters; for example, the minimum χ^2 statistic in normal linear models. Unfortunately, as is well known, many useful statistics are not pivotal, especially with complex models. The Bayesian formulation naturally allows the test statistic to be a function of both data and unknown parameters, and thus allow more direct comparisons between the sample and population characteristics. Gelman *et al.* (1996) call parameter-dependent statistics *discrepancy variables*, to emphasize the goal of assessing the discrepancy between model and data, as oppsed to 'testing' the model's correctness.

Here we discuss model checking using posterior predictive distributions, which was proposed and applied by Guttman (1967) and Rubin (1981, 1984). Gelman *et al.* (1996) provide a general discussion of this method, with special emphasis on using discrepancy variables rather than traditional test statistics. They also present several applications and theoretical examples of the method and make comparisons with the prior predictive test of Box (1980). Meng (1994) presents a similar discussion of the use of posterior predictive p-values for testing hypotheses of parameters within a given model.

Posterior predictive model checking goes as follows. Let y be the observed data, θ be the vector of unknown parameters in the model (including any hierarchical parameters), $P(y|\theta)$ be the likelihood and $P(\theta|y)$ be the posterior distribution. We assume that we have already obtained draws $\theta_1, \ldots, \theta_N$ from the posterior distribution, possibly using Markov chain simulation. We now simulate N hypothetical replications of the data, which we label $y_1^{\text{rep}}, \ldots, y_N^{\text{rep}}$, where y_i^{rep} is drawn from the sampling distribution of y given the simulated parameters θ_i. Thus y^{rep} has distribution $P(y^{\text{rep}}|y) = \int P(y^{\text{rep}}|\theta)P(\theta|y)d\theta$. Creating simulations y^{rep} adapts the old idea of comparing data to simulations from a model, with the Bayesian twist that the parameters of the model are themselves drawn from their posterior distribution.

If the model is reasonably accurate, the hypothetical replications should look similar to the observed data y. Formally, one can compare the data to the predictive distribution by first choosing a discrepancy variable $T(y, \theta)$ which will have an extreme value if the data y are in conflict with the posited model. Then a p-value can be estimated by calculating the proportion of cases in which the simulated discrepancy variable exceeds the realized value:

$$\text{estimated } p\text{-value} = \frac{1}{N} \sum_{i=1}^{N} I(T(y_i^{\text{rep}}, \theta_i) \geq T(y, \theta_i)),$$

where $I(\cdot)$ is the indicator function which takes the value 1 when its argument is true and 0 otherwise. We call $T(y, \theta)$ a 'realized value' because it is *realized* by the observed data y, although it cannot be observed when T depends on unknown parameters. In the special case where the discrepancy variable depends only on data and not on parameters, and thus can be written $T(y)$, it is called a *test statistic*, as in the classical usage.

In practice, we can often visually examine the posterior predictive distribution of the discrepancy variable and compare it to the realized value. If the discrepancy variable depends only on data and not on the parameters θ, we can plot a histogram of posterior predictive simulations of $T(y^{\text{rep}})$ and compare it to the observed value $T(y)$, as in Figure 11.3. A good fit would be indicated by an observed value near the centre of the histogram. If the discrepancy variable is a function of data and parameters, we can plot a scatterplot of the realized values $T(y, \theta_i)$ versus the predictive values $T(y_i^{\text{rep}}, \theta_i)$ on the same scale. A good fit would be indicated by about half the points in the scatterplot falling above the 45° line and half falling below.

The discrepancy variable can be any function of data and parameters. It is most useful to choose a discrepancy variable that measures some aspect of the data that might not be accurately fitted by the model. For example, if one is concerned with outliers in a normal regression model, a sometimes useful discrepancy variable is the proportion of residuals that are more than a specified distance away from zero (see Gelman *et al.*, 1995, Section 8.4). If one is concerned about overall lack of fit in a contingency table model, a χ^2 discrepancy measure can be used (see Gelman *et al.*, 1996). In examples such as hierarchical regression, discrepancy variables are often more easily understood and computed when defined in terms of both data and parameters (for example, residuals from $X\beta$, rather than from $X\hat{\beta}$). Of course, more than one discrepancy variable can be used to check different aspects of the fit. An advantage of Monte Carlo is that the same set of simulations of (θ, y^{rep}) can be used for checking the posterior predictive distribution of many discrepancy variables.

A model does not fit the data if the realized values for some meaningful discrepancy variable are far from the predictive distribution; the discrepancy cannot reasonably be explained by chance if the tail-area probability is close to 0 or 1. The p-values are actual posterior probabilities and can therefore be interpreted directly – but *not* as the posterior probability of 'the model being true'. The role of predictive model checking is to assess the practical fit of a model: we may choose to work with an invalidated model but we should be aware of its deficiencies. On the other hand, a lack of rejection should not be interpreted as 'acceptance' of the model, but rather as a sign that the model adequately fits the aspect of the data being investigated.

Major failures of the model can be addressed by expanding the model, as we discuss in the next section. Lesser failures may also suggest model improvements or might be ignored in the short term if the failure appears not to affect the main inferences. In some cases, even extreme p-values may be ignored if the misfit of the model is substantively small compared to variation within the model. It is important not to interpret p-values as numerical 'evidence'. For example, a p-value of 0.00001 is virtually no stronger, in practice, than 0.001; in either case, the aspect of the data measured by the discrepancy variables is inconsistent with the model. A slight improvement in the model (or correction of a data-coding error!) could bring either p-value to a reasonable range – that is, not too close to 0 or 1 in the typical situation in which we implicitly apply a two-sided test.

11.3 Model improvement via expansion

We now turn to the issue of *expanding* a model that is already set up on the computer and has been estimated from the data. There are three natural reasons to expand a model. First, if the model clearly does not fit the data or prior knowledge, it should be improved in some way, possibly by adding new parameters to allow a better fit. Second, if a modelling assumption is particularly questionable, it may be relaxed. For example, a set of parameters that are assumed equal may be allowed to vary using a random-effects model. Third, a model may be embedded into a larger model to address more general applied questions; for example, a study previously analysed on its own may be inserted into a hierarchical population model (e.g., Rubin, 1981). The goal of our model expansion is not merely to fit the data, but rather to improve the model so as to capture the substantive structures. Thus, when one adds new parameters, a key requirement is that these parameters should have clear substantive meaning. This point is illustrated in our example in the next section.

All these applications of model expansion have the same mathematical structure: the old model, $p(\theta)$, is replaced by a new model, $p(\theta, \phi)$, in which both θ and ϕ may be vectors. In Bayesian notation, the posterior distribution $p(\theta|y)$ is replaced by $p(\theta, \phi|y)$, with an expanded prior distribution, $P(\theta, \phi) = P(\theta|\phi)P(\phi)$. Assuming that the original model in θ has already been programmed for Markov chain simulation, one can use the Gibbs sampler to draw samples from the joint distribution, $p(\theta, \phi|y)$. The step of drawing from $p(\theta|\phi, y)$ is just the problem of drawing from the posterior distribution of θ conditional on the individual model identified by ϕ. The only new step required is sampling from among the models, given data and parameters: $p(\phi|\theta, y)$. If the model class has no simple form, the simulation can be performed using Metropolis–Hastings steps.

For instance, suppose we are interested in the sensitivity of a data analysis with possible outliers to a model with an assumed normal error distribution. A natural expansion would be to replace the normal by a student-t with unknown degrees of freedom, ν. For a Bayesian analysis, a prior distribution must be assumed for ν; a reasonable 'noninformative' prior density might be uniform in $1/\nu$, which after transformation gives $P(\nu) \propto 1/\nu^2$, with the restriction $\nu \geq 1$ so that the limits correspopnd to Cauchy and normal error distributions. (The uniform prior density on ν may seem reasonable at first, but it actually has essentially all its mass 'near' $\nu = \infty$, which corresponds to the normal distribution.) The Markov chain simulation for the student-t model is now obtained by altering the simulations based on the normal model to simulations conditional on ν, and an added step to draw ν. Since the t distributions are not conjugate, it is probably easiest to use the Metropolis–Hastings algorithm to correct for draws based on the normal distribution at each Gibbs step.

A related computational problem is to average over a discrete set of models; see Carlin and Chib (1995), Draper (1995), Gelman and Rubin (1995), Raftery (1995a), and Raftery (1995b: this volume). These computational methods are also useful for sensitivity analysis; see Gelman and Meng (1995).

11.4 Example: hierarchical mixture modelling of reaction times

Neither model checking nor model expansion is enough by itself. Once lack of fit has been found, the next step is to find a model that improves the fit, with the requirement that the new model should be at least as interpretable substantively as the old one. On the other hand, model expansion alone cannot reveal lack of fit of the larger, expanded model. In this example, we illustrate how model checking and expansion can be used in tandem.

11.4.1 The data and the basic model

Belin and Rubin (1990) describe an experiment from psychology measuring 30 reaction times for each of seventeen subjects: eleven non-schizophrenics and six schizophrenics. Belin and Rubin (1995) fit several probability models using maximum likelihood; we describe their approach and fit related Bayesian models. Computation with the Gibbs sampler allows us to fit more realistic hierarchical models to the dataset.

The data are presented in Figures 11.1 and 11.2. It is clear that the response times are higher on average for schizophrenics. In addition, the response times for at least some of the schizophrenic individuals are considerably more variable than the response times for the non-schizophrenic individuals. Current psychological theory suggests a model in which schizophrenics suffer from an attentional deficit on some trials, as well as a general motor reflex retardation; both aspects lead to a delay in the schizophrenics' responses, with motor retardation affecting all trials and attentional deficiency only some.

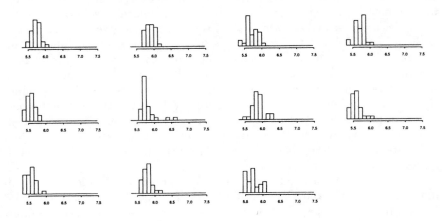

Figure 11.1 *Log response times (in milliseconds) for eleven non-schizophrenic individuals.*

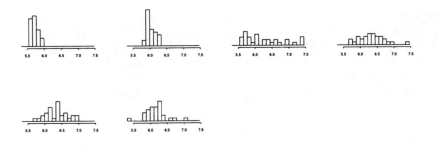

Figure 11.2 *Log response times (in milliseconds) for six schizophrenic individuals.*

To address the questions of psychological interest, the following basic model was fit. Log response times for non-schizophrenics were thought of as arising from a normal random-effects model, in which the responses of person $i = 1, \ldots, 11$ are normally distributed with distinct person mean

α_i and common variance σ_{obs}^2. To reflect the attentional deficiency, the responses for each schizophrenic individual $i = 12, \ldots, 17$ were fit to a two-component mixture: with probability $(1 - \lambda)$ there is no delay and the response is normally distributed with mean α_i and variance σ_{obs}^2, and with probability λ responses are delayed, with observations having a mean of $\alpha_i + \tau$ and the same variance. Because the reaction times are all positive and their distributions are positively skewed, even for non-schizophrenics, the model was fit to the logarithms of the reaction-time measurements. We modify the basic model of Belin and Rubin (1995) to incorporate a hierarchical parameter β measuring motor retardation. Specifically, variation among individuals is modelled by having the person means α_i follow a normal distribution with mean $\nu + \beta S_i$ and variance σ_α^2, where ν is the overall mean response time of non-schizophrenics, and S_i is an observed indicator that is 1 if person i is schizophrenic and 0 otherwise.

Letting y_{ij} be the jth log response time of individual i, the model can be written in the following hierarchical form:

$$
\begin{aligned}
[y_{ij} | \alpha_i, z_{ij}, \phi] &\sim \mathrm{N}(\alpha_i + \tau z_{ij}, \sigma_{obs}^2), \\
[\alpha_i | z, \phi] &\sim \mathrm{N}(\nu + \beta S_i, \sigma_\alpha^2), \\
[z_{ij} | \phi] &\sim \mathrm{Bernoulli}(\lambda S_i),
\end{aligned}
$$

where $\phi = (\sigma_\alpha^2, \beta, \lambda, \tau, \nu, \sigma_{obs}^2)$, and z_{ij} is an unobserved indicator variable that is 1 if measurement j on person i arose from the delayed component and 0 if it arose from the undelayed component. The indicator random variables z_{ij} are not necessary to formulate the model, but allow convenient computation of the modes of (α, ϕ) using the ECM algorithm (Meng and Rubin, 1993) and simulation using the Gibbs sampler. For the Bayesian analysis, the parameters σ_α^2, β, λ, τ, ν, and σ_{obs}^2 are assigned a joint uniform prior distribution, except that λ is restricted to the range $[0.001, 0.999]$ to exclude substantively uninteresting cases, τ is restricted to be positive to identify the model, and σ_α^2 and σ_{obs}^2 are of course restricted to be positive. An alternative non-informative specification, uniform in $\log(\sigma_{obs})$ would give indistinguishable results; in contrast, a uniform prior density on $\log(\sigma_\alpha)$ is well known to yield an improper posterior density (e.g., Tiao and Tan, 1965).

We found the modes of the posterior distribution using the ECM algorithm and then used a mixture of multivariate student-t densities centred at the modes as an approximation to the posterior density. We drew ten samples from the approximate density using importance resampling and then used those as starting points for ten parallel runs of the Gibbs sampler, which adequately converged after 200 steps (in the sense of potential scale reductions $\widehat{R}$ less than 1.1 for all model parameters; see Gelman, 1995: this volume). After discarding the first half of each simulated sequence, we were left with a set of 10×100 samples of the vector of model parameters. Details of the Gibbs sampler implementation and the convergence monitoring appear in Gelman *et al.* (1995) and Gelman and Rubin (1992), respectively.

11.4.2 Model checking using posterior predictive simulation

The model was chosen to fit the unequal means and variances in the two groups of subjects in the study, but there was still some question about the fit to individuals. In particular, the measurements for the first two schizophrenics seem much less variable than the last four. Is this difference 'statistically significant', or could it be explained as a random fluctuation from the model? To compare the observations to the model, we computed s_i, the standard deviation of the 30 log reaction times y_{ij}, for each schizophrenic individual $i = 12, \ldots, 17$. We then defined three test statistics – the smallest, largest, and average of the six values s_i – which we label $T_{\min}(y)$, $T_{\max}(y)$, and $T_{\text{avg}}(y)$, respectively. Examination of Figure 11.2 suggests that $T_{\min}$ is lower and $T_{\max}$ is higher than would be predicted from the model. The third test statistic, T_{avg}, is included as a comparison; we expect it to be very close to the model's predictions, since it is essentially estimated by the model parameters σ_{obs}^2, τ, and λ. For the data in Figure 11.2, the observed values of the test statistics are $T_{\min}(y) = 0.11$, $T_{\max}(y) = 0.58$, and $T_{\text{avg}}(y) = 0.30$.

To perform the posterior predictive model check, we simulate a predictive dataset from the normal-mixture model, for each of the 1 000 simulation draws of the parameters from the posterior distribution. For each of those 1 000 simulated datasets y^{rep}, we compute the test statistics $T_{\min}(y^{\text{rep}})$, $T_{\max}(y^{\text{rep}})$, and $T_{\text{avg}}(y^{\text{rep}})$. Figure 11.3 displays histograms of the 1 000 simulated values of the each statistic, with the observed values, $T(y)$, indicated by vertical lines. The observed data y are clearly atypical of the posterior predictive distribution – $T_{\min}$ is too low and $T_{\max}$ is too high – with estimated p-values within .001 of 0 and 1. In contrast, T_{avg} is well fit by the model, with a p-value of 0.72. More important than the p-values is the poor fit on the absolute scale: the observed minimum and maximum within-schizophrenic standard deviations are off by factors of two compared to the posterior model predictions. Chapter 6 of Gelman *et al.* (1995) presents more examples of evaluating posterior checks and p-values.

11.4.3 Expanding the model

Following Belin and Rubin (1995), we try to fit the data better by including two additional parameters in the model: to allow some schizophrenics to have no attentional delays, and to allow delayed observations to be more variable than undelayed observations. The two new parameters are ω, the probability that each schizophrenic individual has attentional delays, and

σ^2_{obs2}, the variance of attention-delayed measurements. Both these parameters are given uniform prior densities (we use the uniform density on ω because it is proper and the uniform density on σ^2_{obs2} to avoid the singularities at zero variance that occur with uniform prior densities on the log scale for hierarchical and mixture models). For computational purposes, we also introduce another indicator variable for each individual i, W_i, which takes the value 1 if the individual can have attention delays and 0 otherwise. The indicator W_i is automatically 0 for non-schizophrenics and, for each schizophrenic, is 1 with probability ω.

The model we have previously fitted can be viewed as a special case of the new model, with $\omega = 1$ and $\sigma^2_{obs2} = \sigma^2_{obs}$. It is quite simple to fit the new model by just adding three new steps in the Gibbs sampler to update ω, σ^2_{obs2}, and W (parameterized to allow frequent jumps between the states $W_i = 1$ and $W_i = 0$ for each schizophrenic individual i). In addition, the Gibbs sampler steps for the old parameters must be altered somewhat to be conditional on the new parameters. Details appear in Section 16.4 of Gelman *et al.* (1995). We use ten randomly selected draws from the previous posterior simulation as starting points for ten parallel runs of the Gibbs sampler. Because of the added complexity of the model, we ran the simulations for 500 steps, and discarded the first half of each sequence, leaving a set of 10×250 samples from the posterior distribution of the larger model. The potential scale reductions $\hat{R}$ for all parameters were less than 1.1, indicating approximate convergence.

Before performing posterior predictive checks, it makes sense to compare the old and new models in their posterior distributions for the parameters. Table 11.1 displays posterior medians and 95% credible intervals (from the Gibbs sampler simulations) for the parameters of most interest to the psychologists:

- λ, the probability that a response will be delayed, for an individual subject to attentional delays;

- ω, the probability that a schizophrenic will be subject to attentional delays;

- τ, the attentional delay (on the log scale);

- β, the average log response time for the non-delayed observations of schizophrenics minus the average log response time for non-schizophrenics.

Table 11.1 shows a significant difference between the parameter estimates in the two models. Since the old model is nested within the new model, the changes suggest that the improvement in fit is significant. It is a strength of the Bayesian approach, as implemented by iterative simulation, that we can model so many parameters and compute summary inferences for all of them.

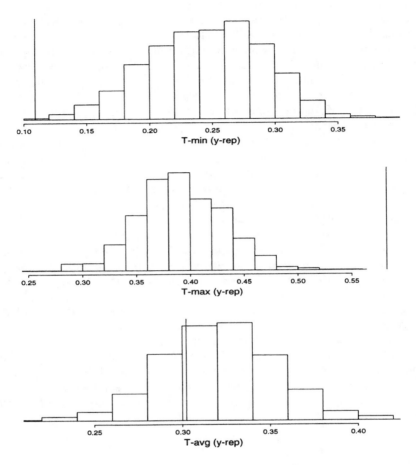

Figure 11.3 *Posterior predictive distributions and observed values for three test statistics: the smallest, largest and average observed within-schizophrenic variances; the vertical line on each histogram represents the observed value of the test statistic.*

11.4.4 Checking the new model

The substantial change in parameter estimates under the new model indicates the importance of model expansion in this example, but how well does the expanded model fit the data? We expect that the new model should show an improved fit to the test statistics considered in Figure 11.3, since the new parameters were added explicitly for this purpose. We emphasize that all the new parameters here have substantive interpretations in psy-

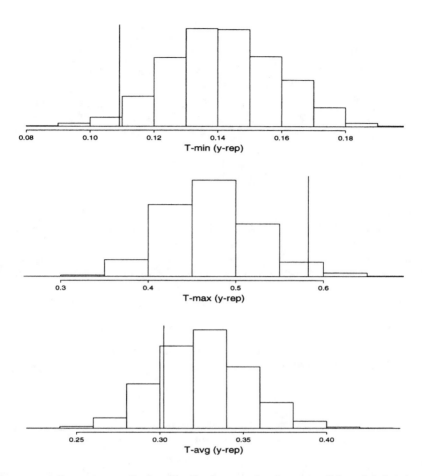

Figure 11.4 *Posterior predictive distributions, under the expanded model, for three test statistics: the smallest, largest and average observed within-schizophrenic variances; the vertical line on each histogram represents the observed value of the test statistic.*

chology. To check the fit of this new model, we use posterior predictive simulation of the same test statistics under the new posterior distribution. The results are displayed in Figure 11.4.

Comparing Figure 11.4 with Figure 11.3, the observed values are in the same place but the posterior predictive distributions have moved closer to them. (The histograms of Figures 11.3 and 11.4 are not plotted on a common scale.) However, the fit is by no means perfect in the new model: the observed values of $T_{\min}$ and $T_{\max}$ are still in the periphery, and the estim-

parameter	old model			new model		
	2.5%	median	97.5%	2.5%	median	97.5%
λ	0.07	0.12	0.18	0.46	0.64	0.88
ω		fixed at 1		0.24	0.56	0.84
τ	0.74	0.85	0.96	0.21	0.42	0.60
β	0.17	0.32	0.48	0.07	0.24	0.43

Table 11.1 *Posterior quantiles for parameters of interest under the old and new mixture models for the reaction time experiment*

ated p-values of the two test statistics are 0.98 and 0.03. (The average statistic, T_{avg}, is still well fitted by the expanded model, with a p-value of 0.81.) The lack of fit is clearly visible in Figure 11.4, and the p-values provide probability summaries. We are left with an improved model that still shows some lack of fit, suggesting directions for further model improvement and sensitivity analysis.

Acknowledgments

We thank the National Science Foundation for grants DMS-9404305, DMS-9457824 and DMS-9204504.

References

Belin, T. R. and Rubin, D. B. (1990) Analysis of a finite mixture model with variance components. *Proc. Am. Statist. Ass. Soc. Statist. Sect.*, 211–215.

Belin, T. R. and Rubin, D. B. (1995) The analysis of repeated-measures data on schizophrenic reaction times using mixture models. *Statist. Med.*, **14**, 747–768.

Box, G. E. P. (1980) Sampling and Bayes' inference in scientific modelling and robustness. *J. R. Statist. Soc.* A, **143**, 383–430.

Carlin, B. P. and Chib, S. (1995) Bayesian model choice via Markov chain Monte Carlo methods. *J. R. Statist. Soc.* B, **57**, 473–484.

Draper, D. (1995) Assessment and propagation of model uncertainty (with discussion). *J. R. Statist. Soc.* B, **57**, 45–98.

Gelman, A. (1995) Inference and monitoring convergence. In *Markov Chain Monte Carlo in Practice* (eds W. R. Gilks, S. Richardson and D. J. Spiegelhalter), pp. 131–143. London: Chapman & Hall.

Gelman, A. and Meng, X.-L. (1995) Discussion on assessment and propagation of model uncertainty (by D. Draper). *J. R. Statist. Soc.* B, **57**, 83.

Gelman, A. and Rubin, D. B. (1992) Inference from iterative simulation using multiple sequences (with discussion). *Statist. Sci.*, **7**, 457–511.

Gelman, A. and Rubin, D. B. (1995) Discussion on Bayesian model selection in social research (by A. E. Raftery). In *Sociological Methodology* (ed. P. V. Marsden). San Francisco: Jossey-Bass (in press).

Gelman, A., Meng, X.-L. and Stern, H. S. (1996) Posterior predictive assessment of model fitness (with discussion). *Statistica Sinica*, (in press).

Gelman, A., Carlin, J. B., Stern, H. S. and Rubin, D. B. (1995) *Bayesian Data Analysis*. London: Chapman & Hall, (in press).

Guttman, I. (1967) The use of the concept of a future observation in goodness-of-fit problems. *J. R. Statist. Soc.* B, **29**, 83–100.

Meng, X.-L. (1994) Posterior predictive p-values. *Ann. Statist.*, **22**, 1142–1160.

Meng, X.-L. and Rubin, D. B. (1993) Maximum likelihood estimation via the ECM algorithm: a general framework. *Biometrika*, **80**, 267–278.

Raftery, A. E. (1995a) Bayesian model selection in social research (with discussion). In *Sociological Methodology* (ed. P. V. Marsden). San Francisco: Jossey-Bass (in press).

Raftery, A. E. (1995b) Hypothesis testing and model selection. In *Markov Chain Monte Carlo in Practice* (eds W. R. Gilks, S. Richardson and D. J. Spiegelhalter), pp. 163–187. London: Chapman & Hall.

Rubin, D. B. (1981) Estimation in parallel randomized experiments. *J. Educ. Statist.*, **6**, 377–401.

Rubin, D. B. (1984) Bayesianly justifiable and relevant frequency calculations for the applied statistician. *Ann. Statist.*, **12**, 1151–1172.

Tiao, G. C. and Tan, W. Y. (1965) Bayesian analysis of random-effect models in the analysis of variance. I: Posterior distribution of variance components. *Biometrika*, **52**, 37–53.

12

Stochastic search variable selection

Edward I George

Robert E McCulloch

12.1 Introduction

When building statistical models, a crucial problem often faced by practitioners is the selection of variables for inclusion. For example, consider the context of multiple linear regression, perhaps the most widely applied statistical methodology. Given a dependent variable Y and a set of potential predictors $X_1, \ldots, X_p$, the problem is to find the 'best' model of the form $Y = X_1^* \beta_1^* + \cdots + X_q^* \beta_q^* + \epsilon$ where $X_1^*, \ldots, X_q^*$ is a 'selected' subset of $X_1, \ldots, X_p$. As described in Miller (1990), the typical solution to such problems has entailed using an omnibus criterion such as AIC, Mallows C_p or BIC to select the 'best' model from a candidate set of tentative models. When p is small enough, this candidate set is the set of all possible models. Otherwise, an *ad hoc* method such as stepwise selection is used first to reduce the number of tentative models to a manageable number.

In this chapter, we describe a new procedure, introduced in George and McCulloch (1993), which stochastically searches for 'promising' subsets of predictors. This procedure, which we call stochastic search variable selection (SSVS), puts a probability distribution on the set of all possible regression models such that 'promising' models are given highest probability, and then uses the Gibbs sampler to simulate a correlated sample from this distribution. The more promising models are then easily identified as those which appear with highest frequency in this sample. This approach can be effective even when the size of the simulated sample is much smaller than the number of possible models (2^p), since the vast majority of the models will have very small probability and can be ignored. We have successfully applied SSVS to problems with hundreds of potential predictors (see George and McCulloch, 1994).

The probability distribution, from which SSVS samples, is the posterior distribution obtained by putting a hierarchical Bayes mixture prior on the regression coefficients. By suitable prespecification of the prior inputs, high posterior probability is allocated to models whose coefficients are substantially different from zero. An attractive feature of this formulation is that it allows for variable selection based on practical significance rather than statistical significance. However, the key to the potential of SSVS is that the Gibbs sampler can be used to simulate an informative sample from the posterior quickly and efficiently. This avoids the overwhelming (and often impossible) burden of calculating the posterior probabilities of all subsets. Related Bayesian variable selection procedures have been proposed by Carlin and Chib (1995), Mitchell and Beauchamp (1988), and Raftery *et al.* (1994).

The SSVS approach turns out to be very powerful in that it can be extended to resolve many related problems. For example, straightforward extensions of SSVS provide systematic approaches for variable selection in generalized linear models (see George *et al.*, 1995), and for hierarchical model building across exchangeable regressions (see George and McCulloch, 1995). Both of these extensions are also described in this chapter.

In Section 12.2, we describe the hierarchical Bayes mixture model which puts high posterior probability on the 'promising' models. In Section 12.3, we show how the Gibbs sampler is used to search the posterior distribution for these promising models. In Section 12.4, we describe the extensions of SSVS for variable selection in generalized linear models and for variable selection across exchangeable regressions. In Section 12.5, we illustrate the application of SSVS to constructing financial portfolios. The example demonstrates how SSVS can be calibrated to incorporate practical significance as a criterion for variable selection.

12.2 A hierarchical Bayesian model for variable selection

The hierarchical Bayesian model for variable selection proposed by George and McCulloch (1993) is as follows. To begin with, the relationship between the dependent variable Y and the set of potential predictors $X_1, \ldots, X_p$ is assumed to be the normal linear model

$$Y \sim N_n(X\beta, \sigma^2 I_n),$$

where Y is $n \times 1$, N_n denotes an n-dimensional multivariate normal distribution, $X = [X_1, \ldots, X_p]$ is $n \times p$, $\beta = (\beta_1, \ldots, \beta_p)^T$, σ^2 is a scalar and I_n denotes the $n \times n$ identity matrix. Both β and σ^2 are considered unknown. For this model, selecting a subset of predictors is equivalent to setting to zero those β_i corresponding to the excluded predictors.

Each component β_i of β is then modelled as having come either from a distribution with most of its mass concentrated about zero, or from a distribution with its mass spread out over plausible values. This is done using a scale mixture of two normal distributions which, using the latent variable $\gamma_i = 0$ or 1, may be conveniently expressed as

$$P(\beta_i \mid \gamma_i) = (1 - \gamma_i)\mathrm{N}(0, \tau_i^2) + \gamma_i \mathrm{N}(0, c_i^2 \tau_i^2)$$

and

$$P(\gamma_i = 1) = 1 - P(\gamma_i = 0) = w_i.$$

(We use $P(\cdot)$ generically to denote distributions on parameters.) The hyperparameters τ_i and c_i are set small and large respectively so that $\mathrm{N}(0, \tau_i^2)$ is concentrated and $\mathrm{N}(0, c_i^2 \tau_i^2)$ is diffuse as in Figure 12.1. The idea is that if $\gamma_i = 0$, β_i would probably be so small that it could be 'safely' estimated by 0, whereas if $\gamma_i = 1$, a nonzero estimate of β_i would probably be required. Based on this interpretation, the hyperparameter $w_i = P(\gamma_i = 1)$ is the prior probability that β_i will require a nonzero estimate, or equivalently that X_i should be included in the model. Recommended strategies for choosing τ_i, c_i, and w_i are described below.

The above mixture prior on each component of β can be obtained using a general multivariate normal mixture prior which is described in George and McCulloch (1993). A special case of this prior which is easy to implement and seems to perform well is the following. Conditionally on the latent vector $\gamma = (\gamma_1, \ldots, \gamma_p)$, this prior on β is multivariate normal

$$P(\beta \mid \gamma) = \mathrm{N}_p(0, D_\gamma^2),$$

where D_γ is a diagonal matrix with i^{th} diagonal element equal to $(1-\gamma_i)\tau_i + \gamma_i c_i \tau_i$, so that the β_i are *a priori* independent given γ. The mixture prior on β is then obtained with any nontrivial prior $P(\gamma)$ on the 2^p possible values of γ. A particularly useful case for the prior for γ is the independent Bernoulli prior

$$P(\gamma) = \prod_{i=1}^{p} w_i^{\gamma_i}(1 - w_i)^{(1-\gamma_i)}, \tag{12.1}$$

especially $P(\gamma) \equiv 1/2^p$. An inverse gamma conjugate prior is put on σ^2, namely

$$P(\sigma^{-2}) = \mathrm{Ga}(\nu/2, \nu\lambda/2) \tag{12.2}$$

which specifies for σ^{-2} a prior mean and variance of λ^{-1} and $2\nu^{-1}\lambda^{-2}$. Choosing ν small can be used to represent vague prior information. Finally, β and σ^2 are treated as *a priori* independent so that $P(\beta, \sigma^2) = P(\beta)P(\sigma^2)$. A modification of this hierarchical prior which offers some computation advantages is described in George and McCulloch (1994).

For this hierarchical mixture model, the latent vector $\gamma = (\gamma_1, \ldots, \gamma_p)$ contains the relevant information for variable selection. If γ were known, then with high probability for suitably chosen $\tau_1, \ldots, \tau_p$ and $c_1, \ldots, c_p$, a desirable model would be obtained by including those X_i for which $\gamma_i = 1$

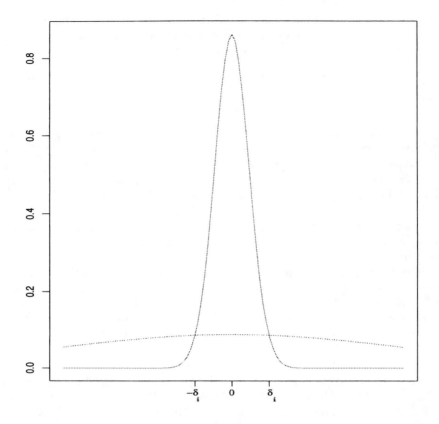

Figure 12.1 *Concentrated density is* $N(0, \tau_i^2)$; *diffuse density is* $N(0, c_i^2\tau_i^2)$; *intersection is at* $\pm\delta_i$.

and excluding those X_i for which $\gamma_i = 0$. Although γ is not known, the posterior distribution of γ, namely

$$P(\gamma \mid Y) \propto P(Y \mid \gamma)P(\gamma),$$

can provide useful variable selection information. Based on Y, the posterior $P(\gamma \mid Y)$ updates the prior probabilities on each of the 2^p possible values of γ. Identifying each γ with a submodel via $(\gamma_i = 1) \Leftrightarrow (X_i$ is included), those γ with higher posterior probability $P(\gamma \mid Y)$ identify the more 'promising' submodels, that is those supported most by the data and the statistician's prior distribution.

The use of high posterior probability values to identify 'promising' models suggests a strategy for selecting τ_i and c_i. Let δ_i be the intersection point of the densities of $N(0, \tau_i^2)$ and $N(0, c_i^2\tau_i^2)$ as shown in Figure 12.1. Note that

$|\beta_i| \leq \delta_i$ corresponds to the region where $N(0, \tau_i^2)$ dominates $N(0, c_i^2\tau_i^2)$, and $|\beta_i| > \delta_i$ corresponds to the region where $N(0, \tau_i^2)$ is dominated by $N(0, c_i^2\tau_i^2)$. Large posterior probability of γ under $P(\gamma \mid Y)$ suggests that $|\beta_i| \leq \delta_i$ for $\gamma_i = 0$, and $|\beta_i| > \delta_i$ for $\gamma_i = 1$. Thus, τ_i and c_i should be selected so that if $|\beta_i| \leq \delta_i$ it would be preferable to set $\beta_i = 0$ and exclude X_i from the model. In particular, when δ_i is the largest value of $|\beta_i|$ for which setting $\beta_i = 0$ would make no practical difference, SSVS will select variables based on 'practical significance' rather than statistical significance. Once δ_i has been selected, we recommend choosing c_i between 10 and 100 to achieve an effective, but not extreme, separation between the two densities. It is also important that $N(0, c_i^2\tau_i^2)$ give reasonable probability to all plausible values of β_i. For a given c_i and δ_i, the implicit value of τ_i is then obtained as $\tau_i = [2\log(c_i)c_i^2/(c_i^2 - 1)]^{-1/2}\delta_i$. An alternative semiautomatic strategy for selecting τ_i and c_i based on statistical significance is described in George and McCulloch (1993).

12.3 Searching the posterior by Gibbs sampling

Implementation of SSVS consists of two stages. The first stage entails setting the prior inputs to the hierarchical model, c_i, τ_i, ν, λ, and $P(\gamma)$, so that γ values corresponding to 'promising' models are assigned higher posterior probability under $P(\gamma|Y)$. The second stage entails the identification of these high posterior probability γ values by Gibbs sampling. This approach avoids the overwhelming difficulties of computing all 2^p posterior probabilities by analytical or numerical methods.

SSVS uses the Gibbs sampler to generate a sequence

$$\gamma^{(1)}, \gamma^{(2)}, \ldots \tag{12.3}$$

which converges in distribution to $\gamma \sim P(\gamma \mid Y)$. The relative frequency of each γ value thus converges to its probability under $P(\gamma|Y)$. In particular, those γ values with highest posterior probability can be identified as those which appear most frequently in this sequence. Note that, in contrast to many applications of the Gibbs sampler, the goal here is not the evaluation of the entire distribution $P(\gamma \mid Y)$. Indeed, most of the 2^p values of γ will have very small probability, will appear very rarely, and so can be ignored. In effect, SSVS uses the Gibbs sampler to search through rather than to evaluate the posterior $P(\gamma \mid Y)$. Consequently, the length of the sequence (12.3) can be much smaller than 2^p, and still serve to identify high probability values.

SSVS generates sequence (12.3) by applying the Gibbs sampler to the complete posterior $P(\beta, \sigma^2, \gamma \mid Y)$. This produces the full sequence of parameter values

$$\beta^{(0)}, \sigma^{(0)}, \gamma^{(0)}, \beta^{(1)}, \sigma^{(1)}, \gamma^{(1)}, \ldots, \beta^{(k)}, \sigma^{(k)}, \gamma^{(k)}, \ldots,$$

a Markov chain converging to $P(\beta, \sigma^2, \gamma \mid Y)$ in which sequence (12.3) is embedded. The Gibbs sampler is implemented as follows. First, $\beta^{(0)}$, $\sigma^{(0)}$ and $\gamma^{(0)}$ are initialized at some reasonable guess such as might be obtained by stepwise regression. Subsequent values are then obtained by successive simulation from the full conditionals

$$P(\beta \mid \sigma^2, \gamma, Y)$$
$$P(\sigma^2 \mid \beta, \gamma, Y) = P(\sigma^2 \mid \beta, Y) \tag{12.4}$$
$$P(\gamma_i \mid \beta, \sigma^2, \gamma_{-i}, Y) = P(\gamma_i \mid \beta, \gamma_{-i}), \quad i = 1, \ldots, p,$$

where $\gamma_{-i} = (\gamma_1, \ldots, \gamma_{i-1}, \gamma_{i+1}, \ldots, \gamma_p)$. At each step, these distributions condition on the most recently generated parameter values. Note that because of the hierarchical structure of the prior, the full conditional for σ^2 depends only on Y and β, and the full conditional for γ_i depends only on β and γ_{-i}.

The success of SSVS is due in large part to the fact that the successive simulations in (12.4) are from standard distributions, and can be carried out quickly and efficiently by routine methods. To begin with, the coefficient vector $\beta^{(j)}$ is obtained by sampling from

$$P(\beta \mid \sigma^2, \gamma, Y) = N_p((X^T X + \sigma^2 D_\gamma^{-2})^{-1} X^T Y, \sigma^2(X^T X + \sigma^2 D_\gamma^{-2})^{-1}). \tag{12.5}$$

The most costly step here is updating $(X^T X + \sigma^2 D_\gamma^{-2})^{-1}$ on the basis of new values of σ^2 and γ. A fast method for performing this step which exploits the Cholesky decomposition is described in George and McCulloch (1994). Next, $\sigma^{(j)}$ is obtained by sampling from

$$P(\sigma^{-2} \mid \beta, Y) = \text{Ga}\left(\frac{n+\nu}{2}, \frac{|Y - X\beta|^2 + \nu\lambda}{2}\right), \tag{12.6}$$

the updated inverse gamma distribution from (12.2).

Finally, the vector γ is obtained componentwise by sampling each γ_i consecutively (and preferably in random order) from the Bernoulli distribution with probability

$$P(\gamma_i = 1 \mid \beta, \gamma_{-i}) = \frac{a}{a+b}, \tag{12.7}$$

where

$$a = P(\beta \mid \gamma_{-i}, \gamma_i = 1)P(\gamma_{-i}, \gamma_i = 1)$$
$$b = P(\beta \mid \gamma_{-i}, \gamma_i = 0)P(\gamma_{-i}, \gamma_i = 0).$$

Under the independent Bernoulli prior $P(\gamma)$ in (12.1), (12.7) simplifies to

$$P(\gamma_i = 1 \mid \beta_i) = \frac{a}{a+b}, \tag{12.8}$$

where

$$a = P(\beta_i \mid \gamma_i = 1)w_i$$
$$b = P(\beta_i \mid \gamma_i = 0)(1 - w_i).$$

The generation of each new γ value proceeds one step at a time where at each step, the value of some γ_i is randomly determined according to the conditional Bernoulli distribution (12.7). Because changing γ_i essentially corresponds to the decision to add or remove X_i from the regression, the generation of sequence (12.3) is equivalent to performing a stochastically directed stepwise procedure.

One may also want to consider using the alternative implementation of SSVS described in George and McCulloch (1994) where the parameters β and σ^2 are integrated out to yield closed form expressions for $P(\gamma|Y)$ up to a normalizing constant. The Gibbs sampler and other Markov chain Monte Carlo methods can then be applied directly to these expressions to generate the sequence (12.3) as a Markov chain converging rapidly to $\gamma \sim P(\gamma|Y)$.

12.4 Extensions

12.4.1 SSVS for generalized linear models

George $et\ al.$ (1995) show how a slight modification of SSVS provides a systematic method for variable selection in the broad class of models of the form $f(y \mid X\beta, \omega)$ where X is the matrix of predictor values and ω represents additional (usually nuisance) parameters. This class includes all generalized linear models in addition to the normal linear model.

To apply SSVS in this context, we use the hierarchical prior of Section 12.2 with $P(\sigma^2)$ replaced by a suitable prior on ω. Using similar considerations as before, the posterior $P(\gamma|Y)$ can be made to put high probability on γ values corresponding to the most 'promising' models. Again, the Gibbs sampler can be used to search for these high probability values by generating a sequence $\gamma^{(1)}, \gamma^{(2)}, \ldots$ which converges to $P(\gamma \mid Y)$. This sequence is here generated by Gibbs sampling from the full posterior $P(\beta, \omega, \gamma \mid Y)$. This requires successive sampling from

$$P(\beta \mid \omega, \gamma, Y)$$

$$P(\omega \mid \beta, Y)$$

$$P(\gamma_i \mid \beta, \gamma_{-i}), \quad i = 1, \ldots, p.$$

It follows from Dellaportas and Smith (1993) that for many generalized linear models, including the binomial, Poisson, normal, gamma and inverse Gaussian link models, $P(\beta \mid \omega, \gamma, Y)$ will be log concave. In all these cases, the generation of β can then be obtained by the adaptive rejection sampling method of Gilks and Wild (1992). In many cases, $P(\omega \mid \beta, Y)$ will belong to a standard family and/or be one dimensional, so that ω can be generated by routine methods. Finally, the distribution of γ_i is again Bernoulli, and so is trivial to generate.

12.4.2 SSVS across exchangeable regressions

George and McCulloch (1995) show how SSVS may be extended to provide a systematic method for variable selection across exchangeable regressions. This setup arises when observations on variables Y and $X_1, \ldots, X_p$ fall into M known groups, $\{Y^1, X_1^1, \ldots, X_p^1\}, \ldots, \{Y^M, X_1^M, \ldots, X_p^M\}$, and within each group Y^m and $X^m = (X_1^m, \ldots, X_p^m)$ are related by the usual linear structure. A reasonable hierarchical model for this general setup entails M different regressions

$$Y^1 \sim \mathrm{N}_{n_1}(X^1\beta^1, \sigma_1^2 I_{n_1}), \ldots, Y^M \sim \mathrm{N}_{n_M}(X^M\beta^M, \sigma_M^2 I_{n_M}), \qquad (12.9)$$

coupled with an exchangeable prior on $\beta^1, \ldots, \beta^M$. For example, in Blattberg and George (1991), weekly sales of bathroom tissue and a set of coincident price and promotional variables were observed for 4 brands in each of 3 supermarket chains. A linear regression was used to relate these variables in each of the M = 12 different brand–chain groups of observations. See also Lindley and Smith (1972).

In the process of building such a hierarchical model, first it is necessary to select the predictor variables which will appear across all the regressions in (12.9). This is where SSVS comes in. Using the hierarchical prior of Section 12.2 with $P(\beta \mid \gamma) = \mathrm{N}_p(0, D_\gamma^2)$ replaced by

$$P(\beta^m \mid \gamma) = \mathrm{N}_p(0, D_\gamma^2), \quad m = 1, \ldots, M, \quad \text{independently}, \qquad (12.10)$$

and $P(\sigma^{-2}) = \mathrm{Ga}(\nu/2, \nu\lambda/2)$ replaced by

$$P(\sigma_m^{-2}) = \mathrm{Ga}(\nu/2, \nu\lambda/2), \quad m = 1, \ldots, M, \quad \text{independently},$$

SSVS can be applied almost exactly as before. The posterior $P(\gamma \mid Y^1, \ldots, Y^M)$ can be specified so as to put highest probability on those γ which correspond to the most promising subset choices according to whether or not X_i should appear in every regression. The Gibbs sampler can then be used to generate a sequence $\gamma^{(1)}, \gamma^{(2)}, \ldots$ which converges in distribution to this posterior by sampling from the full posterior

$$P(\beta^1, \ldots, \beta^M, \sigma_1^2, \ldots, \sigma_M^2, \gamma \mid Y^1, \ldots, Y^M).$$

This requires successive simulation from

$$P(\beta^m \mid \sigma_m^2, \gamma, Y^m), \quad m = 1, \ldots, M$$
$$P(\sigma_m^2 \mid \beta^m, Y^m), \quad m = 1, \ldots, M$$
$$P(\gamma_i \mid \beta^1, \ldots, \beta^M, \gamma_{-i}), \quad i = 1, \ldots, p.$$

Conditional on γ, the M regressions are independent so that the draws of β^m and σ_m^2 are done exactly as in (12.5) and (12.6) with Y, X, β, and σ^2 replaced with Y^m, X^m, β^m, and σ_m^2. The draw of γ uses information from all M regressions and is of the same form as (12.7) with β replaced by $\beta^1, \ldots, \beta^M$. Under the independent Bernoulli prior, we also have the simplification in (12.8) with β_i replaced by $\beta_i^1, \ldots, \beta_i^M$. Again, those values with highest posterior probability can be identified as those which appear most frequently.

SSVS can be further extended to address other problems in this context. Replacing (12.10) by

$$P(\beta^m \mid \beta^*, \gamma) = N_p(\beta^*, D_\gamma^2), \quad m = 1, \ldots, M, \text{ independently}$$

and with $P(\beta^*) = N_p(\beta^{**}, V)$, SSVS can be used to decide which variables should be assigned a common coefficient across the regressions. Finally, this prior also yields shrinkage estimators of the coefficient vectors $\beta^1, \ldots, \beta^M$.

12.5 Constructing stock portfolios with SSVS

In this section, we illustrate the application of SSVS on a real data set. We consider the problem of constructing a portfolio of a small set of stocks which mimics the S&P 500, a broad based market index. Holders of such mimicking portfolios can benefit from the investment advantages of 'holding the market' with a simplified investment strategy and reduced transaction costs. Such a portfolio will be a linear combination of stocks whose returns are highly correlated with the returns of the S&P 500. Thus, this portfolio construction problem is really one of variable selection in linear regression. From a pool of potential stocks, the problem is to find the 'best' regression of the S&P 500 on a small subset of these stocks. This problem is perfectly suited for SSVS.

For this application, our dependent variable Y, was the daily return on the S&P 500, and our pool of potential predictors, $X_1, \ldots, X_{10}$, were daily returns on 10 selected stocks. We used $n = 150$ consecutive daily returns (beginning January 1, 1977) from the Center for Research in Security Prices database (CRSP) at the University of Chicago Graduate School of Business. We have deliberately kept the number of potential predictors small for the purpose of clearly illustrating features of SSVS. For a larger example, see George and McCulloch (1994) where SSVS is used to construct mimicking portfolios from a pool of 200 potential predictors.

To set the prior inputs for SSVS, we first determined that for every X_i, $\delta_i = 0.05$ was the threshold for practical significance, as discussed in Section 12.2. Note, a common threshold was used for all regression coefficients. This determination was based on our sense that when $|\beta_i| \leq 0.05$, the 'effect' of X_i on Y would be negligible so that it would preferable to exclude X_i from the portfolio. We then chose $c_i = 10$ and $\tau_i = 0.023$ to achieve this δ_i. We used the uniform prior on γ, $P(\gamma) \equiv 2^{-10}$ obtained by (12.1) with $w_i = 0.5$. Finally, for the inverse gamma prior (12.2), we set $\lambda = 0.004$, the least-squares estimate of σ^2, and set $\nu = 5$.

Before presenting the results of SSVS as applied to this data set, it is useful to consider the results of the least-squares regression of Y on $X_1, \ldots, X_{10}$. Figure 12.2 displays for each X_i, the classical 95% confidence interval for β_i which is centred at the least-squares estimate $\hat{\beta}_i$. Also displayed are the thresholds of practical significance $\delta_i = \pm 0.05$. Note that a variable X_i is statistically significant if its 95% confidence interval does not contain 0, whereas it is apt to be practically significant if its confidence interval does not appreciably overlap $(-\delta_i, \delta_i)$. For example, X_8 appears to be practically significant, but is not statistically significant. On the other hand, X_9 is not practically significant, but is statistically significant.

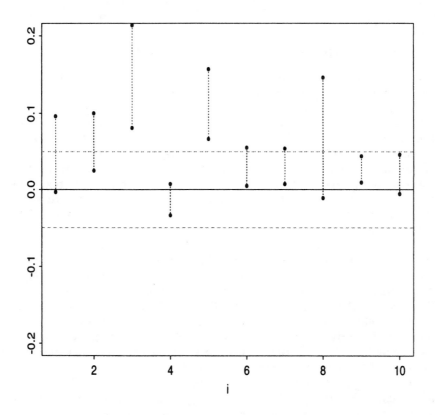

Figure 12.2 *Frequentist 95% confidence intervals for each β_i. Dashed horizontal lines are at $\delta_i = \pm 0.05$.*

Applying SSVS to these data, we simulated a Markov chain of 10 000 values of $\gamma = (\gamma_1, \ldots, \gamma_{10})$ from the posterior distribution $P(\gamma|Y)$. Table 12.1 displays the five most frequent γ values together with the corresponding variable subsets. According to SSVS, these subsets correspond to the five most promising regression models. Note that the variables comprising these subsets, namely X_1, X_2, X_3, X_5 and X_8, are the only reasonable candidates for practical significance as indicated by Figure 12.2. This is a clear demonstration of how SSVS incorporates practical significance into the choice of variables.

γ value	subset	frequency
(0,1,1,0,1,0,0,0,0,0)	X_2, X_3, X_5	1 072
(0,0,1,0,1,0,0,0,0,0)	X_3, X_5	1 020
(0,0,1,0,1,0,0,1,0,0)	X_3, X_5, X_8	537
(0,1,1,0,1,0,0,1,0,0)	X_2, X_3, X_5, X_8	497
(1,1,1,0,1,0,0,0,0,0)	X_1, X_2, X_3, X_5	450

Table 12.1 *The five most frequently selected models*

12.6 Discussion

As the portfolio construction example of Section 12.5 shows, SSVS can identify promising variables on the basis of practical significance. To do this, however, it is necessary to specify the threshold for practical significance δ_i, for each β_i. Of course, such choices will depend on the application. A strategy for selecting δ_i which we have found successful is to first determine ΔY, the largest insignificant change in Y, and ΔX_i, the range of X_i in the population of interest. These quantities are easier to think about than δ_i. Now define $\delta_i = \Delta Y / \Delta X_i$. Any β_i smaller than δ_i would then be practically insignificant in the sense that any possible change in X_i would yield an insignificant change in Y. Note that this strategy may be inappropriate when there are strong dependencies among the β_i.

In applications where the posterior distribution $P(\gamma \mid Y)$ is concentrated on a small set of models, the SSVS frequencies will accumulate there, making the analysis of the output straightforward. However, not all data sets are so obliging. When the data do not strongly support any model, the posterior distribution will be flat (as it should be), and the SSVS frequencies may not accumulate anywhere. A troublesome situation can occur when there is substantial correlation among the covariates. This can lead to an overabundance of good models, which will dilute the posterior probabilities and flatten $P(\gamma|Y)$. One obvious remedy for such a situation, is to eliminate the collinearity by standard methods such as excluding natural proxies before running SSVS. Another remedy is to adjust the prior $P(\gamma)$ to put more weight on parsimonious models. This can force the posterior to concentrate on a smaller set of models.

Finally, we should mention that after identifying a promising subset of variables with SSVS, a remaining issue is the estimation of the regression coefficients β. The least-squares estimates will suffer from selection bias which can be especially severe in large problems. A reasonable alternative may be $\hat{\beta} = E(\beta|Y)$, the posterior mean of the coefficients under the SSVS hierarchical mixture model. We are currently studying the properties of such estimators, and will report on them elsewhere.

Acknowledgement

This research was supported in part by NSF Grant DMS-94-04408.

References

Blattberg, R. C. and George, E. I. (1991) Shrinkage estimation of price and promotional elasticities: seemingly unrelated equations. *J. Am. Statist. Ass.*, **86**, 304–315.

Carlin, B. P. and Chib, S. (1995) Bayesian model choice via Markov chain Monte Carlo methods. *J. R. Statist. Soc.* B, **57**, 473–484.

Dellaportas, P. and Smith, A. F. M. (1993) Bayesian inference for generalized linear and proportional hazards models via Gibbs sampling. *Appl. Statist.*, **42**, 443–460.

George, E. I. and McCulloch, R. E. (1993) Variable selection via Gibbs sampling. *J. Am. Statist. Ass.*, **85**, 398–409.

George, E. I. and McCulloch, R. E. (1994) Fast Bayes variable selection. *CSS Technical Report*, University of Texas at Austin.

George, E. I. and McCulloch, R. E. (1995) Variable selection, coefficient restriction and shrinkage across exchangeable regressions. *CSS Technical Report*, University of Texas at Austin.

George, E. I., McCulloch, R. E. and Tsay, R. (1995) Two approaches to Bayesian model selection with applications. In *Bayesian Statistics and Econometrics: Essays in Honor of Arnold Zellner* (eds D. Berry, K. Chaloner, and J. Geweke). Amsterdam: North-Holland, (in press).

Gilks, W. R. and Wild, P. (1992) Adaptive rejection sampling for Gibbs sampling. *Appl. Statist.*, **41**, 337–348.

Lindley, D. V. and Smith, A. F. M. (1972) Bayes estimates for the linear model (with discussion). *J. R. Statist. Soc.* B, **34**, 1–41.

Miller, A. J. (1990) *Subset Selection in Regression*. New York: Chapman & Hall.

Mitchell, T. J. and Beauchamp, J. J. (1988) Bayesian variable selection in linear regression (with discussion). *J. Am. Statist. Ass.*, **83**, 1023–1036.

Raftery, A. E., Madigan, D. and Hoeting, J. A. (1994) Bayesian model averaging for linear regression. *Technical Report 94/12*, Department of Statistics, Colorado State University.

13

Bayesian model comparison via jump diffusions

David B Phillips
Adrian F M Smith

13.1 Introduction

In recent years, the use of MCMC simulation techniques has made feasible the routine Bayesian analysis of many complex high-dimensional problems. However, one area which has received relatively little attention is that of comparing models of possibly different dimensions, where the essential difficulty is that of computing the high-dimensional integrals needed for calculating the normalization constants for the posterior distribution under each model specification: see Raftery (1995: this volume).

Here, we show how methodology developed recently by Grenander and Miller (1991, 1994) can be exploited to enable routine simulation-based analyses for this type of problem. Model uncertainty is accounted for by introducing a joint prior probability distribution over both the set of possible models and the parameters of those models. Inference can then be made by simulating realizations from the resulting posterior distribution using an iterative jump-diffusion sampling algorithm. The essential features of this simulation approach are that discrete transitions, or *jumps*, can be made between models of different dimensionality, and within a model of fixed dimensionality the conditional posterior appropriate for that model is sampled. Model comparison or choice can then be based on the simulated approximation to the marginal posterior distribution over the set of models. An alternative approach using a reversible-jump Metropolis–Hastings algorithm rather than jump-diffusions has been proposed by Green (1994a,b). At the time of writing, systematic comparisons of the two approaches are not available.

13.2 Model choice

Suppose we wish to carry out inference in a setting where the 'true' model is unknown but is assumed to come from a specified class of parametric models $\{\mathcal{M}_0, \mathcal{M}_1, \ldots\}$, so that the overall parameter space Θ can be written as a countable union of subspaces,

$$\Theta = \bigcup_{k=0}^{\infty} \Theta_k,$$

where Θ_k, a subspace of the Euclidean space $\mathbb{R}^{n(k)}$, denotes the $n(k)$-dimensional parameter space corresponding to model $\mathcal{M}_k$.

Let $\theta \in \Theta$ denote the model parameters; let x denote the observed data vector ($x \in R^n$, for some $n \geq 1$); and let $k \in \{0, 1, \ldots\}$ denote the model labels. Then, given x, the joint posterior probability density π for the model parameters and labels can be written in the form

$$\pi(\theta, k|x) = \frac{p_k}{Z} \exp\{L_k(\theta|x) + P_k(\theta)\} \delta_k(\theta), \qquad \theta \in \Theta, \qquad k = 0, 1, \ldots,$$

where p_k is the prior probability for model $\mathcal{M}_k$ ($\sum_{k=0}^{\infty} p_k = 1$); Z is a normalization constant; $L_k(\theta|x)$ is the conditional log-likelihood for θ given x and model $\mathcal{M}_k$; $P_k(\theta)$ is the conditional log-prior for θ given model $\mathcal{M}_k$; and $\delta_k(\theta) = 1$ if $\theta \in \Theta_k$ and is zero otherwise.

13.2.1 Example 1: mixture deconvolution

Consider the problem of identifying the number of components in a univariate mixture distribution using, for the purposes of illustration, mixtures of normal distributions. A typical data set (see, for example, Escobar and West, 1995) is shown in Figure 13.1. Related data sets and models are discussed by Raftery (1995: this volume) and Robert (1995: this volume).

Assume that the number of components is unknown. For $k = 1, 2, \ldots$, let $\mathcal{M}_k$ denote the model with k components, each with unknown mean μ_i, variance σ_i^2 and mixing weight w_i. (Note that model $\mathcal{M}_0$ makes no sense in this context.) The probability density for an observation $x \in R$ from $\mathcal{M}_k$ takes the form

$$\sum_{i=1}^{k} w_i \frac{1}{\sqrt{2\pi}\sigma_i} \exp\left\{-\frac{1}{2\sigma_i^2}(x - \mu_i)^2\right\},$$

where

$$0 \leq w_i \leq 1, \qquad \sum_{i=1}^{k} w_i = 1. \tag{13.1}$$

Model $\mathcal{M}_k$ has k unconstrained means, k positive variances and k positive weights which sum to one, so that the number of free parameters is $n(k) = 3k - 1$.

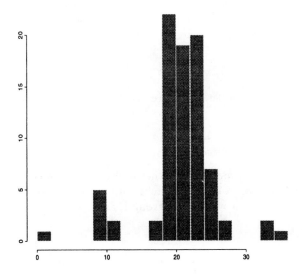

Figure 13.1 *Histogram of* 82 *galaxy velocities from Escobar and West (1995).*

Assume that we observe n data points $x = (x_1, \ldots, x_n)$ independently from the unknown mixture. For model $\mathcal{M}_k$, write

$$\theta = (\mu_1, \sigma_1^2, w_1, \ldots, \mu_k, \sigma_k^2, w_k).$$

Then the log-likelihood function $L_k(\theta|x)$ is given by

$$L_k(\theta|x) = \sum_{j=1}^{n} \log \left(\sum_{i=1}^{k} \frac{w_i}{\sigma_i} \exp \left\{ -\frac{1}{2\sigma_i^2}(x_j - \mu_i)^2 \right\} \right), \qquad \theta \in \Theta_k, \quad (13.2)$$

(ignoring an additive constant).

For the means $\{\mu_i\}$, we assume independent normal $N(\mu, \tau^2)$ priors (with mean μ and variance τ^2); for the inverse variances $\{\sigma^{-2}\}$, we assume independent gamma $Ga(\gamma, \delta)$ priors (with mean γ/δ and variance γ/δ^2); and for the weights $\{w_i\}$ we assume a uniform prior within the region defined by the constraints (13.1). We assume μ, τ^2, γ and δ are known. Then the log-prior $P_k(\theta)$ takes the form

$$P_k(\theta) = a_k - \sum_{i=1}^{k} \left\{ \frac{(\mu_i - \mu)^2}{2\tau^2} + (\gamma + 1) \log \sigma_i^2 + \frac{\delta}{\sigma_i^2} \right\}, \qquad \theta \in \Theta_k, \quad (13.3)$$

where $a_k = k[\gamma \log \delta - \log \tau - \frac{1}{2} \log 2\pi - \log \Gamma(\gamma)] + \log(k-1)!$. Note that although a_k is constant for $\theta \in \Theta_k$, it depends on k, and therefore cannot be ignored for model comparison.

For the prior on the discrete model space, we assume that the number of components k has a Poisson distribution with parameter λ, modified to exclude the case $k = 0$, so that

$$p_k = \frac{\lambda^k}{(e^\lambda - 1)\, k!}, \qquad k = 1, 2, \ldots.$$

We assume λ is known, so the $\{p_k\}$ are also known. Choice of λ is discussed in Section 13.4.

13.2.2 Example 2: object recognition

Suppose an image is composed from the superposition of an unknown number of planar shapes on a uniform background: for example, black discs of equal radius placed on a white background, as depicted in Figure 13.2.

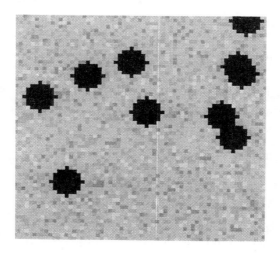

Figure 13.2 *Image showing black discs on a white background.*

Model $\mathcal{M}_k$ corresponds to the image being some arrangement of k discs of radius r in an $m \times m$ lattice of pixels. To simplify the exposition, we assume the disc centres fall within the image scene, and that r is known. We write

$$\theta = (\theta_1, \ldots, \theta_k)$$

to represent the k disc centres for model $\mathcal{M}_k$. Each θ_i is two-dimensional, so $n(k) = 2k$.

We assume that the true image is a two-colour scene, obtained by painting foreground pixels (lying inside at least one of the k discs) black, and background pixels (lying outside all of the discs) white. We assume that the colour of each pixel in the observed image in Figure 13.2 is corrupted by Gaussian noise. Specifically we assume that the greyness x_{ij} of pixel (i, j) in the observed image is independently from a $N(\mu_b, \sigma^2)$ distribution if it is in the foreground, or a $N(\mu_w, \sigma^2)$ distribution if it is in the background, where μ_b, μ_w and σ^2 are assumed known.

Then the log-likelihood for model $\mathcal{M}_k$ is

$$L_k(\theta|x) = -\frac{1}{2\sigma^2} \left(\sum_{(i,j) \in F_k} (x_{ij} - \mu_b)^2 + \sum_{(i,j) \notin F_k} (x_{ij} - \mu_w)^2 \right), \qquad \theta \in \Theta_k,$$

where F_k is the set of foreground pixels.

The prior used for illustration assumes that the disc locations are independently uniformly distributed in the scene, so that

$$P_k(\theta) = -2k \log m, \qquad \theta \in \Theta_k.$$

The Poisson distribution (with parameter λ) can again be used to specify the prior probabilities for each of the models, giving

$$p_k = \frac{\lambda^k}{e^\lambda k!}, \qquad k = 0, 1, 2, \ldots,$$

where λ is assumed known. Here we include the case $k = 0$, to allow image scenes containing no discs.

13.2.3 Example 3: variable selection in regression

Consider the problem of variable selection in regression analysis when a large number m of covariates is available, so that there are 2^m possible models. This problem is also considered by George and McCulloch (1995: this volume). Labelling the models is more complicated in this case. We denote by $\mathcal{M}_{c(k)}$ the regression which includes covariates $c_1, c_2, \ldots, c_k$ only. Thus $c(k) = (c_1, \ldots, c_k)$ is an ordered subsequence of $(1, 2, \ldots, m)$ of length k.

Specifically, denoting the response variable by x_0 and the covariates by $x_1, \ldots, x_m$ (and assuming, for convenience of exposition, that each has been standardized to remove the need for an intercept term), model $\mathcal{M}_{c(k)}$ is given by

$$x_{0,i} = \beta_{c_1} x_{c_1,i} + \ldots + \beta_{c_k} x_{c_k,i} + \epsilon_i,$$

for $i = 1, \ldots, n$, where the $\{\epsilon_i\}$ are independently $N(0, \sigma^2)$ and σ^2 is unknown. Therefore, putting

$$\theta = (\beta_{c_1}, \ldots, \beta_{c_k}, \sigma^2),$$

we have the conditional log-likelihood

$$L_{c(k)}(\theta|x) = -n \log \sigma - \frac{1}{2\sigma^2} \sum_{i=1}^{n} \{x_{0,i} - (\beta_{c_1} x_{c_1,i} + \ldots + \beta_{c_k} x_{c_k,i})\}^2,$$

where $\theta \in \Theta_k$.

For illustration, we assume independent $N(0, \tau^2)$ priors for the regression coefficients, and a $Ga(\gamma, \delta)$ prior for σ^{-2}, where τ^2, γ and δ are known. So the log-prior is given by

$$P_{c(k)}(\theta) = -k(\log \tau + \frac{1}{2} \log 2\pi) - \frac{1}{2\tau^2} \sum_{i=1}^{k} \beta_i^2 - (\gamma+1) \log \sigma^2 - \frac{\delta}{\sigma^2}, \qquad \theta \in \Theta_k.$$

If we assume that each of the $\frac{m!}{k!(m-k)!}$ different models of order k are *a priori* equally likely, and that k has a prior truncated Poisson distribution with parameter λ (truncated because $k \leq m$), then the prior probability for model $\mathcal{M}_{c(k)}$ is

$$p_{c(k)} = \frac{\lambda^k (m-k)!}{\left(e^\lambda - \sum_{j=m+1}^{\infty} \frac{\lambda^j}{j!}\right) m!}, \qquad k = 0, 1, \ldots, m.$$

As before, we assume λ is known.

13.2.4 Example 4: change-point identification

Consider now the problem of identifying the positions of an unknown number of change-points in a univariate time-series. For simplicity, we will suppose that between change-points the observation sequence $x = (x_1, \ldots, x_n)$ is a stationary Gaussian time-series, with unknown mean and variance, and that at a change-point there is a shift in both the mean and the variance. Figure 13.3 shows an example of this type of time series. It will be clear from the following development that more general change-point problems can be handled in a similar way.

Let model $\mathcal{M}_k$ correspond to the existence of k change-points. Let the unknown times of the change-points be denoted by $c_1, \ldots, c_k$. Let μ_i and σ_i^2 denote the mean and variance of x_j for $c_{i-1} < j \leq c_i$ and $i = 1, \ldots, k+1$, where we define $c_0 = 1$ and $c_{k+1} = n$. Then the set of free parameters in model $\mathcal{M}_k$ is

$$\theta = (c_1, \ldots, c_k, \mu_1, \sigma_1^2, \ldots, \mu_{k+1}, \sigma_{k+1}^2),$$

and $n(k) = 3k + 2$.

The likelihood function is then

$$L_k(\theta|x) = -\sum_{i=1}^{k+1} \left\{ (c_i - c_{i-1}) \log \sigma_i + \frac{1}{2\sigma_i^2} \sum_{j=c_{i-1}+1}^{c_i} (x_j - \mu_i)^2 \right\},$$

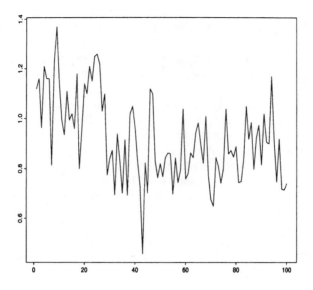

Figure 13.3 *Time-series of the annual discharge from the River Nile at Aswan from 1871 to 1970.*

where $\theta \in \Theta_k$.

The prior adopted for illustration assumes that $(c_1, \ldots, c_k)$ is uniform over all ordered subsequences of $(1, \ldots, n)$ of length k; that $\mu_1, \ldots, \mu_{k+1}$ are independently $N(\mu, \tau^2)$; and that $\sigma_1^{-2}, \ldots, \sigma_{k+1}^{-2}$ are independently $Ga(\gamma, \delta)$, where μ, τ^2, γ and δ are assumed known. Then

$$P_k(\theta) = a_k - \sum_{i=1}^{k+1} \left\{ \frac{(\mu_i - \mu)^2}{2\tau^2} + (\gamma + 1) \log \sigma_i^2 + \frac{\delta}{\sigma_i^2} \right\}, \qquad \theta \in \Theta_k,$$

where $a_k = (k+1)[\gamma \log \delta - \log \tau - \frac{1}{2} \log 2\pi - \log \Gamma(\gamma)] + \log(k - n)! + \log k!$.

We use the truncated Poisson distribution with known parameter λ to specify prior probabilities for the collection of models $\{\mathcal{M}_0, \mathcal{M}_1, \ldots, \mathcal{M}_n\}$, so that

$$p_k = \frac{\lambda^k}{\left(e^\lambda - \sum_{j=n+1}^{\infty} \frac{\lambda^j}{j!} \right) k!}, \qquad k = 0, 1, \ldots, n,$$

as in Section 13.2.3.

13.3 Jump-diffusion sampling

For each of the four examples introduced in the previous section, we have shown how to construct an appropriate probability density over the joint sample space of the models and the parameters. In general, however, these densities are not amenable to analytic or numerical analysis, and so we adopt a simulation approach using MCMC with target density $\pi(\theta, k|x)$. Throughout the remainder of this section, for notational convenience, we shall suppress dependence on x, so for example $\pi(\theta, k|x)$ will be denoted $\pi(\theta, k)$.

We construct a Markov process $(\theta, k)^{(t)}$ in continuous time t which produces samples from the target density $\pi(\theta, k)$. This stochastic process is generated from a jump component which makes discrete transitions between models at random times, and a diffusion component which samples values for the model-specific parameters between the jumps. Under appropriate conditions the stationary distribution of $(\theta, k)^{(t)}$ has density π, so that pseudo-samples from π may be produced by taking evenly spaced states from a realization of the process, and ergodic averages may be used to estimate expectations, with respect to π, of integrable functions of (θ, k).

We now informally describe the sampling algorithm; see Grenander and Miller (1994) and Amit *et al.* (1993) for details of the theorems which underpin the methodology.

13.3.1 The jump component

Suppose that at time t the process is in state (θ, k). The probability of jumping out of model $\mathcal{M}_k$ is specified through a *jump intensity* $q((\theta, k), (\phi, h))$, where $\phi \in \Theta_h$ and $h \neq k$. Informally, $q((\theta, k), (\phi, h))dt$ can be thought of as the probability density of jumping from (θ, k) to $\phi \in \Theta_h$ in an infinitesimally small interval of time dt.

Let $\mathcal{J}(\theta, k)$ denote the set of all (ϕ, h) which can be reached from (θ, k) in a single jump. This is called the *jump set* of (θ, k). Often it is convenient to choose $\mathcal{J}(\theta, k)$ such that

$$\mathcal{J}(\theta, k) \subset \Theta_{k-1} \cup \Theta_{k+1}, \qquad (13.4)$$

that is, from model $\mathcal{M}_k$, we can only jump to models $\mathcal{M}_{k-1}$ or $\mathcal{M}_{k+1}$. Regularity conditions given in Grenander and Miller (1994) require that $\mathcal{J}(\theta, k)$ must be chosen so that jumps are always *local*, i.e. that for some suitably defined metric D and constant B:

$$D((\theta, k), (\phi, h)) = D((\phi, h), (\theta, k)) < B, \qquad (13.5)$$

for all states (ϕ, h) in $\mathcal{J}(\theta, k)$. For example, in Sections 13.4–13.7 we have set

$$D((\theta, k), (\phi, h)) = \|\theta - \theta_0\| + \|\phi - \phi_0\|,$$

where θ_0 and ϕ_0 are roughly centrally located fixed points in Θ_k and Θ_h respectively, and $\|\dots\|$ denotes Euclidean (L_2) distance. However, condition (13.5) is really more a theoretical concern than a practical one, because

B can be chosen large enough to make (13.5) largely ineffective. We require that jump sets are *reversible*, i.e.

$$\text{if } (\phi, h) \in \mathcal{J}(\theta, k) \text{ then } (\theta, k) \in \mathcal{J}(\phi, h). \tag{13.6}$$

The jump sets used in Sections 13.4–13.7 satisfy both (13.5) and (13.4), which jointly satisfy (13.6).

To decide when to jump, the marginal jump intensity must be calculated. This is

$$r(\theta, k) = \sum_{h} \int_{(\phi, h) \in \mathcal{J}(\theta, k)} q((\theta, k), (\phi, h)) d\phi. \tag{13.7}$$

The marginal probability of jumping out of model $\mathcal{M}_k$ in time interval $(t, t + dt)$ is then $r(\theta, k)dt$. Note that this will in general depend on (θ, k). Jump times $t_1, t_2, \ldots$ can be sampled by generating unit exponential random variables $e_1, e_2, \ldots$, and setting t_i to be such that

$$\int_{t_{i-1}}^{t_i} r((\theta, k)^{(t)}) \, dt \geq e_i,$$

where $t_0 = 0$. In practice, the algorithm must be followed in discrete time-steps of size Δ.

If a jump must be made at time t (i.e. if $t = t_i$ for some i), then the new state (ϕ, h) must be selected according to the transition kernel:

$$Q((\theta, k), (\phi, h)) = \frac{q((\theta, k), (\phi, h))}{r(\theta, k)}, \qquad (\phi, h) \in \mathcal{J}(\theta, k). \tag{13.8}$$

Informally, $Q((\theta, k), (\phi, h))$ is the conditional probability density of jumping from (θ, k) to $\phi \in \Theta_h$, given that a jump out of model $\mathcal{M}_k$ must be made immediately.

Jump intensity $q((\theta, k), (\phi, h))$ must be chosen so that a balance condition is satisfied, to ensure that the stationary distribution of the process is $\pi(\theta, k)$ (Grenander and Miller, 1994). We now describe two strategies proposed by Grenander and Miller (1994) and Amit *et al.* (1993), which ensure that the balance condition is satisfied.

Gibbs jump dynamics

Under this choice of dynamics, the process jumps from one subspace to another with transition intensity given by

$$q((\theta, k), (\phi, h)) = Cp_h \exp\{L_h(\phi) + P_h(\phi)\}, \tag{13.9}$$

for $(\phi, h) \in \mathcal{J}(\theta, k)$, where constant C can be chosen arbitrarily. The choice of this constant will affect the frequency of jumps. In practice, some experimentation with C may be required to achieve good mixing.

For most problems, calculation of the marginal jump intensity (13.7) and direct sampling from the complex, high-dimensional posterior distribution

π restricted to $\mathcal{J}(\theta, k)$ will be impractical. To tackle this problem, Amit *et al.* (1993) show that the transition intensity function q can be concentrated on jump sets $\mathcal{J}(\theta, k)$ of reduced dimension. In particular, $\mathcal{J}(\theta, k)$ might constrain some of the elements of ϕ to equal some of those in θ. Suppose $\theta = (\theta', \theta'')$, where θ' are the parameters which are held constant in $\mathcal{J}(\theta, k)$, i.e. $\phi = (\theta', \phi'')$. Then (13.9) becomes

$$q((\theta', \theta'', k), (\theta', \phi'', h)) = C p_h \exp\left\{ L_h(\theta', \phi'') + P_h(\theta', \phi'') \right\}.$$

The parameters of θ which are to be held constant may depend on k and h, but we suppress this in the notation. However, the reversibility condition (13.6) must be preserved.

For example, suppose model $\mathcal{M}_k$ comprises the parameters $(\theta_1, \ldots, \theta_k)$, for $k \geq 1$, so the models are nested. Define $\mathcal{J}(\theta, k)$ to be the set

$$\{h = k - 1 \quad \text{and} \quad (\phi_1, \ldots, \phi_{k-1}) = (\theta_1, \ldots, \theta_{k-1})\}$$
$$\text{or}$$
$$\{h = k + 1 \quad \text{and} \quad (\phi_1, \ldots, \phi_k) = (\theta_1, \ldots, \theta_k)\},$$

subject to (13.5). Thus, if $h = k - 1$, ϕ can only be the point obtained by dropping the last component of θ. If $h = k + 1$, ϕ can be any point whose first k components are identical to θ, subject to (13.5). Clearly, this $\mathcal{J}(\theta, k)$ satisfies the reversibility condition (13.6).

With jump sets of reduced dimension, the marginal jump intensity $r(\theta, k)$ is calculated from

$$r(\theta, k) = \sum_h \int_{(\theta', \phi'', h) \in \mathcal{J}(\theta, k)} q((\theta', \theta'', k), (\theta', \phi'', h)) d\phi''. \quad (13.10)$$

Thus in the above example

$$\begin{aligned}
r(\theta, k) \quad \propto \quad & p_{k-1} \exp\left\{ L_{k-1}(\theta_1, \ldots, \theta_{k-1}) + P_{k-1}(\theta_1, \ldots, \theta_{k-1}) \right\} \\
+ \quad & p_{k+1} \int \exp\left\{ L_{k+1}(\theta_1, \ldots, \theta_k, \phi_{k+1}) + \right. \\
& \left. P_{k+1}(\theta_1, \ldots, \theta_k, \phi_{k+1}) \right\} d\phi_{k+1},
\end{aligned}$$

where the integral is over all ϕ_{k+1} subject to (13.5). The transition kernel $Q((\theta, k), (\phi, h))$ in (13.8) is then

$$\frac{p_{k-1}}{r(\theta, k)} \exp\left\{ L_{k-1}(\phi) + P_{k-1}(\phi) \right\}$$

for $\phi = (\theta_1, \ldots, \theta_{k-1})$ and $h = k - 1$, and

$$\frac{p_{k+1}}{r(\theta, k)} \exp\left\{ L_{k+1}(\phi) + P_{k+1}(\phi) \right\}$$

for $\phi = (\theta_1, \ldots, \theta_k, \phi_{k+1})$ and $h = k+1$. We can sample from this as follows.

First sample h, setting $h = k - 1$ with probability

$$\frac{p_{k-1}}{r(\theta, k)} \exp\left\{L_{k-1}(\theta_1, \ldots, \theta_{k-1}) + P_{k-1}(\theta_1, \ldots, \theta_{k-1})\right\}.$$

Then, if $h = k - 1$, set $\phi = (\theta_1, \ldots, \theta_{k-1})$. Otherwise, sample ϕ_{k+1} from the density proportional to

$$\exp\left\{L_{k+1}(\theta_1, \ldots, \theta_k, \phi_{k+1}) + P_{k+1}(\theta_1, \ldots, \theta_k, \phi_{k+1})\right\}.$$

This is proportional to the full conditional distribution for ϕ_{k+1} in model $\mathcal{M}_{k+1}$.

Using jump sets of reduced dimension will render the calculation of r and the sampling from Q more amenable to available methods. In some applications, it may be possible to evaluate r analytically; otherwise techniques of numerical or Monte-Carlo integration may be used. The variable selection problem introduced in Section 13.2.3 is analysed in Section 13.6 using Gibbs jump dynamics on reduced-dimension jump sets.

Metropolis jump dynamics

Under this second choice of dynamics, the process jumps from one subspace to another with transition intensity $q((\theta, k), (\phi, h))$ given by

$$q((\theta, k), (\phi, h)) = C \min\left(1, \exp\left\{L_h(\phi) - L_k(\theta)\right\}\right) p_h \exp\left\{P_h(\phi)\right\}, \quad (13.11)$$

for $(\phi, h) \in \mathcal{J}(\theta, k)$. The process can be simulated as follows. Calculate jump times from (13.7) using a modified jump intensity:

$$q((\theta, k), (\phi, h)) = C p_h \exp\left\{P_h(\phi)\right\}. \quad (13.12)$$

At each jump time, generate a candidate state (ϕ, h) from the prior restricted to $\mathcal{J}(\theta, k)$, i.e. with density proportional to $p_h \exp\{P_h(\phi)\}$, and accept this point with probability

$$\min\left(1, \exp\left\{L_h(\phi) - L_k(\theta)\right\}\right); \quad (13.13)$$

otherwise reject this point and set (ϕ, h) equal to the current point (θ, k). This is just the Metropolis–Hastings algorithm; the modified jump intensity (13.12) generates more frequent jump times than would (13.11), but some of these jump times do not result in a jump.

This choice of jump dynamics is more easily implemented than Gibbs jump dynamics, since generating new states from the prior is generally straightforward. However, it may again be necessary to concentrate the jump intensity (13.11) on jump sets of reduced dimension. For this, (13.12) is replaced by

$$q((\theta', \theta'', k), (\theta', \phi'', h)) = C p_h \exp\left\{P_h(\theta', \phi'')\right\}, \quad (13.14)$$

using the notation described above, and the marginal jump intensity is calculated from (13.10). The examples described in Sections 13.4, 13.5 and 13.7 use Metropolis jump dynamics on reduced-dimension jump sets.

Choice of jump dynamics

We have described two examples of transition kernels which may be used to generate the jumps. For Gibbs jump dynamics, we need to be able to integrate over, and sample directly from, the posterior (or a conditional form thereof). For Metropolis jump dynamics, we need to be able to integrate over, and sample directly from, the prior. In most applications, Metropolis jump dynamics is likely to be more easily implemented, although it may require longer runs if a high proportion of proposed jumps are rejected. In the examples we have looked at (Sections 13.4–13.7), we have mostly used Metropolis jump dynamics. However, for the example in Section 13.6 where we used Gibbs jump dynamics, the algorithm was clearly efficient.

13.3.2 Moving between jumps

At times between jumps, i.e. for $t_{i-1} < t < t_i$, $i = 1, 2, \ldots$, the process follows a Langevin diffusion in the subspace to which the process jumped at the previous jump time. Within the fixed subspace Θ_k, this is given by

$$d\theta^{(t)} = \frac{dt}{2} \left[\frac{d}{d\theta} \left(L_k(\theta) + P_k(\theta) \right) \right]_{\theta^{(t)}} + dW_{n(k)}^{(t)}, \qquad \theta \in \Theta_k,$$

where $W_{n(k)}^{(t)}$ is standard Brownian motion in $n(k)$ dimensions. This can be approximated, using a discrete time-step Δ, by

$$\theta^{(t+\Delta)} = \theta^{(t)} + \frac{\Delta}{2} \frac{d}{d\theta} \left(L_k(\theta^{(t)}) + P_k(\theta^{(t)}) \right) + \sqrt{\Delta} z_{n(k)}^{(t)}, \qquad (13.15)$$

where $z_{n(k)}^{(t)}$ is a vector of $n(k)$ independent standard normal variates. See Gilks and Roberts (1995: this volume) for futher discussion of this simulation.

In the following sections, we illustrate the use of the methodology discussed in this section by considering in turn the four applications introduced in Section 13.2.

13.4 Mixture deconvolution

To apply the above methodology, we first return to the example of Section 13.2.1, where the problem is to estimate the number of components in a normal mixture. Recall that the parameters for the model $\mathcal{M}_k$ are $\{\mu_i, \sigma_i^2, w_i; \ i = 1, \ldots, k\}$, i.e. component means, variances and mixing weights, so that $\theta = (\theta_1, \ldots, \theta_k) \in \Theta_k$ where $\theta_i = (\mu_i, \sigma_i^2, w_i)$.

Within any subspace Θ_k, the diffusion (13.15) is straightforward to simulate, requiring only the calculation of derivatives with respect to θ of $L_k(\theta|x)$ in (13.2) and $P_k(\theta)$ in (13.3). To avoid complications with boundary conditions, we simulate the diffusion on the unconstrained transformed variables

$$\mu_i' = \mu_i, \qquad \sigma_i^{2'} = \log(\sigma_i^2), \qquad \text{for } i = 1, \ldots, k,$$

and

$$w_i' = \log\left(\frac{w_i}{1 - \sum_{j=1}^i w_j}\right), \qquad \text{for } i = 1, \ldots, k-1.$$

For the jump part of the Markov process we use Metropolis dynamics, due to the difficulty in directly sampling from π or any of the conditionals of π. We limit discussion here to the case where $\mathcal{J}(\theta, k) \subset \Theta_{k-1} \cup \Theta_{k+1}$, i.e. we can only increase or decrease the number of components by one at a time. Assume that the current model is $\mathcal{M}_k$, and that the θ_i have been reordered so that $\mu_1 < \mu_2 < \ldots < \mu_k$ (note that π is invariant to such reorderings). The essence of the jump process is as follows.

When adding a new component, i.e. when model $\mathcal{M}_{k+1}$ is under consideration, there are $k+1$ 'positions' where a new component may be added; before component 1, between components 1 and 2, and so on. A mean μ_{k+1} and variance σ_{k+1}^2 for the new component are sampled from the prior, so that $\mu_{k+1} \sim N(\mu, \tau^2)$ and $\sigma_{k+1}^{-2} \sim \text{Ga}(\gamma, \delta)$. To sample a weight w_{k+1} we must first downweight some of the existing weights (which sum to unity). If the component is to be added in the jth position, i.e. $\mu_{j-1} < \mu_{k+1} < \mu_j$, then we sample a triple (w_{j-1}, w_{k+1}, w_j) from their prior conditioned on the remaining $k-2$ weights (w_{j-1} and w_j are the weights of the components positioned to the immediate left and right of the new component). If the new component is to be added before μ_1 or after μ_k, then respectively the pairs (w_{k+1}, w_1) or (w_k, w_{k+1}) are sampled conditioning on the remaining $k-1$ weights. This ensures that $\sum_{i=1}^{k+1} w_i = 1$.

To make a transition to model $\mathcal{M}_{k-1}$ we must remove a component. If component j is to be deleted, then we simply resample (w_{j-1}, w_{j+1}) from the prior conditioned on the remaining $k-3$ weights, where w_{j-1} and w_{j+1} are the weights of the components positioned to the immediate left and right of the deleted component. If the first or k^{th} components are to be deleted, we set $w_2 = 1 - \sum_{i=3}^k w_i$ or $w_{k-1} = 1 - \sum_{i=1}^{k-2} w_i$, respectively. For both types of jump move, we have used reduced-dimensional jump sets.

Integrating the jump intensity (13.14) over the jump sets described above gives $r(\theta, k)$ proportional to

$$(k-1)!\frac{p_{k-1}}{k}\left[s_{k-1,1}w_1 + s_{k-1,k}w_k + \frac{1}{2}\sum_{j=2}^{k-1} s_{k-1,j}(w_{j-1} + w_j + w_{j+1})^2\right]$$

$$+ k!s_k p_{k+1}\left[\Psi_{-\infty,\mu_1} + \Psi_{\mu_k,\infty} + \sum_{j=2}^k \Psi_{\mu_{j-1},\mu_j}(w_{j-1} + w_j)\right],$$

where $s_k = \exp P_k(\theta)$ in (13.3); $s_{k-1,j}$ is of the form of $\exp P_{k-1}(\theta)$, except that the summation in (13.3) goes over $i = 1, \ldots, j-1, j+1, \ldots, k$; $\Psi_{a,b}$ denotes the integral from a to b of a $N(\mu, \tau^2)$ distribution; and all parameters take their values from θ. This expression ignores the restriction (13.5).

Having calculated $r(\theta, k)$ we can generate jump times as described in Section 13.3.1. At the jump times, ϕ can be sampled as described above, and accepted with probability (13.13).

13.4.1 Dataset 1: galaxy velocities

The first dataset we consider is taken from Escobar and West (1995), and is shown in Figure 13.1. The data consists of the velocities at which 82 distant galaxies are diverging from our own galaxy.

Recall that the continuous diffusion must be approximated using a discrete time-step Δ. An appropriate choice of Δ is very much dependent on the specific application, and for this example (and the others to be described) this was chosen on a trial-and-error basis. In the following simulation, we used a value of $\Delta = 0.01$. Kloeden and Platen (1992) offer some theoretical and practical guidelines to assist in this choice in more general settings.

The results from running the jump-diffusion algorithm for an initial period of 100 'burn-in' iterations (time-steps) followed by a further 1 000 iterations are shown in Figure 13.4, where we have plotted sample predictive distributions spaced 50 iterations apart. Figure 13.5 gives the predictive distribution corresponding to the set of parameters with maximum posterior probability.

The sample output of the Markov chain is concentrated on the five models $\mathcal{M}_6$, $\mathcal{M}_7$, $\mathcal{M}_8$, $\mathcal{M}_9$, and $\mathcal{M}_{10}$, for which the frequencies (i.e. estimates of posterior model probabilities) are given in Table 13.1. These differ from the results obtained by Escobar and West (1995) in that more components are favoured here. If the presence of components with small variance and weight (see Figure 13.5) is undesirable, then the prior can be modified appropriately. Our simulations used the hyperparameters $\mu = 20.0$, $\tau^2 = 100.0$, $\gamma = 2.0$, $\delta = 0.5$, and $\lambda = 3.0$. Therefore, we could decrease δ to favour larger variances, and decrease λ to further penalize the addition of extra components. Also, we could assume something other than a uniform prior on the weights.

13.4.2 Dataset 2: length of porgies

For the second demonstration of this technique applied to mixtures, we consider a dataset studied by Shaw (1988), which consists of the lengths of 13 963 porgies (a species of fish), tabulated in Table 13.2.

Notice that the data are only available in grouped form, which creates an additional complication. Let c_j denote the jth class limit (so that $c_1 = 9$,

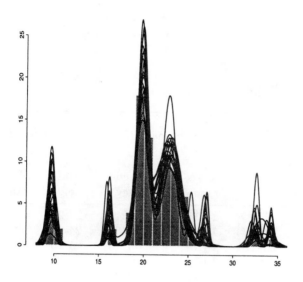

Figure 13.4 *Overlay of 20 sampled predictive densities for galaxy dataset.*

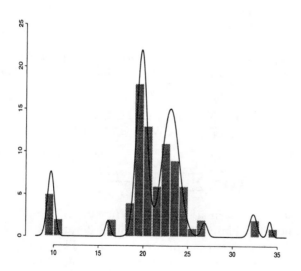

Figure 13.5 *Overlay of predictive density corresponding to maximum posterior estimate for galaxy dataset.*

number of components	estimated posterior probability
6	0.025
7	0.394
8	0.321
9	0.217
10	0.043

Table 13.1 *Estimated model probabilities for galaxy dataset*

class range	observed frequency	class range	observed frequency	class range	observed frequency
9-10	509	16-17	921	23-24	310
10-11	2240	17-18	448	24-25	228
11-12	2341	18-19	512	25-26	168
12-13	623	19-20	719	26-27	140
13-14	476	20-21	673	27-28	114
14-15	1230	21-22	445	28-29	64
15-16	1439	22-23	341	29-30	22

Table 13.2 *Length of porgies, taken from Shaw (1988)*

$c_2 = 10$, and so on), m the number of classes, and n_j the number of lengths in the range $[c_j, c_{j+1})$. Then, the modified likelihood function (Shaw, 1988) takes the form

$$L_k(\theta|x) = \sum_{i=1}^{m} n_i \log \left\{ \frac{F_{k,i+1} - F_{k,i}}{F_{k,n+1} - F_{k,1}} \right\},$$

with

$$F_{k,i} = \sum_{j=1}^{k} w_j \Phi \left(\frac{c_i - \mu_j}{\sigma_j} \right),$$

where Φ is the standard normal distribution function.

To illustrate how the jump-diffusion sampling iteratively refines an initially poor set of parameters (and incorrect model), Figure 13.6 shows the sample predictive densities corresponding to the first 12 iterations. Note that even in as few iterations as this, the algorithm has captured the main structure in the data.

The results from running the jump-diffusion algorithm for 1 000 iterations, after an initial 'burn-in' period of 100 iterations, are shown in Figures 13.7 and 13.8, where we have plotted sample predictive distributions

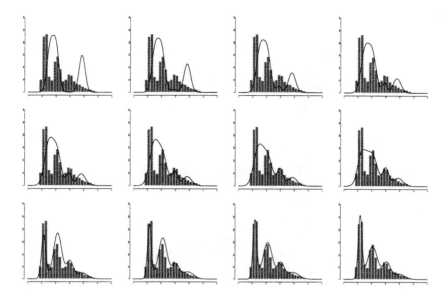

Figure 13.6 *Predictive densities for the first 12 iterations of the jump-sampling algorithm overlaid on the porgie data histogram.*

(spaced 50 iterations apart) and the predictive distribution corresponding to the set of parameters with maximum posterior probability.

Notice that model uncertainty arises in the right-hand tails of the histogram where there is less data. The sample output is concentrated on the four models $\mathcal{M}_4$, $\mathcal{M}_5$, $\mathcal{M}_6$, and $\mathcal{M}_7$, for which the frequencies are given in Table 13.3. This confirms that the model with 5 components is heavily favoured, a result in agreement with the conclusions of Shaw (1988) based on background knowledge (probably five age-groups) of the fish population under study.

number of components	estimated posterior probability
4	0.294
5	0.685
6	0.015
7	0.006

Table 13.3 *Estimated model probabilities for porgie dataset*

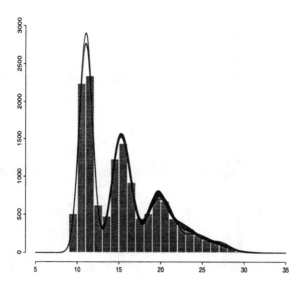

Figure 13.7 *Overlay of* 20 *sampled predictive densities for porgie dataset.*

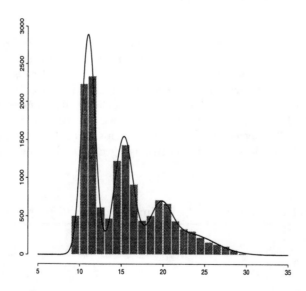

Figure 13.8 *Overlay of predictive density corresponding to maximum posterior estimate for porgie dataset.*

13.5 Object recognition

We now return to the example of Section 13.2.2, in which the inference problem is to identify the positions of an unknown number of objects in an image (a simplified form of the illustrative example used by Grenander and Miller, 1994). Recall that $\mathcal{M}_k$ represents a model with k randomly positioned discs, with centres given by $\theta = (\theta_1, \ldots, \theta_k) \in \Theta_k$. The posterior distribution on $\Theta = \bigcup_{k=0}^{\infty} \Theta_k$ is as specified in Section 13.2.2.

The details of the jump-diffusion sampling algorithm are much as for the previous application. We use Metropolis-type jump dynamics (again due to the difficulty in directly simulating shapes from π), so that at jump times a new model (ϕ, h) is proposed from the prior, and accepted with probability depending on the posterior. We allow the model to change from order k to $k-1$ or $k+1$, which in this context means that a new disc may be added at a random location, or one of the existing discs may be deleted. Note that both the addition and deletion of single discs need not affect the remaining discs. Therefore we define the set $\mathcal{J}(\theta, k)$ as

$$\{h = k - 1 \quad \text{and} \quad (\phi_1, \ldots, \phi_{k-1}) = (\theta_1, \ldots, \theta_{j-1}, \theta_{j+1}, \ldots, \theta_k)\},$$
$$\text{for some } j \in \{1, \ldots, k\}$$

$$\text{or}$$

$$\{h = k + 1 \quad \text{and} \quad (\phi_1, \ldots, \phi_k) = (\theta_1, \ldots, \theta_k)\}.$$

Between jumps, the fixed number of objects are modified by changing their locations according to the diffusion (13.15). Given a time step Δ, this is straightforward to simulate, since the gradient terms may be easily approximated using standard image-processing algorithms (see, for example, Pratt, 1991).

13.5.1 Results

We demonstrate the above procedure by the analysis of a particular synthesised image. To illustrate how the jump-diffusion sampling iteratively refines an initially poor set of parameters (and incorrect model), Figure 13.9 shows a sequence of iterates corresponding to the first 12 jumps. Note how the initially placed discs are quickly discarded, and how correctly-positioned discs are subsequently introduced.

Figure 13.10 shows the picture corresponding to the set of parameters (and corresponding model) with maximum posterior probability in a sample of 1 000 extracted from the sample output of the Markov chain. The sample is concentrated on the two models $\mathcal{M}_8$ and $\mathcal{M}_9$, with frequencies 0.779 and 0.221, which seems reasonable given that the actual number of discs in the simulated image is 8. We used a value of $\lambda = 6$, corresponding to a prior mean of 6.

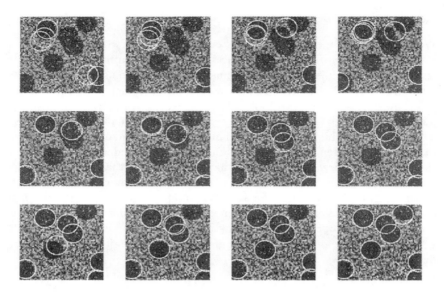

Figure 13.9 *Sequence of iterates showing the jumps from the sample output of the jump-sampling algorithm.*

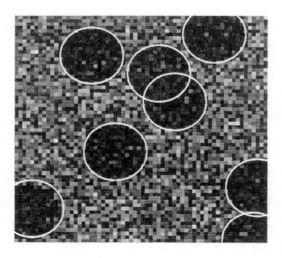

Figure 13.10 *Picture corresponding to the maximum posterior estimate of the number of discs and their locations.*

13.6 Variable selection

We now return to the example of Section 13.2.3, concerning variable selection in regression models. Recall that we have a response variable x_0 and a number of covariates x_1 to x_m, and we want to decide which combination of the x_i best predicts x_0.

This application provides a good illustration of a situation where Gibbs is preferable to Metropolis jump dynamics, since given the model $\mathcal{M}_{c(k)}$, it is straightforward to integrate over, and sample directly from, restricted forms of the posterior.

We choose $\mathcal{J}(\theta, c(k))$ to limit the available jump moves from model $\mathcal{M}_{c(k)}$ to include only those models $\mathcal{M}_{c'(h)}$ for which $c'(h)$ can be obtained from $c(k)$ by adding or deleting a single entry. For example, with $m = 3$, we can only jump from $\mathcal{M}_{(1,2)}$ to $\mathcal{M}_{(1)}$, $\mathcal{M}_{(2)}$, and $\mathcal{M}_{(1,2,3)}$. We also fix the variance σ^2 for the jump moves.

We consider a dataset analysed by George and McCulloch (1993), relating to the effect of 4 factors on the contentment of a group of 39 workers. Table 13.4 shows the posterior frequencies for each model visited, again derived as ergodic averages from a run of length 1 000 of the jump-diffusion algorithm (with hyperparameter settings $\mu = 0.0$, $\tau^2 = 10.0$, $\gamma = 5.0$, $\delta = 5.0$ and $\lambda = 2.0$). The two most likely models are $\mathcal{M}_{(3,4)}$ and $\mathcal{M}_{(1,3,4)}$, which is in agreement with the results obtained by George and McCulloch (1993).

model variables	estimated posterior probability
(3)	0.018
(1,3)	0.044
(2,3)	0.006
(3,4)	0.267
(1,2,3)	0.015
(1,3,4)	0.399
(2,3,4)	0.078
(1,2,3,4)	0.173

Table 13.4 *Estimated model probabilities for worker contentment dataset*

13.7 Change-point identification

As a final application, we return to the example of Section 13.2.4, which involves the identification of an unknown number of change-points in a sequence. Our analysis extends that of Stephens (1994), who conditions on a known maximum number of change-points. Recall that $\mathcal{M}_k$ represents the model with k randomly positioned change-points, and that

$$\theta = (c_1, \ldots, c_k, \mu_1, \sigma_1^2, \ldots, \mu_{k+1}, \sigma_{k+1}^2).$$

i.e. change-points and means and variances within homogeneous segments. Details of the posterior density are given in Section 13.2.4.

The implementation of the jump-diffusion sampling algorithm follows along similar lines to that described for the analysis of mixtures in Section 13.4. We allow moves to increase or decrease the number of change-points by one. Adding a change-point involves splitting a subsection of the series, say $\{j; \ c_i < j \leq c_{i+1}\}$, into two and replacing $(\mu_{i+1}, \sigma_{i+1}^2)$ by parameters suitable for each part. Deleting a change-point involves merging two subsections, say $\{j; \ c_i < j \leq c_{i+1}\}$ and $\{j; \ c_{i+1} < j \leq c_{i+2}\}$, into one and replacing $(\mu_{i+1}, \sigma_{i+1}^2)$ and $(\mu_{i+2}, \sigma_{i+2}^2)$ by a single mean and variance parameter appropriate for $(x_{c_i+1}, \ldots, x_{c_{i+2}})$. As for the case of mixtures, we use Metropolis rather than Gibbs jump dynamics (due to the difficulty in the latter of directly sampling from either π or its conditionals), and define $\mathcal{J}(\theta, k)$ to fix those parts of θ not directly affected by the jumps.

13.7.1 Dataset 1: Nile discharge

We begin by analysing the Nile data, shown in Figure 13.3. The sample output from a run of length $1\,000$ of the jump-diffusion algorithm is concentrated on the two models $\mathcal{M}_0$ and $\mathcal{M}_1$, corresponding to none versus one change-point. The respective probabilities are 0.054 and 0.946, therefore heavily favouring one change-point. These results are in agreement with those obtained by other authors (see for example Freeman, 1986), who found evidence for exactly one change-point.

Figure 13.11 shows the predictive distribution corresponding to those parameters with maximum posterior probability overlayed on the underlying series. Conditional on the one change-point model, Figure 13.12 shows the marginal posterior histogram of the change-point position, based on an approximate posterior sample of size 100 derived from the sample output by retaining every 10th iteration.

13.7.2 Dataset 2: facial image

This approach to identifying change-points in a sequence can be adapted to find edges in an image, simply by treating each image line as a separate sequence to be processed (see Stephens and Smith, 1993). We analysed the image shown in the left-hand panel of Figure 13.13, and obtained the results shown in the right-hand panel, after running the jump-diffusion algorithm independently for 50 iterations on each image row and column. This produces an acceptable edge map, and compares favourably with the results obtained by Stephens and Smith (1993), who conditioned on a fixed

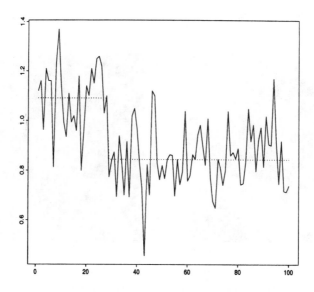

Figure 13.11 *Maximum posterior estimate of change-point position and mean parameters (dotted line).*

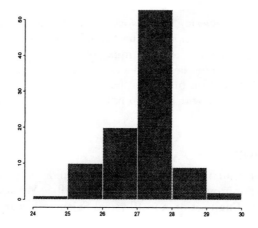

Figure 13.12 *Marginal posterior histogram for change-point position conditioned on the model $\mathcal{M}_1$.*

number of change-points.

Figure 13.13 *Image (left-hand panel) and corresponding estimated change-points (right-hand panel).*

13.8 Conclusions

We have illustrated a Bayesian framework for model choice problems using jump-diffusion sampling, demonstrating the effectiveness of this approach for a range of applications. There are many other problems which may be analysed using the methods described here. In particular, we can approximate unknown distribution functions 'non-parametrically' using suitably constructed mixtures, for which we do not have to specify in advance the number of components.

Acknowledgements

We have benefited from discussions with Ulf Grenander and Michael Miller, and from editorial improvements made by Wally Gilks. We acknowledge the support of the UK Engineering and Physical Science Research Council.

References

Amit, Y., Grenander, U. and Miller, M. I. (1993) Ergodic properties of jump-diffusion processes. *Technical Report 361*, Department of Statistics, University of Chicago.

Escobar, M. D. and West, M. (1995) Bayesian density estimation and inference using mixtures. *J. Am. Statist. Ass.*, **90**, 577–588.

Freeman, J. M. (1986) An unknown change point and goodness of fit. *Statistician*, **35**, 335–344.

George, E. I. and McCulloch, R. E. (1993) Variable selection via Gibbs sampling. *J. Am. Statist. Ass.*, **88**, 881–889.

George, E. I. and McCulloch, R. E. (1995) Stochastic search variable selection. In *Markov Chain Monte Carlo in Practice* (eds W. R. Gilks, S. Richardson and D. J. Spiegelhalter), pp. 203–214. London: Chapman & Hall.

Gilks, W. R. and Roberts, G. O. (1995) Strategies for improving MCMC. In *Markov Chain Monte Carlo in Practice* (eds W. R. Gilks, S. Richardson and D. J. Spiegelhalter), pp. 89–114. London: Chapman & Hall.

Green, P. J. (1994a) Discussion on representations of knowledge in complex systems (by U. Grenander and M. I. Miller). *J. R. Statist. Soc.* B, **56**, 589–590.

Green, P. J. (1994b) Reversible jump MCMC computation and Bayesian model determination. *Technical Report 5-94-03*, University of Bristol.

Grenander, U. and Miller, M. I. (1991) Jump-diffusion processes for abduction and recognition of biological shapes. Technical report, Electronic Signals and Systems Research Laboratory, Washington University.

Grenander, U. and Miller, M. I. (1994) Representations of knowledge in complex systems, *J. R. Statist. Soc.* B, **56**, 549–603.

Kloeden, P. E. and Platen, E. (1992) *Numerical Solution of Stochastic Differential Equations*. Berlin: Springer-Verlag.

Pratt, W. K. (1991) *Digital Image Processing*, 2nd edn. New York: Wiley.

Raftery, A. E. (1995) Hypothesis testing and model selection. In *Markov Chain Monte Carlo in Practice* (eds W. R. Gilks, S. Richardson and D. J. Spiegelhalter), pp. 163–187. London: Chapman & Hall.

Robert, C. P. (1995) Mixtures of distributions: inference and estimation. In *Markov Chain Monte Carlo in Practice* (eds W. R. Gilks, S. Richardson and D. J. Spiegelhalter), pp. 441–464. London: Chapman & Hall.

Shaw, J. E. H. (1988) Aspects of numerical integration and summarisation. In *Bayesian Statistics 3* (eds J. M. Bernardo, M. H. DeGroot, D. V. Lindley, and A. F. M. Smith), pp. 411–428. Oxford: Oxford University Press.

Stephens, D. A. (1994) Bayesian retrospective multiple-changepoint identification. *Appl. Statist.*, **43**, 159–178.

Stephens, D. A. and Smith, A. F. M. (1993) Bayesian edge-detection in images via changepoint methods. In *Computing Intensive Methods in Statistics* (eds W. Härdle and L. Simar), pp. 1–29. Heidelberg: Physica-Verlag.

14

Estimation and optimization of functions

Charles J Geyer

14.1 Non-Bayesian applications of MCMC

Non-Bayesian applications of MCMC are highly varied and will become increasingly important as time goes on. Perhaps there are important applications yet to be invented. MCMC is a powerful numerical method of doing probability calculations when analytic procedures fail. So it would seem that MCMC should eventually have applications in all areas of statistics. Anyone who needs to calculate a probability or expectation that cannot be done analytically should ponder how MCMC might help.

So far, the major non-Bayesian applications have been to hypothesis testing (Besag and Clifford, 1989, 1991; Guo and Thompson, 1992; Diaconis and Sturmfels, 1993) and to likelihood inference (Ogata and Tanemura, 1981, 1984, 1989; Penttinen, 1984; Strauss, 1986; Younes, 1988; Moyeed and Baddeley, 1991; Geyer and Thompson 1992, 1995; Geyer 1993, 1994; Thompson and Guo, 1991; Guo and Thompson, 1991, 1994; Geyer and Møller, 1994). A new area of application to optimal decisions in sequential control problems has been opened by Chen *et al.* (1993).

Though all of these applications are important, they are not closely related. Hence this chapter will only cover the estimation and maximization or minimization of functions by Monte Carlo, as arises in likelihood inference and decision theory.

14.2 Monte Carlo optimization

There are two distinct notions that might be called 'Monte Carlo optimization'. One is *random search optimization* and the other is optimization in which the objective function is calculated by Monte Carlo. It is the latter that will be called 'Monte Carlo optimization' here.

In random search optimization, the problem is to maximize or minimize some objective function $w(x)$ which is known and can be calculated exactly. Because the function may have many local maxima, it is feared that simple hill-climbing strategies like Newton–Raphson will fail to find the global maximum, hence the objective function is evaluated at a large number of points x_1, x_2, ... and the x_i with the largest value of $w(x_i)$ is taken to approximate the global maximizer. In random search optimization, the x_i are generated by some random process. Examples are simulated annealing and genetic algorithms. The randomness in the search pattern is important, since random search tends to beat deterministic search in high-dimensional spaces, but the exact sampling distribution of the x_i is unimportant. Any distribution which puts appreciable probability in neighbourhoods of the global maximum will do.

In what we are calling 'Monte Carlo optimization', the problem is to maximize or minimize an objective function $w(x)$ whose exact form is unknown. It is specified by an expectation*

$$w(x) = E_x g(x, Y) = \int g(x, y) \, dP_x(y), \tag{14.1}$$

where $g(x, y)$ is a known function, the distribution P_x of the random variable Y may depend on x, and the expectation (14.1) cannot be calculated exactly.

There are two ways to go about solving such a problem using Monte Carlo evaluation of the integral. The first involves obtaining a new Monte Carlo sample for each evaluation of (14.1). That is, whenever we want to evaluate $w(x)$ for some x, we obtain a sample $Y_1, Y_2, \ldots$ from P_x and calculate

$$w_n(x) = \frac{1}{n} \sum_{i=1}^{n} g(x, Y_i). \tag{14.2}$$

Here and elsewhere in this chapter, the sample need not be independent; it may be an MCMC sample. The second involves *importance sampling*. Suppose there exists a distribution Q which does not depend on x, such

---

* Readers should not be frightened by the measure-theoretic notation here. Integration with respect to the probability measure P_x is used here only to avoid one formula for the discrete case where it denotes a finite sum $\sum_y g(x, y) p_x(y)$ and another formula for the continuous case where it denotes an integral $\int g(x, y) p_x(y) \, dy$. In some problems, y is a vector with both discrete and continuous components, and (14.1) represents a sum over the discrete components and integration over the continuous components.

that each of the P_x has a density $f_x(y)$ with respect to $Q^\dagger$. Then

$$w(x) = \int g(x,y)f_x(y)\,dQ(y). \tag{14.3}$$

So if Y_1, Y_2, ... is a sample from Q,

$$w_n(x) = \frac{1}{n}\sum_{i=1}^{n} g(x,Y_i)f_x(Y_i) \tag{14.4}$$

is an approximation to $w(x)$. Differentiating (14.4) also gives approximations to derivatives of $w(x)$ if (14.3) can be differentiated under the integral sign:

$$\frac{\partial w(x)}{\partial x_k} = \int \left(\frac{\partial g(x,y)}{\partial x_k}f_x(y) + g(x,y)\frac{\partial f_x(y)}{\partial x_k}\right)\,dQ(y)$$

is approximated by

$$\frac{\partial w_n(x)}{\partial x_k} = \frac{1}{n}\sum_{i=1}^{n}\left(\frac{\partial g(x,Y_i)}{\partial x_k}f_x(Y_i) + g(x,Y_i)\frac{\partial f_x(Y_i)}{\partial x_k}\right), \tag{14.5}$$

and similarly for higher-order derivatives. Note that it makes no sense to differentiate (14.2) to approximate derivatives, since there the distribution of the Y_i depends on x.

We call the first method a 'many samples' method because it requires one sample per evaluation of the objective function. It is very inefficient compared to the second method, which we call a 'single-sample' method because it uses the same sample for all evaluations of the objective function. For a fixed sample Y_1, Y_2, ... from Q, (14.4) approximates $w(x)$ for all x: the function w_n defined by (14.4) approximates the function w defined by (14.1). The same cannot be said of (14.2) because the distribution P_x of the sample Y_1, Y_2, ... depends on x. So each sample can only be used for the evaluation of $w(x)$ at one point x. This is what makes the many-samples method inefficient.

Despite their inefficiency, many-samples methods can be used. Diggle and Gratton (1984) proposed many-samples methods for maximum likelihood estimation in complex models. The approximation to the objective function given by (14.2) is exceedingly rough since the Monte Carlo errors in $w_n(x_1)$ and $w_n(x_2)$ are independent no matter how close x_1 is to x_2. Hence there is no hope of approximating derivatives. Thus Diggle and Gratton (1984) proposed evaluating $w(x)$ on a grid, smoothing the evaluates using a kernel smoother, and then maximizing the resulting smooth approximation.

A method that is somewhere between 'many samples' and 'single-sample' is Monte Carlo Newton–Raphson proposed by Penttinen (1984) in which

---

$\dagger$ This means $f_x(y) = p_x(y)/q(y)$ where p_x and q are the density functions for P_x and Q respectively. This assumes that $p_x(y) = 0$ whenever $q(y) = 0$; otherwise P_x does not have a density with respect to Q.

the gradient and Hessian (second-derivative matrix) of the objective function are evaluated by Monte Carlo using (14.5) and its higher-order analogue in each iteration, but a new sample is used for each iteration. Though this method is more efficient than that of Diggle and Gratton (1984), it is still not as efficient as single-sample methods. Besides the inefficiency, neither of these procedures is susceptible to theoretical analysis. It is difficult to prove convergence of the algorithms or estimate the size of the Monte Carlo error.

A third method attempts to maximize the objective function without evaluating it. This is the method of *stochastic approximation*, also called the Robbins–Munro algorithm (Wasan, 1969). Explaining this method would take us away from our main theme, so we just note that this method has been used for maximum likelihood by Younes (1988) and by Moyeed and Baddeley (1991). This method is also inefficient compared to the single-sample method. Moreover, it provides only a maximum likelihood estimate (MLE), not estimates of likelihood ratios or Fisher information. Hence it provides no help in calculating tests or confidence intervals. Still, this method may be used to obtain an initial crude MLE that can then be improved by the single-sample method (Younes, 1992).

14.3 Monte Carlo likelihood analysis

As many of the other chapters of this book attest, MCMC has become a miraculous tool of Bayesian analysis. Once a likelihood and prior are accepted, any desired Bayesian calculation can be accomplished by MCMC. This gives workers a new freedom in choice of models and priors and calls for a 'model liberation movement' (Smith, 1992) away from models 'assumed for reasons of mathematical convenience' to models of actual scientific interest (and also for a 'prior liberation movement' to priors that reflect actual subjective judgement).

It has been less widely appreciated, perhaps because of the limitations of early work, that likelihood analysis is similarly liberated. It does not matter if densities contain unknown normalizing constants, if there are missing data, or if there is conditioning on ancillary statistics. Likelihood analysis can still be done. Maximum likelihood estimators and likelihood ratio statistics can be calculated. Their asymptotic distributions can be calculated, or their exact sampling distributions can be approximated by a parametric bootstrap. Profile likelihoods, support regions (level sets of the likelihood), and other tools of likelihood analysis can also be calculated. In the single-sample approach, all of this is done using a single run of the Markov chain.

14.4 Normalizing-constant families

Statistical models with complex dependence have become increasingly important in recent years. Such models arise in spatial statistics, statistical genetics, biostatistics, sociology, psychological testing, and many other areas (Geyer and Thompson, 1992). Even for fairly simple models, missing data or conditioning can produce complex dependence. Many of these models are included in the following general framework.

Let $\mathcal{H} = \{h_\theta : \theta \in \Theta\}$ be a family of *unnormalized densities* with respect to some measure μ on some sample space, indexed by an unknown parameter θ. That is, each h_θ is a nonnegative function on the sample space such that

$$c(\theta) = \int h_\theta(x)\, d\mu(x) \qquad (14.6)$$

is nonzero and finite, and

$$f_\theta(x) = \frac{1}{c(\theta)} h_\theta(x) \qquad (14.7)$$

is a probability density with respect to μ. Then $\mathcal{F} = \{f_\theta : \theta \in \Theta\}$ is a (normalized) family of densities with respect to μ. The function $c(\theta)$ is called in rather inappropriate terminology the 'normalizing constant' of the family. One should keep in mind that it is not a constant but a function on the parameter space. One should think of the unnormalized densities as being known functions and the normalizing constant as being an unknown function, the integral (14.6) being analytically intractable.

Some familiar statistical models are special cases of normalizing-constant families. The most familiar are exponential families where $h_\theta(x)$ takes the form $e^{t(x) \cdot \theta}$ for some vector valued statistic $t(x)$ and parameter θ. Another special case comprises Gibbs distributions, which gave the Gibbs sampler its name, used in spatial statistics and image processing. This is really the same case under a different name since finite-volume Gibbs distributions are exponential families. A slightly more general case comprises Markov random fields, more general because the unnormalized densities can be zero at some points, as in hard-core point processes.

A rather different case arises from conditioning. Suppose $g_\theta(x, y)$ is a family of bivariate densities, and we wish to perform conditional likelihood inference, conditioning on the statistic y. The conditional family has the form

$$f_\theta(x|y) = \frac{g_\theta(x, y)}{\int g_\theta(x, y)\, dx}$$

which has the form (14.7) with $h_\theta(x) = g_\theta(x, y)$, the variable y being considered fixed at the observed value.

The notion of a normalizing-constant family is much larger than these specific examples. It permits an enormous scope for modelling. Just make up any family of nonnegative functions of the data and parameter. If they are integrable, they specify a model. It might be thought that the scope is too large. How can any inference be done when there is so little mathematical structure with which to work?

For any fixed parameter value θ, we can sample from the distribution P_θ having density f_θ with respect to μ using the Metropolis–Hastings algorithm, without any knowledge of the normalizing constant. That is what the Metropolis–Hastings algorithm does: sample from densities known up to a constant of proportionality.

Since we can simulate any normalizing-constant family, we also want to carry out inference. A Monte Carlo approximation to the likelihood can be constructed (Geyer and Thompson, 1992) using MCMC simulations. Let ψ be any point in the parameter space Θ such that h_ψ dominates h_θ for all θ, that is $h_\psi(x) = 0$ implies $h_\theta(x) = 0$ for almost all x. Let $X_1, X_2, \ldots, X_n$ be simulations from P_ψ obtained by MCMC. The log likelihood, or more precisely the log likelihood ratio between the points θ and ψ, corresponding to the observation x is

$$
\begin{aligned}
l(\theta) &= \log \frac{f_\theta(x)}{f_\psi(x)} \\[2mm]
&= \log \frac{h_\theta(x)}{h_\psi(x)} - \log \frac{c(\theta)}{c(\psi)} \\[2mm]
&= \log \frac{h_\theta(x)}{h_\psi(x)} - \log E_\psi \frac{h_\theta(X)}{h_\psi(X)}
\end{aligned} \qquad (14.8)
$$

because

$$
\frac{c(\theta)}{c(\psi)} = \frac{1}{c(\psi)} \int h_\theta(x)\, d\mu(x) = \int \frac{h_\theta(x)}{h_\psi(x)} f_\psi(x)\, d\mu(x).
$$

A Monte Carlo approximation to (14.8) is obtained by replacing the expectation with its empirical approximation

$$
l_n(\theta) = \log \frac{h_\theta(x)}{h_\psi(x)} - \log \left(\frac{1}{n} \sum_{i=1}^{n} \frac{h_\theta(X_i)}{h_\psi(X_i)} \right). \qquad (14.9)
$$

The general heuristic for Monte Carlo likelihood analysis is to treat l_n as if it were the exact log likelihood l, following the program laid out in Section 14.2. The Monte Carlo MLE $\hat{\theta}_n$ is the maximizer of l_n. It approximates the exact MLE $\hat{\theta}$, the maximizer of l. The Monte Carlo approximation of the observed Fisher information $-\nabla^2 l(\theta)$ is $-\nabla^2 l_n(\theta)$[‡]. Similarly for other objects: we replace profiles of l with profiles of l_n, level sets of l (likelihood-based confidence sets) with level sets of l_n. Geyer (1994) gives details.

[‡] $\nabla l(\theta)$ denotes the gradient vector $\frac{\partial l(\theta)}{\partial \theta}$, and $\nabla^2 l(\theta)$ denotes the Hessian matrix whose $(i,j)^{th}$ element is $\frac{\partial^2 l(\theta)}{\partial \theta_i \partial \theta_j}$.

Typically $\hat{\theta}_n$ is found by solving $\nabla l_n(\theta) = 0$, then an approximation to the Monte Carlo error in $\hat{\theta}_n$ is found as follows. Let

$$\sqrt{n}\nabla l_n(\theta) \xrightarrow{\mathcal{D}} N\big(0, A(\theta)\big) \qquad (14.10)$$

be the Markov chain central limit theorem for the score (see Tierney, 1995: this volume). Equation (14.10) says that, as the MCMC sample size n increases, the vector $\sqrt{n}\nabla l_n(\theta)$ converges in distribution to a multivariate normal distribution, with mean vector zero and variance-covariance matrix $A(\theta)$. Let $B(\theta) = -\nabla^2 l(\theta)$ be the observed Fisher information. Then, under suitable regularity conditions (Geyer, 1994),

$$\sqrt{n}(\hat{\theta}_n - \hat{\theta}) \xrightarrow{\mathcal{D}} N\big(0, B(\hat{\theta})^{-1}A(\hat{\theta})B(\hat{\theta})^{-1}\big). \qquad (14.11)$$

Thus the Monte Carlo MLE $\hat{\theta}_n$ converges to $\hat{\theta}$ as $n \to \infty$, and the variance of $\hat{\theta}_n$ is approximately $n^{-1}B(\hat{\theta})^{-1}A(\hat{\theta})B(\hat{\theta})^{-1}$. A consistent estimator of $B(\hat{\theta})$ is $-\nabla^2 l_n(\hat{\theta}_n)$.

There are as many different ways of estimating $A(\theta)$ as there are methods of estimating the variance of the sample mean of a time series or Markov chain. Geyer (1992) gives a review. For chains with a regeneration point, it is also possible to use regeneration methods (Mykland *et al.*, 1995; Geyer and Thompson, 1995).

Space does not permit an exhaustive treatment of variance estimation, so only a few comments will be given on methods using direct estimation of autocorrelations. We provide an estimate of the variance matrix $A(\theta)$ in (14.10). From (14.9),

$$\nabla l_n(\theta) = \frac{\frac{1}{n}\sum_{i=1}^n \left[\big(t_\theta(x) - t_\theta(X_i)\big)\frac{h_\theta(X_i)}{h_\psi(X_i)}\right]}{\frac{1}{n}\sum_{k=1}^n \frac{h_\theta(X_k)}{h_\psi(X_k)}}, \qquad (14.12)$$

where $t_\theta(X) = \nabla \log h_\theta(X)$. The denominator converges to a constant $c(\theta)/c(\psi)$. The numerator is the sample mean of a mean-zero time series $Z_1, Z_2, \ldots, Z_n$, where

$$Z_i = \big(t_\theta(x) - t_\theta(X_i)\big)\frac{h_\theta(X_i)}{h_\psi(X_i)}. \qquad (14.13)$$

Note that the definition of Z_i depends on θ though this is not indicated in the notation. The asymptotic variance of $\sqrt{n}$ times the sample mean of this time series is typically the sum of the autocovariances

$$\Sigma(\theta) = \text{var}(Z_i) + 2\sum_{k=1}^{\infty} \text{cov}(Z_i, Z_{i+k}) \qquad (14.14)$$

where the variances and autocovariances here refer to the stationary chain and are interpreted as matrices if Z_i is a vector. Equation (14.14) can be estimated by a weighted combination of the empirical variances and autocovariances

$$\hat{\Sigma}(\theta) = \widehat{\mathrm{var}}(Z_i) + 2\sum_{k=1}^{\infty} w(k)\widehat{\mathrm{cov}}(Z_i, Z_{i+k}) \qquad (14.15)$$

where a $\hat{\ }$ denotes the usual empirical estimate and $w(k)$ is a weight function decreasing from 1 to 0 as k increases. Many weight functions have been proposed in the time-series literature (Priestley, 1981, pp. 437 and 563). Thus $A(\theta)$ in (14.10) is estimated by

$$\left(\frac{c(\psi)}{c(\theta)}\right)^2 \hat{\Sigma}(\theta) \approx \left(\frac{1}{n}\sum_{k=1}^{n}\frac{h_\theta(X_k)}{h_\psi(X_k)}\right)^{-2}\hat{\Sigma}(\theta), \qquad (14.16)$$

(Geyer, 1994).

The main point of (14.12–14.15) is not their specific form, since other schemes are possible. What is important is the recognition that the size of the Monte Carlo error can be estimated. It must be emphasized that

$$n^{-1}B(\theta)^{-1}A(\theta)B(\theta)^{-1}$$

is the estimated variance of the Monte Carlo error in calculating $\hat{\theta}$, i.e. the difference between the Monte Carlo approximation to the MLE $\hat{\theta}_n$ and the exact MLE $\hat{\theta}$ which we are unable to calculate. It is an entirely different problem to estimate the difference between the exact MLE $\hat{\theta}$ and the true parameter value θ_0. Under the usual regularity conditions for maximum likelihood, this is the inverse Fisher information, which is estimated by $B(\theta)^{-1}$.

For complex models, the usual asymptotics of maximum likelihood may not hold and $B(\theta)^{-1}$ may not be a reasonable estimator (or at least may not be *known* to be reasonable). Geyer and Møller (1994) give an example. In that case, the sampling distribution of the MLE may be approximated by a parametric bootstrap (Geyer, 1991; Geyer and Møller, 1994). Here too, the single-sample method provides efficient sampling:

Step 1: Generate a MCMC sample $X_1, X_2, \ldots, X_n$ from P_ψ. Calculate $\hat{\theta}_n$, the maximizer of (14.9).

Step 2: Generate bootstrap samples $X_1^*, X_2^*, \ldots, X_m^*$ from $P_{\hat{\theta}_n}$. For each $k = 1, \ldots, m$, calculate $\hat{\theta}_{n,k}^*$, the maximizer of $l_n(\,\cdot\,; X_k^*)$, where $l_n(\,\cdot\,; X_k^*)$ denotes $l_n(\theta)$ in (14.9) evaluated at $x = X_k^*$.

Then $\hat{\theta}_{n,1}^*, \ldots, \hat{\theta}_{n,m}^*$ approximates the sampling distribution of the MLE. Importance sampling has made it possible to calculate a parametric bootstrap using only two Monte Carlo samples, rather than one for each $\hat{\theta}_{n,k}^*$. With further use of the importance sampling formula the entire calculation could be done using only one sample, but for once we refrain from insisting on a single sample. The point is that m bootstrap samples can be calculated without obtaining m Monte Carlo samples.

This ability to bootstrap efficiently completes our toolkit for Monte Carlo likelihood analysis. Of course, when the asymptotics of the MLE are unknown, the usual asymptotic justification of the bootstrap is unavailable. So we can only claim that the procedure is a reasonable application of the bootstrap heuristic. Questions of whether the bootstrap 'works' will remain open until the asymptotics are understood.

14.5 Missing data

Missing data has always been a problem for likelihood inference. The EM algorithm (Dempster *et al.*, 1977) provides a method for finding maximum likelihood estimates in some problems, but not all. It requires closed-form expressions for the conditional expectations in the 'E step'. Moreover, since EM in its basic form does not evaluate the likelihood or its derivatives, it does not calculate the Fisher information. There are methods (Sundberg, 1974; Louis, 1982; Meng and Rubin, 1991) that can be used in conjunction with EM to do this, but they require more conditional expectations to be known in closed form or are slow. Monte Carlo EM (Wei and Tanner, 1990; Guo and Thompson, 1991) can be used when conditional expectations are not known in closed form, but is very slow. EM, though useful, does not do all missing data problems and does does not give us all we want in the problems it does.

Monte Carlo maximum likelihood can be used to do any missing data problem in which the complete data likelihood is known (Thompson and Guo, 1991) or known up to a normalizing constant (Gelfand and Carlin, 1993). Moreover, as was the case with normalizing-constant families, it can be used to calculate all useful objects: the Fisher information, likelihood ratios, profile likelihoods, support regions, parametric bootstraps, etc.

Consider first the case where the normalizing constant is known: $f_\theta(x, y)$ is the joint density with x missing and y observed. Again, let ψ be some point in the parameter space such that $f_\psi(x, y) = 0$ implies $f_\theta(x, y) = 0$ for almost all x and the observed y. Then the log likelihood is

$$l(\theta) = \log \frac{f_\theta(y)}{f_\psi(y)} = \log E_\psi \left\{ \frac{f_\theta(X, Y)}{f_\psi(X, Y)} \,\bigg|\, Y = y \right\}. \qquad (14.17)$$

The natural Monte Carlo approximation of the log likelihood is now

$$l_n(\theta) = \log \left(\frac{1}{n} \sum_{i=1}^{n} \frac{f_\theta(X_i, y)}{f_\psi(X_i, y)} \right) \qquad (14.18)$$

where $X_1, X_2, \ldots, X_n$ are realizations from the conditional distribution of X given $Y = y$ and parameter value ψ, typically simulated using the Metropolis–Hastings algorithm, since the normalizing constant $f_\psi(y)$, being the likelihood we want to approximate, is unknown.

All of the procedures discussed in Section 14.4 can also be applied here. The maximizer of l_n approximates the maximizer of l, the Monte Carlo error can be derived from a central limit theorem as in (14.12–14.16), and so forth. In particular, the observed Fisher information is approximated by $-\nabla^2 l_n(\hat{\theta}_n)$.

If we shift our perspective slightly, the same scheme is seen to do any ordinary (non-Bayes) empirical Bayes problem. Let $f(x, y|\theta)$ be the likelihood (as before x is missing and y is observed) and $\pi(\theta|\phi)$ the prior for θ, where ϕ is a hyperparameter. We want to estimate ϕ by maximizing the marginal likelihood $p(y|\phi)$. But this is just a missing data problem in disguise. What was the complete data likelihood $f_\theta(x, y)$ above is now the joint density of the data and parameter $f(x, y|\theta)\pi(\theta|\phi)$; what was the missing data x is now the pair (x, θ) consisting of both missing data and parameter; and what was the parameter θ is now the hyperparameter ϕ. With these identifications, we see that the problem is to find the ϕ that maximizes the analogue of the complete data likelihood with the analogue of the missing data integrated out.

Thus we see that missing data is no obstacle to Monte Carlo likelihood methods. This is even true in normalizing-constant families (Gelfand and Carlin, 1993), though the analysis becomes more complicated. Now the joint density up to an unknown normalizing constant is $h_\theta(x, y)$ with x missing and y observed. Again, fix a parameter value ψ so that $h_\psi(x, y) = 0$ implies $h_\theta(x, y) = 0$ for almost all x and y and also for almost all x when y is fixed at the observed value. Now the log likelihood is

$$l(\theta) = \log E_\psi \left(\frac{h_\theta(X, Y)}{h_\psi(X, Y)} \,\bigg|\, Y = y \right) - \log E_\psi \frac{h_\theta(X, Y)}{h_\psi(X, Y)} \tag{14.19}$$

and its natural Monte Carlo approximation is

$$l_n(\theta) = \log \left(\frac{1}{n} \sum_{i=1}^{n} \frac{h_\theta(X_i^\star, y)}{h_\psi(X_i^\star, y)} \right) - \log \left(\frac{1}{n} \sum_{j=1}^{n} \frac{h_\theta(X_j, Y_j)}{h_\psi(X_j, Y_j)} \right) \tag{14.20}$$

where $X_1^\star$, $X_2^\star$, ..., X_n^* are samples from the conditional distribution of X given $Y = y$ and (X_1, Y_1), (X_2, Y_2), ..., (X_n, Y_n) are samples from the unconditional distribution (both for the parameter value ψ).

The necessity for two samples makes this situation more complex than the preceding two. Moreover the difference of two Monte Carlo expectations in (14.20) requires larger Monte Carlo sample sizes. Still this method must work given large enough sample sizes, and all of the associated methods from Section 14.4 can also be used.

14.6 Decision theory

Likelihoods are not the only objective functions defined by expectations that statisticians maximize or minimize. Many decision problems have no closed-form solutions, and Monte Carlo optimization can be used to obtain approximate solutions. This is particularly true in optimal control problems (Chen *et al.*, 1993) where the predictive distribution of the next observation tends to be complex, and the optimal control setting cannot be found in closed form. The general situation is as follows. There is some probability distribution P and a loss function $L(w, x)$, a function of an action w and a random variable X having distribution P. The problem is to find the action w which minimizes the expected loss

$$r(w) = EL(w, X) = \int L(w, x) \, dP(x) \qquad (14.21)$$

having the natural Monte Carlo approximation

$$r_n(w) = \frac{1}{n} \sum_{i=1}^{n} L(w, X_i), \qquad (14.22)$$

where $X_1, X_2, \ldots, X_n$ are samples from P. In the problems studied by Chen *et al.* (1993), the distribution P, the one-step-ahead predictive distribution of the data, did not depend on the action w, the control setting at the next time period. Hence (14.22) can be calculated for different values of w using the sample $X_1, X_2, \ldots, X_n$, without an importance sampling correction as in the likelihood analyses. In a problem where P in (14.21) was replaced by P_w depending on the action w, an importance sampling correction would be needed. If for some fixed value u of w we sampled from P_u, the formulae would involve importance weights $p_w(x)/p_u(x)$.

Though there are important differences from likelihood analyses in that the analogues of Fisher information, bootstrapping, and so forth are un-interesting, the basic heuristic is unchanged. The Monte Carlo objective function $r_n(w)$ approximates the exact objective function $r(w)$, the minimizer $\hat{w}_n$ of $r_n(w)$ approximates the minimizer $\hat{w}$ of $r(w)$. A central limit theorem for $\sqrt{n}(\hat{w}_n - \hat{w})$ can be obtained from the Taylor series expansion of r_n very similar to the one in Section 14.4, and so forth. Chen *et al.* (1993) give details.

14.7 Which sampling distribution?

One important issue has been ignored in the preceding discussion. Which parameter value ψ is to be used for the simulation? Although the methods 'work' for any ψ, there can be severe loss of efficiency if ψ is chosen badly. There are several ways to see this. One way is that the method relies on reweighting P_ψ to P_θ using 'importance weights' proportional to $h_\theta(x)/h_\psi(x)$. If h_θ is very large where h_ψ is very small, there will be a few very large importance weights and many very small ones, and the reweighted distribution will not be accurate. Another way in which this inefficiency is automatically revealed is in estimating the Monte Carlo error using (14.12–14.16). The importance weights enter the formula (14.12) for the Monte Carlo approximation to the score, hence large variations in importance weights will inflate the variance and autocovariances of the Z_i defined by (14.13). This must be the case, of course. When the variance is inflated by a bad choice of ψ, the variance estimate captures this.

So it is necessary to make a good choice of ψ. The ideal choice would be to use $\psi = \hat{\theta}$, which would make all of the importance weights equal to one, but this is impossible. We are using simulations from P_ψ to find $\hat{\theta}$, so we do not know the value $\hat{\theta}$.

One solution to this problem, used by Geyer and Thompson (1992), is to iterate the algorithm. Start at some value ψ_0. Simulate from P_{ψ_0}. Find the Monte Carlo MLE $\hat{\theta}_0$. If ψ_0 is very far from the exact MLE $\hat{\theta}$, the Monte Carlo likelihood approximation will be bad, and $\hat{\theta}_0$ may be far from $\hat{\theta}$. It may be that the Monte Carlo likelihood does not even achieve its maximum, so there is no maximizer. In this case, it is necessary to do a constrained maximization, letting $\hat{\theta}_0$ be the maximizer of the Monte Carlo likelihood in some neighbourhood of ψ_0 in which the Monte Carlo approximation is trusted (Geyer and Thompson, 1992). This 'trust region' approach is a good idea whether or not the Monte Carlo likelihood has an unconstrained maximizer. It is never a good idea to trust the approximation too far away from ψ_0. Having decided on $\hat{\theta}_0$, the next iteration sets $\psi_1 = \hat{\theta}_0$, simulates from P_{ψ_1}, finds the maximizer $\hat{\theta}_1$ of the Monte Carlo likelihood (possibly restricted to a trust region), and so forth. It is not necessary to iterate to 'convergence'. As soon as $\hat{\theta}_i$ is reasonably close to ψ_i, it is time to stop the iteration, perhaps taking one more very large sample from the distribution for $\psi_{i+1} = \hat{\theta}_i$ to obtain a high precision estimate. Although this scheme is workable in practice, it lacks elegance, so there is reason to consider other methods of finding a 'good' ψ.

Younes (1992) proposed stochastic approximation (the Robbins–Munro algorithm) to find an approximation to $\hat{\theta}$ following Younes (1988) and Moyeed and Baddeley (1991). This approximation can then be used as the ψ in the scheme of Geyer and Thompson (1992). There does not seem to be a direct comparison of iteration versus stochastic approximation in the literature so it is not clear which of the two methods is the best way to find a 'good' ψ. Ultimately, it may turn out that both methods are inferior to importance sampling, the subject of the next section.

14.8 Importance sampling

In a sense, all of the likelihood approximations in Sections 14.4 and 14.5 are special cases of importance sampling, but they are not in the usual spirit of importance sampling, which is to choose the sampling distribution so as to minimize the Monte Carlo error. The ψ that does this is the exact MLE $\hat{\theta}$. When we use any other ψ we are 'anti-importance' sampling – using the same formulae, but for a different purpose.

We can return to the original purpose of importance sampling, i.e. variance reduction, by choosing a sampling distribution that is not in the model. The notation used in Sections 14.4 and 14.5, with h_ψ indicating the sampling density, gives a certain symmetry to the formulae, but is misleading in that the sampling density need not be a distribution in the model. The only requirement is that it dominates h_θ for all θ, the conditions more precisely stated just before (14.8), (14.17), and (14.19).

We could choose any unnormalized density satisfying the domination condition for sampling, replacing h_ψ everywhere by this new function. If this new sampling density puts probability in all regions of interest in the sample space, that is, everywhere h_θ is large for all 'interesting' θ, then the Monte Carlo errors will be much smaller than any sampler using an importance sampling distribution h_ψ in the model. Green (1992) called this the principle of sampling from the 'wrong' model. There is always a better importance sampling distribution outside the model of interest. Green (1992) gives a specific proposal for spatial lattice models with 'intrinsic' priors.

There seems to be no good method for finding the optimal importance sampling distribution in all problems. One suggestion (Torrie and Valleau, 1977) is to use an arbitrary importance sampling density, that is defined by table look-up, with the entries in the table determined by trial and error. This too is inelegant, but can be made to work.

Another suggestion (Geyer, 1993) is to use only importance sampling distributions of a special form, finite mixtures of distributions in the model of interest. Torrie and Valleau (1977) also do the same in some of their examples. That is, we use an importance sampling density of the form

$$h_{\mathrm{mix}}(x) = \sum_{j=1}^m \frac{p_j}{c(\theta_j)} h_{\theta_j}(x) \qquad (14.23)$$

where the p_j are nonnegative and sum to one and $c(\theta)$ is given by (14.6). This method requires no trial and error. If we have computer code to sample from h_θ for any θ in the model, then we can also sample from the mixture density (14.23) by the simple expedient of obtaining a sample $\{X_{ij}, i = 1, \ldots, n_j\}$ from each of the h_{θ_j} and considering the combined samples to be a sample from the mixture, where $p_j = n_j / \sum_k n_k$. Thus no new code need be written to implement the sampler. At the expense of a bit of additional code, the chains can be 'Metropolis-coupled' to accelerate mixing

(Geyer, 1991; see also Gilks and Roberts, 1995: this volume).

We do, however, need a method to estimate the $c(\theta_j)$, which are typically unknown and are required to calculate the mixture density (14.23) and the importance weights in formulae like (14.9). The following curious scheme accomplishes this. Write $p_j/c(\theta_j) = e^{\eta_j}$ and consider the probability that a point x in the pooled sample came from the component of the mixture with density h_{θ_j}. This is

$$p_j(x, \eta) = \frac{h_{\theta_j}(x)e^{\eta_j}}{\sum_k h_{\theta_k}(x)e^{\eta_k}}. \tag{14.24}$$

This reparameterization of the problem has been chosen to put it in the form of a logistic regression. It now seems sensible to estimate the η_i by maximizing the log likelihood for the logistic regression

$$l_{n_1,\ldots,n_m}(\eta) = \sum_{j=1}^{m} \sum_{i=1}^{n_j} \log p_j(X_{ij}, \eta). \tag{14.25}$$

This is what Geyer (1993) calls 'reverse logistic regression'. (Note that (14.25) is a quasi-likelihood: it is constructed as if group membership j is sampled conditionally for each observation X; in fact each X is sampled conditioning on its group membership j.) The η_j are only identifiable up to an additive constant: we can add an arbitrary constant to each η_j without altering p_j in (14.24). Thus the normalizing constants are only determined up to a constant of proportionality. Thus the mixture density (14.23) is estimated up to a constant of proportionality as

$$h_{\text{mix}}(x) = \sum_{j=1}^{m} h_{\theta_j}(x)e^{\hat{\eta}_j}.$$

This h_{mix} can then replace h_ψ, and the pooled sample $\{X_{ij}\}$ then replaces the $\{X_i\}$ in (14.9) and in other uses of importance sampling to evaluate expectations. For any unnormalized density $h_\theta(x)$ dominated by $h_{\text{mix}}(x)$ and any function $g(x)$ integrable with respect to P_θ

$$\frac{\sum_{j=1}^{m} \sum_{i=1}^{n_j} g(X_{ij}) \frac{h_\theta(X_{ij})}{h_{\text{mix}}(X_{ij})}}{\sum_{j=1}^{m} \sum_{i=1}^{n_j} \frac{h_\theta(X_{ij})}{h_{\text{mix}}(X_{ij})}} \tag{14.26}$$

approximates $E_\theta g(X)$. Similar formulae apply for the missing data likelihood methods. The formulae for the asymptotic variance of $\hat{\theta}_n$ in Section 14.4 are still valid. This method works well and provides a stable method for importance sampling (Geyer 1993; Geyer and Møller, 1994).

Before leaving logistic regression, it should be noted that the key procedure, estimating the normalizing constants by maximizing (14.25) can be used where normalizing constants are the primary object of interest. This occurs in likelihood analysis with missing data, formula (14.18), where the likelihood $f_\theta(y)$ (with the missing data integrated out) is the normalizing constant for the distribution $f_\theta(x|y) = f_\theta(x, y)/f_\theta(y)$ of the missing data given the observed data. Hence logistic regression can be used to calculate likelihood ratios for pairs (θ_i, θ_j) that are components of the sampling mixture. This has been used by Thompson *et al.* (1993).

Yet another method of mixture sampling is 'simulated tempering' (Marinari and Parisi, 1992; Geyer and Thompson, 1995). Here the normalizing constants are determined by preliminary experimental runs and the chain simulates the mixture distribution directly. The state of the chain at iteration t consists of the pair of variables (X_t, I_t) such that when $I_t = i$ then X_t is 'from' the distribution with density h_{θ_i}. More precisely the scheme implements a Metropolis–Hastings algorithm for the unnormalized density

$$h_{\text{mix}}(x, i) = h_{\theta_i}(x)\pi_i, \qquad (14.27)$$

the π_i being the tuning constants that must be adjusted using evidence from preliminary runs to make the sampler mix well. The algorithm updates the variables X and I one at a time, updating x using the same Gibbs or Metropolis updates for h_{θ_i} used by the other schemes and updating i with a simple Metropolis update that proposes to change i to a new value j and rejects according to the value of the odds ratio

$$\frac{h_{\theta_j}(x)\pi_j}{h_{\theta_i}(x)\pi_i}.$$

The π_i can be determined automatically by stochastic approximation (Geyer and Thompson, 1995). For good mixing, the π_i should be approximately the inverse normalizing constants $\pi_i \approx 1/c(\theta_i)$. The output of a simulated tempering sampler can be used for a slightly different mixture estimation scheme. Summing over i in (14.27) gives

$$h_{\text{mix}}(x) = \sum_i h_{\theta_i}(x)\pi_i \qquad (14.28)$$

for the marginal density of x. This can be used in place of (14.23), both in likelihood approximations and in the calculation of expectations by (14.26).

14.9 Discussion

MCMC is a powerful method of calculating analytically intractable probabilities and expectations. Averaging over a single run of the Markov chain only calculates expectations with respect to the stationary distribution of the chain, but expectations of many different functions can be calculated from the same run. By using the importance sampling formula, reweighting the stationary distribution to other distributions of interest, expectations with respect to many different distributions can be calculated from the same run. This enables us to calculate a Monte Carlo approximation of the entire objective function of an optimization problem from one run.

This approximation to the objective function is maximized to yield an approximation to the solution of the exact optimization problem. Any optimization algorithm can be used. For very complicated optimization problems involving many variables and constraints (Geyer and Thompson, 1992) sophisticated modern optimization software is necessary. For simpler problems, Newton–Raphson or iteratively reweighted least squares might do.

No matter what optimization algorithm is chosen, it is important to keep the two parts conceptually separated. First we approximate the objective function, then we maximize this approximation. As Sections 14.7 and 14.8 describe, one may need some preliminary experimentation or rather sophisticated importance sampling schemes to make single-sample MCMC optimization work well, but with these improvements single-sample schemes are conceptually simpler, more efficient, and more widely applicable than other methods.

Conflating optimization and estimation, with steps of the Monte Carlo intermingled with steps of the optimization, complicates analysis of the algorithm without improving performance. In fact, conflated algorithms generally have worse performance because they are so difficult to analyse and improve. When the Monte Carlo and the optimization are conceptually separated, things are much simpler. Under suitable regularity conditions, the optimizer of the approximate objective function converges to the optimizer of the exact objective function, and the errors are approximately normal with variances that can be estimated. Moreover, the optimization algorithm may be varied as appropriate to the specific problem without affecting these asymptotics.

References

Besag, J. and Clifford, P. (1989) Generalized Monte Carlo significance tests. *Biometrika*, **76**, 633–642.

Besag, J. and Clifford, P. (1991) Sequential Monte Carlo p-values. *Biometrika*, **78**, 301–304.

Chen, L. S., Geisser, S. and Geyer, C. J. (1993) Monte Carlo minimization for sequential control. *Technical Report 591*, School of Statistics, University of Minnesota.

Dempster, A. P., Laird, N. M. and Rubin, D. B. (1977) Maximum likelihood from incomplete data via the EM algorithm (with discussion). *J. R. Statist. Soc B*, **39**, 1–37.

Diaconis, P. and Sturmfels, B. (1993) Algebraic algorithms for sampling from conditional distributions. Technical report, Department of Mathematics, Harvard University.

Diggle, P. J. and Gratton, R. J. (1984) Monte Carlo methods of inference for implicit statistical models (with discussion). *J. R. Statist. Soc* B, **46**, 193–227.

Gelfand, A. E. and Carlin, B. P. (1993) Maximum likelihood estimation for constrained or missing data models. *Canad. J. Statist.*, **21**, 303–311.

Geyer, C. J. (1991) Markov chain Monte Carlo maximum likelihood. In *Computing Science and Statistics: Proceedings of the 23rd Symposium on the Interface* (ed. E. M. Keramidas), pp. 156–163. Fairfax: Interface Foundation.

Geyer, C. J. (1992) Practical Markov chain Monte Carlo (with discussion). *Statist. Sci.*, **7**, 473–511.

Geyer, C. J. (1993) Estimating normalizing constants and reweighting mixtures in Markov chain Monte Carlo. *Technical Report 568*, School of Statistics, University of Minnesota.

Geyer, C. J. (1994) On the convergence of Monte Carlo maximum likelihood calculations. *J. R. Statist. Soc* B, **56**, 261–274.

Geyer, C. J. and Møller, J. (1994) Simulation procedures and likelihood inference for spatial point processes. *Scand. J. Statist.*, **21**, 359–373.

Geyer, C. J. and Thompson, E. A. (1992) Constrained Monte Carlo maximum likelihood for dependent data (with discussion). *J. R. Statist. Soc.* B, **54**, 657–699.

Geyer, C. J. and Thompson, E. A. (1995) Annealing Markov chain Monte Carlo with applications to pedigree analysis. *J. Am. Statist. Ass.*, (in press).

Gilks, W. R. and Roberts, G. O. (1995) Strategies for improving MCMC. In *Markov Chain Monte Carlo in Practice* (eds W. R. Gilks, S. Richardson and D. J. Spiegelhalter), pp. 89–114. London: Chapman & Hall.

Green, P. J. (1992) Discussion on Constrained Monte Carlo maximum likelihood for dependent data (by C. J. Geyer and E. A. Thompson). *J. R. Statist. Soc.* B, **54**, 683–684.

Guo, S. W. and Thompson, E. A. (1991) Monte Carlo estimation of variance component models for large complex pedigrees. *IMA J. Math. Appl. Med. Biol.*, **8**, 171–189.

Guo, S. W. and Thompson, E. A. (1992) Performing the exact test of Hardy–Weinberg equilibrium for multiple alleles. *Biometrics*, **48**, 361–372.

Guo, S. W. and Thompson, E. A. (1994) Monte Carlo estimation of mixed models for large complex pedigrees. *Biometrics*, **50**, 417–432.

Louis, T. A. (1982) Finding the observed information matrix when using the EM algorithm. *J. R. Statist. Soc* B, **44**, 226–233.

Marinari, E. and Parisi, G. (1992) Simulated tempering: a new Monte Carlo scheme. *Europhys. Lett.*, **19**, 451–458.

Meng, X.-L. and Rubin, D. B. (1991) Using EM to obtain asymptotic variance-covariance matrices: the SEM algorithm. *J. Am. Statist. Ass.*, **86**, 899–909.

Moyeed, R. A. and Baddeley, A. J. (1991) Stochastic approximation of the MLE for a spatial point pattern. *Scand. J. Statist.*, **18**, 39–50.

Mykland, P., Tierney, L. and Yu, B. (1995) Regeneration in Markov chain samplers. *J. Am. Statist. Ass.*, **90**, 233–241.

Ogata, Y. and Tanemura, M. (1981) Estimation of interaction potentials of spatial point patterns through the maximum likelihood procedure. *Ann. Inst. Statist. Math.* B, **33**, 315–338.

Ogata, Y. and Tanemura, M. (1984) Likelihood analysis of spatial point patterns. *J. R. Statist. Soc* B, **46**, 496–518.

Ogata, Y. and Tanemura, M. (1989) Likelihood estimation of soft-core interaction potentials for Gibbsian point patterns. *Ann. Inst. Statist. Math.*, **41**, 583–600.

Penttinen, A. (1984) *Modelling Interaction in Spatial Point Patterns: Parameter Estimation by the Maximum Likelihood Method.* Jyväskylä Studies in Computer Science, Economics, and Statistics, **7**. University of Jyväskylä.

Priestley, M. B. (1981) *Spectral Analysis and Time Series.* London: Academic Press.

Smith, A. F. M. (1992) Discussion on constrained Monte Carlo maximum likelihood for dependent data (by C. J. Geyer and E. A. Thompson). *J. R. Statist. Soc.* B, **54**, 684–686.

Strauss, D. (1986) A general class of models for interaction. *SIAM Review*, **28**, 513–527.

Sundberg, R. (1974) Maximum likelihood theory for incomplete data from an exponential family. *Scand. J. Statist.*, **1**, 49–58.

Thompson, E. A. and Guo, S. W. (1991) Evaluation of likelihood ratios for complex genetic models. *IMA J. Math. Appl. Med. Biol.*, **8**, 149–169.

Thompson, E. A., Lin, S., Olshen, A. B. and Wijsman, E. M. (1993) Monte Carlo analysis on a large pedigree. *Genet. Epidem.*, **10**, 677–682.

Tierney, L. (1995) Introduction to general state space Markov chain theory. In *Markov Chain Monte Carlo in Practice* (eds W. R. Gilks, S. Richardson and D. J. Spiegelhalter), pp. 59–74. London: Chapman & Hall.

Torrie, G. M. and Valleau, J. P. (1977) Nonphysical sampling distributions in Monte Carlo free-energy estimation: umbrella sampling. *J. Comput. Phys.*, **23**, 187–199.

Wasan, M. T. (1969) *Stochastic Approximation.* Cambridge: Cambridge University Press.

Wei, G. C. G. and Tanner, M. A. (1990) A Monte Carlo implementation of the EM algorithm and poor man's data augmentation. *J. Am. Statist. Ass.*, **85**, 699–704.

Younes, L. (1988) Estimation and annealing for Gibbsian fields. *Annales de l'Institut Henri Poincaré. Section B, Probabilites et statistiques*, **24**, 269–294.

Younes, L. (1992) Discussion on Constrained Monte Carlo maximum likelihood for dependent data (by C. J. Geyer and E. A. Thompson). *J. R. Statist. Soc* B, **54**, 694–695.

15

Stochastic EM: method and application

Jean Diebolt
Eddie H S Ip

15.1 Introduction

There are many missing value or incomplete data problems in the statistical literature. The *EM* (expectation-maximization) algorithm, formalized by Dempster *et al.* (1977), has become a popular tool for handling such problems. Surprisingly, many important inference problems in statistics, such as latent variable and random parameter models, turn out to be solvable by EM when they are formulated as missing value problems. In these applications, it is not the data collected from the field that are in a real sense missing. Rather, it is the hidden parameters or variables which are systematically not observed. Here, we deal mainly with this kind of structural missing data.

The widespread use of EM since Dempster *et al.* (1977) is phenomenal. The main appeals of this iterative algorithm are that it is relatively easy to program and that it produces maximum likelihood estimates (MLEs). However, there are also well documented shortcomings of EM. For example, it can converge to local maxima or saddle points of the log-likelihood function, and its limiting position is often sensitive to starting values. In some models, the computation of the E-step involves high dimensional integrations over subsets of k-dimensional Euclidean space $\mathbb{R}^k$, which may be intractable.

Stochastic EM (Celeux and Diebolt, 1985) provides an attractive alternative to EM. This involves iterating two steps. At the S-step, the missing data are imputed with plausible values, given the observed data and a current estimate of the parameters. At the M-step, the MLE of the parameters is computed, based on the pseudo-complete data. The M-step is often easy to perform. Unlike the deterministic EM algorithm, the final output from stochastic EM is a sample from a stationary distribution whose mean is close to the MLE and whose variance reflects the information loss due to missing data.

Stochastic EM generally converges reasonably quickly to its stationary regime. Such a behaviour can partially be explained by the duality principle (Diebolt and Robert, 1994). The deterministic M-step and the stochastic S-step generate a Markov Chain which converges to its stationary distribution faster than the usual long chains of conditional stochastic draws used in many Gibbs sampling schemes.

Stochastic EM is particularly useful in problems where the E-step of EM involves numerical integration. In stochastic EM, the E-step is replaced by simulation. Stochastic EM also gives fast and reliable results when the aim of the analysis is to restore the missing data. Computer code for stochastic EM can be reused to obtain approximate estimates of standard errors. We illustrate these ideas with examples in Section 15.4.

In Section 15.2, we explain the EM algorithm. Section 15.3 describes stochastic EM. Two applications of stochastic EM are discussed in Section 15.4. A simple example involving censored Weibull data illustrates the implementation of stochastic EM. An empirical Bayes probit regression example illustrates how intractable numerical integrations in EM can be replaced by a more tractable simulation step. A real data set for cognitive diagnosis in psychological and educational testing is used.

15.2 The EM algorithm

Since stochastic EM works similarly to EM in a missing data context, we describe in this section the missing-data set-up of EM.

For simplicity, suppose that the complete data $x = \{x_1, \ldots, x_n\}$ are realizations of random variables from some distribution having density $f(x \mid \theta)$. We write $x = (y, z)$, where z represents the unobserved samples and y represents the observed samples. With this notation, the observed data log-likelihood function is

unobserved data

$$\ell_{obs}(\theta; y) = \log \int f((y, z) \mid \theta) \, dz. \tag{15.1}$$

The computation and maximization of $\ell_{obs}(\theta; y)$ is difficult in many situations of interest. In contrast, the complete-data log-likelihood function, $\ell_c(\theta; x) = \log f((y, z) \mid \theta)$, often admits a closed-form expression for the corresponding MLE, $\hat{\theta}_{ML}(y, z)$. Under model (15.1), information about the missing data is completely embedded in the conditional density of z given

y,

$$k\left(z \mid \theta, y\right) = f\left(\left(y, z\right) \mid \theta\right) / \int f\left(\left(y, z'\right) \mid \theta\right) \, \mathrm{d}z'. \tag{15.2}$$

The basic idea of the EM algorithm is to replace maximization of the log-likelihood of the observed data, $\ell_{obs}(\theta; y)$, with successive maximizations of the conditional expectation $Q\left(\theta \mid \theta^{(m)}\right)$ of the complete-data log-likelihood function $\ell_c(\theta; x)$ given the observations y and a current fit $\theta^{(m)}$ of the parameter:

$$Q\left(\theta \mid \theta^{(m)}\right) = E_{\theta^{(m)}}(\ell_c(\theta; x) \mid y). \tag{15.3}$$

The expectation here is with respect to $k(z \mid \theta^{(m)}, y)$.

Given $\theta^{(m)}$ at the mth iteration of EM, the $(m+1)^{th}$ iteration is conducted in two steps:

1. E-step : Compute $Q\left(\theta \mid \theta^{(m)}\right)$.

2. M-step : Update $\theta^{(m)}$ by computing the maximizer $\theta^{(m+1)}$ of $Q\left(\theta \mid \theta^{(m)}\right)$.

This iterative process is repeated until convergence is apparent. For details about convergence properties of the sequence $\theta^{(m)}$, see Dempster *et al.* (1977) and Wu (1983).

As pointed out by Efron (1977) in the discussion of Dempster *et al.* (1977), the underlying identity driving EM is Fisher's identity:

$$\dot{\ell}_{obs}(\theta; y) = E_\theta(\dot{\ell}_c(\theta; x) \mid y) \tag{15.4}$$

where $\dot{\ell}$ denotes $\partial \ell / \partial \theta$, and expectation is with respect to $k(z \mid \theta, y)$. Solving $\dot{\ell}_{obs}(\theta; y) = 0$ is therefore equivalent to solving $E_\theta(\dot{\ell}_c(\theta; x) \mid y) = 0$. We do not know the true value of θ, so we cannot calculate the expectation over the distribution $k(z \mid \theta, y)$. However, if we have a guess for θ, then $k(z \mid \theta, y)$ is completely determined. We shall see how this fact is utilized in the stochastic EM algorithm in the next section.

15.3 The stochastic EM algorithm

15.3.1 Stochastic imputation

Stochastic EM (Broniatowski *et al.*, 1983; Celeux and Diebolt, 1985) provides us with a flexible and powerful tool to handle complex models, especially those for which EM is difficult to implement. It is particularly appealing in situations where inference based on complete data is easy.

The main idea of stochastic EM is, at each iteration m, to 'fill-in' for the missing data z with a single draw from $k(z \mid \theta^{(m)}, y)$, where $\theta^{(m)}$ is the current estimate of θ. This imputation of z is based on all our current information about θ, and hence provides us with a plausible pseudo-complete sample. Once we have a pseudo-complete sample, we can directly maximize its log-likelihood to obtain an updated MLE, $\theta^{(m+1)}$. The whole process is iterated.

Alternately imputing pseudo-complete data (the S-step) and performing maximization (the M-step) generates a Markov chain $\{\theta^{(m)}\}$ which converges to a stationary distribution $\pi(.)$ under mild conditions (Ip, 1994a). The stationary distribution is approximately centred at the MLE of θ and has a variance that depends on the rate of change of $\theta^{(m)}$ in the EM iterations. In most situations where we have used stochastic EM, convergence of $\theta^{(m)}$ has been reasonably fast. In practice, a number of iterations would be required as a burn-in period, to allow $\{\theta^{(m)}\}$ to approach its stationary regime. For example, in mixture models, 100–200 iterations are often used for burn-in (see for example Diebolt and Celeux, 1993). Further iterations generate a sequence $\{\theta^{(m)}\}$ that is approximately stationary. This sequence of points represents a set of good guesses for θ with respect to various plausible values of the missing data.

The Markov chain output $\{\theta^{(m)}\}$ generated by stochastic EM can be used in a variety of ways. First, it provides a region of interest for θ which we call the plausible region. The plausible region can be explored by exploratory data analysis (EDA) tools, such as two-dimensional graphical displays. We now illustrate this idea with an example.

15.3.2 Looking at the plausible region

The following data set is adapted from Murray (1977) and is also considered by Tanner (1991). It consists of 12 bivariate observations:

1	1	-1	-1	2	2	-2	-2	*	*	*	*
1	-1	1	-1	*	*	*	*	2	2	-2	-2

where * indicates a missing value. Overlooking the evident discreteness in these data, we assume that the data come from a bivariate normal distribution with mean $(0,0)$, correlation ρ and common variance σ^2. Let $\theta = (\rho, \sigma)$. The log-likelihood surface of this data has a saddle point at $\rho = 0$, $\sigma^2 = 2.5$ and two maxima of equal log-likelihood at $\rho = 0.5$ and $\rho = -0.5$, $\sigma^2 = 2.667$. Clearly, using MLE naively would be misleading.

Figure 15.1(a) displays graphically the values of 1 000 stochastic EM iterates, each point corresponding to the MLE of (ρ, σ^2) from a pseudo-complete data set. The slightly smoothed histogram of $\{\rho^{(m)}\}$ is shown in Figure 15.1(b). This shows that stochastic EM 'spends more time' in more probable regions, close to $\rho = -0.7$ and $\rho = 0.7$. Thus stochastic EM can provide clear evidence of bimodality.

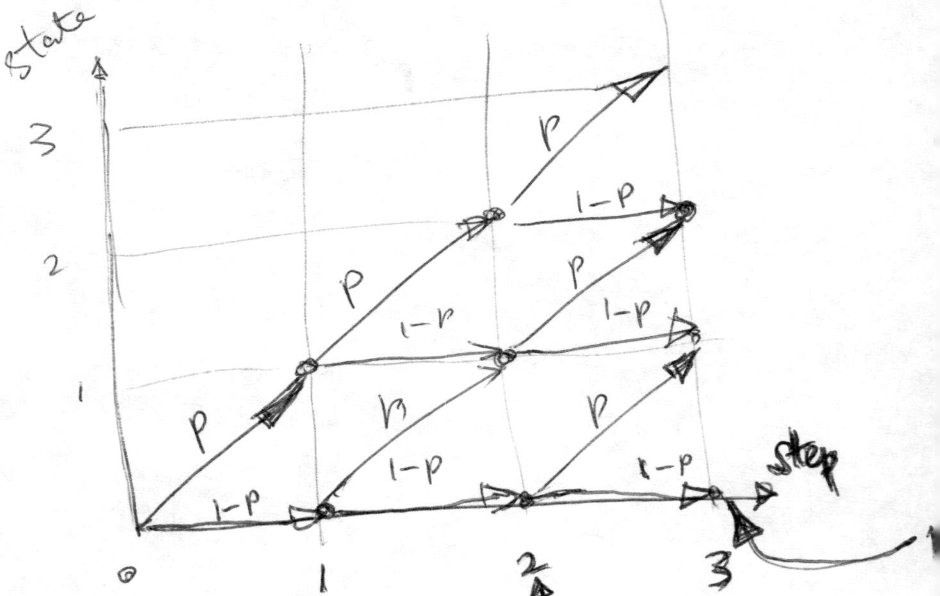

state

3

2

1

p

$1-p$

p

$1-p$

p

$1-p$

p

$1-p$

p

$1-p$

p

$1-p$

h

$1-p$

$1-p$

$1-p$

step

0   1   2   3

$p_0(1) = 1-p$ $p_0(2) = (1-p)^2$

$p_0(0) = 1$ $p_1(1) = p$ $p_1(2) = (1-p)p + p(1-p)$
                                                          $= 2p(1-p)$

$p_2(2) = p^2$

+2

$\frac{09}{48}$

$\frac{5}{45}$

$\frac{1}{6}$

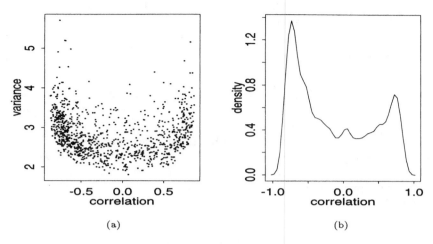

Figure 15.1 *Stochastic EM on data from Murray (1977): (a) plot of variance versus correlation; (b) histogram of ρ.*

15.3.3 Point estimation

To make further use of the array of points generated by stochastic EM, we consider the mean of the stationary distribution $\pi(.)$ as an estimate for θ. We call this mean the *stochastic EM estimate* and denote it by $\tilde{\theta}_n$. This estimate does not agree with the MLE in general. In the exponential family case, which is the usual setting for EM, $\tilde{\theta}_n$ differs from the MLE by $O(1/n)$ (Ip, 1994a). In some simple examples, $\tilde{\theta}_n$ actually coincides with the MLE.

If at each S-step more than one sample is drawn from $k(z|\theta^{(m)}, y)$, and the mean of these multiple draws is imputed for the missing data z, then biased estimates of θ may result (Ip, 1994a). Loosely speaking, the *single* draw makes stochastic EM converge to the right distribution, and covers the MLE in an almost unbiased way. The elegance of stochastic EM actually lies in its simplicity.

Results for consistency and asymptotic normality of the estimator $\tilde{\theta}_n$ have been established for specific examples (Diebolt and Celeux, 1993). These results are also discussed in Ip (1994a).

Besides $\tilde{\theta}_n$, another estimate of θ can be derived from the stochastic EM iterates, namely, the point with the largest log-likelihood $\ell_{obs}(\theta^{(m)}, y)$ in the plausible region. From our experience of stochastic EM, this point is usually adequately close to the MLE for most practical purposes, especially when the region of interest exhibits only one cluster. However, to obtain this point requires the extra effort of evaluating $\ell_{obs}(\theta^{(m)}, y)$ for each m. This function evaluation can hardly be avoided in EM if one wants a tight criterion for convergence based on the difference of the values of the log-likelihood between successive iterates.

15.3.4 Variance of the estimates

Estimating the variance of MLEs is an important issue using EM. Various authors have devised different solutions. Efforts have been made to avoid extra work in writing new code. Louis (1982), Carlin (1987), Meilijson (1989) and Meng and Rubin (1991) give useful results.

The variance of $\tilde{\theta}_n$ can be estimated by the inverse of the observed information matrix $-\ddot{\ell}_{obs}(\theta; y)$ evaluated at $\theta = \tilde{\theta}_n$, where $\ddot{\ell}$ denotes $\partial^2\ell/\partial\theta\partial\theta^{\mathrm{T}}$. Direct computation of $-\ddot{\ell}_{obs}(\theta; y)$ will not in general be convenient, as it involves integration over $k(z|\theta, y)$. However, Louis (1982) derives a useful identity relating the observed data log-likelihood and the complete data log-likelihood:

$$-\ddot{\ell}_{obs}(\theta; y) = E_\theta\{-\ddot{\ell}_c(\theta; x) \mid y\} - \text{cov}_\theta\{\dot{\ell}_c(\theta; x) \mid y\}, \qquad (15.5)$$

where expectations are with respect to $k(z|\theta, y)$.

We propose a conditional parametric boostrap method (Efron, 1992) for estimating the variance of $\tilde{\theta}_n$ by exploiting (15.5). The idea is to replace the theoretical mean and covariance on the right-hand side of (15.5) with a sample mean and covariance, based on a sample $z^{(1)}, z^{(2)}, \ldots, z^{(M)}$ generated independently from $k(z|\theta, y)$, where θ is fixed at $\tilde{\theta}_n$. We illustrate the above procedure with an example in Section 15.4.

Efron(1992) points out that the methods proposed by Meng and Rubin (1991), Meilijson (1989) and Carlin (1987) are basically delta methods and often underestimate the variance. The above scheme gives an approximation to the observed information matrix and may also be subject to the same kind of warning.

15.4 Examples

In this section, we illustrate the application of stochastic EM to two real data examples.

15.4.1 Type-I censored data

In the first example, the data is assumed to be generated from a Type-I censored Weibull model. For a discussion of MLE in censored Weibull data and applications in reliability see, for example, Meeker *et al.* (1992).

Point estimation

Suppose that $x = \{x_i, i = 1, \ldots, n\}$ is an ordered sample from a Weibull distribution with density

$$f(x|\theta) \;=\; \frac{\beta}{\alpha}\left(\frac{x}{\alpha}\right)^{\beta-1}\exp\left[-\left(\frac{x}{\alpha}\right)^{\beta}\right], \qquad x > 0,$$

where $\theta = (\alpha, \beta)$ denotes the scale and shape parameters.

With Type-I censoring at a fixed time c, we only observe a censored sample $\{y_1, \ldots, y_r\}$, $y_i < c$ for $i = 1, \ldots, r$. We treat the problem as a typical missing value problem where x is the complete data and $y = \{y_1 = x_1, \ldots, y_r = x_r, y_{r+1} = c, \ldots, y_n = c\}$ are the observed data. Let $z = \{z_{r+1}, \ldots, z_n\}$ denote the $n - r$ survival times greater than c, and for convenience let $w_i = z_i - c$ denote residual lifetime, for $i = r + 1, \ldots, n$.

The complete-data log-likelihood is

$$\ell_c(\theta|x) = n\log\frac{\beta}{\alpha} + \sum_{i=1}^{n}(\beta - 1)\log\left(\frac{x_i}{\alpha}\right) - \sum_{i=1}^{n}\left(\frac{x_i}{\alpha}\right)^{\beta}.$$

Finding the MLE involves solving the likelihood equations:

$$\frac{\sum_i x_i^{\beta}\log x_i}{\sum_i x_i^{\beta}} - \frac{1}{\beta} - \frac{1}{n}\sum_i \log x_i = 0, \tag{15.6}$$

$$\alpha = \left(\frac{1}{n}\sum_i x_i^{\beta}\right)^{\frac{1}{\beta}}. \tag{15.7}$$

When the data are censored at a fixed time c, the observed data likelihood is $\prod_i f(y_i|\theta)^{\delta_i} S(c|\theta)^{1-\delta_i}$, where

$$\delta_i = \begin{cases} 1 & \text{if } x_i \leq c, \\ 0 & \text{if } x_i > c, \end{cases}$$

and $S(c|\theta) = \exp\left[-(c/\alpha)^{\beta}\right]$.

Stochastic EM consists of two steps, the stochastic imputation step (S-step) and the maximization step (M-step). We look at these two steps separately.

At the mth iterate, given the current estimate $\theta^{(m)}$, the S-step samples $n - r$ observations $\{w_i^{(m)}, i = r + 1, \ldots, n\}$ independently from the conditional density

$$k(w|\theta^{(m)}, c) \;=\; \frac{f(c + w \mid \theta^{(m)})}{S(c \mid \theta^{(m)})}.$$

Then $z_i^{(m)} = c + w_i^{(m)}$ is calculated to 'fill-in' for the censored data.

The M-step is straightforward. With the pseudo-complete sample

$$\left\{y_1, \ldots, y_r, z_{r+1}^{(m)}, \ldots, z_n^{(m)}\right\}$$

available, we solve the log-likelihood equations (15.6)-(15.7), updating the parameter θ to $\theta^{(m+1)}$. A Newton–Raphson type algorithm is generally good enough to obtain the MLE within ten iterations when started at a position not too far from the maximum.

To initialize stochastic EM, a starting value $\theta^{(0)}$ must be provided. Stochastic EM is generally very robust to the starting values, in contrast to deterministic EM. After a burn-in period of m_o iterations, the sequence $\theta^{(m)}$ should be close to its stationary regime and we stop after a sufficiently large number of iterations T.

In the following application, $m_o = 200$ and $T = 300$. The stochastic EM estimate of θ is given by

$$\tilde{\theta}_n = (T - m_o)^{-1} \sum_{m=m_o+1}^{T} \theta^{(m)}.$$

We consider a real data set consisting of 50 points measuring the modulus of rupture of a sample of fir woods. A two-parameter Weibull distribution fits the data well. The Cramer–Von Mises statistics is 0.0495 and the Anderson–Darling statistics is 0.346. Both statistics show no significant deviation from the fitted Weibull model (Lawless, 1982). We artificially censor the data at 1.6, corresponding to a censoring rate of 34%. The stochastic EM and the EM algorithms are both applied to the censored data.

Table 15.1 summarizes the results for the two estimation methods. The MLE for the original uncensored data are also included as a reference.

	estimate		standard error	
	α	β	α	β
uncensored MLE	1.616	4.651	0.052	0.497
$\tilde{\theta}_n$	1.579	5.431	0.050	0.793
EM	1.575	5.435	0.051	0.839
bootstrap	-	-	0.052	0.790

Table 15.1 *Results for fir wood rupture data*

Figure 15.2 shows the points visited by the Monte Carlo chain $\{\theta^{(m)}\}$ and the location of the various estimates. Another run of stochastic EM with a different random seed would produce slightly different estimates. We tried several runs and found that differences were negligible.

Standard errors

The standard error for the stochastic EM estimates is obtained using the Louis identity (15.5), as described in Section 15.3.4. To compute the expected complete data information matrix $E_\theta\{-\ddot{\ell}_c(\theta; x) \mid y\}$ in (15.5), we need the second derivatives of the complete-data log-likelihood:

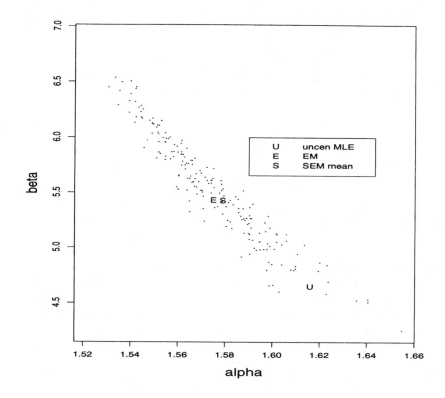

Figure 15.2 *Markov chain generated by stochastic EM for fir wood rupture data showing* 300 *points.*

$$\frac{\partial^2 \ell_c}{\partial \alpha^2} = n\frac{\beta}{\alpha^2} - \left(\frac{\beta+1}{\alpha}\right)\left(\frac{\beta}{\alpha}\right)\sum_i \left(\frac{x_i}{\alpha}\right)^\beta, \qquad (15.8)$$

$$\frac{\partial^2 \ell_c}{\partial \beta^2} = \frac{-n}{\beta^2} - \sum_i \left[\log\left(\frac{x_i}{\alpha}\right)\right]^2 \left(\frac{x_i}{\alpha}\right)^\beta, \qquad (15.9)$$

$$\frac{\partial^2 \ell_c}{\partial \alpha \partial \beta} = \frac{-n}{\alpha} + \frac{1}{\alpha}\sum_i \left(\frac{x_i}{\alpha}\right)^\beta + \frac{\beta}{\alpha}\sum_i \left[\log\left(\frac{x_i}{\alpha}\right)\right]\left(\frac{x_i}{\alpha}\right)^\beta. \qquad (15.10)$$

We create pseudo-complete samples $\{x_i^{(1)}\}, \ldots, \{x_i^{(M)}\}$ by appending to the observed data M samples generated from the conditional distribution $k(z \mid \tilde{\theta}_n, y)$. For each sample, the complete data information matrix $-\ddot{\ell}_c(\theta; x)$ is obtained from (15.8–15.10). By averaging over the M complete data information matrices, we obtain an empirical approximation to the first term on the right-hand side of (15.5). The second term of (15.5),

$$\text{cov}_\theta\{\dot{\ell}_c(\theta; x) \mid y\},$$

is approximated in a similar fashion by computing the sample covariance matrix of the M vectors of first partial derivatives of the complete data log-likelihood, which are given by the following two equations:

$$\frac{\partial \ell}{\partial \alpha} = n\frac{\beta}{\alpha} + \frac{\beta}{\alpha} \sum_i \left(\frac{x_i}{\alpha}\right)^\beta, \tag{15.11}$$

$$\frac{\partial \ell}{\partial \beta} = \frac{n}{\beta} + \sum_i \log\left(\frac{x_i}{\alpha}\right) - \sum_i \left[\log\left(\frac{x_i}{\alpha}\right)\right] \left(\frac{x_i}{\alpha}\right)^\beta. \tag{15.12}$$

In this application, $M = 100$. Standard errors calculated by this method are given in the second row of Table 15.1. Standard errors using the observed information matrix at the MLE are also given in the third row of Table 15.1.

To assess the performance of these estimated standard errors, we also calculated bootstrap standard error estimates. These were calculated from 1 000 datasets, generated by resampling from the empirical distribution of y. The results are shown in the last row of Table 15.1. The standard error estimates in the last three rows of Table 15.1 are similar. It seems that the stochastic EM estimate is quite accurate in approximating the standard error.

We also performed a small simulation study to compare the stochastic EM and the EM estimates for highly censored Weibull data. Thirty data sets were generated independently from a Weibull distribution with $\alpha = 1.0$ and $\beta = 2.0$. Both stochastic EM and EM were applied to the data sets. The censoring point was fixed at 0.8 so that the censoring rate was about 50% on average. The boxplots for the two parameters are shown in Figure 15.3. The stochastic EM estimates and the censored MLE look comparable.

15.4.2 Empirical Bayes probit regression for cognitive diagnosis

We show in this example how stochastic EM can be applied in an empirical Bayes analysis of binary data where EM can be quite difficult to implement. This example is taken from Ip (1994b).

The data are responses to 60 mathematics items from a sample of 2 000 scholastic aptitude test (SAT) test-takers. In the following analysis, we use a random sample of 1 000 test-takers from the data base. All 60 items are coded 1 or 0 according to whether or not an item is testing a certain cognitive attribute (Tatsuoka, 1990). Examples of attributes include arithmetic skill, algebra, geometry, application of simple rules and theorems, logical

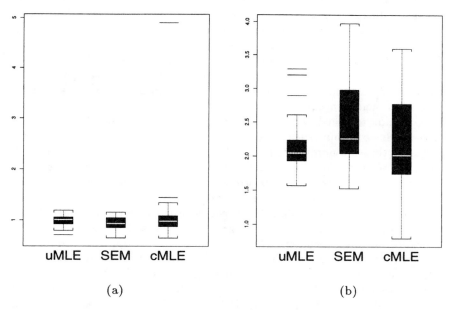

Figure 15.3 *Boxplots for estimates of (a) α and (b) β (right panel) from 30 simulated data sets generated with $\alpha = 1.0$ and $\beta = 2.0$, showing uncensored MLE (uMLE), stochastic EM (SEM) and censored MLE (cMLE) estimates.*

thinking, cognitive restructuring, etc. There are 14 attributes altogether and each item generally tests three to six attributes.

Model and inference

Suppose J items $Y_{i1}, \ldots, Y_{iJ}$ are recorded on each of $i = 1, \ldots, n$ test-takers, where the $\{Y_{ij}\}$ are independent Bernoulli random variables with probabilities $P(Y_{ij} = 1) = p_{ij}$, $j = 1, \ldots, J$. Assume that the ith test-taker is characterized by a $p \times 1$ parameter vector β_i, called the ability vector, signifying his or her proficiency in each attribute. We propose the following probit model:

$$P(Y_{ij} = 1 \mid \beta_i) = \Phi(x_j^{\mathrm{T}} \beta_i), \qquad (15.13)$$

where $x_j = (x_{j1}, \ldots, x_{jp})$ is a vector of known binary covariates flagging the attributes tested by the jth test item and $\Phi(.)$ is the cumulative normal distribution function. Let X denote the $n \times p$ matrix with j^{th} row x_j.

To model heterogeneity among individuals, we use a parametric empirical Bayes model (Morris, 1983) by assuming that the β_i follow a multivariate normal distribution

$$\beta_i \sim \mathrm{N}_p(\mu, \Sigma). \qquad (15.14)$$

Thomas *et al.* (1992) propose a normal mixing model similar to the one above for hospital effects on mortality in a medical context. The mean μ is set to 0 in their application. Under the parametric empirical Bayes framework, $\theta = (\mu, \Sigma)$ in (15.14) must be replaced by an estimate from the data. In the SAT subsample analysed here: $n = 1\,000$, $p = 14$ and $J = 60$.

Because the parameters β_i are regarded as a random sample from a universe of test-takers and not observed, they can be regarded as missing data. Although conceptually EM is quite straightforward, the amount of computation involved here can be prohibitive. To estimate the number of high-dimensional numerical integrations required for EM in this problem, first note that the number of parameters to be estimated is $p + p(p+1)/2$, which is of order 100. The number of test-takers is $1\,000$. For each EM iteration, we are required to perform a 14-dimensional integral numerically $1\,000 \times 100$ times. Assuming the number of EM iterations is of the order 100 (in practice it could range from several dozen to several hundred), the total number of 14-dimensional integrals required is of the order 10^7.

Let us see how stochastic EM can be applied to avoid these numerical integrations. First, we exploit the Gaussian structure of the probit link in (15.13) and add an extra 'layer' of missing values by introducing n latent continuous variables $Z_1, \ldots, Z_n$, each of which is a $J \times 1$ vector. For a fixed i, Z_{ij} is distributed as $N(x_j^T \beta_i, 1)$. Define $Y_{ij} = 1$ if $Z_{ij} > 0$ and $Y_{ij} = 0$ if $Z_{ij} \leq 0$. This formulation is equivalent to the empirical Bayes formulation described above. It is helpful, although not necessary, to think of Z_{ij} as a continuous score obtained by the ith test-taker on the jth item: we only observe either 1 or 0 according to whether this score is above or below zero.

To invoke stochastic EM, we treat the Z_i and the β_i as missing values. Thus we need to draw a sample from the joint distribution of (β, Z), conditioning on the observed data y. This joint distribution is quite intractable, but there are a number of ways of sampling from it. We describe one method which we call the *short Gibbs sampler*. This method exploits the fact that it is easy to sample from the distribution of each variable conditional on the other. The short Gibbs sampler alternately simulates:

$$\beta_i \sim N\left((X^T X + \Sigma^{-1})^{-1}(\Sigma^{-1}\mu + X^T Z_i), (X^T X + \Sigma^{-1})^{-1}\right) \quad (15.15)$$

given y, Z, μ, Σ; and

$$Z_{ij} \sim N(x_j' \beta_i, 1), \quad (15.16)$$

given β, y, μ, Σ, where Z_{ij} is truncated on the left by 0 when $y_{ij} = 1$, and is truncated on the right by 0 when $y_{ij} = 0$.

The short Gibbs sampler is similar to the sampler described in Albert and Chib (1993). At each iteration of the S-step of stochastic EM, the short Gibbs sampler is burnt-in through G iterations of steps (15.15–15.16). The values of $\{\beta_i, i = 1, \ldots, n\}$ from the Gth iterate of the S-step are used in the usual way in the M-step to update parameters μ and Σ. This M-step is straightforward because the distribution is Gaussian. As before, these S- and M-steps are repeated T times.

There are two immediate advantages in using stochastic EM here. First, it avoids the huge number of 14-dimensional integrations required by the E-step of EM. Second, it makes the restoration of β easy: Monte Carlo samples of β from the empirical Bayes posterior distribution of β_i given y and $\tilde{\theta}_n = (\tilde{\mu}, \tilde{\Sigma})$ are readily available using short Gibbs, requiring no new computer code.

To avoid an identification problem, we take one of the variances Σ_{11} (the variance of the first ability parameter) to be known. The programs are written in C and take about five hours to run on a DEC 5000 machine.

Results

To assess the effect of the number of short-Gibbs iterates G and stochastic EM iterates T, we performed experiments on a smaller simulated data set with $p = 3$, $J = 10$ and $n = 200$, and with G ranging from 10 to 1 000. We found that G need not be large to obtain good samples from $k(\beta, Z|y, \mu, \Sigma)$, and that results from these settings for G were quite similar. We also tried ranging T between 300 to 3 000. Again, differences in the results were negligible. In the above SAT data analysis, we set $G = 20$.

We do not give the details of the numerical results here but just report some of the findings by inspecting the stochastic EM estimate of Σ. Several low-level cognitive attributes, for instance arithmetic skill and knowledge in substantive areas (algebra, advanced algebra) have correlations between 0.5 and 0.6. Geometry is the only attribute which stands out, being slightly negatively correlated to most other attributes. High-order cognitive skills, such as logical thinking and ability to solve problems involving more than one step, do not appear to correlate with low-order cognitive skills.

The histograms in Figure 15.4 give an approximation to the empirical Bayes posterior distributions of two components ('arithmetic' and 'cognitive restructuring') of β_i given y, μ and Σ, for two subjects i. The first subject had an overall score of 55%; the second subject had all items correct. These distributions were calculated from 300 iterations of short Gibbs, with $\tilde{\theta}_n = (\tilde{\mu}, \tilde{\Sigma})$.

The empirical Bayes model (15.14) works better than simply treating all individuals independently. In particular, without (15.14), the 'ability' coefficients for subjects having all responses correct would tend to infinity.

Acknowledgement

This work was done while the authors were at the Statistics Department, Stanford University; the first author as a visitor, and the second as a graduate student.

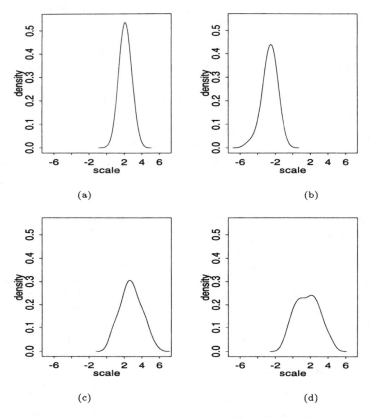

Figure 15.4 *Distribution of attribute arithmetic (panels a,c) and cognitive restructuring (panels b,d). Panels a,b: individual with 55 percent correct responses; panels c,d: individual with 100% correct responses.*

References

Albert, J and Chib, S. (1993) Bayesian analysis of binary and polychotomous response data. *J. Am. Statist. Ass.*, **88**, 669–679.

Broniatowski, M., Celeux, G. and Diebolt, J. (1983) Reconnaissance de Melanges de densites par un algorithme d'apprentissage probabiliste. *Data Anal. Informatics*, **3**, 359–374.

Carlin, J. B. (1987) Seasonal analysis of economic time series. *Doctoral Thesis*, Department of Statistics, Harvard University.

Celeux, G. and Diebolt, J. (1985) The SEM algorithm: a probabilistic teacher algorithm derived from the EM algorithm for the mixture problem. *Comp. Statist. Quart.*, **2**, 73–82.

Dempster, A. P., Laird, N. M. and Rubin, D. B. (1977) Maximum likelihood from incomplete data via the EM algorithm (with discussion). *J. R. Statist. Soc.* B, **39**, 1–38.

Diebolt, J. and Celeux, G. (1993) Asymptotic properties of a stochastic EM algorithm for estimating mixing proportions. *Comm. Statist. B: Stoch. Mod.* **9**, 599–613.

Diebolt, J. and Robert, C. P. (1994) Estimation of finite mixture distributions through Bayesian sampling. *J. R. Statist. Soc.* B, **56**, 363–375.

Efron, B. (1977) Discussion on maximum likelihood from incomplete data via the EM algorithm (by A. Dempster, N. Laird, and D. Rubin). *J. R. Statist. Soc.* B, **39**, 1–38.

Efron, B. (1992) Missing data, imputation and the bootstrap. Technical report, Division of Biostatistics, Stanford University.

Ip, E. H. S. (1994a) A stochastic EM estimator in the presence of missing data – theory and applications. Technical report, Department of Statistics, Stanford University.

Ip, E. H. S. (1994b) Using the stochastic EM algorithm in multivariate hierarchical models. Technical report, Department of Statistics, Stanford University.

Lawless, J. (1982) *Statistical Models and Methods for Lifetime Data.* New York: Wiley.

Louis, T. (1982) Finding the observed information matrix when using the EM algorithm. *J. R. Statist. Soc.* B, **44**, 226–233.

Meeker, W. Q., Escobar, W. and Hill, J. (1992) Sample sizes for estimating the Weibull hazard. *IEEE Trans. Reliability*, **41**, 133–138.

Meilijson, I. (1989) A fast improvement to the EM algorithm on its own terms. *J. R. Statist. Soc.* B, **51**, 127–138.

Meng, X. and Rubin, D. (1991) Using EM to obtain asymptotic variance-covariance matrices: the SEM algorithm. *J. Am. Statist. Ass.*, **86**, 899–909.

Morris, C. N. (1983) Parametric empirical Bayes inference: theory and applications. *J. Am. Statist. Ass.*, **8**, 47–59.

Murray, G. (1977) Comment on maximum likelihood from incomplete data via the EM algorithm (by A. Dempster, N. Laird, and D. Rubin). *J. R. Statist. Soc.* B, **39**, 1–38.

Tanner, M. (1991) *Tools for Statistical Inference.* Hayward: IMS.

Tatsuoka, K. K. (1990) Toward an integration of item-response theory and cognitive error diagnosis. In *Diagnostic Monitoring of Skills and Knowledge Acquisition* (eds N. Frederikson, R. Glacer, A. Lesgold, M. C. Shafto), pp. 453–488. Hillsdale: Lawrence Erlbaum.

Thomas, N., Longford, N. and Rolph, J. (1992) A statistical framework for severity adjustment of hosiptal mortality. *Technical Report N-3501-HCFA*, The Rand Corporation.

Wu, C. F. J. (1983) On the convergence of the EM algorithm. *Ann. Statist.*, **11**, 95–103.

16

Generalized linear mixed models

David G Clayton

16.1 Introduction

The development of MCMC methods has opened up the possibility that statistical methods can be freshly tailored for each application, and that it will no longer be necessary to force an analysis into ill-fitting procedures solely for reasons of computational tractability. Attractive though such a scenario undoubtedly is, the need will remain for flexible off-the-peg methods to meet many of the exigencies of day-to-day data analysis. This chapter deals with such a method: Bayesian analysis using generalized linear mixed models (GLMMs).

Since their formal definition by Nelder and Wedderburn (1972), generalized linear models (GLMs) have had an immense impact on the theory and practice of statistics. These models, together with closely related models for point processes and survival analysis have dominated the applied literature for two decades. More recently there has been considerable interest in extending these models to include random effects, in much the same way that the Gaussian linear model has been extended (Harville, 1977). Such generalization stems primarily from the need to accommodate more complex error structures such as those arising in multilevel models (Bryk and Raudenbush, 1992). However, away from the world of Gaussian responses, inference in random-effects models runs into computational problems. Several approximate approaches, recently reviewed by Breslow and Clayton (1993), have been proposed. In this chapter, a subclass of GLMMs is described for which Bayesian inference may conveniently be achieved using MCMC methods, specifically Gibbs sampling.

Section 16.2 briefly reviews the classical ideas which underly the GLM. Section 16.3 discusses the problem of Bayesian estimation in such models and Section 16.4 describes application of the Gibbs sampler. In Section 16.5, the formal extension to the class of GLMMs is defined, and it is shown how the Gibbs sampling algorithm introduced earlier may be extended. Section 16.6 deals in detail with the parameterization of main effects and interaction terms involving categorical variables (or *factors*) and with the specification of appropriate multivariate prior distributions, and Section 16.7 discusses the choice of prior distributions for the hyperparameters of these distributions. Section 16.8 briefly illustrates the diversity of applications of this class of models in biostatistics. Finally Section 16.9 discusses the implementation of these methods for wider use.

16.2 Generalized linear models (GLMs)

The definition of GLMs is now well known. A set of observed responses, $y_i (i = 1, \ldots, n)$ are assumed to come from exponential family distributions with means μ_i and variances $\phi V(\mu_i)/w_i$. The *variance function*, $V()$, is determined by the choice of the appropriate member of the exponential family. In some cases the *scale factor*, ϕ, is an unknown parameter while in others (notably the important Poisson and binomial cases) it simply takes the value 1. The *prior weights*, w_i, are known constants whose main use is to deal with data records which are summaries of more extensive raw datasets.

The mean responses are related to the elements of a *linear predictor* vector, η, via a *link function*, $g()$, the linear predictor being given by a linear regression model involving a vector of explanatory variables x_i, as follows

$$g(\mu_i) = \eta_i = \beta^{\mathrm{T}} x_i.$$

Writing the matrix of explanatory explanory variables as $X = (x_1, \ldots, x_i, \ldots, x_n)^{\mathrm{T}}$, the linear model is

$$\eta = X\beta.$$

The log likelihood for the regression coefficients, β, takes simple forms (see, for example, McCullagh and Nelder, 1989) and the first derivative vector is

$$\sum_{i=1}^{n} w_i^* r_i x_i,$$

where r_i is a residual projected onto the scale of the linear predictor

$$r_i = (y_i - \mu_i)g'(\mu_i)$$

and

$$w_i^* = w_i \left[\phi V(\mu_i) \{g'(\mu_i)\}^2 \right]^{-1}.$$

For canonical link functions, $g'(\mu) = [V(\mu)]^{-1}$ and the expression

$$-\sum_{i=1}^{n} w_i^* x_i x_i^{\mathrm{T}},$$

gives the second derivative of the log likelihood for all β. For other link functions, this expression is only equal to the second derivative at $\hat{\beta}$, the maximum likelihood estimate (MLE) of β. However, for the computations described in this chapter, it is only ever necessary to have a good approximation to the second derivative of the log likelihood and the above expression is always adequate in practice.

From the above results, it follows that $\hat{\beta}$ can be calculated by iterative repetition of conventional weighted least-squares calculations. Its location does not depend on the value of the scale factor but estimates of standard errors do, so that (where necessary) a subsidiary procedure is required to estimate ϕ. Here full maximum likelihood estimation may involve unpleasant functions (such as digamma and trigamma functions) and more usually we calculate the moment estimate of ϕ obtained by equating the Pearsonian χ^2 statistic

$$\sum_{i=1}^{n} \hat{w}_i^* \hat{r}_i^2 = \sum_{i=1}^{n} w_i \frac{(y_i - \hat{\mu}_i)^2}{\phi V(\hat{\mu}_i)}$$

to the residual degrees of freedom, where $\hat{\ }$ denotes evaluation at $\hat{\beta}$.

16.3 Bayesian estimation of GLMs

For Bayesian estimation in GLMs, there is no general family of prior distributions for the parameters β which leads to analytically tractable posteriors in all cases. However, MCMC methods avoid the need for this and, for flexibility, the prior distribution for β will be assumed to be multivariate normal with variance-covariance matrix Λ^{-1}. The reason for parameterizing the prior in terms of the precision matrix, Λ, will become clear later, and its detailed structure will be discussed in Section 16.6.

The first derivative of the log posterior distribution is

$$-\Lambda\beta + \sum_{i=1}^{n} w_i^* r_i x_i$$

and, with the same provisos as before, the expression

$$-\Lambda - \sum_{i=1}^{n} w_i^* x_i x_i^{\mathrm{T}}$$

can be used for the second derivatives. Thus the posterior mode may be calculated by a simple modification of the conventional iteratively reweighted least-squares algorithm. For the improper uniform prior, $\Lambda = 0$ and the posterior mode is the MLE. In other cases, the posterior mode is a compromise between the MLE and the prior mean so that the estimates are shrunk towards zero. An interesting special case is when $\Lambda = \lambda I$, where the posterior mode estimate may be obtained using iteratively reweighted ridge regression calculations. The posterior distribution of β is log-concave and, given sufficient data, is approximately Gaussian with mean equal to the posterior mode and variance equal to (minus) the inverse of the second derivative matrix at the mode. With large samples and uniform priors, inference is therefore identical to that which would normally follow from conventional use of the GLIM program (Francis *et al.*, 1993).

When the scale factor ϕ is not identically 1, its estimation from a Bayesian perspective is rather more difficult. It is convenient to assume that, *a priori*, ϕ^{-1} is distributed independently of β according to a gamma distribution. In this way, with appropriate choices of prior, Bayesian inference coincides with classical inference in the important special case of the Gaussian linear model. In other cases, the posterior for ϕ is intractable but the moment estimator can be justified as an approximate Bayes estimate by approximating the likelihood for ϕ and integrating out the nuisance parameters, β, using a Laplace approximation.

16.4 Gibbs sampling for GLMs

Small sample inference for β is more difficult, since the posterior distribution is intractable. However, Dellaportas and Smith (1993) showed how the Gibbs sampler can be used to sample from this posterior distribution, thus allowing estimation of posterior moments, probabilities, and credible intervals for the parameters. As usual, the idea is to sample each parameter in turn from its distribution conditional upon data and current values of all other parameters (its *full conditional* distribution); see Gilks *et al.* (1995: this volume). Dellaportas and Smith (1993) considered only the case of *diagonal* Λ and in this case the regression coefficients are conditionally independent of each other given data and ϕ, leading to a particularly simple implementation. However, the algorithm is easily extended to the more general case, as will be shown below.

It follows from well-known properties of the multivariate normal distribution (Dempster, 1972; Whittaker, 1990) that the prior distribution of one regression coefficient, β_i, conditional upon the other coefficients is Gaussian with mean

$$-\sum_{j \neq i} \frac{\Lambda_{ij}}{\Lambda_{ii}} \beta_j$$

and variance $1/\Lambda_{ii}$. The full conditional distribution of β_i is obtained by multiplying this by the likelihood function for β_i. This is log-concave and can be reliably and efficiently sampled using the adaptive rejection sampling algorithm of Gilks and Wild (1992); see Gilks (1995: this volume). Each cycle of the Gibbs sampling algorithm samples a new value of each β_i given current values of all other parameters. Finally, if ϕ is unknown, a new value should be sampled from its full conditional distribution. Since the applications described in this chapter have involved only Poisson and binomial likelihoods (for which $\phi = 1$), this step will not be discussed in any further detail.

This algorithm is an effective use of MCMC methods to deal with a useful but intractable model. It performs extremely well in practice, provided that the problem is not so ill-conditioned as to lead to high correlations in the posterior distribution of β. Ill-conditioned problems should first be regularized by linear transformation of the X matrix; see Gilks and Roberts (1995: this volume).

16.5 Generalized linear mixed models (GLMMs)

Generalized linear mixed models (GLMMs) are GLMs which include one or more *random effects*. Such models have the potential to deal with data involving multiple sources of random error, such as repeated measures within subjects. In this section, GLMMs will first be defined from a frequentist perspective. The necessary extensions for Bayesian inference will then be discussed.

16.5.1 Frequentist GLMMs

Following Harville (1977), the extension is usually indicated by rewriting the regression equations for the linear predictor as

$$\eta = X\beta + Zb.$$

The matrix X defines the fixed-effect part of the linear model and the matrix Z defines the random-effect part. The fixed effects β are unknown constants, but the random effects b are random variables drawn from a distribution. The parameters of the distribution of random effects, commonly termed *hyperparameters*, must also be estimated from the data. These will be denoted by θ. Usually the random effects are assumed to be drawn from a Gaussian distribution and the hyperparameters, θ, have an interpretation in terms of variance components.

The usual strategy is to attempt to estimate β, ϕ and θ by maximum likelihood using the integrated likelihood

$$[\text{Data}|\beta, \phi, \theta] \propto \int [\text{Data}|\beta, b, \phi][b|\theta]db, \qquad (16.1)$$

(where $[u|v]$ generically denotes the conditional distribution of u given v). If possible, improved estimation of variance components (ϕ, θ) may be achieved by conditioning on the sufficient statistics for β. (In the fully Gaussian case, this approach leads to 'restricted' maximum likelihood (REML) estimation for θ, ϕ.) The random effects b are estimated by empirical Bayes estimates, defined as the mean of the distribution

$$[b|\text{Data}, \beta = \hat{\beta}, \phi = \hat{\phi}, \theta = \hat{\theta}] \propto [\text{Data}|\beta = \hat{\beta}, b, \phi = \hat{\phi}][b|\theta = \hat{\theta}],$$

where $\hat{\ }$ denotes a MLE obtained from (16.1).

While this approach is quite feasible for the Gaussian mixed model, the computations are intractable in other cases. Breslow and Clayton (1993) review approaches to approximate these calculations. There are two main approaches and both reduce to iteratively reweighted versions of the calculations for the Gaussian linear mixed model. Although these perform quite well in a wide range of circumstances, they are generally quite laborious since they typically involve, at each iteration, inversion of a square matrix of size equal to the total number of β and b parameters. When the random effects have a nested, or 'multilevel' structure, inversion by partitioning can dramatically reduce the computational burden, but there are many useful models for which this is not possible.

The Gibbs sampling approach to the GLM requires only a minor extension to accommodate the introduction of random effects. A particularly attractive feature is that the amount of computation depends only linearly upon the total number of parameters. This approach requires a Bayesian formulation of the model.

16.5.2 Bayesian GLMMs

From the Bayesian standpoint, there is no need to partition the vector of explanatory variables as (x, z) with a corresponding partition of the parameter vector into fixed and random effects (β, b). At this level in the model, *all* the parameters are random variables drawn from a multivariate Gaussian distribution with zero mean. Thus, the distinction between the two sorts of effect may be dropped and the linear model may simply be specified as for a simple GLM:

$$\eta = X\beta.$$

The difference between the two sorts of effect lies in the specification of the precision matrix, Λ, in the 'prior' model

$$\beta \sim N(0, \Lambda).$$

For fixed effects, the relevant elements of this matrix are known constants expressing subjective prior knowledge, while for random effects they depend on unknown hyperparameters, θ, which are estimated from the data. To complete the Bayesian formulation of the problem it remains only to specify a *hyperprior* distribution for θ.

Further simplification is possible if we are prepared to adopt improper uniform priors for all the fixed effects. In this case, the partition of the model into fixed and random effects corresponds to a partitioning of the prior precision matrix

$$\Lambda = \left[\begin{array}{cc} 0 & 0 \\ 0 & \Lambda_1(\theta) \end{array} \right]. \tag{16.2}$$

This simplification will be assumed for the remainder of this chapter.

Extension of the Gibbs sampling algorithm to deal with random effects requires only the addition of a further step at the end of each cycle to sample the hyperparameters θ. It is clear from Figure 16.1 that this distribution only depends on current values of β. Figure 16.1(a) shows the model as a directed acyclic graph (DAG), as discussed in Spiegelhalter *et al.* (1995: this volume). Figure 16.1(b) shows the corresponding undirected conditional-independence graph (Whittaker, 1990), from which it is clear that θ is independent of x and y given β.

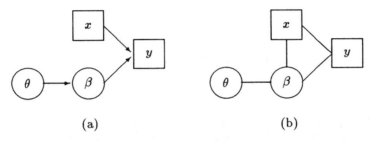

(a)   (b)

Figure 16.1 *A GLMM as a graphical model.*

Although there may be many parameters in such a model, the algorithm for sampling the β parameters may be implemented efficiently because
1. the X matrix will usually be sparse and the likelihood calculations for any one parameter β_i will only involve the rows in which i^{th} column vector of X is non-zero, and
2. the prior mean of β_i depends only on the values β_j corresponding to values of j for which $\Lambda_{ij} \neq 0$.

Thus if both the X and Λ matrices are sparse and if they are stored in such a way that sequences of non-zero elements can be efficiently retrieved, the computational burden is not excessive.

Efficient sampling of the conditional posterior for θ can be achieved if the hyperprior is conjugate when β is taken as data. The choice of hyperprior will be discussed in Section 16.7, after more detailed consideration of the form of the prior precision matrix, Λ.

16.6 Specification of random-effect distributions

16.6.1 Prior precision

The general form of Λ is determined by the structure of terms in the linear model specification. Consider, for example, a model for additive effects of two categorical variables such as *age-band* and geographical *area*, with $p+1$ and $q+1$ levels respectively. In the notation of Wilkinson and Rogers (1973), such a model might be indicated by the expression

$$1 + \mathsf{Age} + \mathsf{Area}$$

This model has $1 + p + q$ parameters in total, falling into three distinct groups:

   1   the single intercept, or 'grand mean', parameter,

   Age   the p parameters representing age effects, and

   Area   the q parameters representing area effects.

The terms of the model formula correspond to a partitioning of X and β so that

$$\eta = \left(X^{[1]}, X^{[2]}, \ldots, X^{[j]}, \ldots, \right) \begin{pmatrix} \beta^{[1]} \\ \beta^{[2]} \\ \vdots \\ \beta^{[j]} \\ \vdots \end{pmatrix}.$$

The first parameter becomes the intercept by letting $X^{[1]} = 1_n$, the unit vector of length n. (This explains the use of the name 1 in model formulae.)

Corresponding to this partitioning of the β vector is a similar partitioning of Λ into a block structure. It is useful to start from the assumption that there may, *a priori*, be correlation between parameters within the same term but not between parameters in different terms. Thus, in the example above, it is likely that the parameters which together estimate the age relationship will be interrelated *a priori* and likewise for the region effects. However, there would seem to be no reason for prior correlation between two parameters drawn from different groups. In this case, the prior precision matrix is block-diagonal. A further assumption which holds in all the applications to be discussed in this chapter is that each block is known up to a constant multiple, so that

$$\Lambda = \begin{bmatrix} \theta^{[1]}S^{[1]} & & & \\ & \theta^{[2]}S^{[2]} & & \\ & & \cdots & \\ & & & \theta^{[j]}S^{[j]} & \\ & & & & \cdots \end{bmatrix}.$$

The block in Λ which describes the interrelationship between parameters in the jth term is given by the product $\theta^{[j]}S^{[j]}$, where $\theta^{[j]}$ is a scalar and $S^{[j]}$ is a known matrix which determines the structure of the relationship. These S matrices will be referred to as *prior structure matrices*. When using the Gibbs sampler, it is desirable for such matrices to be sparse.

If term j represents a set of fixed effects, then $\theta^{[j]} = 0$, by assumption (16.2). For a random-effect term, $\theta^{[j]}$ is an unknown parameter to be estimated. The following subsections discuss further problems in the parameterization of random-effect terms. For clarity, the $[j]$ superscript will be omitted although the discussion will concern single terms rather than the entire linear model.

16.6.2 Prior means

Without loss of generality, the prior mean of β may be taken as zero. If this were not so and $E(\beta) = \mu$, a known vector, it follows from the identity

$$X\beta = X(\beta - \mu) + X\mu$$

that the model could be recast in terms of parameters $(\beta - \mu)$, which have zero mean, by including the known 'offset' vector $X\mu$ in the linear model. Similarly, if μ is a linear function of unknown fixed parameters γ with $\mu = T\gamma$, the linear model may be written

$$X\beta = X(\beta - \mu) + XT\gamma$$

so that γ may be incorporated as fixed-effect parameters by inclusion of the block XT in the design matrix. The random effects will then have zero mean.

16.6.3 Intrinsic aliasing and contrasts

Mixed models arise most commonly in simple multi-level problems. In the simplest case, the observational units fall into groups, where the groups may be regarded as being sampled from some larger universe. For example, with repeated-measures data on a collection of subjects, a 'unit' is an individual measurement, and a 'group' comprises the measurements on single subject. Statistical models for such data commonly assume units to be statistically independent within groups, but that some of the parameters of the within-group model vary randomly from group to group.

If there are g groups and the $n \times g$ matrix G codes group membership, with $G_{ik} = 1$ if unit i falls in group k and 0 otherwise, the usual way of specifying such independent random group effects is to take $X = G$ and $S = I_g$, the $g \times g$ identity matrix. This is the *full dummy variable* parameterization of the model term. However, if the model also contains an intercept term, there is *intrinsic aliasing* owing to the linear relationship $G1_g = 1_n$. This problem is clarified by reparameterizing the group effects by post-multiplying G by a $g \times g$ matrix whose first column is the unit vector:

$$X = G(1_g, C) = (1_n, GC).$$

The column vector 1_n will then appear twice in the X matrix and the model will have two intercept parameters: one fixed, the other random. The parameterization of the remaining effects is controlled by C, a $g \times (g-1)$ *contrast matrix*[†]. This will be referred to as the *contrast* parameterization of the term.

Unlike models with intrinsically aliased fixed effects, models with intrinsic aliasing in the random effects *can* be fitted, since the prior distribution on the random effect ensures that there is a best fit model. It is clear that the best estimate of the random intercept will be at its prior mean: zero. The point estimates of the parameters will, therefore, be exactly the same as if the second intercept were simply omitted from the model: the usual method of dealing with intrinsic aliasing in fixed-effect models. However, the standard errors of the fixed-effect intercept will differ between these two strategies. This reflects the different interpretations of this parameter between the two approaches. If the random intercept is allowed to remain in the model, the fixed intercept is interpreted in terms of the *population* of groups which might have been observed while, if it is omitted, the fixed intercept must be interpreted in terms of the groups actually observed. In practice, the distinction will rarely be of importance and the choice will be made according to ease of implementation.

Reparameterization of the group effects as contrasts has implications for the precision structure of the random-effect parameters. Writing the full dummy variable parameters as δ, the contrast parameters are given by the equation

$$\beta = (1_g, C)^{-1}\delta$$

and, if the δ parameters are independent with equal variances, it follows that the variance-covariance matrix of β is proportional to

$$\begin{pmatrix} 1_g^T 1_g & 1_g^T C \\ C^T 1_g & C^T C \end{pmatrix}^{-1}$$

If the first (intercept) parameter is omitted, the variance-covariance matrix for the remaining $g-1$ parameters is proportional to the lower right-hand block of the above matrix. The precision structure matrix for the contrast parameterization is, therefore,

$$S = C^T C - \frac{1}{g} C^T 1_g 1_g^T C.$$

For the *corner* parameterization, as used in the GLIM program, groups $2, \ldots, g$ are each contrasted with group 1, so $\beta_k = \delta_k - \delta_1$. The corresponding

---

[†] This terminology follows that used for the implementation of linear models in the S language for statistical computing (Chambers and Hastie, 1992).

contrast matrix is

$$C = \begin{bmatrix} 0 & 0 & 0 & \cdot \\ 1 & 0 & 0 & \cdot \\ 0 & 1 & 0 & \cdot \\ 0 & 0 & 1 & \cdot \\ \cdot & \cdot & \cdot & \cdot \end{bmatrix}$$

and

$$S = I_{g-1} - \frac{1}{g} 1_{g-1} 1_{g-1}^{\mathrm{T}}.$$

The *Helmert* contrasts compare the second group with the first, the third with the mean of the first two, and so on. A contrast matrix which defines such contrasts is

$$C = \begin{bmatrix} -1 & -1 & -1 & \cdot \\ 1 & -1 & -1 & \cdot \\ 0 & 2 & -1 & \cdot \\ 0 & 0 & 3 & \cdot \\ \cdot & \cdot & \cdot & \cdot \end{bmatrix}.$$

This specifies the contrasts

$$\beta_k = \frac{1}{k}\left(\delta_k - \frac{1}{k-1}\sum_{l=1}^{k-1}\delta_l\right), \qquad (k > 1).$$

These are *a priori* independent of one another, since

$$S = \begin{bmatrix} 2 & & & \\ & 6 & & \\ & & 12 & \\ & & & \cdot \end{bmatrix}.$$

At first sight it might be thought that requiring S to be sparse would suggest parameterizing the model in terms of Helmert rather than corner contrasts. However, the X matrix would then be less sparse and this would usually impose a more serious computational burden. Furthermore, a minor elaboration of the computations replaces the requirement that S be sparse with the less stringent condition that a large number of its elements are equal (to the value a, let us say). That is,

$$S = P + a11^{\mathrm{T}}$$

where P is a sparse matrix. In this case, the full conditional for a single parameter, β_k, is obtained by multiplying its likelihood function by a Gaussian density with mean

$$-\frac{1}{a + P_{kk}}\left[a(\beta_+ - \beta_k) + \sum_{l \neq k} P_{kl}\beta_l\right]$$

(which depends on current parameter values), and variance

$$\frac{1}{\theta(a + P_{kk})},$$

where β_+ is the sum of parameter values in the group. If P is sparse and if β_+ is stored, the full conditional for β_k may be calculated quickly. After sampling a new value, the stored value of β_+ is easily updated.

It follows from the above discussion that the corner parameterization leads to more efficient Gibbs sampling than other choices of contrasts. However, poor choice of the corner category may lead to an ill-conditioned parameterization, as there may be strong correlations in the posterior distribution of β. This could lead to relatively poor performance of the Gibbs sampler. With this proviso, in general it will be advantageous to choose the parameterization for sparseness of the X matrix rather than of the S matrix.

16.6.4 Autocorrelated random effects

The remainder of this section is concerned with random effects which are interdependent. These most commonly arise when the factor levels correspond to locations in time and/or space, since in such situations there may be a tendency for the effects of two levels which are close together to be similar.

As before, the form of the precision structure matrix, S, will depend on the choice of contrasts. It will usually be most convenient to define the interdependence structure for the case of the full (i.e. original) parameterization and to derive the appropriate choice of S for contrast parameterizations. If the inverse variance-covariance matrix of the g random effects in the full parameterization is proportional to the $g \times g$ matrix U, then an extension of the argument presented above for independent effects shows that the precision structure matrix for a contrast parameterization defined by the contrast matrix C is

$$S = C^\mathrm{T} U C - \frac{C^\mathrm{T} U 1_g 1_g^\mathrm{T} U C}{1_g^\mathrm{T} U 1_g}.$$

In all the examples to be discussed below, U is not of full rank and $U1_g = 0$. In these circumstances, the second term in the above expression vanishes. In the case of the corner parameterization, S is obtained by simply deleting the first row and column of U.

In choosing a parameterization, it is important to ensure that S is of full rank. Otherwise, the posterior distribution of the random effects will be improper and the Gibbs sampler will generally fail.

16.6.5 The first-difference prior

When the factor levels $1, \ldots, g$ fall in some natural order, it will usually be more appropriate to adopt a prior in which effects for neighbouring categories tend to be alike. The simplest of such priors is the *random-walk* or *first-difference* prior in which each effect is derived from the immediately preceding effect plus a random perturbation:

$$
\begin{aligned}
[\beta_k | \beta_1, \beta_2, \ldots, \beta_{k-1}] &= \beta_{k-1} + \epsilon_k \\
\epsilon_k &\sim N\left(0, \tfrac{1}{\theta}\right)
\end{aligned}
\qquad (k = 2, \ldots, g).
$$

This is illustrated as a DAG in Figure 16.2. The undirected conditional-independence graph has the same structure (but with undirected links) and the conditional mean for each effect is the mean of the two immediately neighbouring effects.

Figure 16.2 *The random-walk (first-difference) prior.*

Note that this prior does not specify a distribution for β_1, and this is reflected in a loss of rank of the inverse variance-covariance structure:

$$
U = \begin{bmatrix}
1 & -1 & & & & \\
-1 & 2 & -1 & & & \\
& -1 & 2 & -1 & & \\
& & \cdot & \cdot & \cdot & \\
& & & -1 & 2 & -1 \\
& & & & -1 & 1
\end{bmatrix}.
$$

In the corner parameterization, β_1 is set to zero and S is derived from U simply by deleting its first row and column.

An equivalent model is obtained by parameterizing the factor effect in terms of the $g - 1$ first-difference comparisons $\beta_k = \delta_k - \delta_{k-1}, (k > 1)$, by taking

$$
C = \begin{bmatrix}
0 & 0 & 0 & \cdot \\
1 & 0 & 0 & \cdot \\
1 & 1 & 0 & \cdot \\
1 & 1 & 1 & \cdot \\
\cdot & \cdot & \cdot & \cdot
\end{bmatrix},
$$

and S as the identity matrix. However, X is then less sparse and computation at each step of the Gibbs sampler will be increased.

16.6.6 The second-difference prior

A more elaborate model, the *stochastic-trend* or *second-difference* prior, derives each effect from a linear extrapolation from the two effects immediately preceding it, plus a random perturbation:

$$[\beta_k|\beta_1, \beta_2, \ldots, \beta_{k-1}] = 2\beta_{k-1} - \beta_{k-2} + \epsilon_k$$
$$\epsilon_k \sim N\left(0, \tfrac{1}{\theta}\right)$$

$$(k = 3, \ldots, g).$$

This is illustrated as a DAG in Figure 16.3. The undirected conditional-independence graph takes a similar form so that the conditional prior mean for each effect is obtained by (cubic) interpolation from the four nearest neighbours.

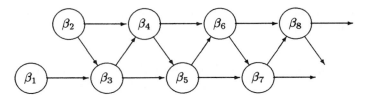

Figure 16.3 *The stochastic-trend (second-difference) prior.*

Since the distributions of both β_1 and β_2 are unspecified, U only has rank $g - 2$ and it will be necessary to parameterize the random effect as $g - 2$ contrasts and, in addition, to fit a fixed effect for linear trend across categories. The matrix for the full dummy variable parameterization is

$$U = \begin{bmatrix} 1 & -2 & 1 & & & & & & \\ -2 & 5 & -4 & 1 & & & & & \\ 1 & -4 & 6 & -4 & 1 & & & & \\ & 1 & -4 & 6 & -4 & 1 & & & \\ & & \cdot & \cdot & \cdot & \cdot & \cdot & & \\ & & & 1 & -4 & 6 & -4 & 1 & \\ & & & & 1 & -4 & 6 & -4 & 1 \\ & & & & & 1 & -4 & 5 & -2 \\ & & & & & & 1 & -2 & 1 \end{bmatrix}.$$

The simplest parameterization as $g - 2$ contrasts is to extend the corner constraint to fix both the first and last group effects to be zero. The corresponding prior structure matrix is as above, but with first and last rows and columns omitted. An equivalent but computationally inefficient alternative is to expresses the random effect as $g - 2$ *a priori* independent second differences.

The present discussion of first- and second-difference priors has implicitly assumed that the factor levels are, in some sense, equally spaced. This can be relaxed and leads to prior structure matrices of a similar form but with elements whose values depend on the spacing of the observation points. See Berzuini (1995: this volume) for an application of stochastic-trend priors to patient monitoring.

16.6.7 General Markov random field priors

When the factor levels represent positions in some higher-dimensional space, a Markov random field (MRF) prior may be defined in terms of their adjacencies, A_{kl}. Adjacency may be coded as 1 or 0 (for adjacent or not adjacent respectively), or they may code the *degree* of adjacency between levels. The most important use of MRF priors is in the analysis of spatial data.

The model with

$$
\begin{aligned}
U_{kl} &= -A_{kl} && (k \neq l) \\
U_{kk} &= \sum_{l \neq k} A_{kl}
\end{aligned}
$$

leads to a conditional prior mean which is the mean of the effects of adjacent levels (weighted by the degree of their adjacency). As in the case of the random-walk prior, the rank of U is only $g - 1$ and it will be necessary to use a non-aliased parameterization such as the corner parameterization. If the pattern of adjacency is such that the levels fall into several distinct groups or 'islands', then there will be further loss of rank, leading to an improper posterior distribution. A possible remedy is to re-express the model in terms of fixed effects to model differences between islands, and nested random effects for levels within islands. See Mollié (1995: this volume) for an application of MRF priors to disease mapping.

16.6.8 Interactions

A major aim of this chapter is to suggest a class of generalized linear models which are generally useful and which may be easily specified by the user. Interaction terms in the model present a problem.

Certain rules are clear. An interaction term which involves only simple covariates and factors whose main effects are to be modelled as fixed effects (which we shall call *F-factors*) must be a fixed-effect term. Conversely, any interaction which contains a factor whose main effects are modelled as random effects (an *R-factor*) must itself be a random-effect term. However, the appropriate precision structures for random interactions are not so obvious and they lead to the possibility of off-diagonal blocks in the Λ matrix. In this section, some tentative suggestions are made for general rules for specifying Λ when there are interactive random effects.

Interaction between an R-factor and a fixed covariate

The first case to be considered is interaction between a random group effect and a fixed covariate effect. The presence of such interaction implies that the within-group regression lines have different intercepts and slopes. The simplest model of this type assumes that the group-specific slopes and intercepts are independently sampled from a bivariate Gaussian distribution.

This model has g random intercepts and g random slopes. The earlier discussion of intrinsic aliasing now applies to both terms, and it may be preferable to reparameterize each in terms of $g-1$ contrasts. The usual rule for parameterizing linear models is that the section of each x vector which codes for an interaction term is constructed by taking the direct product of the sections which code for the main effects. That is, if for unit i the sections of the covariate vector for model terms j and k take on values $x_i^{[j]}$ and $x_i^{[k]}$, then the section which codes for interaction takes the value $x_i^{[j]} \otimes x_i^{[k]}$, where $\otimes$ denotes the (left) direct product. A rule for deriving the prior precision structure for an interaction between a random factor and a fixed covariate may be constructed by assuming that it is possible to 'centre' the covariate by subtracting a vector of constants in such a way that the random intercepts and slopes are independent of one another. Some algebra then shows that the appropriate prior precision matrix for the model

$$1 + \mathbf{Covariate} + \mathbf{Group} + \mathbf{Group \cdot Covariate}$$

is

$$\Lambda = \begin{bmatrix} 0 & 0 & 0 & 0 \\ 0 & 0 & 0 & 0 \\ 0 & 0 & \theta^{[3]}S & \theta^{[3,4]}S \\ 0 & 0 & \theta^{[3,4]}S & \theta^{[4]}S \end{bmatrix}$$

where S is the prior structure matrix appropriate for the parameterization of group contrasts. Only the prior structure matrix for the random main effect needs to be specified.

Interaction between an R-factor and an F-factor

The model for interaction between a fixed-effect factor and a random-effect factor follows as a straightforward generalization of the above argument, since the main effects of a fixed level factor on g levels are equivalent to $g - 1$ separate covariate effects. For the model

$$1 + \mathbf{Fixed} + \mathbf{Random} + \mathbf{Fixed \cdot Random}$$

the prior precision matrix is

$$\Lambda = \begin{bmatrix} 0 & & \\ & 0 & \\ & & S \otimes \Theta \end{bmatrix}$$

where S is the prior structure matrix for the random-effects factor and Θ is a $g \times g$ symmetric matrix of hyperparameters. Note that Θ_{11} is the hyperparameter for the main effect of the random factor.

Interaction between two R-factors

The final type of interaction which must be considered is between two factors which are both modelled by random effects. Let S_A and S_B be the prior structure matrices for the main effect terms for two such factors, **A** and **B**. The interaction term **A** · **B** must also define a set of random effects and here it is suggested that a sensible rule for deriving the prior precision structure for this term is

$$S_{A \cdot B} = S_A \otimes S_B.$$

It is further suggested that this term should be treated as independent of the main effects terms so that the prior precision matrix for the model

$$1 + A + B + A \cdot B$$

is

$$\Lambda = \begin{bmatrix} 0 & & & \\ & \theta^{[2]} S_A & & \\ & & \theta^{[3]} S_B & \\ & & & \theta^{[4]} S_A \otimes S_B \end{bmatrix}.$$

The rationale for this proposal stems from the observation that, by suitable choice of contrast matrix, it is possible to reparameterize any factor main effect so that its prior structure matrix is the identity matrix. The rule proposed above is equivalent to suggesting that, if both factors are reparameterized in this way so that $S_A = I$ and $S_B = I$, then it should follow that $S_{A \cdot B} = I$ also.

This rule seems to generate plausible prior structure matrices. For example, if the priors for the main effects of **A** and **B** are both random-walk priors as illustrated in Figure 16.2, then this rule generates a prior structure matrix for the **A** · **B** interaction term which may be represented by the undirected conditional-independence graph of the form illustrated in Figure 16.4. By choosing contrast matrices so that the main effects are coded as first differences, this model represents interaction *a priori* in a table, y, cross-classified by the two factors as a series of independent contrasts of the form

$$y_{k+1,l+1} - y_{k,l+1} - y_{k+1,l} + y_{k,l}.$$

16.7 Hyperpriors and the estimation of hyperparameters

In the Gibbs sampling algorithm, values of the hyperparameters, $\theta^{[j]}$, are sampled from their full conditional distributions. These are equivalent to

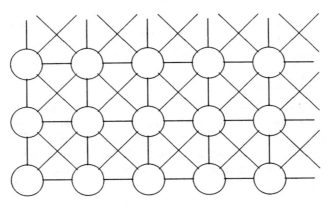

Figure 16.4 *Undirected conditional-independence structure for* **A · B** *interaction.*

Bayesian posteriors derived by treating the current value of the corresponding vector of parameters, $\beta^{[j]}$, as observed data. Since $[\beta|\theta]$ is Gaussian, this is a standard problem.

In most cases, the vector $\beta^{[j]}$ is conditionally independent of other parameters given the data and the hyperparameter $\theta^{[j]}$. The most convenient choice of hyperprior for $\theta^{[j]}$ is then a gamma distribution with scale parameter $\alpha^{[j]}$ and shape parameter (degrees of freedom) $\nu^{[j]}$. With this choice, the full conditional distribution of $\theta^{[j]}$ is also a gamma distribution, with scale parameter $\alpha^{[j]} + (\beta^{[j]})^{\mathrm{T}} S^{[j]} \beta^{[j]}$ and shape parameter $\nu^{[j]} + p^{[j]}$, where $p^{[j]}$ is the number of parameters in the jth term.

It is also useful and desirable in practice to be able to specify a finite range of support for the hyperprior. This is trivial to implement since all that is necessary is to discard any value which falls outside the specified range and resample.

When there are off-diagonal blocks in the Λ matrix, as in Section 16.6.8, the position is rather more complicated. The natural choice of hyperprior for the Θ matrix is a Weibull distribution, since this leads to another Weibull distribution for the full conditional distribution.

The most straightforward method for sampling the hyperparameters is the Gibbs sampler, in which each hyperparameter is visited in turn and sampled from its full conditional distribution. However, it is likely that other MCMC methods might perform rather better for this purpose, since there may be relatively strong posterior correlations between different hyperparameters; see Gilks and Roberts (1995: this volume) and Besag *et al.* (1995).

16.8 Some examples

This section describes some of the many uses of GLMMs in biostatistics.

16.8.1 Longitudinal studies

An important application of random-effects models is to longitudinal studies in which measurements are available for a group of subjects at each of a series of time points. Covariates may vary between subjects but also within subjects over time. Such data may be modelled using GLMMs with random subject effects.

Thall and Vail (1990) describe such a study in which people with epilepsy were randomly assigned to two treatment groups and seizures counted for each of four post-treatment periods. The pre-treatment seizure rate (**Pre**) was used as a covariate. Breslow and Clayton (1993) discuss various GLMMs for these data. Conditional upon subject-level random effects, it seems most natural to adopt a GLM with Poisson errors and log link. The simplest model is

$$1 + \texttt{Pre} + \texttt{Treatment} + \texttt{Period} + \texttt{Subject}$$

in which **Treatment** and **Period** are fixed-effect factors and **Subject** is a factor modelled as a random effect using one of the approaches outlined in Section 16.6.3. A more elaborate model allows for a linear effect of time since start of treatment (**T**), randomly varying between subjects, by inclusion of two new terms:

$$+\texttt{T} + \texttt{Subject} \cdot \texttt{T}.$$

16.8.2 Time trends for disease incidence and mortality

The first- and second-difference priors described in Sections 16.6.5 and 16.6.6 may be used for semiparametric smooth curve fitting. A particularly fertile area for such models is in the study of time trends. For example, in epidemiology a frequent requirement is to analyse counts of cases (individuals with disease) and population denominators in rectangular tables cross-classified by age and calendar period. The GLM approach to such data uses log-linear Poisson models such as

$$\log(\texttt{Rate}) = 1 + \texttt{Age} + \texttt{Period},$$

a model in which the time trends for different age groups are parallel. Commonly, **Period** is grouped rather coarsely and fitted as a fixed-effect factor. Alternatively, polynomial models can be used but these are usually criticized on the grounds that a change of a polynomial coefficient to adapt the model to one data point can have very strong influence on fitted values remote from it. A better alternative is to subdivide finely into periods and fit this term as a random effect with a first- or second-difference prior. In the latter case, it will also be necessary to fit a linear trend term since the second difference parameterization represents deviations from a linear trend.

For many diseases, it is sufficient to code age in broad (5- or 10-year) bands and fit fixed effects. However, for diseases such as childhood leukaemia which have a more irregular age dependency, it may be necessary to group into much finer age bands and to fit random effects with first- or second-difference priors as for the Period effects. In other cases, trends may be better explained in terms of Cohort effects: differences between groups defined by date of birth.

A more general model for trends in disease rates is the age–period–cohort model

$$1 + \text{Age} + \text{Period} + \text{Cohort},$$

which has received considerable attention in the epidemiological literature. This model is not fully identifiable owing to the linear dependency between age, present date, and date of birth (see, for example, Clayton and Schifflers, 1987). As a result of this dependency, it is only possible to estimate two parameters for the linear trend components of the Age, Period, and Cohort effects. Second differences are, however, identifiable. Berzuini and Clayton (1994) investigated the use of an age–period–cohort GLMM for modelling disease trends on a very fine time grid. Identifiability considerations lead naturally to the choice of stochastic trend (second-difference) priors for the random effects of Age, Period, and Cohort, supplemented by fixed linear trend effects for two of these.

Keiding (1990) has discussed smooth non-parametric estimation of hazard rates in two-dimensional time planes such as the age-period space, and proposed the use of bivariate kernel smoothing. However, there are difficulties in selecting values for the parameters which define the kernel function since, for a bivariate kernel, there should be at least three such parameters. The ideas outlined in Section 16.6.8 suggest a more formal approach by fitting the model

$$1 + \text{Age} + \text{Period} + \text{Age} \cdot \text{Period}$$

to data tabulated on a very fine grid. The three hyperparameters for the three sets of random effects are analogous to the three parameters which would properly be needed to specify a bivariate kernel smoother.

16.8.3 Disease maps and ecological analysis

A particularly fruitful area for application of GLMMs has been geographical epidemiology. Two problems are of interest:

1. the estimation of *maps* of geographical variation of disease incidence and mortality (closely related to the identification of *disease clusters*); and

2. the investigation of *ecological relationships* between disease rates and population mean exposures taking geographically defined population groups as the data points.

These problems are discussed at length in Mollié (1995: this volume).

Clayton and Kaldor (1987) first considered the disease mapping problem using random-effects models. The problem amounts to the analysis of tables with rows defined by age groups and a very large number of columns defined by geographical area. The data in each cell are counts of cases of disease together with the corresponding person-years observation. It has been known for some time that the classical procedure of indirect age standardization is equivalent to fitting the model

$$1 + \texttt{Age} + \texttt{Area}$$

with log link and Poisson errors. However, Clayton and Kaldor drew attention to serious deficiencies of the MLEs of **Area** effects for mapping disease. Instead they suggested fitting **Area** as random effects and computing empirical Bayes estimates for mapping purposes. A similar suggestion was made by Manton *et al.* (1989).

This suggestion works perfectly well so long as there remains an appreciable amount of data for each area but, when the data concern small areas, this may not always be the case. However, if the geographical variation is such that there is a tendency for neighbouring areas to have similar rates, the use of a MRF prior will allow estimation of the map. Clayton and Kaldor suggested the use of a spatially autocorrelated prior with a spatial autocorrelation parameter whose value corresponds to a smooth gradation from total *a priori* independence of areas at one extreme to strong dependence on neighbouring rates (as in Section 16.6.7) at the other.

A practical problem with this model is that the spatial autocorrelation parameter is difficult to estimate. Even when using Gibbs sampling, the conditional posterior for this parameter involves the determinant of a $p \times p$ matrix, where p is the number of areas. Besag *et al.* (1991) suggested decomposing the **Area** effect into two parts. In the terminology introduced by Clayton and Bernardinelli (1991) this suggestion corresponds to the model

$$1 + \texttt{Age} + \texttt{Heterogeneity} + \texttt{Clustering}$$

where **Heterogeneity** refers to an independent random area effect (as described in Section 16.6.3), and **Clustering** refers to a completely spatially-autocorrelated random area effect (as described in Section 16.6.7). Note that the relative magnitude of these two components of the area effect is controlled by the corresponding hyperparameters and these are estimated from the data.

Clayton and Bernardinelli (1991) extended this model further by the inclusion of covariates and discussed its uses in the study of ecological relationships.

16.8.4 Simultaneous variation in space and time

The preceding two applications may be regarded as simple special cases of a general concern of epidemiology: the simultaneous variation of disease rates over space *and* time. This section considers the use of GLMMs for this more general problem. Since it describes work in progress, it is more speculative than earlier sections.

The data for such studies will consist of a three-dimensional table of rates defined by age, geographical area, and calendar period. In principle the joint variation of (age-standardized) rates in space and time is described by the model

$$1 + \text{Area} + \text{Period} + \text{Area} \cdot \text{Period}.$$

where **Area** is modelled by random effects; **Period** may be modelled by fixed effects if the effect may be assumed to be linear within areas and/or if there are relatively few periods; and the **Area · Period** interaction is also modelled as a random effect with a prior structure derived as suggested in Section 16.6.8. The **Area** effect can be modelled using either prior independence assumptions or Markov random field priors. Alternatively the suggestion of Besag *et al.* (1991) may be extended, leading to the model

$$
\begin{aligned}
1 \quad &+ \quad \text{Heterogeneity} + \text{Clustering} + \text{Period} \\
&+ \quad \text{Heterogeneity} \cdot \text{Period} + \text{Clustering} \cdot \text{Period}.
\end{aligned}
$$

Models like this could be used, for example, to investigate the hypothesis that the disease has an infectious aetiology. As before, models can be enriched by including covariate effects, thus allowing investigation of effects of differential intervention, such as the introduction of screening programmes for early detection and treatment of disease.

16.8.5 Frailty models in survival analysis

There is a close connection between counting-process models for analysis of survival times and event histories, and generalized linear models (see for example Clayton, 1988). When events are assumed to arise from an exponential survival time distribution or from a time-homogeneous Poisson process, the likelihood contributions for each observation are Poisson in form and conventional GLM software can be used to model the relationship between covariates and the probability rate for the event of interest. When there is a requirement to allow rates to vary with time, a likelihood factorization permits the partitioning of an observation of a subject through time into a series of observations through different parts of the time scale. This is illustrated in Figure 16.5. The line at the top of the figure represents observation of a subject until eventual failure, and the three subsequent lines represent the partitioning of this observation into three sections according to a stratification of the time-scale; the first two observations end in censoring, while the third ends in failure.

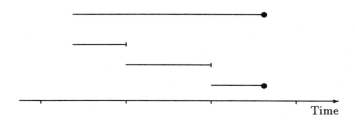

Figure 16.5 *The follow-up record in survival analysis.*

Proportional hazard models for such data are equivalent to GLMs with Poisson errors, log link function, and models of the form

$$1 + \texttt{Time} + \texttt{Covariate}.$$

For large-scale epidemiological analyses, **Time** is usually grouped rather coarsely into 5- or 10-year bands but, in the version of the model proposed by Cox (1972), time is infinitely subdivided. Even in this latter case, however, **Time** is modelled as a fixed effect and the MLE of the time effect is zero everywhere except for the points at which failures occurred. In the work of Berzuini and Clayton (1994), it was pointed out that an alternative is to subdivide the time-scale rather finely and use priors such as those described in Sections 16.6.5 and 16.6.6 to obtain smooth non-parametric estimates of the relationship between event rate and time.

Perhaps a more significant possibility is opened up when there is some relationship between groups of observations. For example, the data may represent repeated (censored) observations of a failure time within the same subject, or survival times of different members of a family. Random effect models are then appropriate. For example, for repeated observations within the same subject, one might fit the model

$$1 + \texttt{Time} + \texttt{Subject} + \texttt{Covariate}$$

where **Subject** is fitted as random effects. Such models are often called 'frailty' models. Clayton (1991) discussed implementation via Gibbs sampling of the non-parametric time-effect model with a log-gamma distribution for the random effects. However, the assumption of Gaussian random effects is just as easily implemented and then frailty models are a special case of the GLMMs described here. The Gaussian assumption allows greater flexibility than the more usual gamma assumption in its capacity to model multiple, correlated random effects. For example, in twin and other family studies, the assumption that frailty is genetically inherited suggests a model in which the pattern of intercorrelation between random effects for related individuals is a function of the degree of kinship: see Thomas and Gauderman (1995: this volume).

16.9 Discussion

The work described in this chapter has been carried out using a package of C functions developed by the author. For each application, it has been necessary to write a main program to read the data and to set up the model. Only block-diagonal prior precision matrices are currently implemented. Despite these limitations, experience has been sufficient to demonstrate that the class of models described above are generally useful.

One of the aims of this chapter has been to present a family of models which could be implemented reasonably simply. In particular, it is important that models can be specified using the syntax of Wilkinson and Rogers (1973) as used in the GLIM program, thus allowing expansion of the X matrix by indicator variables which ensure correct parameterization of effects of factors and of interaction terms. A more recent implementation in the S language (Chambers and Hastie, 1992) adds the ability to attach a contrast matrix, C, as an attribute of a factor. This chapter suggests that further extension of the syntax to allow specification of a broad class of GLMMs requires only the ability to define a prior structure as a further attribute of any factor to be modelled by random effects. The most natural way to do this is to attach, as an attribute of a random factor, the matrix U representing the prior precision structure for the full dummy variable parameterization of the model term.

Gibbs sampling for GLMMs has proved highly successful in a number of diverse applications. The most important conditions for the method to work in practice are that:

1. the X matrix is not too ill-conditioned; and

2. proper priors are used for the hyperparameters, θ, when there are insufficient data to yield a concentrated likelihood.

Conditioning of the X matrix can be improved by careful parameterization of the model. For example, covariates should be centred (see Gilks and Roberts, 1995: this volume) and any strong dependencies between covariates removed by the construction of new indices (for example, replacing height and weight by height and Quetelet's index: weight/(height)^2). The problem of specifying suitable priors for the hyperparameters is a more difficult one. In some cases, improper priors can be used, but this course of action is risky since improper posteriors may be obtained in some circumstances. A useful device for ensuring proper posteriors is to limit the support for the hyperpriors, as indicated in Section 16.7; a uniform prior may be adopted for a variance component, but specification of upper and lower bounds is advisable.

The use of MCMC methods for GLMMs has several advantages over the alternative iteratively reweighted least-squares (IRLS) methods reviewed by Breslow and Clayton (1993):

1. IRLS methods often require inversion of extremely large matrices. The exception to this rule is nested-random-effect models in which the prior structure matrices have special structures (such as diagonal or tridiagonal) since the relevant matrix inversion may then be carried out by partitioning. In other cases, such as spatial problems and problems involving crossed random effects, the computations may be very laborious. In contrast, Gibbs sampling requires only that the prior structure matrices are sparse.

2. IRLS methods do not compute interval estimates for random effects or for hyperparameters. These are obtained as a natural by-product of MCMC methods.

However, the restriction of the GLMMs to the class described in this chapter does have some undesirable consequences. In particular, the following models (which must properly considered to be GLMMs) are not easily dealt with:

1. models in which the prior distributions of random effects are better described in terms of variance-covariance matrices and for which the corresponding precision matrices are not sparse;

2. random effects whose prior precision matrices depend upon several hyperparameters (that is, the case where $\theta^{[j]}$ is a vector rather that a scalar).

The problem in this latter case is that the full conditional distribution for $\theta^{[j]}$ depends on the determinant of the matrix $\Lambda^{[j]}(\theta^{[j]})$ and the repeated evaluation of this determinant generally imposes a heavy computational burden.

Despite these limitations, MCMC methods have provided a technical breakthrough for estimation in this broad and useful class of models.

Acknowledgement

This work was supported in part by United States Department of Health and Human Services grant number CA61042–1.

References

Berzuini, C. (1995) Medical monitoring. In *Markov Chain Monte Carlo in Practice* (eds W. R. Gilks, S. Richardson and D. J. Spiegelhalter), pp. 321–337. London: Chapman & Hall.

Berzuini, C. and Clayton, D. G. (1994) Bayesian analysis of survival on multiple time-scales. *Statist. Med.*, **13**, 823–838.

Besag, J., York, J. and Mollié, A. (1991) Bayesian image restoration with two applications in spatial statistics. *Ann. Inst. Statist. Math.*, **43**, 1–59.

Besag, J., Green, P. J., Higdon, D. and Mengersen, K. (1995) Bayesian computation and stochastic systems. *Statist. Sci.*, **10**, 3–41.

Breslow, N. and Clayton, D. (1993) Approximate inference in generalized linear mixed models. *J. Am. Statist. Ass.*, **88**, 9–25.

Bryk, A. and Raudenbush, S. (1992) *Hierarchical Linear Models: Applications and Data Analysis Methods.* Newbury Park: Sage.

Chambers, J. and Hastie, T. (eds) (1992) *Statistical Models in S.* Pacific Grove: Wadsworth and Brooks/Cole.

Clayton, D. (1988) The analysis of event history data: a review of progress and outstanding problems. *Statist. Med.*, **7**, 819–41.

Clayton, D. (1991) A Monte Carlo method for Bayesian inference in frailty models. *Biometrics*, **47**, 467–85.

Clayton, D. and Bernardinelli, L. (1991) Bayesian methods for mapping disease risk. In *Small Area Studies in Geographical and Environmental Epidemiology* (eds J. Cuzick and P. Elliot), pp. 205–220. Oxford: Oxford University Press.

Clayton, D. and Kaldor, J. (1987) Empirical Bayes estimates of age-standardized relative risks for use in disease mapping. *Biometrics*, **43**, 671–81.

Clayton, D. and Schifflers, E. (1987) Models for temporal variation in cancer rates. II: Age-period-cohort models. *Statist. Med.*, **6**, 469–81.

Cox, D. R. (1972) Regression models and life tables (with discussion). *J. R. Statist. Soc.* B, **34**, 187–220.

Dellaportas, P. and Smith, A. F. M. (1993) Bayesian inference for generalized linear and proportional hazards models via Gibbs sampling. *Appl. Statist.*, **42**, 443–60.

Dempster, A. (1972) Covariance selection. *Biometrics*, **28**, 157–75.

Francis, B., Green, M. and Payne, C. (eds) (1993) *The GLIM System: Release 4 Manual.* Oxford: Oxford University Press.

Gilks, W. R. (1995) Full conditional distributions. In *Markov Chain Monte Carlo in Practice* (eds W. R. Gilks, S. Richardson and D. J. Spiegelhalter), pp. 75–88. London: Chapman & Hall.

Gilks, W. R. and Roberts, G. O. (1995) Strategies for improving MCMC. In *Markov Chain Monte Carlo in Practice* (eds W. R. Gilks, S. Richardson and D. J. Spiegelhalter), pp. 89–114. London: Chapman & Hall.

Gilks, W. R. and Wild, P. (1992) Adaptive rejection sampling for Gibbs sampling. *Appl. Statist.*, **41**, 1443–63.

Gilks, W. R., Richardson, S. and Spiegelhalter, D. J. (1995) Introducing Markov chain Monte Carlo. In *Markov Chain Monte Carlo in Practice* (eds W. R. Gilks, S. Richardson and D. J. Spiegelhalter), pp. 1–19. London: Chapman & Hall.

Harville, D. (1977) Maximum likelihood approaches to variance component estimation and to related problems. *J. Am. Statist. Ass.*, **72**, 320–40.

Keiding, N. (1990) Statistical inference in the Lexis diagram. *Phil. Trans. Roy. Soc. Lon.* A, **332**, 487–509.

Manton, K., Woodbury, M. A., Stallard, E., Riggan, W., Creason, J. and Pellom, A. (1989) Empirical Bayes procedures for stabilizing maps of US cancer mortality rates. *J. Am. Statist. Ass.*, **84**, 637–50.

McCullagh, P. and Nelder, J. (1989) *Generalized Linear Models (2nd edn)*. London: Chapman & Hall.

Mollié, A. (1995) Bayesian mapping of disease. In *Markov Chain Monte Carlo in Practice* (eds W. R. Gilks, S. Richardson and D. J. Spiegelhalter), pp. 359–379. London: Chapman & Hall.

Nelder, J. and Wedderburn, R. (1972) Generalized linear models. *J. R. Statist. Soc.* A, **135**, 370–84.

Spiegelhalter, D. J., Best, N. G., Gilks, W. R. and Inskip, H. (1995) Hepatitis B: a case study in MCMC methods. In *Markov Chain Monte Carlo in Practice* (eds W. R. Gilks, S. Richardson and D. J. Spiegelhalter), pp. 21–43. London: Chapman & Hall.

Thall, P. and Vail, S. (1990) Some covariance models for longitudinal count data with overdispersion. *Biometrics*, **46**, 657–71.

Thomas, D. C. and Gauderman, W. J. (1995) Gibbs sampling methods in genetics. In *Markov Chain Monte Carlo in Practice* (eds W. R. Gilks, S. Richardson and D. J. Spiegelhalter), pp. 419–440. London: Chapman & Hall.

Whittaker, J. (1990) *Graphical Models in Applied Multivariate Statistics*. New York: Wiley.

Wilkinson, G. and Rogers, C. (1973) Symbolic description of factorial models. *Appl. Statist.*, **22**, 392–9.

17

Hierarchical longitudinal modelling

Bradley P Carlin

17.1 Introduction

Many datasets arising in statistical and biostatistical practice consist of repeated measurements (usually ordered in time) on a collection of individuals. Data of this type are referred to as *longitudinal data*, and require special methods for handling the correlations that are typically present among the observations on a given individual. More precisely, suppose we let Y_{ij} denote the jth measurement on the ith individual in the study, $j = 1, \ldots, s_i$ and $i = 1, \ldots, n$. Arranging each individual's collection of observations in a vector $Y_i = (Y_{i1}, \ldots, Y_{is_i})^{\mathrm{T}}$, we might attempt to fit the usual fixed-effects model

$$Y_i = X_i \alpha + \epsilon_i \,,$$

where X_i is a $s_i \times p$ design matrix of covariates and α is a $p \times 1$ parameter vector. While the usual assumptions of normality, zero mean, and common variance might still be appropriate for the components of ϵ_i, their independence is precluded by the aforementioned correlations within the Y_i vector. Hence we would be forced to adopt a general covariance matrix Σ, though we might assume that this matrix follows some simplifying pattern, such as compound symmetry or a first-order autoregressive structure. Another difficulty is that data from living subjects frequently exhibit more variability than can be adequately explained using a simple fixed-effects model.

A popular alternative is provided by the *random-effects* model

$$Y_i = X_i \alpha + W_i \beta_i + \epsilon_i \,, \tag{17.1}$$

where W_i is a $s_i \times q$ design matrix (q typically less than p), and β_i is a $q \times 1$ vector of subject-specific random effects, usually assumed to be normally distributed with mean vector 0 and covariance matrix V. The β_i capture any subject-specific mean effects, and also enable the model to reflect any extra-normal variability in the data accurately. Furthermore, it is now realistic to assume that, given β_i, the components of ϵ_i *are* independent, and hence we might set $\Sigma = \sigma^2 I_{s_i}$; marginalizing over β_i, the Y_i components are again correlated, as desired. Models of this type have been very popular for longitudinal data since their appearance in the paper of Laird and Ware (1982).

Notice that models like (17.1) can accommodate the case where not every individual has the same number of observations (unequal s_i), perhaps due to missed visits, data transmission errors, or study withdrawal. In fact, features far broader than those suggested above are possible in principle. For example, heterogeneous error variances σ_i^2 may be employed, removing the assumption of a common variance shared by all subjects. The Gaussian errors assumption may be dropped in favour of other symmetric but heavier-tailed densities, such as the student t-distribution. Covariates whose values change and growth curves that are nonlinear over time are also possible within this framework (see Bennett *et al.*, 1995: this volume).

In practice, however, the benefits described in the preceding paragraph would only be attainable using traditional statistical methods if it were possible to integrate all of the β_i nuisance parameters out of the likelihood in closed form, leaving behind a low-dimensional marginal likelihood for the fixed effects α. This would, in turn, limit consideration to fairly special cases (e.g. balanced data with a normal error distribution for both the data and the random effects). An increasing number of popular commercially available computer packages (e.g. SAS Proc Mixed and BMDP 5V) allow the fitting of models of this type. Still, to be of much use in practice, our computational approach must not be subject to these limitations.

As a result, we adopt a Markov chain Monte Carlo (MCMC) approach to perform the integrations in our high-dimensional (and quite possibly analytically intractable) likelihood. Several previous papers (e.g. Zeger and Karim, 1991; Gilks *et al.*, 1993) have employed MCMC methods in longitudinal modelling settings as general as ours or more so. Longitudinal models appear throughout this volume, and in particular in the chapters by Spiegelhalter *et al.*, Berzuini, Bennett *et al.* and Clayton.

Given this computational setting, it becomes natural to adopt a fully Bayesian point of view, though several authors have shown the applicability of MCMC methods to certain likelihood analyses (see for example Geyer and Thompson, 1992; Kolassa and Tanner, 1994; Carlin and Gelfand, 1993; Geyer, 1995: this volume). Hence we require prior distributions for α, σ^2, and V to complete our model specification. Bayesians often think of the distribution on the random effects β_i as being part of the prior (the 'first stage', or *structural* portion), with the distribution on V forming another part (the 'second stage', or *subjective* portion). Together, these two pieces are sometimes referred to as a *hierarchical prior*, motivating the term *hierarchical model* to describe the entire specification.

The remainder of this chapter is organized as follows. Section 17.2 briefly describes the clinical background to our study of the progression of human immunodeficiency virus (HIV). Section 17.3 presents the precise form of model (17.1) appropriate for our dataset and outlines the diagnostic and analytic tools used in the implementation of our MCMC algorithm. The results of applying this methodology to data arising from a clinical trial established to compare the effect of two antiretroviral drugs on the immune systems of patients with late-stage HIV infection are given in Section 17.4. Finally, Section 17.5 summarizes our findings and discusses their implications for the routine use of MCMC Bayesian methods in longitudinal data analysis.

17.2 Clinical background

The absolute number of CD4 lymphocytes in the peripheral blood is used extensively as a prognostic factor and surrogate marker for progression of disease and for death consequent to infection with HIV. As a prognostic tool, the CD4 cell count is used to initiate and modify antiretroviral treatment, for initiation of primary prophylaxis against various opportunistic infections, and for selection and stratification of patients in clinical trials. It has also been extensively recommended as a surrogate endpoint for progression of disease events and for death, in order to facilitate earlier conclusions about the efficacy of antiretroviral drugs. Indeed, an increase in the average CD4 count of patients receiving the drug didanosine (ddI) in a clinical trial was the principal argument used in its successful bid for limited licensing in the United States. In a pre-licensing expanded access trial of another such drug, zalcitabine (ddC), using patients unable to tolerate either zidovudine (AZT) or ddI, those patients with higher CD4 counts at baseline had a CD4 rise within the first 12 weeks, apparently sustained for some time.

A recently concluded US study, funded by the National Institutes of Health and conducted by the Community Programs for Clinical Research on AIDS (CPCRA), has provided additional data relevant to the issue of CD4 response in patients taking ddI and ddC. This study was a direct randomized comparison of these two drugs in patients intolerant or unresponsive to AZT; lack of effective response to AZT will be termed *AZT failure*, below. While several recent papers have questioned the appropriateness of CD4 count as a surrogate endpoint for clinical disease progression or death (Choi *et al.*, 1993; Goldman *et al.*, 1993; Lin *et al.*, 1993), in this analysis we use the CPCRA trial data to test whether ddI and ddC produce an increase in average CD4 count, and whether the CD4 patterns for the HIV-infected individuals in the two drug groups differ significantly.

The details of the conduct of the ddI/ddC study are described by Abrams *et al.* (1994); only the main relevant points are given here. HIV-infected patients were eligible if they had two CD4 counts of 300 or fewer cells or had AIDS, and if they fulfilled specific criteria for AZT intolerance or AZT failure. The randomization was stratified by clinical unit and by AZT intolerance versus failure, into treatment groups of either 500 mg per day ddI or 2.25 mg per day ddC. The study opened January, 1991 and enrolment of $n = 467$ patients was completed the following September. CD4 counts were measured at baseline (study entry) and again at 2, 6, 12, and, in some cases, 18 months post-baseline, so that $s_i = 5$ for an individual with no missing data. Any drug-induced improvement in CD4 count should manifest itself clinically by the 2-month visit.

17.3 Model detail and MCMC implementation

Since we wish to detect a possible boost in CD4 count two months after baseline, we attempt to fit a model that is linear but with possibly different slopes before and after this time. Thus the subject-specific design matrix W_i for patient i in (17.1) has j^{th} row

$$w_{ij} = (1, \, t_{ij}, \, (t_{ij} - 2)^+),$$

where $t_{ij} \in \{0, 2, 6, 12, 18\}$ and $z^+ = \max(z, 0)$. The three columns of W_i correspond to individual-level intercept, slope, and change in slope following the change-point, respectively. We account for the effect of covariates by including them in the fixed-effect design matrix X_i. Specifically, we set

$$X_i = (W_i, \, d_i W_i, \, a_i W_i), \qquad (17.2)$$

where d_i is a binary variable indicating whether patient i received ddI ($d_i = 1$) or ddC ($d_i = 0$), and a_i is another binary variable telling whether the patient was diagnosed as having AIDS at baseline ($a_i = 1$) or not ($a_i = 0$). Notice from (17.2) that $p = 3q = 9$; the two covariates are being allowed to affect any or all of the intercept, slope, and change in slope of the overall population model. The corresponding elements of the α vector then quantify the effect of the covariate on the form of the CD4 curve. In particular, our interest focuses on the α parameters corresponding to drug status, and whether they differ from 0. The β_i are of secondary importance, though they would be critical if our interest lay in individual-level prediction.

Adopting the usual exchangeable normal model for the random effects, we obtain a likelihood of the form

$$\prod_{i=1}^{n} N_{s_i}(Y_i | X_i \alpha + W_i \beta_i, \sigma^2 I_{s_i}) \prod_{i=1}^{n} N_3(\beta_i | 0, V), \qquad (17.3)$$

where $N_k(\cdot | \mu, \Sigma)$ denotes the k-dimensional normal distribution with mean vector μ and covariance matrix Σ. To complete our Bayesian model specification, we adopt the prior distributions $\alpha \sim N_9(c, D)$, $\tau \sim Ga(a, b)$, and

$U \sim \mathrm{W}(\rho, (\rho R)^{-1})$, where $\tau = \sigma^{-2}$, $U = V^{-1}$, $\mathrm{Ga}(a, b)$ denotes a Gamma distribution with mean a/b and variance a/b^2 and $\mathrm{W}(\rho, (\rho R)^{-1})$ denotes a Wishart distribution with degrees of freedom ρ and expectation R^{-1}. (Thus U has a distribution proportional to

$$|U|^{\frac{\rho-m-1}{2}} \exp\left[-\frac{\rho}{2}\mathrm{trace}(RU)\right],$$

where $m = 3$ and the proportionality constant is a function of m, R and ρ only.) See Bernardo and Smith (1994) for further discussion of the Wishart distribution and its use as a prior. These distributions are *conjugate* with the likelihood (17.3) in the sense that they lead to full conditional distributions for α, τ, and U that are again normal, gamma, and Wishart, respectively, thus facilitating implementation of the Gibbs sampler (see Gilks *et al.*, 1995: this volume). While occasionally criticized as too restrictive, conjugate priors offer sufficient flexibility in our highly parameterized model setting, as we demonstrate below.

Values for the parameters of our priors (sometimes called *hyperparameters*) may sometimes be elicited from clinical experts (Chaloner *et al.*, 1993) or past datasets (Lange *et al.*, 1992). In our case, since little is reliably known concerning the CD4 trajectories of late-stage patients receiving ddI or ddC, we prefer to choose hyperparameter values that lead to fairly vague, minimally informative priors. Care must be taken, however, that this does not lead to impropriety in the posterior distribution. For example, taking ρ extremely small and D extremely large would lead to confounding between the fixed and random effects. (To see this, consider the simple but analogous case of the one-way layout, where $Y_i \sim \mathrm{N}(\mu + \alpha_i, \sigma^2)$ independently for $i = 1, \ldots, n$, with non-informative priors on both μ and α_i.) To avoid this while still allowing the random effects a reasonable amount of freedom, we adopt the rule of thumb wherein $\rho = n/20$ and $R = \mathrm{Diag}((r_1/8)^2, (r_2/8)^2, (r_3/8)^2)$, where r_i is total range of plausible parameter values across the individuals. Since R is roughly the prior mean of V, this gives a ± 2 prior standard deviation range for β_i that is half of that plausible in the data. Since in our case Y_{ij} corresponds to square root CD4 count in late-stage HIV patients, we take $r_1 = 16$ and $r_2 = r_3 = 2$, and hence our rule produces $\rho = 24$ and $R = \mathrm{Diag}(2^2, (0.25)^2, (0.25)^2)$.

Turning to the prior on σ^2, we take $a = 3$ and $b = 200$, so that σ^2 has both mean and standard deviation equal to 10^2, a reasonably vague (but still proper) specification. Finally, for the prior on α we set $c = (10, 0, 0, 0, 0, 0, -3, 0, 0)$ and $D = \mathrm{Diag}(2^2, 1^2, 1^2, (0.1)^2, 1^2, 1^2, 1^2, 1^2, 1^2)$, a more informative prior but biased strongly away from 0 only for the baseline intercept, α_1, and the intercept adjustment for a positive AIDS diagnosis, α_7. These values correspond to our expectation of mean baseline square root CD4 counts of 10 and 7 (100 and 49 on the original scale) for the AIDS-negative and AIDS-positive groups, respectively. This prior also insists that the drug intercept be very small in magnitude, since patients were assigned to drug group at random (α_4 is merely a placeholder in our model). It also allows the data to determine the degree to which the CD4 trajectories depart from horizontal, both before and after the change-point. Similarly, while both α and the β_i have been assumed uncorrelated *a priori*, both specifications are vague enough to allow the data to determine the proper amount of correlation *a posteriori*.

As mentioned above, we adopt the Gibbs sampler as our MCMC method due to the conjugacy of our prior specification with the likelihood. The required full conditional distributions for our model are presented in Section 2.4 of Lange *et al.* (1992), who used them in a change-point analysis of longitudinal CD4 cell counts for a group of HIV-positive men in San Francisco; the reader is referred to that paper for details. Alternatively, our model could be described graphically as in Figure 17.1 (see also Spiegelhalter *et al.*, 1995: this volume), and subsequently analysed in BUGS (Gilks *et al.*, 1992). This graphical modelling and Bayesian sampling computer package frees the user from computational details, though the current BUGS restriction to univariate nodes complicates our longitudinal model specification substantially.

The MCMC Bayesian approach also facilitates choosing among various competing models through calculation of the distribution of each datapoint given the remaining points in the dataset, sometimes called the *cross-validation predictive* distribution. For example, writing $Y_{(ij)}$ as the collection of all the observed data except Y_{ij}, we could subtract the mean of this distribution from each datapoint, obtaining the Bayesian cross-validation residual $e_{ij} = y_{ij} - E(Y_{ij}|y_{(ij)})$; here Y denotes a random variable, and y denotes its observed value. Better-fitting models will have generally smaller residuals, as measured by the sum of their squares or absolute values, or perhaps a graphical display such as a histogram. We also might plot each e_{ij} versus its standard deviation $\sqrt{\text{var}(Y_{ij}|y_{(ij)})}$, to measure model accuracy relative to precision. Alternatively, we might look directly at the value of the cross-validation predictive density at the observed datapoints, $f(y_{ij}|y_{(ij)})$, sometimes called the *conditional predictive ordinate*, or CPO. Since they are nothing but observed likelihoods, larger CPOs suggest a more likely model. A suitable summary measure here might be the log of the product of the CPOs for each datapoint. Notice that using either of these two criteria, the competing models need not use the same parametric structure, or even the same covariates. This is in stark contrast to traditional statistical methods, where two models cannot be compared unless one is a special case of the other (i.e., the models are 'nested'). Estimates of $E(Y_{ij}|y_{(ij)})$, $\text{var}(Y_{ij}|y_{(ij)})$, and $f(y_{ij}|y_{(ij)})$ are easily obtained using the output from the Gibbs sampler, as described in Gelfand *et al.* (1992) and Gelfand (1995: this volume).

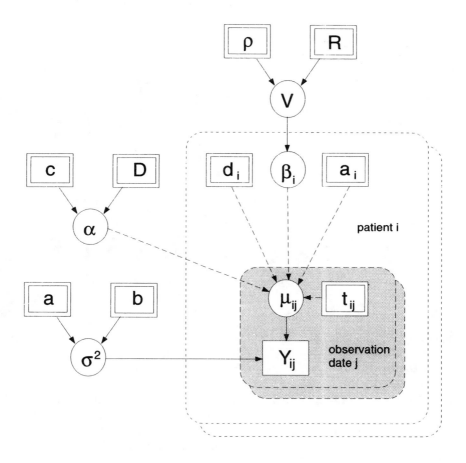

Figure 17.1 *Graphical representation of the model.*

17.4 Results

Before undertaking the formal analyses described in the previous section, we consider some exploratory and summary plots of our dataset. Boxplots of the individual CD4 counts for the two drug groups, shown in Figures 17.2(a) and 17.2(b), indicate a high degree of skewness toward high CD4 values. This, combined with the count nature of the data, suggests a square root transformation for each group. As Figures 17.2(c) and 17.2(d) show, this transformation improves matters considerably; while upper tails are still present, plots on the log scale (not shown) indicate that this transformation goes a little too far, creating lower tails. Since many authors

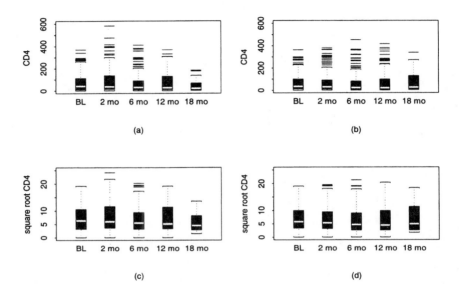

Figure 17.2 *Exploratory plots of CD4 counts over time, ddI/ddC data: (a) CD4 counts, ddI treatment group; (b) CD4 counts, ddC treatment group; (c) square root CD4 counts, ddI group; (d) square root CD4 counts, ddC group.*

(e.g. Lange *et al.*, 1992; McNeil and Gore, 1994) have used the square root transformation with CD4 counts, we employ it in all subsequent analyses. The sample medians, shown as white horizontal bars on the boxplots, offer reasonable support for our assumption of a linear decline in square root CD4 after two months. The sample sizes at the five time points, namely (230, 182, 153, 102, 22) and (236, 186, 157, 123, 14) for the ddI and ddC groups, respectively, indicate an increasing degree of missingness as the study wears on. There are only 1 405 total observations, an average of roughly 3 per study participant.

We fit the model described in Section 17.3 by running 5 parallel Gibbs sampler chains for 500 iterations each. These calculations were performed using Fortran 77 with IMSL subroutines on a single SPARCstation 2, and took approximately 30 minutes to execute. A crudely overdispersed starting distribution for V^{-1} was obtained by drawing 5 samples from a Wishart distribution with the R matrix given above but having $\rho = \dim(\beta_i) = 3$, the smallest value for which this distribution is proper. The gth chain for each of the other parameters was initialized at its prior mean plus $(g - 3)$

prior standard deviations, $g = 1, \ldots, 5$. While monitoring each of the 1417 parameters individually is not feasible, Figure 17.3 shows the resulting Gibbs chains for the first 6 components of α, the 3 components of β_8 (a typical random effect), the model standard deviation σ, and the (1,1) and (1,2) elements of V^{-1}, respectively. Also included are the point estimate and 95th percentile of Gelman and Rubin's (1992) scale reduction factor, which measures between-chain differences and should be close to 1 if the sampler is close to the target distribution (see also Gelman, 1995: this volume), and the autocorrelation at lag 1 as estimated by the first of the 5 chains. The figure suggests convergence for all parameters with the possible exception of the two V^{-1} components. While we have little interest in posterior inference on V^{-1}, caution is required since a lack of convergence on its part could lead to false inferences concerning other parameters that *do* appear to have converged (Cowles, 1994). In this case, the culprit appears to be the large positive autocorrelations present in the chains, a frequent cause of degradation of the Gelman and Rubin statistic in otherwise well-behaved samplers. Since the plotted chains appear reasonably stable, we stop the sampler at this point, concluding that an acceptable degree of convergence has been obtained.

Based on the appearance of Figure 17.3, we discarded the first 100 iterations from each chain to complete sampler 'burn-in'. For posterior summarization, we then used all of the remaining 2 000 sampled values despite their autocorrelations, in the spirit of Geyer (1992). For the model variance σ^2, these samples produce an estimated posterior mean and standard deviation of 3.36 and 0.18, respectively. By contrast, the estimated posterior mean of V_{11} is 13.1, suggesting that the random effects account for much of the variability in the data. The dispersion evident in exploratory plots of the β_i posterior means (not shown) confirms the necessity of their inclusion in the model.

Figure 17.4 displays prior densities and corresponding posterior mixture density estimates for all 9 components of the fixed-effect parameter vector α. Posterior modes and 95% equal-tail credible intervals based on these displays are given in Table 17.1. The overlap of prior and posterior in Figure 17.4(d) confirms that the randomization of patients to drug group has resulted in no additional information concerning α_4 occurring in the data. Figures 17.4(e) and 17.4(f) (and the corresponding entries in Table 17.1) suggest that the CD4 trajectories of persons receiving ddI do have slopes that are significantly different from those of the ddC patients, both before and after the two-month change-point. AIDS diagnosis also emerges as a significant predictor of the two slopes and the intercept, as seen Figures 17.4(g), 17.4(h) and 17.4(i).

To quantify these differences, Figure 17.5 plots the fitted population trajectories that arise from (17.1) by setting β_i and ϵ_i to zero and replacing each α component by its estimated posterior mode. These fitted models,

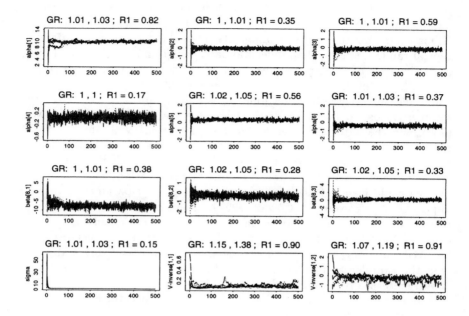

Figure 17.3 *Convergence monitoring plots, 5 independent chains: GR = Gelman and Rubin's (1992) convergence diagnostics; R1 = lag 1 autocorrelation; horizontal axis is iteration number; vertical axes are* α_1, α_2, α_3, α_4, α_5, α_6, $\beta_{8,1}$, $\beta_{8,2}$, $\beta_{8,3}$, σ, $(V^{-1})_{11}$ *and* $(V^{-1})_{12}$.

			mode	95% interval	
baseline :	intercept	α_1	9.938	9.319	10.733
	slope	α_2	−0.041	−0.285	0.204
	change in slope	α_3	−0.166	−0.450	0.118
drug :	intercept	α_4	0.004	−0.190	0.198
	slope	α_5	0.309	0.074	0.580
	change in slope	α_6	−0.348	−0.671	−0.074
AIDS diagnosis :	intercept	α_7	−4.295	−5.087	−3.609
	slope	α_8	−0.322	−0.588	−0.056
	change in slope	α_9	0.351	0.056	0.711

Table 17.1 *Point and interval estimates, fixed-effect parameters, full model*

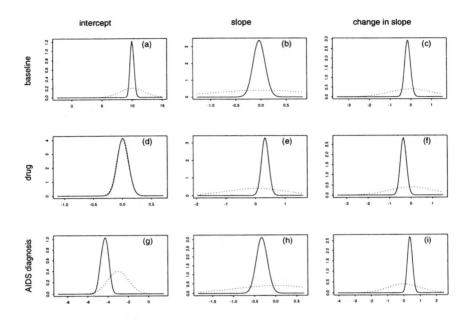

Figure 17.4 *Fixed-effects (α) priors (dashed lines) and estimated posteriors (solid lines).*

plotted for each of the four possible drug–diagnosis combinations, show that an improvement in (square root) CD4 count typically occurs only for AIDS-negative patients receiving ddI. However, there is also some indication that the ddC trajectories 'catch up' the corresponding ddI trajectories by the end of the observation period. Moreover, while the trajectories of ddI patients may be somewhat better, this improvement appears clinically insignificant compared to that provided by a negative AIDS diagnosis at baseline.

The relatively small differences between the pre- and post-change-point slopes indicated in Figure 17.5 raise the question of whether these data might be adequately explained by a simple linear decay model. To check this, we ran 5 sampling chains of 500 iterations each for this model, simply dropping the portions of the change-point model and prior specification that were no longer needed. We compare these two models using the cross-validation ideas described near the end of Section 17.3, using a 5% subsample of 70 observations, each from a different randomly selected individual. Figure 17.6 offers a side-by-side comparison of histograms and normal empirical quantile-quantile (Q-Q) plots of the resulting residuals for Model 1 (change-point) and Model 2 (linear). The Q-Q plots indicate a reasonable

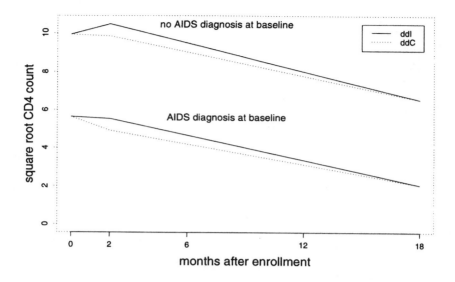

Figure 17.5 *Fitted population model by treatment group and baseline prognosis.*

degree of normality in the residuals for both models. The absolute values
of these residuals for the two models sum to 66.37 and 70.82, respectively,
suggesting almost no degradation in fit using the simpler model, as antici-
pated. This similarity in quality of fit is confirmed by Figure 17.7(a), which
plots the residuals versus their standard deviations. While there is some
indication of smaller standard deviations for residuals in the full model (see
for example the two clumps of points on the right-hand side of the plot),
the reduction appears negligible.

Finally, a pointwise comparison of CPO is given in Figure 17.7(b), along
with a reference line marking where the values are equal for the two models.
Since larger CPO values are indicative of better model fit, the predominance
of points below the reference line implies a preference for Model 1 (the full
model). Still, the log(CPO) values sum to −130.434 and −132.807 for the
two models respectively, in agreement with our previous finding that the
improvement offered by Model 1 is very slight. In summary, while the more
complicated change-point model was needed to address the specific clinical
questions raised by the trial's designers concerning a CD4 boost at two
months, the observed CD4 trajectories can be adequately explained using
a simple linear model with our two covariates. Point and interval estimates
for the remaining fixed-effect parameters in this reduced model are shown
in Table 17.2. Interestingly, we see that the CD4 trajectories of patients in

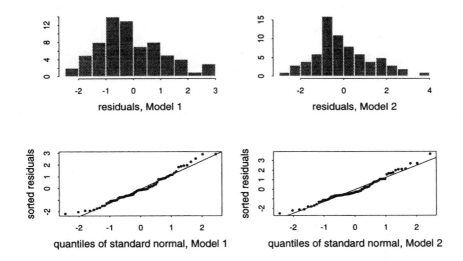

Figure 17.6 *Bayesian model assessment based on 70 random observations from the 1 405 total. Model 1: change-point model; Model 2: linear decay model.*

the two drug groups no longer differ significantly under this simple linear decline model.

			mode	95% interval	
baseline :	intercept	α_1	10.153	9.281	10.820
	slope	α_2	−0.183	−0.246	−0.078
drug :	intercept	α_4	0.030	−0.165	0.224
	slope	α_5	0.041	−0.060	0.102
AIDS diagnosis :	intercept	α_7	−4.574	−5.280	−3.530
	slope	α_8	−0.004	−0.099	0.091

Table 17.2 *Point and interval estimates, fixed-effect parameters, reduced model*

17.5 Summary and discussion

Thanks to the ever-increasing speed and declining cost of personal computers and workstations, MCMC methods have made Bayesian analysis of nontrivial models and datasets a reality for the first time. As the diversity

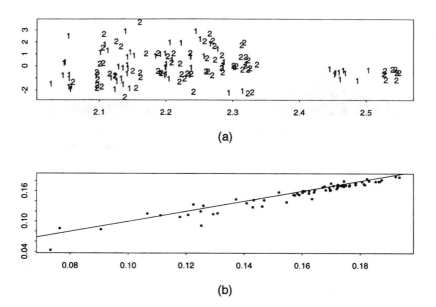

(a)

(b)

Figure 17.7 *Bayesian model choice comparing Model 1 (change-point model) and Model 2 (linear decay model); based on 70 random observations from the 1 405 total. (a) Cross-validation residuals $e_r = y_r - E(Y_r|y_{(r)})$ (horizonatal axis) versus s.d.$(Y_r|y_{(r)})$; plotting character indicates model number. (b) Comparison of CPO values, model 1 (horizontal axis) versus model 2.*

of the recent literature (and this volume) suggests, Bayesian models of virtually unlimited complexity may now be seriously contemplated. However, with this freedom have come unanticipated dangers, such as models with hidden identifiability defects, and sampling algorithms that violate conditions necessary for their own convergence. Since virtually all practising statisticians learned the art of data analysis prior to the MCMC Bayesian revolution, most should probably receive a certain amount of retraining in order to avoid these pitfalls and take full advantage of this exciting new methodology.

In the case of models for longitudinal data, several fairly specific guidelines can be given. First, users must think carefully about the number of random effects to include per individual, and restrict their variability sufficiently so that reasonably precise estimation of the fixed effects is still possible. In our dataset, while there was some evidence that heterogeneous variance parameters σ_i^2 might be called for, this would have brought the number of subject-specific parameters per individual to four – one more than the average number of CD4 observations per person in our study. Fortunately, accidental model over-parameterization can often be detected by ultrapoor performance of multichain convergence plots and diagnostics of the variety shown in Figure 17.3. In cases where reducing the total dimension of the model is not possible, recent work by Gelfand *et al.* (1994) indicates that the relative prior weight assigned to variability at the model's first stage (σ^2) versus its second stage (V) can help to suggest an appropriate transformation to improve algorithm convergence. More specifically, had our prior put more credence in small σ^2 values and larger random effects, a sampler operating on the transformed parameters $\alpha^* = \alpha$, $\beta_i^* = \beta_i + \alpha$ would have been recommended. See Gilks and Roberts (1995: this volume) for further details of this reparameterization.

The question of data missingness is also an important one that arises frequently in clinical trials work. Our approach effectively assumes that all missing data are ignorable, which, while less restrictive than 'missing completely at random', is still a fairly strong assumption. In our case, missing counts have likely arisen because the patient refused or was too ill to have CD4 measured. Since our patients have been unable to tolerate or not responded to AZT therapy, they are indeed quite ill. Thus our estimates are very likely biased toward higher CD4 counts for the later time periods; we are essentially modelling CD4 progression among persons still able to have CD4 measured. In situations where explicit modelling of the missingness mechanism is critical, tobit-type models can be employed; see Cowles *et al.* (1993) for an application to estimating compliance in a clinical trial setting.

Acknowledgements

This work was supported in part by National Institute of Allergy and Infectious Diseases (NIAID) FIRST Award 1-R29-AI33466. In addition, the dataset analysed herein was collected by the CPCRA as sponsored by NIAID Contract N01-AI05073, whose cooperation is gratefully acknowledged. The author thanks Prof. Anne Goldman for access to the ddI/ddC dataset and several helpful discussions, and the editors for numerous comments that led to substantial improvements in the presentation.

References

Abrams, D. I., Goldman, A. I., Launer, C., Korvick, J. A., Neaton, J. D., Crane, L. R., Grodesky, M., Wakefield, S., Muth, K., Kornegay, S., Cohn, D. L., Harris, A., Luskin-Hawk, R., Markowitz, N., Sampson, J. H., Thompson, M., Deyton, L. and the Terry Beirn Community Programs for Clinical Research on AIDS (1994) Comparative trial of didanosine and zalcitabine in patients with human immunodeficiency virus infection who are intolerant of or have failed zidovudine therapy. *New Engl. J. Med.*, **330**, 657–662.

Bennett, J. E., Racine-Poon, A. and Wakefield, J. C. (1995) MCMC for nonlinear hierarchical models. In *Markov Chain Monte Carlo in Practice* (eds W. R. Gilks, S. Richardson and D. J. Spiegelhalter), pp. 339–357. London: Chapman & Hall.

Bernardo, J. M. and Smith, A. F. M. (1994) *Bayesian Theory.* Chichester: Wiley.

Berzuini, C. (1995) Medical monitoring. In *Markov Chain Monte Carlo in Practice* (eds W. R. Gilks, S. Richardson and D. J. Spiegelhalter), pp. 321–337. London: Chapman & Hall.

Carlin, B. P. and Gelfand, A. E. (1993) Parametric likelihood inference for record breaking problems. *Biometrika*, **80**, 507–515.

Chaloner, K., Church, T., Louis, T. A. and Matts, J. (1993) Graphical elicitation of a prior distribution for a clinical trial. *Statistician*, **42**, 341–353.

Choi, S., Lagakos, S. W., Schooley, R. T. and Volberding, P. A. (1993) CD4+ lymphocytes are an incomplete surrogate marker for clinical progression in persons with asymptomatic HIV infection taking zidovudine. *Ann. Intern. Med.*, **118**, 674–680.

Clayton, D. G. (1995) Generalized linear mixed models. In *Markov Chain Monte Carlo in Practice* (eds W. R. Gilks, S. Richardson and D. J. Spiegelhalter), pp. 275–301. London: Chapman & Hall.

Cowles, M. K. (1994) Practical issues in Gibbs sampler implementation with application to Bayesian hierarchical modeling of clinical trial data. Ph.D. dissertation, Division of Biostatistics, University of Minnesota.

Cowles, M. K., Carlin, B. P. and Connett, J. E. (1993) Bayesian tobit modeling of longitudinal ordinal clinical trial compliance data. *Research Report 93-007*, Division of Biostatistics, University of Minnesota.

Gelfand, A. E. (1995) Model determination using sampling-based methods. In *Markov Chain Monte Carlo in Practice* (eds W. R. Gilks, S. Richardson and D. J. Spiegelhalter), pp. 145–161. London: Chapman & Hall.

Gelfand, A. E., Dey, D. K. and Chang, H. (1992) Model determination using predictive distributions with implementation via sampling-based methods (with discussion). In *Bayesian Statistics 4* (eds. J. M. Bernardo, J. O. Berger, A. P. Dawid and A. F. M. Smith), pp. 147–167. Oxford: Oxford University Press.

Gelfand, A. E., Sahu, S. K. and Carlin, B. P. (1994) Efficient parametrizations for normal linear mixed models. *Research Report 94-01*, Division of Biostatistics, University of Minnesota.

Gelman, A. (1995) Inference and monitoring convergence. In *Markov Chain Monte Carlo in Practice* (eds W. R. Gilks, S. Richardson and D. J. Spiegelhalter), pp. 131–143. London: Chapman & Hall.

Gelman, A. and Rubin, D. B. (1992) Inference from iterative simulation using multiple sequences (with discussion). *Statist. Sci.*, **7**, 457–511.

Geyer, C. J. (1992) Practical Markov chain Monte Carlo (with discussion). *Statist. Sci.*, **7**, 473–511.

Geyer, C. J. (1995) Estimation and optimization of functions. In *Markov Chain Monte Carlo in Practice* (eds W. R. Gilks, S. Richardson and D. J. Spiegelhalter), pp. 241–258. London: Chapman & Hall.

Geyer, C. J. and Thompson, E. A. (1992) Constrained Monte Carlo maximum likelihood for dependent data (with discussion). *J. R. Statist. Soc.*, B, **54**, 657–699.

Gilks, W. R. and Roberts, G. O. (1995) Strategies for improving MCMC. In *Markov Chain Monte Carlo in Practice* (eds W. R. Gilks, S. Richardson and D. J. Spiegelhalter), pp. 89–114. London: Chapman & Hall.

Gilks, W. R., Richardson, S. and Spiegelhalter, D. J. (1995) Introducing Markov chain Monte Carlo. In *Markov Chain Monte Carlo in Practice* (eds W. R. Gilks, S. Richardson and D. J. Spiegelhalter), pp. 1–19. London: Chapman & Hall.

Gilks, W. R., Thomas, A. and Spiegelhalter, D. J. (1992) Software for the Gibbs sampler. In *Computing Science and Statistics, Vol. 24* (ed. H. J. Newton), pp. 439–448. Fairfax Station: Interface.

Gilks, W. R., Wang, C. C., Yvonnet, B. and Coursaget, P. (1993) Random-effects models for longitudinal data using Gibbs sampling. *Biometrics*, **49**, 441–453.

Goldman, A. I., Carlin, B. P., Crane, L. R., Launer, C., Korvick, J. A., Deyton, L. and Abrams, D. I. (1993) Response of $CD4^+$ and clinical consequences to treatment using ddI or ddC in patients with advanced HIV infection. *Research Report 93–013*, Division of Biostatistics, University of Minnesota.

Kolassa, J. E. and Tanner, M. A. (1994) Approximate conditional inference in exponential families via the Gibbs sampler. *J. Am. Statist. Ass.*, **89**, 697–702.

Laird, N. M. and Ware, J. H. (1982) Random effects models for longitudinal data. *Biometrics*, **38**, 963–974.

Lange, N., Carlin, B. P. and Gelfand, A. E. (1992) Hierarchical Bayes models for the progression of HIV infection using longitudinal CD4 T-cell numbers (with discussion). *J. Am. Statist. Ass.*, **87**, 615–632.

Lin, D. Y., Fischl, M. A. and Schoenfeld, D. A. (1993) Evaluating the role of CD4-lymphocyte counts as surrogate endpoints in human immunodeficiency virus clinical trials. *Statist. Med.*, **12**, 835–842.

McNeil, A. J. and Gore, S. M. (1994) Statistical analysis of zidovudine effect on CD4 cell counts in HIV disease. Technical report, Medical Research Council Biostatistics Unit, Cambridge.

Spiegelhalter, D. J., Best, N. G., Gilks, W. R. and Inskip, H. (1995) Hepatitis B: a case study in MCMC methods. In *Markov Chain Monte Carlo in Practice* (eds W. R. Gilks, S. Richardson and D. J. Spiegelhalter), pp. 21–43. London: Chapman & Hall.

Zeger, S. L. and Karim, M. R. (1991) Generalized linear models with random effects: a Gibbs sampling approach. *J. Am. Statist. Ass.*, **86**, 79–86.

18

Medical monitoring

Carlo Berzuini

18.1 Introduction

The term 'medical monitoring' refers to the process of observing a patient through time, with the aim of predicting the likely course of his (or her) disease. Medical monitoring may generate quite complex time-dependent data. Typically, it generates serial measurements of markers reflecting the evolution of a patient's disease, e.g. chemical parameters measured in urine or blood. In addition, it may generate a chronological record of key failure events occurring to the patient, e.g. infection episodes or death. Patient progression is thus observed in terms of two parallel processes: a time series of marker measurements and a history of failure events. Joint modelling of these two processes is of primary interest in many areas of medical research. In patients with HIV infection, for example, interest may focus on the relationship between the evolution of immunological markers (e.g. CD4 count) and time to AIDS or death; see also Carlin (1995: this volume).

Cox's proportional hazards model (Cox, 1972) may be used to study the relationship between marker evolution and failure. However, to apply this model, precise knowledge of the patient's entire marker history is necessary. Unfortunately, because markers are usually measured only periodically and with substantial measurement error, use of Cox's model in this context presents a number of serious difficulties (Tsiatis *et al.*, 1992). We present a rather general class of models for predicting failure in individual patients. Computations are conveniently performed via MCMC methods.

The most complex model discussed in this chapter represents marker observations generated by a latent discrete-time stochastic process. This avoids strong assumptions about the shape of individual marker profiles, which may consist of long and irregular series of measurements. We discuss approaches to recursive Monte Carlo updating of predictions, for use in real-time forecasting, and present a method for comparing the predictive ability

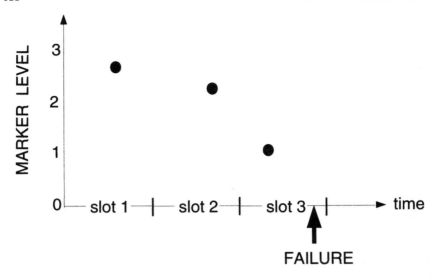

Figure 18.1 *Observations on fictitious patient k, monitored between time 0 and failure. Dots represent marker observations (which may be missing in any time slot). In our notation, these data are* $y_{k1} = 2.7$, $y_{k2} = 2.3$, $y_{k3} = 1.0$, $d_{k1} = d_{k2} = 0$, $d_{k3} = 1$.

of competing models. Finally, we illustrate the methods with an analysis of data from transplant patients affected by cytomegalovirus infection.

18.2 Modelling medical monitoring

We concentrate on the following simple problem. We want to predict when a patient will experience a defined failure event, using serial marker observations that are predictive of failure. For simplicity, we assume the failure event to be nonrepeatable. Extension to recurrent failure events, i.e. events which a patient can experience more than once, is straightforward.

18.2.1 Nomenclature and data

Figure 18.1 shows observations on a fictitious patient taken over a period of time. These comprise a series of marker measurements taken at irregularly spaced time points, and a single failure event, after which the patient is withdrawn from observation. Patient-specific covariate information, such as treatment descriptors, may also be present. Choice of the origin of the time-scale will depend on the application. One possibility is to define it as the time of onset of the patient's disease.

We divide the time-scale into a series of intervals called *slots*, such that the true marker level and the hazard of failure (the probability of failure per unit time) of each patient can be assumed constant within each slot. Without loss of generality, we assume the slots have unit length.

The data for patient k consist of triples (y_{kj}, d_{kj}, Z_{kj}), where

$$y_{kj} = \text{(possibly missing) inaccurate marker measurement in slot } j,$$

$$d_{kj} = \begin{cases} 0 & \text{if patient } k \text{ was not observed to fail in slot } j \\ 1 & \text{otherwise}, \end{cases}$$

$$Z_{kj} = \text{vector containing values of covariates in slot } j.$$

18.2.2 Linear growth model

We take the $\{d_{kj}\}$ to be conditionally independent Bernoulli variables with $P(d_{kj} = 1) = 1 - \exp(-\lambda_{kj})$, where λ_{kj} is the conditional hazard of failure for patient k in slot j. To fix ideas we assume (with some loss of generality) that the $\{y_{kj}\}$ are conditionally independent normal variables with common measurement-error variance τ_y and means ξ_{kj}. Thus:

$$y_{kj} \sim \text{N}(\xi_{kj}, \tau_y),$$
$$d_{kj} \sim \text{Bernoulli}\left(1 - e^{-\lambda_{kj}}\right).$$

We complete the model with structural assumptions about the marker and hazard evolution. One possibility is:

$$\xi_{kj} = \beta_{k0} + \beta_{k1}(j - \bar{j}) + \beta_2 Z_{kj}, \tag{18.1}$$

$$\log \lambda_{kj} = \log \lambda_{0j} + \theta_0 + \theta_1 \xi_{kj} + \theta_2 Z_{kj}. \tag{18.2}$$

Equation (18.1) assumes that the natural evolution of each marker profile is linear in time, each patient k being individually assigned an intercept β_{k0} and slope β_{k1}. To reduce posterior correlation in slopes and intercepts, time j is expressed relative to $\bar{j}$: the slot containing the mean follow-up time over all patients (see Gilks and Roberts, 1995: this volume). A similar device could be applied to the covariates. Note that, in the above model, covariates Z_{kj} have a direct effect on the failure rate λ_{kj}, and an indirect effect via the marker ξ_{kj}.

Based on a proportional hazards assumption (Cox, 1972), (18.2) models the log hazard as the sum of an intercept parameter θ_0, a time-slot effect $\log \lambda_{0j}$, a marker effect $\theta_1 \xi_{kj}$ and a covariate effect $\theta_2 Z_{kj}$. Note that λ_{0j} is the conditional hazard of failure in slot j for a hypothetical patient with $Z_{kj} = \xi_{kj} = 0$: this is a discrete-time approximation to the baseline hazard function of Cox (1972). Equation (18.2) assumes that the risk of patient k failing in slot j is independent of his past history, given the true values of his marker and covariates in that slot. More general models of failure can be developed within the framework described here.

We adopt the following prior specifications; parameters not mentioned here are assumed to have flat (improper) priors:

$$\beta_{k0} \sim N(\mu_0, \tau_{\beta_0}), \tag{18.3}$$

$$\beta_{k1} \sim N(\alpha\beta_{k0} + \rho, \tau_{\beta_1}), \tag{18.4}$$

$$\tau_y^{-1} \sim Ga(r_y, s_y),$$

$$\tau_{\beta_0}^{-1} \sim Ga(r_0, s_0),$$

$$\tau_{\beta_1}^{-1} \sim U(r_1, s_1),$$

$$\lambda_{0j} = \theta_3 \, j^{\,\theta_3-1}, \tag{18.5}$$

where $Ga(r, s)$ generically denotes a gamma distribution with mean r/s and variance r/s^2, and $U(r, s)$ denotes a uniform distribution with support in the interval $[r, s]$. Here r_y, s_y, r_0, s_0, r_1 and s_1 are known constants.

To introduce an element of similarity among marker profiles for different patients, equations (18.3) and (18.4) assume that the (β_{k0}, β_{k1}) pairs are drawn from a common bivariate normal population distribution. In standard Bayesian growth curve analysis, one would express prior correlation of the intercepts β_{k0} and slopes β_{k1} through an inverse Wishart prior for their variance-covariance matrix. With the Wishart prior, uncertainty is represented by a single scalar parameter: the degrees of freedom (see Carlin, 1995: this volume). Our approach is different: by modelling the slopes $\{\beta_{k1}\}$ conditionally on the intercepts $\{\beta_{k0}\}$ in (18.4), we allow uncertainty about the slopes and intercepts to be expressed through two parameters, τ_{β_0} and τ_{β_1}, resulting into a more flexible prior. Parameters α and ρ are unkown.

Equation (18.5) assumes that the log baseline hazard function is linear in the logarithm of time, as in the Weibull model commonly used in survival analysis. Thus, the dependence of risk on time is described by a single scalar parameter, θ_3. In some situations, a less restrictive model of baseline hazard evolution may be more appropriate. One possibility, explored by Berzuini and Clayton (1994), is to treat the $\{\log\lambda_{0j}\}$ as free parameters subject to an autoregressive prior probability structure. This allows smooth estimates of the underlying baseline hazard function to be obtained without strong assumptions about its shape; see Clayton (1995: this volume).

18.2.3 Marker growth as a stochastic process

In many situations, a linear trend for marker evolution (18.1) will be unrealistic, and we might prefer a more flexible representation. One possibility might be to replace the linear growth model with a polynomial function of time, but this will often be inappropriate, especially in the context of forecasting (Diggle, 1990: page 190).

Another solution is to model marker evolution as a stochastic process. This is achieved in the model below by including in (18.1) a component ϵ_{kj}, such that the sequence $\epsilon_{k1}, \epsilon_{k2}, \ldots$ is a discrete-time stochastic process. The model below also allows for Gaussian *i.i.d.* measurement error in the covariates: specifically the $\{Z_{kj}\}$ are interpreted as inaccurate measurements of unobserved time-constant covariates ω_k. The model is as follows:

$$y_{kj} \sim \mathrm{N}(\xi_{kj}, \tau_y), \tag{18.6}$$
$$d_{kj} \sim \mathrm{Bernoulli}(1 - e^{-\lambda_{kj}}),$$
$$Z_{kj} \sim \mathrm{N}(\omega_k, \tau_z),$$
$$\xi_{kj} = \beta_1(j - 1) + \beta_2\omega_k + \epsilon_{kj}, \tag{18.7}$$
$$\log \lambda_{kj} = \log \theta_3 + (\theta_3 - 1)\log j + \theta_0 + \theta_1\xi_{kj} + \theta_2\omega_k, \tag{18.8}$$
$$\epsilon_{k1} \sim \mathrm{N}(\mu_0, \tau_{\beta_0}),$$
$$\tau_y^{-1} \sim \mathrm{Ga}(r_y, s_y),$$
$$\tau_z^{-1} \sim \mathrm{Ga}(r_z, s_z),$$
$$\tau_{\beta_0}^{-1} \sim \mathrm{Ga}(r_0, s_0),$$

where r_y, s_y, r_z, s_z, r_0 and s_0 are known constants.

In (18.7), the stochastic process ϵ_{kj} represents marker fluctuations around a linear time trend with slope β_1. We define the stochastic process ϵ_{kj} within patient k to have the following Gaussian autoregressive (order 2) structure:

$$\epsilon_{k2} \sim \mathrm{N}(\epsilon_{k1}, \eta),$$
$$\epsilon_{kj} \sim \mathrm{N}\left((1 + \gamma)\epsilon_{k,j-1} - \gamma\epsilon_{k,j-2}, (1 - \gamma^2)\eta\right), \qquad j > 2, \tag{18.9}$$
$$\eta^{-1} \sim \mathrm{Ga}(r_\eta, s_\eta),$$
$$\gamma \sim \text{truncated } \mathrm{N}(r_\gamma, s_\gamma),$$

where $0 \leq \gamma < 1$ and r_η, s_η, r_γ and s_γ are known constants. When forecasting at the subject level is of interest, we recommend replacing β_1 with a subject-specific parameter β_{k1}.

First differences $\nu_{kj} = \xi_{k,j+1} - \xi_{kj}$ can be shown to have the same marginal prior distribution

$$\nu_{kj} \sim \mathrm{N}(\beta_1, \eta), \tag{18.10}$$

so that η is simply interpreted as the marginal variance of first differences. Large values of η allow the slope of each individual marker profile to deviate substantially from the population slope β_1, whereas small values of η cause marker profiles to be pulled towards a line of slope β_1. Values for r_η and s_η for the prior distribution of η can be set by considering the interpretation of η: that there is a 95% probability that the increase in a patient's marker level, from one time slot to the next, will be within $\beta_1 \pm 1.96\sqrt{\eta}$.

Parameter γ controls the smoothness of covariate profiles. When γ is close to 0, the $\{\epsilon_{k1}, \epsilon_{k2}, \ldots\}$ sequence for patient k will approximate a random walk with rapid and unpredictable changes in slope. When γ is

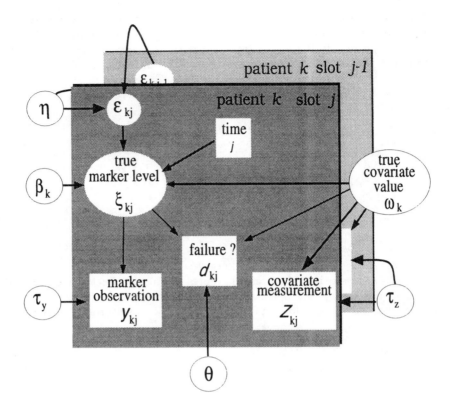

Figure 18.2 *Graphical representation of the stochastic process model (detail).*

close to 1, first differences (18.10) will tend to be highly correlated within each patient, leading to smooth estimates of marker profiles. When the data contain little information about γ, usually when the individual series of marker measurements are short, long-term predictions will usually be sensitive to prior knowledge about γ. This suggests that we develop ways of eliciting prior information about this parameter, and that we analyse the sensitivity of model-based inferences to changes in the prior for γ.

Figure 18.2 shows a graphical representation of the above model, for two generic time slots for a single generic patient. See Spiegelhalter *et al.* (1995: this volume) for an introduction to graphical models.

The model can be extended to deal with repeatable failure events, for example recurrent infections in a patient. Then d_{kj}, the observed count of failures for subject k in slot j, can be modelled as a Poisson variable with mean λ_{kj}. Equations (18.2) and (18.8) might then include additional terms representing dependence of λ_{kj} on aspects of the patient's failure history, such as the time elapsed since the patient's previous failure.

Finally, note that (18.9) can be expressed as a nonlinear bivariate autoregressive process of order one:

$$\nu_{kj} \sim \mathrm{N}\left(\gamma\nu_{k,j-1}, (1-\gamma^2)\eta\right), \qquad \epsilon_{kj} = \epsilon_{k,j-1} + \nu_{k,j-1}.$$

With this formulation of the underlying stochastic process, the model closely resembles the dynamic hierarchical models described in Gamerman and Migon (1993).

18.3 Computing posterior distributions

We adopt a Bayesian approach to facilitate estimation of the joint probability distribution of predicted quantities. The integrals required for a full Bayesian treatment of the most general model described so far, have no closed analytical solution. One approach is to approximate these integrals by a Laplace approximation, leading to penalised quasi-likelihood maximization. However, this method does not allow forecasts fully to reflect uncertainty about the unknown parameters. It also requires matrix inversions that become problematic when time is very finely subdivided into slots. Therefore we use Gibbs sampling to perform the computations, not least because this readily allows further generalization of the models. Other authors, e.g. Gilks *et al.* (1993) and McNeil and Gore (1994), have applied Gibbs sampling to similar models.

When using Gibbs sampling, one particularly convenient feature of the models we have described is that the full conditional distributions of all the parameters are log-concave under a constant or log-concave prior. Therefore the extremely efficient adaptive rejection sampling algorithm of Gilks and Wild (1992) can be used for sampling from full conditional distributions (see Gilks, 1995: this volume). Introduction of nonlinearities (e.g. piecewise-linear marker profiles with unknown change-points, or a nonlinear relationship between $\log \lambda_{kj}$ and ξ_{kj}) will usually lead to non-log-concave (sometimes piece-wise log-concave) full conditional distributions. For sampling from non-log-concave full conditional distributions, a Metropolis–Hastings algorithm might be adopted; alternatively, adaptive rejection Metropolis sampling (ARMS) might be used (see Gilks, 1995: this volume).

18.3.1 Recursive updating

Patient monitoring data arise sequentially, and with each data acquisition new forecasts may be required. In their pure form, MCMC methods would need to be restarted with each data acquisition, making their implementation in real time impractical. However, an approximate method which we now describe allows the posterior distribution of parameters of interest to be updated *recursively* (Berzuini *et al.*, 1994).

We concentrate on the stochastic process model of Section 18.2.3. We use the notation $[a \mid b]$ generically to denote the distribution of a given b, and $\{\theta^{(i)}\}_M$ to denote a collection of samples of parameter θ: $\{\theta^{(1)}, \ldots, \theta^{(M)}\}$.

Suppose that patient k at time t has just completed his j^{th} slot, making new data (y_{kj}, d_{kj}, Z_{kj}) available for processing. Let $D_{k,j-1}$ denote all data available just before t, excluding (y_{kj}, d_{kj}, Z_{kj}), and let Δ denote all model parameters except $(\eta, \epsilon_{k1}, \epsilon_{k2}, \ldots)$. Suppose that from calculations made before time t we have a sample of size M, reflecting the posterior distribution of the model parameters prior to the new data:

$$\{\epsilon_{k1}^{(i)}, \ldots, \epsilon_{k,j-1}^{(i)}, \eta^{(i)}, \Delta^{(i)}\}_M \approx [\epsilon_{k1}, \ldots, \epsilon_{k,j-1}, \eta, \Delta \mid D_{k,j-1}], \quad (18.11)$$

where '$\approx$' means 'approximately sampled from'. Upon arrival of the new data (y_{kj}, d_{kj}, Z_{kj}) this sample is not discarded, but is updated (for use in subsequent calculations) into a new sample of size Q:

$$\{\tilde{\epsilon}_{k1}^{(i)}, \ldots, \tilde{\epsilon}_{kj}^{(i)}, \tilde{\eta}^{(i)}, \tilde{\Delta}^{(i)}\}_Q \approx [\epsilon_{k1}, \ldots, \epsilon_{kj}, \eta, \Delta \mid D_{k,j-1}, y_{kj}, d_{kj}, Z_{kj}],$$
$$(18.12)$$

which reflects the change in the posterior due to the new data. This updating is performed recursively whenever a patient completes a slot. Parameters specific to patients no longer being monitored at time t, need not be included in Δ or involved in the updating.

The updating may be carried out by importance resampling (Rubin, 1987). This comprises two steps. In the first step, we augment sample (18.11) into a sample

$$\{\epsilon_{k1}^{(i)}, \ldots, \epsilon_{kj}^{(i)}, \eta^{(i)}, \Delta^{(i)}\}_M \approx [\epsilon_{k1}, \ldots, \epsilon_{kj}, \eta, \Delta \mid D_{k,j-1}] \quad (18.13)$$

by including samples $\epsilon_{kj}^{(i)}$ generated from their prior (18.9), i.e.

$$\epsilon_{kj}^{(i)} \sim N\left((1 + \gamma^{(i)})\epsilon_{k,j-1}^{(i)} - \gamma^{(i)}\epsilon_{k,j-2}^{(i)}, (1 - (\gamma^{(i)})^2)\eta^{(i)}\right).$$

The second step consists of generating the target sample (18.12) by resampling with replacement from sample (18.13), where the rth element of (18.13), $(\epsilon_{k1}^{(r)}, \ldots, \epsilon_{kj}^{(r)}, \eta^{(r)}, \Delta^{(r)})$, is sampled with probability proportional to the importance weight:

$$[y_{kj}, d_{kj}, Z_{kj} \mid \epsilon_{kj}^{(r)}, \eta^{(r)}, \Delta^{(r)}].$$

This importance weight is entirely specified by model (18.6)–(18.8).

Elements of (18.13) which are in the tails of the current posterior will tend to have relatively small importance weights, and consequently a high probability of being 'killed' in the resampling. Consequently, samples drawn in future steps of the updating process will only be generated on the basis of the most effective of the available samples. Liu and Chen (1995) demonstrate that this leads to more precise estimates of predictive quantities. Unfortunately, if the new data carry substantial information, the output sample (18.12) may be markedly impoverished as a consequence of the resampling. This might be partly remedied within the above scheme by drawing $\epsilon_{kj}^{(i)}$ from its conditional posterior distribution given

$$\gamma^{(i)}, \ \epsilon_{k,j-1}^{(i)}, \ \epsilon_{k,j-2}^{(i)}, \ \eta^{(i)}$$

and the new data, but this might make calculation of the importance weights difficult. In such circumstances, it may be better to abandon the above scheme in favour of a general method described by Berzuini *et al.* (1994) for recursive updating of posteriors using a combination of importance sampling and MCMC.

18.4 Forecasting

Real-time implementation of our methodology demands that we are able to update the failure-time forecast for any patient k whenever he completes a time slot j. Under the stochastic process model, conditional on the updated sample (18.12), we draw for $i = 1, \ldots, Q$ and recursively for $h = j+1, \ldots, J$:

$$\tilde{\epsilon}_{kh}^{(i)} \ \sim \ \mathrm{N}\left((1 + \gamma^{(i)})\tilde{\epsilon}_{k,h-1}^{(i)} - \gamma^{(i)}\tilde{\epsilon}_{k,h-2}^{(i)} \ , \ (1 - (\gamma^{(i)})^2)\tilde{\eta}^{(i)}\right).$$

Jointly with (18.12) this yields an updated sample

$$\{\tilde{\epsilon}_{k1}^{(i)}, \ldots, \tilde{\epsilon}_{kJ}^{(i)}, \tilde{\eta}^{(i)}, \tilde{\Delta}^{(i)}\}_Q \ \sim \ [\epsilon_{k1}, \ldots, \epsilon_{kJ}, \eta, \Delta \mid D_{k,j-1}, y_{kj}, d_{kj}, Z_{kj}],$$

from which we may compute samples of the hazard of failure (18.8) for patient k in future slots:

$$\lambda_{kh}^{(i)} \ = \ \tilde{\theta}_3^{(i)} \exp\left((\tilde{\theta}_3^{(i)} - 1)\log h + \tilde{\theta}_0^{(i)} + \tilde{\theta}_1^{(i)}\tilde{\xi}_{kh}^{(i)} + \tilde{\theta}_2^{(i)}\tilde{\omega}_k^{(i)}\right), \quad (18.14)$$

for $i = 1, \ldots, Q$ and $h = j + 1, \ldots, J$.

Suppose we wish to express the forecast in terms of the probability p_{kjh} that patient k, now failure-free at the end of slot j, will still be failure-free at the end of a specified future slot h, for $(j < h \leq J)$. Letting T_k denote the patient's future failure time and t_q denote the right extreme of the qth slot, this probability is formally given by

$$p_{kjh} \ = \ \int_{H \in L(t_j, t_h)} [T_k > t_h \mid D_{kj}, H] \, [H \mid D_{kj}] \ dH,$$

where $L(t_j, t_h)$ represents the set of all possible marker paths for patient k in time-interval (t_j, t_h); H is a particular marker path; and D_{kj} represents all the data available at t_j. Note that p_{kjh} is the conditional expectation, given D_{kj}, of the probability that patient k will survive up to t_h. The expectation involves an integration over the set of all possible paths of the marker in the time interval (t_j, t_h). An (approximate) Monte Carlo estimator for p_{kjh} based on samples (18.14) is

$$\hat{p}_{kjh} = \frac{1}{Q} \sum_{i=1}^{Q} \exp\left(-\sum_{s=j+1}^{h} \lambda_{ks}^{(i)}\right). \tag{18.15}$$

18.5 Model criticism

We will normally want to check whether our model predicts failure adequately. For this we propose a method based on the predictive distribution of the observed failure indicators, $\{d_{kj}\}$.

Let D denote the complete dataset, and $D_{(kj)}$ the same dataset after deleting data item d_{kj}. By analogy with leave-one-out cross-validation, we argue that we should check each observed failure indicator d_{kj} against its univariate predictive distribution $[d_{kj} \mid D_{(kj)}]$. For example we may compute, for each k and j, the discrepancies

$$R_{kj} = d_{kj} - E(d_{kj} \mid D_{(kj)}).$$

A Monte Carlo estimate for R_{kj} is given by the *predictive residual*:

$$\hat{R}_{kj} = d_{kj} - 1 + \frac{1}{M} \sum_{i=1}^{M} \exp\left(-\lambda_{kj}^{(i)}\right),$$

where the sample $\{\lambda_{kj}^{(i)}\}_M$ should be drawn from $[\lambda_{kj} \mid D_{(kj)}]$. In practice, we may draw this sample from $[\lambda_{kj} \mid D]$ for all k and j using a single MCMC run, provided we subsequently adjust for the fact that the intended posterior is $[\lambda_{kj} \mid D_{(kj)}]$. This adjustment can be performed by importance resampling; see Gelfand (1995: this volume).

Many large $|\hat{R}_{kj}|$, for different values of k and j, cast doubt on the model's ability to predict failure, and the sign of $\hat{R}_{kj}$ allows patterns of optimistic (or pessimistic) prediction to be revealed. Many $\hat{R}_{kj}$ close to 1 indicates a tendency towards unpredicted failures, whereas large negative values of $\hat{R}_{kj}$ indicate that the model tends to predict failures which do not occur. A useful summary is the kernel density plot of the collection of predictive residuals $\{\hat{R}_{kj}; k = 1, 2, \ldots; j = 1, 2, \ldots\}$. Roughly speaking, we prefer models with relatively less dispersed predictive residuals.

18.6 Illustrative application

18.6.1 The clinical problem

Morbidity and mortality after organ transplantation is mainly due to infection by cytomegalovirus (CMV). This virus proliferates in the tissues of a transplant recipient during immuno-suppressive treatment undertaken to

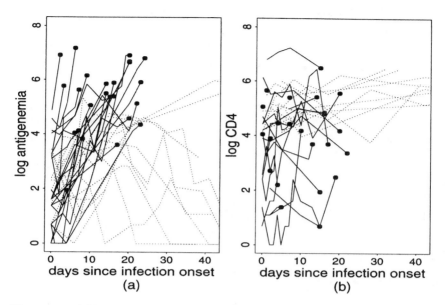

Figure 18.3 *(a)* $\log_e$ *CMV-antigenemia profiles for 37 patients, (b)* $\log_e$ *CD4-count profiles for the same patients. Observed deterioration events are represented by dots. Solid lines – patients who were observed to deteriorate; broken lines – patients not observed to deteriorate.*

prevent graft rejection. An important indicator of the intensity of a CMV infection is CMV antigenemia. During the infection some patients develop *deterioration*, i.e. the onset of life-threatening symptoms. Aggressive anti-viral treatment can reverse the deterioration process, but at the price of severe side-effects.

It would be unwise to impose the anti-viral treatment and its associated side-effects on all patients, since some patients can defeat the virus by themselves. We propose a *predictive* approach to treatment in which each patient is monitored through serial measurements of CMV antigenemia and CD4 cell count (indicators of immune system vitality). The anti-viral treatment should be started only when these measurements are predictive of imminent deterioration. The prediction can be performed by using the model described in the next section.

Figure 18.3(a) shows $\log_e$ CMV-antigenemia profiles for 37 heart trans-plant recipients during CMV infection, plotted versus time since onset of in-fection. Figure 18.3(b) shows $\log_e$ CD4 count profiles for the same patients. Times of measurement were irregular. Data obtained after deterioration or during non-standard treatment were disregarded in this analysis.

18.6.2 The model

We analysed the antigenemia/CD4 data using both the linear growth model of Section 18.2.2 and the stochastic process model of Section 18.2.3, with 3-day time slots. In these analyses, marker y_{kj} was observed antigenemia; covariate Z_{kj} was observed CD4 count; and failure d_{kj} was deterioration.

The underlying value of CD4, ω_k, for each patient was treated as a time-constant covariate, because observed CD4 counts were roughly constant for each patient. To avoid problems of identifiability, we took θ_3 to be unity, so that no direct effect of time on risk of deterioration was considered. When using the stochastic process model, γ was set to 0.05, specifying a second-order random-walk prior process for ϵ_{kj} (although fixing γ was not necessary).

The calculations were performed using the Gibbs sampling program BUGS (Spiegelhalter *et al.*, 1994, 1995: this volume). We used estimates from the linear growth model to provide good initial values for the stochastic process model.

18.6.3 Parameter estimates

Table 18.1 compares prior and posterior marginal distributions for selected parameters for each of the two models considered. Note that the effect of the data was to shift the posterior means of four crucial hyperparameters, η^{-1}, τ_y^{-1}, $\tau_{\beta_0}^{-1}$ and τ_z^{-1}, markedly away from their prior means, suggesting that the results are not dominated by prior specifications.

parameter	prior mean	s.d.	linear growth posterior mean	s.d.	stochastic process posterior mean	s.d.
η^{-1}	0.08	0.28	2.38	1.00	1.60	0.21
τ_y^{-1}	1.00	1.00	0.82	0.08	2.60	0.47
$\tau_{\beta_0}^{-1}$	0.11	0.35	0.09	0.04	2.45	1.17
τ_z^{-1}	1.00	1.00	1.50	0.18	1.50	0.18
θ_0	–	–	−5.79	1.66	−4.77	1.19
θ_1	–	–	1.24	0.29	0.98	0.22
θ_2	–	–	−0.48	0.18	−0.51	0.16
β_2	–	–	−0.16	0.14	0.06	0.15

Table 18.1 *Prior specifications and posterior estimates for selected parameters from the linear growth and stochastic process models. A '–' denotes a flat prior.*

In terms of parameters θ_1, θ_2 and β_2, there is substantial agreement between the two models. Both models strongly support $\theta_1 > 0$, in agreement with the accepted clinical view that high levels of antigenemia are associated with high risk of deterioration. Similarly, both models suggest $\theta_2 < 0$, i.e. that CD4 protects against infection, as is well known. This suggests that some patients with CMV infection might fare better if CMV anti-viral therapy were accompanied by a reduction in immunosuppressive therapy. Neither model provides strong evidence of an effect of CD4 on antigenemia. Overall, the above estimates suggest that on-line monitoring of antigenemia and CD4 data can provide a useful basis for predicting a patient's residual time to deterioration.

18.6.4 Predicting deterioration

Figure 18.4 illustrates the dynamic forecasting method of Section 18.4 for a randomly selected patient, using the linear growth and stochastic process models. Forecasts were made initially when only 2 antigenemia measurements were available for this patient (first row). The second and third rows show the revised forecasts after 5 and 8 antigenemia measurements.

In each diagram of Figure 18.4, a vertical line marks the time at which the forecast took place, and black dots represent the patient's antigenemia data available at that time. The solid curve running in the middle of the shaded band represents an estimate for the underlying log-antigenemia path ξ_{kj} up to the time of the forecast. The shaded band shows a pointwise 90% Bayesian credible region for that path. Both the log-antigenemia path and the shaded band were calculated straightforwardly from the generated posterior samples.

The dashed curve on the right of the vertical line is a survival curve predicting the patient's residual time to deterioration, calculated according to (18.15). For any given future time, this curve yields the probability that the patient will still be deterioration-free at that time (being unity initially).

The forecasts made after slot 2 (day 6) of the patient's follow-up were based mainly on population (between-patient) information, whereas the subsequent two forecasts were largely determined by data specific to the patient. Both models reacted to the increase in antigenemia between slots 2 and 5 by shortening the predicted median residual time to deterioration. However, the subsequent antigenemia decrease caused both models to yield a substantially more favourable prognosis. The patient chosen for this illustration ultimately did not develop deterioration, perhaps helped by his high CD4 count. Note that, in Figure 18.4(f), the predicted log antigenemia path between slots 10 and 13 follows the mean population trend, ignoring the previous negative trend in log antigenemia observations for this patient. This is a consequence of having set $\gamma = 0.05$, which implies low correlation between first differences ν_{kj}.

Figure 18.4 suggests that the stochastic process model fits the antigenemia data better than the linear growth model, and this impression was

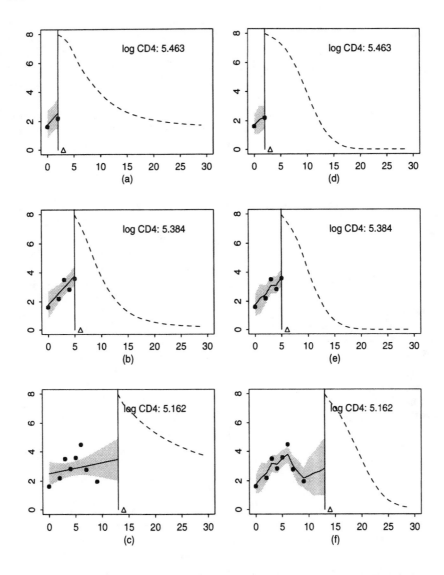

Figure 18.4 *Illustrating dynamic forecasting for a randomly selected patient, using the linear growth model (panels (a), (b) and (c)) and the stochastic process model (panels (d), (e) and (f)). Forecasts were made after slot 2 (panels (a) and (d)), slot 5 (panels (b) and (e)) and slot 13 (panels (c) and (f)). Horizontal axes are time since onset of CMV infection (days/3); dots represent log antigenemia data available at the forecast time; solid curves are underlying log antigenemia trajectories; dashed curves are forecasted deterioration-free survival curves; and 'log CD4' denotes the mean log CD4 count for the patient at the forecast time.*

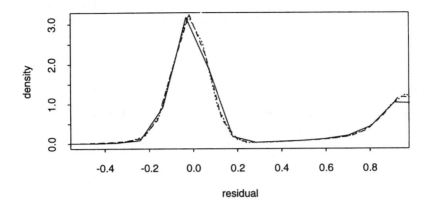

Figure 18.5 *Comparison of predictive residual density plots for the linear growth model (solid line) and the stochastic process model (dotted line: $\gamma = 0.1$; broken line: $\gamma = 0.9$).*

confirmed by approximate prequential testing (see Dawid, 1984).

Figure 18.5 compares predictive residual density plots (see Section 18.5) for the linear growth and stochastic process models, the latter with two different values of γ. The marginal densities of the residuals from the three models are nearly identical, indicating that the they provide similar performance in predicting failure.

18.7 Discussion

We have combined time-series and failure-time models to represent the natural history of post-transplant CMV infection. The model seems to yield interesting and plausible estimates for parameters of biological interest, and clinically useful patient predictions. The technique for recursive updating, described in Section 18.3.1, allows models such as those described here to be used in on-line clinical forecasting systems.

In our illustrative application, the sophisticated stochastic process model did not appear to predict patient failure better than the simpler linear growth model. In other situations, perhaps involving longer series of marker measurements, the more sophisticated model may well lead to better predictions, especially when the data contain substantial information about parameter γ.

Several extensions to the proposed models are possible. For example, normal distributions might be replaced by t-distributions for added robustness. A nonlinear relationship between the log hazard of failure and marker levels might be considered; this will usually lead to non-log-concave full conditional distributions. Also, one might wish to allow β_2, θ_1 and θ_2 to depend on time, or the risk of failure to depend on multiple time scales (Berzuini and Clayton, 1994).

Acknowledgements

This research was partially funded by the Italian M.U.R.S.T. I am particularly indebted to Walter Gilks for suggestions on earlier drafts of this chapter; to Cristiana Larizza, Andrew Thomas, Jeremy Taylor and David Spiegelhalter for insightful discussions; to Dr Grossi of the IRCCS Policlinico San Matteo, Pavia, for providing the data; and to L. Bernardinelli and C. Gobbi for suggestions and help in analysing the data.

References

Berzuini, C. and Clayton, D. G. (1994) Bayesian analysis of survival on multiple time scales. *Statistics in Medicine*, **13**, 823–838.

Berzuini, C., Best, N., Gilks, W. R. and Larizza, C. (1994) Dynamic graphical models and Markov chain Monte Carlo methods. Technical report, Department of Informatics and Systematics, Faculty of Engineering, University of Pavia.

Carlin, B. P. (1995) Hierarchical longitudinal modelling. In *Markov Chain Monte Carlo in Practice* (eds W. R. Gilks, S. Richardson and D. J. Spiegelhalter), pp. 303–319. London: Chapman & Hall.

Clayton, D. G. (1995) Generalized linear mixed models. In *Markov Chain Monte Carlo in Practice* (eds W. R. Gilks, S. Richardson and D. J. Spiegelhalter), pp. 275–301. London: Chapman & Hall.

Cox, D. R. (1972) Regression models and life tables. *J. R. Statist. Soc.* B, **34**, 187–220.

Dawid, A. P. (1984) Statistical theory: the prequential approach (with discussion). *J. R. Statist. Soc.* A, **147**, 278–292.

Diggle, P. J. (1990) *Time Series: A Biostatistical Introduction*. Oxford: Oxford Science Publications.

Gamerman, D. and Migon, H. S. (1993) Dynamic hierarchical models. *J. R. Statist. Soc.* B, **55**, 629–642.

Gelfand, A. E. (1995) Model determination using sampling-based methods. In *Markov Chain Monte Carlo in Practice* (eds W. R. Gilks, S. Richardson and D. J. Spiegelhalter), pp. 145–161. London: Chapman & Hall.

Gilks, W. R. (1995) Full conditional distributions. In *Markov Chain Monte Carlo in Practice* (eds W. R. Gilks, S. Richardson and D. J. Spiegelhalter), pp. 75–88. London: Chapman & Hall.

Gilks, W. R. and Roberts, G. O. (1995) Strategies for improving MCMC. In *Markov Chain Monte Carlo in Practice* (eds W. R. Gilks, S. Richardson and D. J. Spiegelhalter), pp. 89–114. London: Chapman & Hall.

Gilks, W. R. and Wild, P. (1992) Adaptive rejection sampling for Gibbs sampling. *Applied Statistics*, 41, 337–348.

Gilks, W. R., Clayton, D. G., Spiegelhalter, D. J., Best, N. G., McNeil, A. J., Sharples, L. D. and Kirby, A. J. (1993) Modelling complexity: applications of Gibbs sampling in medicine. *J. R. Statist. Soc.* B, 55, 39–52.

Liu, J. S. and Chen, R. (1995) Blind deconvolution via sequential imputations. *J. Am. Statist. Ass.*, 90, 567–576.

McNeil, A. J. and Gore, S. M. (1994) Statistical analysis of Zidovudine effect on CD4 cell counts in HIV disease. Technical report, Medical Research Council Biostatistics Unit, Cambridge.

Rubin, D. B. (1987) Discussion on the calculation of posterior distributions by data augmentation (by M. A. Tanner and W. H. Wong). *J. Am. Statist. Ass.*, 52, 543–546.

Spiegelhalter, D. J., Best, N. G., Gilks, W. R. and Inskip, H. (1995) Hepatitis B: a case study in MCMC methods. In *Markov Chain Monte Carlo in Practice* (eds W. R. Gilks, S. Richardson and D. J. Spiegelhalter), pp. 21–43. London: Chapman & Hall.

Spiegelhalter, D. J., Thomas, A., Best, N. G. and Gilks, W. R. (1994) *BUGS: Bayesian Inference Using Gibbs Sampling.* Cambridge: Medical Research Council Biostatistics Unit.

Tsiatis, A. A., DeGruttola, V., Strawderman, R. L., Dafni, U., Propert, K. J. and Wulfsohn, M. (1992) The relationship of CD4 counts over time to survival in patients with AIDS: is CD4 a good surrogate marker? In *AIDS Epidemiology: Methodological Issues*, (eds N. P. Jewell, K. Dietz and V. T. Farewell), pp. 256–274. Boston: Birkhauser.

19

MCMC for nonlinear hierarchical models

James E Bennett
Amy Racine-Poon
Jon C Wakefield

19.1 Introduction

Nonlinear hierarchical models are used in a wide variety of applications, for example in growth curve analysis (Berkey, 1982) and in pharmacokinetic applications (Steimer *et al.*, 1985). The implementation of such models is the subject of much research interest; see Lindstrom and Bates (1990) for a non-Bayesian perspective.

Consider the following three-stage hierarchical model. At the first stage of the model we assume that $f(\theta_i, t_{ij})$ represents the predicted response at the jth observation on the ith individual at time t_{ij} where θ_i is a vector of p regression parameters, f is a nonlinear function and $i = 1, \ldots, M$; $j = 1, \ldots, n_i$. Let y_{ij} represent the observed response. The distribution of the responses is assumed to be normal or lognormal, that is

$$y_{ij} = f(\theta_i, t_{ij}) + \varepsilon_{ij} \tag{19.1}$$

or

$$\log y_{ij} = \log f(\theta_i, t_{ij}) + \varepsilon_{ij} \tag{19.2}$$

where the ε_{ij} are independent and identically distributed normal random variables with zero mean and variance τ^{-1} (i.e. precision τ). We denote this distributional form (the first-stage distribution) by $p_1(y_{ij} \mid \theta_i, \tau^{-1})$.

At the second stage of the model we have

$$E[\theta_i \mid \mu] = g(\mu) \tag{19.3}$$

where μ is a vector of parameters for the population mean. We assume that the 'random effects'

$$b_i = \theta_i - E[\theta_i \mid \mu]$$

are independent identically distributed random variables with zero mean and scale matrix Σ, following either a multivariate normal distribution or a multivariate student t-distribution with degrees of freedom ν (see Bernardo and Smith, 1994, for a definition of the multivariate t-distribution). We generically denote this distributional form for θ_i (the second-stage distribution) by $p_2(\theta_i \mid \mu, \Sigma)$. At the third stage of the model, we assume that τ has a gamma distribution $\text{Ga}\left(\frac{\nu_0}{2}, \frac{\tau_0 \nu_0}{2}\right)$, μ is multivariate normal $\text{N}(c, C)$ and Σ^{-1} is Wishart $\text{W}(\rho, [\rho R]^{-1})$ for known v_0, τ_0, c, C, R and ρ. Here we use the parameterizations of the normal, gamma and Wishart distributions as given in Carlin (1995: this volume).

With a linear model at the first stage, the implementation of the Gibbs sampler is straightforward since all conditional distributions assume standard forms; see Gelfand et $al.$ (1990) for the normal second-stage case and Wakefield et $al.$ (1994) for the student-t second-stage case. The analysis of related random effects frailty models via the Gibbs sampler is considered by Clayton (1991). Lange et $al.$ (1992) also consider a nonlinear hierarchical structure, though the form of their models results in conditional distributions from which sampling is straightforward.

In this chapter, we describe and compare various MCMC strategies for implementing the model defined in (19.1–19.3). We concentrate on the sampling of the individual-specific parameters θ_i, as these are the only parameters which have non-standard full-conditional distributions (see below). Wakefield et $al.$ (1994) sample individual elements of the θ_i vectors using the extended ratio-of-uniforms method (Wakefield et $al.$, 1991). This approach, though successful, requires familiarity with the ratio-of-uniforms method and involves three maximizations per random variate generation, a potentially daunting process for the uninitiated. We show that convergence is accelerated if the complete θ_i vector is updated simultaneously instead of one element at a time. Zeger and Karim (1991) consider the Gibbs analysis of hierarchical models in which the first stage is a generalized linear model. They overcome the problem of generating from the awkward full conditionals for μ and θ_i by rejection sampling from an approximate rejection envelope. This approach is unreliable as the effects of a nondominating 'envelope' will be difficult to determine, though Tierney (1994) provides a Metropolis–Hastings remedy for this. Dellaportas and Smith (1993) exploit the log-concavity of the full conditional distributions in a non-hierarchical generalized linear model by using the adaptive rejection sampling method of Gilks and Wild (1992). Gilks et $al.$ (1995) propose a univariate 'Metropolized' adaptive rejection sampling scheme which can be used for models such as those considered here though this method, like the ratio-of-uniforms method, involves univariate sampling.

Section 19.2 describes a new scheme for generation from the full conditional distributions of the random effects and gives the forms of all of the other full conditionals. Section 19.3 compares the various methods suggested in Section 19.2, using an example analysed in a classical framework by Lindstrom and Bates (1990). Section 19.4 contains an analysis of a complex pharmacokinetic-pharmacodynamic dataset, and Section 19.5 concludes with a discussion.

19.2 Implementing MCMC

For the hierarchical model outlined in Section 19.1, the conditional independence structure suggests a Gibbs sampling approach (Gelfand and Smith, 1990; see also Spiegelhalter *et al.*, 1995: this volume). Let $\theta = (\theta_1, \ldots, \theta_M)$ and $y = (y_1, \ldots, y_M)$. Let $[\theta_i \mid .]$ denote the full conditional distribution of θ_i, that is, the distribution of θ_i given all other quantities in the model (see Gilks, 1995: this volume). Unfortunately, the nonlinear model $f(\theta_i, t_{ij})$ leads to a form for $[\theta_i \mid .]$ that is non-standard and is known only up to a normalizing constant. We first give, for completeness, the form of the other conditional distributions. For further details, see Gelfand *et al.* (1990) and Wakefield *et al.* (1994).

For normal first- and second-stage distributions, defining $\overline{\theta} = \frac{1}{M} \sum_{i=1}^{M} \theta_i$ and $V^{-1} = M\Sigma^{-1} + C^{-1}$, we have the following full conditional distributions:

$$[\mu \mid .] = \text{N}\left(V\left\{M\Sigma^{-1}\overline{\theta} + C^{-1}c\right\}, V\right),$$

$$[\Sigma^{-1} \mid .] = \text{W}\left(M + \rho, \left\{\sum_{i=1}^{M}(\theta_i - \mu)^{\text{T}}(\theta_i - \mu) + \rho R\right\}^{-1}\right),$$

$$[\tau \mid .] = \text{Ga}\left(\frac{1}{2}\left\{\nu_0 + \sum_{i=1}^{M} n_i\right\},\right.$$
$$\left.\frac{1}{2}\left\{\sum_{i=1}^{M}\sum_{j=1}^{n_i}[y_{ij} - f(\theta_i, t_{ij})]^2 + \nu_0\tau_0\right\}\right).$$

Extension to a second-stage multivariate t-distribution with fixed degrees of freedom ν and variance-covariance matrix $\frac{\nu}{\nu-2}\Sigma$ is straightforward if we exploit the fact that such a distribution can be expressed as a scale mixture of normals (Racine-Poon, 1992; Bernardo and Smith, 1994). For this, M additional parameters $\lambda = (\lambda_1, \ldots, \lambda_M)$ are introduced with second-stage distribution $\theta_i \sim \text{N}(\mu, \lambda_i^{-1}\Sigma)$, where the univariate λ_i parameters are assigned a $\text{Ga}(\frac{\nu}{2}, \frac{\nu}{2})$ distribution at the third stage. Defining

$U^{-1} = \Sigma^{-1} \sum_{i=1}^{M} \lambda_i + C^{-1}$, we then have

$$[\mu \mid .] \;=\; \mathrm{N}\left(U\left\{\Sigma^{-1}\sum_{i=1}^{M}\lambda_i\theta_i + C^{-1}c\right\},\, U\right),$$

$$[\Sigma^{-1} \mid .] \;=\; \mathrm{W}\left(M+\rho\,,\left\{\sum_{i=1}^{M}\lambda_i(\theta_i-\mu)^{\mathrm{T}}(\theta_i-\mu)+\rho R\right\}^{-1}\right),$$

$$[\nu_i\lambda_i \mid .] \;=\; \mathrm{Ga}\left(\frac{1}{2}\{\nu+p\},\, \frac{1}{2}\right),$$

where

$$\nu_i = (\theta_i - \mu)^{\mathrm{T}}\Sigma^{-1}(\theta_i - \mu) + \nu.$$

Sampled values for λ_i substantially less than 1.0 indicate that individual i may be outlying.

We now describe various methods for sampling from $[\theta_i \mid .]$, $i = 1, \ldots, M$.

19.2.1 Method 1: Rejection Gibbs

The rejection sampling method of random variate generation is described in Ripley (1987); see also Gilks (1995: this volume). Smith and Gelfand (1992) describe how this technique can be straightforwardly applied in Bayesian applications, using the prior as the density from which candidate points are generated.

We implement rejection sampling in the context of Gibbs sampling as follows. Let $\ell_i(\theta_i, \tau)$ denote the likelihood for the ith individual and suppose that each of these functions can be maximized with respect to θ_i. For the normal likelihood (19.1),

$$\log \ell_i(\theta_i, \tau) = -\frac{\tau}{2}\sum_{j=1}^{n_i}[y_{ij} - f(\theta_i, t_{ij})]^2 , \qquad (19.4)$$

ignoring terms which do not depend on θ. We propose using the prior $p_2(\theta_i \mid \mu, \Sigma)$ as a rejection envelope for $[\theta_i \mid .]$. We define $\hat{\theta}_i$ to be the value of θ_i that maximizes $\ell_i(\theta_i, \tau)$, for fixed τ, and refer to it as the maximum likelihood estimate (MLE). Note that, for both the normal likelihood (19.1) and the lognormal likelihood (19.2), $\hat{\theta}_i$ does not depend on τ, so $\hat{\theta}_i$ need be evaluated only once, prior to the MCMC run. If $n_i = 1$, the maximized likelihood (19.4) will equal unity since, in all but pathological cases, the single point can be fitted exactly. Note that 1 always provides an upper bound for $\ell_i(\hat{\theta}_i, \tau)$ in (19.4). This may be useful for individuals for whom few data are available and whose likelihood is conseqently difficult to maximize. However, this could result in a highly inefficient algorithm, as the dominating envelope will be larger than it needs to be and will therefore lead to more points being generated before one is accepted. Note also that it is simple to check, during a MCMC run, that no violations of the bound occur, if it is uncertain that a true maximum has been determined. If such violations occur then the bound can be adjusted accordingly, though technically the chain is effectively restarted at this time.

Since $[\theta_i \mid .] \propto \ell_i(\theta_i, \tau) \times p_2(\theta_i \mid \mu, \Sigma)$, generation from the full conditional distributions proceeds according to the following rejection-sampling scheme:

> *Step 1:* Generate θ_i from $p_2(\theta_i \mid \mu, \Sigma)$ and U from a uniform distribution on $[0, 1]$.
>
> *Step 2:* If $U \leq \ell_i(\theta_i, \tau)/\ell_i(\hat{\theta}_i, \tau)$ then accept θ_i, otherwise return to *Step 1*.

Generation from $p_2(. \mid \mu, \Sigma)$ will be straightforward since this distribution is typically multivariate normal or multivariate student-t. The great advantage of this approach is its simplicity. We make no claims for its efficiency in a given context, though the fact that the whole vector θ_i is generated will lead to a Gibbs chain that will converge more quickly than one in which each component of θ_i is generated separately. The method is likely to be more efficient when the data are sparse, since the likelihood function will then be fairly flat.

19.2.2 Method 2: Ratio Gibbs

Wakefield *et al.* (1994) generate separately each of the p elements of θ_i using an extended ratio-of-uniforms method; see Gilks (1995: this volume) for an account of the basic ratio-of-uniforms method.

19.2.3 Method 3: Random-walk Metropolis

Hybrid MCMC algorithms allow some parameters to be updated by sampling from full conditionals as in Gibbs sampling, and others to be updated via Metropolis–Hastings steps (Tierney, 1994). A natural choice of Metropolis–Hastings for updating the θ_i parameters is random-walk Metropolis (see Roberts, 1995: this volume). For this, we use a multivarate normal proposal distribution centred at the current value of θ_i, with covariance matrix given by a constant c times the inverse information matrix:

$$\Omega_i = -\left[\frac{\partial^2 \log \ell_i(\theta_i, \tau)}{\partial \theta_i \partial \theta_i^T} \right]_{\theta_i = \hat{\theta}_i}^{-1}, \tag{19.5}$$

where τ is fixed at its current value. We assume here and below that each individual i supplies enough information to allow $\hat{\theta}_i$ and Ω_i to be calculated.

19.2.4 Method 4: Independence Metropolis–Hastings

Another natural choice of Metropolis–Hastings for updating θ_i is an independence sampler (see Roberts, 1995: this volume). For this we choose a multivariate normal proposal density $N(\hat{\theta}_i, \Omega_i)$, where Ω_i is given by (19.5).

19.2.5 Method 5: MLE/prior Metropolis–Hastings

For fixed τ, we can approximate the likelihood function $\ell_i(\theta_i, \tau)$ by a multivariate normal distribution for θ_i: $N(\hat{\theta}_i, \Omega_i)$. If the second-stage distribution is normal, we may approximate the full conditional distribution $[\theta_i \mid .]$ by the product of the approximate likelihood for θ_i and the prior. This yields an approximate full conditional that is multivariate normal with mean

$$\hat{\theta}_i - \Omega_i(\Omega_i + \Sigma)^{-1}(\hat{\theta}_i - \mu)$$

and covariance

$$(\Omega_i^{-1} + \Sigma^{-1})^{-1},$$

where μ, Σ and τ (and hence Ω_i) take their current values. This provides another Metropolis–Hastings proposal distribution. A similar argument in the context of a Bayesian EM algorithm for a mixed-effects model was used by Racine-Poon (1985).

Many other implementations of Metropolis–Hastings for sampling the $\{\theta_i\}$ are possible; see Gilks and Roberts (1995: this volume).

19.3 Comparison of strategies

Here we compare the strategies outlined in Section 19.2 on a previously published dataset. To compare strategies, account should be taken of the time involved in implementing the algorithm, the computer time required per iteration, and the total number of iterations needed to achieve adequate precision in ergodic averages. We formally compare the required number of iterations using the method of Raftery and Lewis (1992), whilst for the other two considerations we comment informally.

The method of Raftery and Lewis (1992) is described in Raftery and Lewis (1995: this volume). The method estimates, on the basis of a pilot run, the number of iterations required to estimate the qth posterior quantile of some function of the model parameters to within $\pm r$ with probability s. The length of burn-in is controlled by another variable, denoted by ε in Raftery and Lewis (1995: this volume). Our experience with this method is that, for well-behaved posteriors and 'good' parameterizations, it is reasonably reliable, but it can be unreliable in other circumstances; see Wakefield (1993).

For each of Methods 1 to 5 described above, we conducted a pilot run of 1 000 iterations, and performed the Raftery and Lewis calculations separately for each model parameter, setting $q = 0.025$, $r = 0.005$, $s = 0.95$ and $\varepsilon = 0.01$. The same initial starting points were used for each method and were chosen to be close to the posterior means for each parameter.

19.3.1 Guinea pigs data

This example was analysed by Lindstrom and Bates (1990) and concerns the prediction of uptake volume of guinea pig tissue by concentration of B-methylglucoside. There are ten responses for each of eight guinea pigs and we assume the model

$$\log y_{ij} = \log \left[\frac{\psi_{1i} B_{ij}}{\psi_{2i} + B_{ij}} + \psi_{3i} \right] + \varepsilon_{ij},$$

where y_{ij} is the uptake volume and B_{ij} the concentration of B-methylgluco-side for the jth sample from the ith individual, $i = 1, \ldots, 8$; $j = 1, \ldots, 10$. The ε_{ij} are independent normal random variables with zero mean and common variance. Here each of the random effects ψ_1, ψ_2 and ψ_3 are strictly positive so we assume that $\theta_{ki} = \log \psi_{ki}$, $k = 1, 2, 3$, arise from a trivariate normal distribution with mean μ and variance-covariance matrix Σ.

Table 19.1 gives the total number of iterations (in thousands) recommended for each of the parameters by the method of Raftery and Lewis (1992), for each of Methods 1 to 5. We see that the rejection Gibbs method (Method 1) performs well. For this method, the average number of rejection-sampling iterations needed per acceptance is, by guinea pig, 98, 27, 32, 23, 38, 116, 19, 26, with an overall average of 47. The ratio-of-uniforms Gibbs method (Method 2) has recommended numbers of iterations up to 3 times larger. This slower convergence is due to the element-wise updating of each random-effect vector, introducing additional serial dependence in the Markov chain.

Table 19.2 gives the empirical acceptance rates for the three Metropolis–Hastings methods. We first consider the random-walk Metropolis method (Method 3). As expected, the acceptance rate increases as the variance of the proposal distribution decreases (i.e. as c decreases). This is because points close to the current point are more likely to be accepted. However, a high acceptance rate is not the only requirement: for rapid mixing the chain must move about the non-negligible support of the stationary distribution as rapidly as possible. Therefore, there is a trade-off between small and large c: small c produces generally small movements at each iteration, whilst large c causes occasional large moves but frequently no move at all. With reference to Table 19.1, the $c = \frac{1}{4}$ version is the best here, although results for $c = \frac{1}{2}$ and $c = 1$ are almost as good. Thus for random-walk Metropolis, the optimal acceptance rate is around 55%, although acceptance rates around 30% produce comparable results. These results agree moderately well with theoretical results for random-walk Metropolis in

	reject.	ratio	random-walk Metropolis					indep.	MLE/
	Gibbs	Gibbs			c			M-H	prior
			$\frac{1}{4}$	$\frac{1}{2}$	1	2	4		M-H
θ_{11}	4.5	8.2	16.4	12.7	16.4	13.0	22.8	6.2	4.1
θ_{12}	4.1	5.3	10.4	12.3	12.3	43.9	28.4	6.6	7.0
θ_{13}	3.9	5.2	6.6	15.2	21.7	61.0	26.5	7.6	4.5
θ_{14}	3.9	6.2	11.8	8.2	9.9	31.3	28.5	9.9	4.8
θ_{15}	4.1	6.2	16.4	12.3	17.7	21.7	16.4	30.9	9.0
θ_{16}	3.8	7.0	14.2	24.7	26.5	17.7	97.6	57.0	437.6
θ_{17}	4.1	7.6	12.3	9.0	14.1	19.7	26.5	5.5	4.3
θ_{18}	4.1	14.3	14.1	9.9	16.3	19.7	80.3	7.0	5.5
θ_{21}	4.1	5.8	20.3	14.8	13.3	16.3	17.2	7.6	4.3
θ_{22}	4.1	5.5	24.7	14.8	16.4	32.9	28.5	8.2	5.2
θ_{23}	4.3	4.5	11.8	12.3	16.4	61.7	24.7	7.1	4.5
θ_{24}	3.8	4.7	9.9	16.4	13.3	12.3	62.2	19.0	6.6
θ_{25}	4.3	9.4	35.4	14.2	41.1	19.7	24.7	41.1	8.2
θ_{26}	3.9	4.9	14.1	26.0	19.7	102.9	49.8	31.8	38.0
θ_{27}	4.5	6.6	14.1	15.2	17.4	14.1	54.4	5.8	6.3
θ_{28}	3.8	5.5	11.0	13.3	12.7	26.5	19.7	9.0	4.7
θ_{31}	4.3	4.7	32.9	32.9	19.0	31.3	61.0	57.0	13.3
θ_{32}	4.7	7.0	16.4	43.4	27.7	22.8	32.9	21.2	7.0
θ_{33}	3.9	5.5	12.3	19.7	28.5	15.2	96.9	43.4	15.2
θ_{34}	3.8	5.5	17.7	17.7	24.7	52.2	20.3	8.2	8.9
θ_{35}	4.5	5.8	42.2	19.7	17.3	13.6	51.4	11.7	6.2
θ_{36}	4.1	5.8	15.2	46.3	16.4	14.1	24.1	8.2	7.1
θ_{37}	3.9	6.2	12.3	32.9	15.2	19.7	35.1	10.4	10.6
θ_{38}	4.3	9.6	14.1	11.0	28.5	73.8	69.4	27.8	10.6
τ_{12}	4.5	4.3	3.9	3.8	3.9	4.3	4.5	4.1	4.1
μ_1	4.9	4.3	4.3	4.5	5.2	4.5	3.9	5.2	4.9
μ_2	4.7	4.1	3.9	14.3	6.2	11.4	3.9	5.8	4.3
μ_3	4.3	13.4	3.9	5.2	4.1	14.9	3.9	4.1	4.3
Σ_{11}	4.1	4.1	4.5	4.3	4.3	3.9	4.7	4.3	4.3
Σ_{12}	4.7	4.3	4.7	4.7	4.1	4.1	4.1	4.3	4.1
Σ_{13}	4.3	5.5	4.5	3.9	11.4	4.5	4.3	4.5	4.1
Σ_{22}	4.1	4.5	3.8	4.3	4.7	4.7	4.5	4.3	4.5
Σ_{23}	4.5	5.5	4.7	4.5	4.5	4.3	4.3	4.3	3.9
Σ_{33}	4.1	4.5	4.5	4.9	4.0	3.9	4.1	4.5	4.3

Table 19.1 *Recommended numbers of iterations (in thousands) for the Guinea pigs data, for Methods 1 to 5*

high-dimensional symmetric problems, for which the optimal acceptance rate has been shown to be 24% (Gelman *et al.*, 1995; Roberts, 1995: this volume).

individual	random-walk Metropolis					independence Metropolis– Hastings	MLE/prior Metropolis– Hastings
	c						
	$\frac{1}{4}$	$\frac{1}{2}$	1	2	4		
1	56	44	27	21	8	27	87
2	61	45	30	16	9	44	89
3	58	44	31	20	10	36	86
4	57	46	33	17	11	41	88
5	56	46	32	19	10	35	88
6	46	36	24	14	7	20	38
7	60	47	32	18	12	47	91
8	60	48	33	22	10	45	91

Table 19.2 *Metropolis–Hastings acceptance rates (percentages) for the Guinea pigs example, by individual and type of Metropolis–Hastings*

The independence and MLE/prior Metropolis–Hastings methods (Methods 4 and 5) perform moderately well, although less well than the best random-walk Metropolis. Note the single extremely high recommendation (437 600 iterations) for the MLE/prior method; this value occurs for individual 6 who has a correspondingly low acceptance rate.

In terms of ease of implementation, the Metropolis–Hastings methods (Methods 3-5) are the most favourable. In general, no maximizations are required for the random-walk Metropolis but in our case we use the asymptotic variance-covariance matrix in the proposal distribution and this is evaluated at the MLEs. The random-walk Metropolis also requires some tuning to find the optimal value of c.

The Metropolis–Hastings methods are by far the most computationally efficient, with just a single evaluation of $[\theta_i \mid .]$ being required per iteration. The computer time per iteration for the rejection Gibbs method (Method 1) is far more than for the Metropolis–Hastings methods, since sampling must be repeated until acceptance. The ratio-of-uniforms method (Method 2) is also computationally expensive since 3 maximizations per generated variate are required.

The non-expert user is likely to find large non-interactive computing time requirements more acceptable than potentially lower computing times but which involve more work on tuning the Markov chain.

19.4 A case study from pharmacokinetics-pharmacodynamics

In this section, we carry out a re-analysis of a population pharmacokinetic-pharmacodynamic dataset previously analysed by Wakefield and Racine-Poon (1995).

Pharmacokinetic (PK) models predict the time course of a drug's concentration profile following the administration of a known dose. Pharmacodynamic (PD) models relate the response measurements to the dose or to some function of the plasma drug concentration. Population PK data consist of dose histories, individual covariates and measured drug concentrations with associated sampling times. Population PD data consist of dose histories, covariates and some response measure. Population analyses, whether they be PK or PD, attempt to explain the variability observed in the recorded measurements and are increasingly being seen as an important aid in drug development. It is natural to model population data hierarchically and hence the Markov chain techniques described in the previous sections are relevant. We now describe the dataset and the algorithm we utilize for our analysis.

The drug administered in this trial was the antithrombotic agent: recombinant Hirudin. The part of the study we concentrate on here consisted of four groups of four volunteers, each of whom received on day one of the study an intravenous bolus injection of recombinant Hirudin. Each group received one of the doses: $0.01, 0.03, 0.05$ or 0.1 mg/kg. Blood samples were taken at 0 hours, immediately before the bolus of recombinant Hirudin, and between 0.08 hours and 24 hours subsequently. Sixteen blood samples were taken in total from each individual. The response measure we shall concentrate on is the coagulation parameter: activated partial thromblastic time (APTT). APTT generally increases following administration of the drug.

Let y_{ij} denote the jth measured concentration for individual i measured at time t_{ij} and let z_{ij} be the reciprocal of the measured APTT at time t_{ij}, $i = 1, \ldots, 16; j = 1, \ldots, 16$. We now write down the first-stage likelihood for a particular individual. Let θ_i denote a vector of individual-specific PK parameters and ϕ_i a vector of individual-specific PD parameters. Also let τ_θ and τ_ϕ denote intra-individual precision parameters for concentration and response data respectively. Finally, let y_i and z_i denote the concentration and response data of individual i. The likelihood function for the PK/PD parameters of a particular individual is then given by

$$
\begin{aligned}
l(\theta_i, \phi_i, \tau_\theta, \tau_\phi) &= f(y_i, z_i \mid \theta_i, \phi_i, \tau_\theta, \tau_\phi) \\
&= f(y_i \mid \theta_i, z_i, \phi_i, \tau_\theta) f(z_i \mid \theta_i, \phi_i, \tau_\phi) \\
&= f(y_i \mid \theta_i, \tau_\theta) f(z_i \mid \theta_i, \phi_i, \tau_\phi),
\end{aligned}
$$

since the concentrations depend on the PK parameters only. Note that the PK parameter estimation may be influenced by the PD data via the second term of the above expression.

The pharmacokinetics of recombinant Hirudin are described by a 'sum of three exponentials' model. That is, the predicted concentration at time t_{ij} is given by

$$f_1(\theta_i, t_{ij}) = D_i(A_i e^{-\alpha_i t_{ij}} + B_i e^{-\beta_i t_{ij}} + C_i e^{-\gamma_i t_{ij}})$$

where D_i is the dose in mg/kg and

$$\theta_i = (\log A_i, \log B_i, \log C_i, \log(\alpha_i - \beta_i), \log(\beta_i - \gamma_i), \log \gamma_i).$$

The constraints $A_i, B_i, C_i > 0$ ensures positivity and $\alpha_i > \beta_i > \gamma_i > 0$ ensures identifiability. It is common when analysing PK data, to use a model of the form

$$\log y_{ij} = \log f_1(\theta_i, t_{ij}) + \varepsilon_{ij},$$

where ε_{ij} denotes a zero-mean error term, since assay techniques are frequently observed to produce measurements whose precision decreases with increasing concentration. In our case, the assay validation data confirmed this. On examining the data, we found that some measurements appeared to be outlying. With normal errors such points would overly influence parameter estimates. To guard against this we used a student-t error distribution with 4 degrees of freedom and precision τ_θ. This distribution has more weight in the tails and hence accommodates outliers better than the normal alternative.

The lower limit of detection for the drug concentration was $L = 0.1$ nmol/litre. The PK data contained a large number of concentration measurements that were labelled as below this limit of detection. Rather than leave these censored readings out of the analysis, which would lead to serious bias, we incorporated them via a likelihood term

$$P(y_{ij} < L \mid \theta_i, \tau_\theta) = \int_0^L f(y_{ij} \mid \theta_i, \tau_\theta) \, dy_{ij},$$

where P denotes probability and f denotes the student-t distribution with 4 degrees of freedom and precision τ_θ. This expression is the likelihood of θ_i for an observation lying at an unknown point somewhere between 0 and L. To evaluate this term we numerically calculate the cumulative density function of the student-t_4 distribution.

At the second stage of the PK model we assume that the θ_i vectors arise from a multivariate student-t_ν distribution with mean μ_θ and variance-covariance matrix $\frac{\nu}{\nu-2}\Sigma_\theta$. The use of a multivariate t-distribution at the second stage robustifies the analysis against outlying individuals. Implementation of a second-stage multivariate t-distribution via a scale mixture of normals is discussed in Section 19.2. We used $\nu = 4$ degrees of freedom.

We turn now to the PD relationship. In Wakefield and Racine-Poon (1995), APTT at time t was assumed to be linearly related to the square root of the concentration at time t. An alternative model that is more physiologically reasonable is the following. The relationship between the actual concentration $f_1(\theta, t)$ and the actual inhibition effect (reduction in clotting) $f_2(\phi, \theta, t)$ is given by the inhibition sigmoid Emax model (Holford and Sheiner, 1981)

$$f_2(\phi_i, \theta_i, t_{ij}) = \phi_{1i} \left(1 - \frac{f_1(\theta_i, t_{ij})^{\frac{1}{2}}}{f_1(\theta_i, t_{ij})^{\frac{1}{2}} + \phi_{2i}^{\frac{1}{2}}} \right),$$

where ϕ_{1i} is the reciprocal of baseline APTT (that is $\phi_1 = APTT_0^{-1}$ where $APTT_0$ is the amount of APTT when no drug is present), ϕ_{2i} is the concentration required to produce 50% inhibition and $\phi_i = (\phi_{1i}, \phi_{2i})$. As for the PK data a lognormal error distribution with precision τ_ϕ was chosen. The second-stage model for the PD parameters ϕ_i was assumed to be normal $N(\mu_\phi, \Sigma_\phi)$. We could also have specified a multivariate t-distribution here but no individual's PD parameters were outlying. Figure 19.1 shows the directed graph which represents this situation (see Spiegelhalter $et\ al.$, 1995: this volume).

Finally, vague conjugate priors with large variances were assumed for μ_θ and μ_ϕ with inverse Wishart priors for Σ_θ, Σ_ϕ and gamma priors for τ_θ, τ_ϕ.

The PK parameters θ_i were sampled using the random-walk Metropolis algorithm described in Section 19.2.3, although instead of using the asymptotic variance-covariance matrix Ω_i given in (19.5), we used the asymptotic correlation matrix, as some of the elements of Ω_i were very large due to the poor behaviour of the six-dimensional likelihood surface. By converting to a correlation matrix, the principle characteristics of the surface were retained. The constant $c = c_1$ for this method was chosen empirically so that the Metropolis algorithm gave reasonable acceptance probabilities. For the PD parameters ϕ_i, a random-walk Metropolis, with variance $c = c_2$ times the asymptotic variance-covariance matrix, was constructed. Table 19.3 gives the Metropolis acceptance rates for each of the individuals and for the PK and PD parameter vectors. Table 19.4 gives posterior summaries of the population PK and PD parameters. Figure 19.2 shows predicted mean profiles for the two PD models: the inhibition model described here and the linear model that was used in Wakefield and Racine-Poon (1995). There is little difference between the two predicted profiles.

19.5 Extensions and discussion

We return now to the model and notation of Section 19.1. So far we have assumed that each component of the θ-vector is a random effect and that there are no covariates at the second stage. In this section, we relax these assumptions. We also discuss hybrid MCMC strategies.

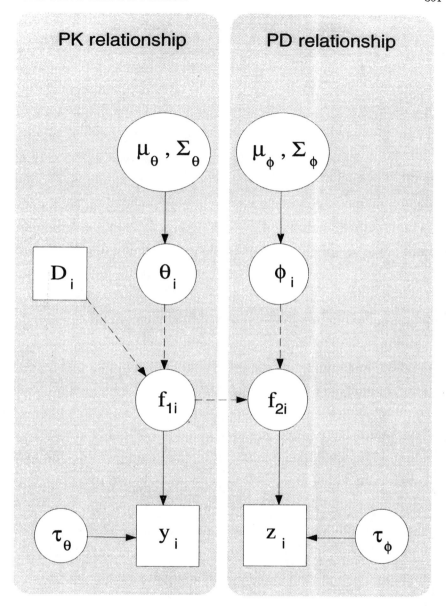

Figure 19.1 *Directed graph for PK/PD model.*

individual	PK parameters $(c_1 = 0.005)$	PD parameters $(c_2 = 0.01)$
1	48.0	31.2
2	51.6	30.1
3	38.1	36.3
4	58.2	31.7
5	43.2	45.6
6	41.7	46.0
7	48.9	46.1
8	48.0	46.3
9	32.7	53.8
10	32.0	51.7
11	43.7	43.7
12	40.6	54.5
13	31.9	55.4
14	31.2	53.1
15	32.4	58.4
16	32.4	49.7

Table 19.3 *Random-walk Metropolis acceptance rates (percentages) for PK and PD parameters*

Second-stage dependence on covariates

If we have covariate relationships of the form

$$E(\theta_i \mid \mu, x_i) = g(\mu, x_i)$$

that is, relationships involving covariates x_i that are *constant* within an individual, then the approaches outlined in Section 19.2 can be extended in an obvious fashion. For example, the rejection Gibbs strategy (Method 1) merely samples the θ_i vector via a normal distribution with mean $g(\mu, x_i)$ determined by the current value of μ. If the covariate values are changing within an individual during the course of the data collection, there is no straightforward solution since the prior distribution for the random effects, $N(g(\mu, x_{ij}), \Sigma)$, is now changing at each time point. The obvious remedy would be to set up a Metropolis–Hastings scheme, for example by taking the average value of the covariate within an individual, and then using the kernel so defined as the basis of a Metropolis–Hastings step (Wakefield, 1992).

Hybrid MCMC strategies

For the example considered in Section 19.3, we started each chain from a 'good' initial value. In other examples, we have started from 'bad' positions

parameter	posterior mean	posterior s.d.
PK parameters		
$\mu_{\theta 1}$	7.32	0.116
$\mu_{\theta 2}$	5.86	0.0769
$\mu_{\theta 3}$	2.82	0.0919
$\mu_{\theta 4}$	1.60	0.123
$\mu_{\theta 5}$	-0.253	0.171
$\mu_{\theta 6}$	-1.70	0.263
$\Sigma_{\theta 11}$	0.0940	0.0482
$\Sigma_{\theta 22}$	0.0633	0.0342
$\Sigma_{\theta 33}$	0.122	0.0540
$\Sigma_{\theta 44}$	0.137	0.0765
$\Sigma_{\theta 55}$	0.436	0.189
$\Sigma_{\theta 66}$	1.032	0.478
PD parameters		
$\mu_{\phi 1}$	-3.41	0.0295
$\mu_{\phi 2}$	4.42	0.101
$\Sigma_{\phi 11}$	0.0138	0.00611
$\Sigma_{\phi 22}$	0.127	0.0663

Table 19.4 *Posterior summaries for population PK and PD parameters*

and have encountered very slow convergence with the Metropolis–Hastings algorithm. Slow convergence even occurred for the population parameters μ and Σ and was due to random effects becoming stuck or moving very slowly. Tierney (1994) suggests a cyclical Markov chain strategy. In our context, this might be implemented using an n-iteration cycle, comprising $n-1$ Metropolis–Hastings steps followed by a single Gibbs step using the rejection algorithm of Method 1. This may prevent the chain becoming stuck. One advantage of a Gibbs sampling approach is that there is no need to have initial estimates of the random-effects vectors since the conditional distributions only depend on the population parameters and the precision τ.

In the examples considered above, the Gibbs chain in which all the random effects were sampled as vectors (Method 1) produced very good convergence characteristics. The greatest disadvantage of the rejection Gibbs method is its computational inefficiency which may increase with the dimensionality of θ_i. The Metropolis–Hastings methods are potentially far more efficient though usually at the expense of some tailoring to the problem in hand. It is possible to use different strategies for different individuals. For example, the Gibbs rejection algorithm is likely to be inefficient for a

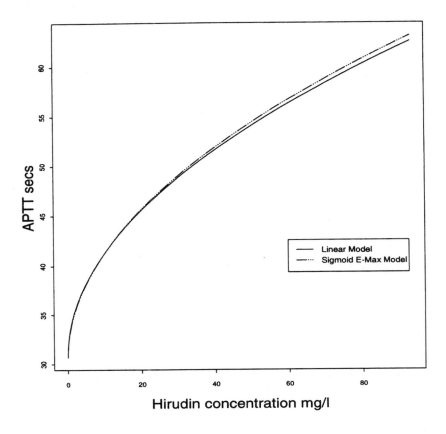

Figure 19.2 *Concentration/effect curves for two alternative PK/PD models.*

particular individual, if the individual has an abundance of data since, in
this case, the likelihood will be peaked. With sparse data (and consequently
a flat likelihood) we would expect to obtain a more efficient algorithm. For
the Metropolis–Hastings methods, a reliable estimate of the information
matrix is required, a requirement which, in other examples we have looked
at, has sometimes been difficult to achieve. Such a matrix will be easy to
determine if the likelihood is peaked. This suggests that it may be prof-
itable to use Gibbs for those individuals with sparse data and Metropolis
for those individuals with an abundance of data. It is unlikely that this
strategy would be followed, however, since the 'set-up' time in deciding
which algorithm is to be used for each individual would be prohibitive.

The practitioner may find the array of possible MCMC strategies and implementational details somewhat daunting. Such details for the Metropolis–Hastings algorithm include location, scale and family of generating distribution, and the number n of Metropolis–Hastings steps to take within a cycle. We hope we have given some insight into the relative merits of some of these possibilities in the context of nonlinear hierarchical models.

Acknowledgements

The first author's research was funded by Glaxo Group Research. We have had useful discussions with David Phillips, Gareth Roberts, Adrian Smith and David Stephens.

References

Berkey, C. S. (1982) Comparison of two longitudinal growth models for preschool children. *Biometrics*, **38**, 221–234.

Bernardo, J. M. and Smith, A. F. M. (1994) *Bayesian Theory*. Chichester: Wiley.

Carlin, B. P. (1995) Hierarchical longitudinal modelling. In *Markov Chain Monte Carlo in Practice* (eds W. R. Gilks, S. Richardson and D. J. Spiegelhalter), pp. 303–319. London: Chapman & Hall.

Clayton, D. G. (1991) A Monte Carlo method for Bayesian inference in frailty models. *Biometrics*, **47**, 467–485.

Dellaportas, P. and Smith, A. F. M. (1993) Bayesian inference for generalized linear and proportional hazards models via Gibbs sampling. *Appl. Statist.*, **41**, 443–459.

Gelfand, A. E. and Smith, A. F. M. (1990) Sampling based approaches to calculating marginal densities. *J. Am. Statist. Ass.*, **85**, 398–409.

Gelfand, A. E., Hills, S. E., Racine-Poon, A. and Smith, A. F. M. (1990) Illustration of Bayesian inference in normal data models using Gibbs sampling. *J. Am. Statist. Ass.*, **85**, 972–985.

Gelman, A. Roberts, G. O. and Gilks, W. R. (1995) Efficient Metropolis jumping rules. In *Bayesian Statistics 5* (eds J. M. Bernardo, J. Berger, A. P. Dawid and A. F. M. Smith). Oxford: Oxford University Press (in press).

Gilks, W. R. (1995) Full conditional distributions. In *Markov Chain Monte Carlo in Practice* (eds W. R. Gilks, S. Richardson and D. J. Spiegelhalter), pp. 75–88. London: Chapman & Hall.

Gilks, W. R. and Roberts, G. O. (1995) Strategies for improving MCMC. In *Markov Chain Monte Carlo in Practice* (eds W. R. Gilks, S. Richardson and D. J. Spiegelhalter), pp. 89–114. London: Chapman & Hall.

Gilks, W. R. and Wild, P. (1992) Adaptive rejection sampling for Gibbs sampling. *Appl. Statist.*, **41**, 337–348.

Gilks, W. R., Best, N. G. and Tan, K. K. C. (1995) Adaptive rejection Metropolis sampling within Gibbs sampling. *Applied Statistics* (in press).

Holford, N. H. G. and Sheiner, L. B. (1981) Kinetics of pharmacological reponse. *Pharm. Therap.*, **16**, 143–166.

Lange, N., Carlin, B. P. and Gelfand, A. E. (1992) Hierarchical Bayes models for the progression of HIV infection using longitudinal CD$_4$ counts. *J. Am. Statist. Ass.*, **87**, 615–632.

Lindstrom, M. J. and Bates, D. M. (1990) Nonlinear mixed effects models for repeated measures data. *Biometrics*, **46**, 673–687.

Racine-Poon, A. (1985) A Bayesian approach to nonlinear random effects models. *Biometrics*, **41**, 1015–1024.

Racine-Poon, A. (1992) SAGA: sample assisted graphical analysis. In *Bayesian Statistics 4* (eds J. M. Bernardo, J. O. Berger, A. P. Dawid and A. F. M. Smith). Oxford: Oxford University Press.

Raftery, A. E. and Lewis, S. (1992) How many iterations in the Gibbs sampler? In *Bayesian Statistics 4*, (eds J. M. Bernardo, J. O. Berger, A. P. Dawid and A. F. M. Smith). Oxford: Oxford University Press.

Raftery, A. E. and Lewis, S. M. (1995) Number of iterations, convergence diagnostics and generic Metropolis algorithms. In *Markov Chain Monte Carlo in Practice* (eds W. R. Gilks, S. Richardson and D. J. Spiegelhalter), pp. 115–130. London: Chapman & Hall.

Ripley, B. (1987) *Stochastic Simulation.* New York: Wiley.

Roberts, G. O. (1995) Introduction to Markov chains for simulation. In *Markov Chain Monte Carlo in Practice* (eds W. R. Gilks, S. Richardson and D. J. Spiegelhalter), pp. 45–57. London: Chapman & Hall.

Smith, A. F. M. and Gelfand, A. E. (1992) Bayesian statistics without tears: a sampling resampling perspective. *Am. Statist.*, **46**, 84–88.

Spiegelhalter, D. J., Best, N. G., Gilks, W. R. and Inskip, H. (1995) Hepatitis B: a case study in MCMC methods. In *Markov Chain Monte Carlo in Practice* (eds W. R. Gilks, S. Richardson and D. J. Spiegelhalter), pp. 21–43. London: Chapman & Hall.

Steimer, J. L., Mallet, A., Golmard, J. and Boisieux, J. (1985) Alternative approaches to estimation of population pharmacokinetic parameters: comparison with the nonlinear mixed effect model. *Drug Metab. Rev.*, **15**, 265–292.

Tierney, L. (1994) Markov chains for exploring posterior distributions (with discussion). *Ann. Statist.*, **22**, 1701–1762.

Wakefield, J. C. (1992) Bayesian analysis of population pharmacokinetics. Technical report, Department of Mathematics, Imperial College, London.

Wakefield, J. C. (1993) Contribution to the discussion on the Gibbs sampler and other Markov chain Monte Carlo methods. *J. R. Statist. Soc. B*, **55**, 3–102.

Wakefield, J. C. and Racine-Poon, A. (1995) An application of Bayesian population pharmacokinetic/pharmacodynamic models to dose recommendation. *Statist. Med.*, to appear.

Wakefield, J. C., Gelfand, A. E. and Smith, A. F. M. (1991) Efficient generation of random variates via the ratio-of-uniforms method. *Statist. Comput.*, **1**, 129–133.

Wakefield, J. C., Smith, A. F. M., Racine-Poon, A. and Gelfand, A. E. (1994) Bayesian analysis of linear and nonlinear population models using the Gibbs sampler. *Appl. Statist.*, **41**, 201–221.

Zeger, S. L. and Karim, M. R. (1991) Generalized linear models with random effects: a Gibbs sampling approach. *J. Am. Statist. Ass.*, **86**, 79–86.

20

Bayesian mapping of disease

Annie Mollié

20.1 Introduction

The analysis of geographical variation in rates of disease incidence or mortality is useful in the formulation and validation of aetiological hypotheses. Disease mapping aims to elucidate the geographical distribution of underlying disease rates, and to identify areas with low or high rates. The two main conventional approaches are maps of standardized rates based on Poisson inference, and maps of statistical significance. The former has the advantage of providing estimates of the parameters of interest, the disease rates, but raises two problems.

First, for rare diseases and for small areas, variation in the observed number of events exceeds that expected from Poisson inference. In a given area, variation in the observed number of events is due partly to Poisson sampling, but also to extra-Poisson variation arising from variability in the disease rate within the area, which results from heterogeneity in individual risk levels within the area.

These considerations have led several authors to develop Bayesian approaches to disease mapping. This consists of considering, in addition to the observed events in each area, prior information on the variability of disease rates in the overall map. Bayesian estimates of area-specific disease rates integrate these two types of information. Bayesian estimates are close to the standardized rate, when based on a large number of events or person-years of exposure. However, with few events or person-years, prior information on the overall map will dominate, thereby shrinking standardized rates towards the overall mean rate.

The second problem in using the conventional approach based on Poisson inference, is that it does not take account of any spatial pattern in disease, i.e. the tendency for geographically close areas to have similar disease rates. Prior information on the rates, allowing for local geographical dependence, is pertinent. With this prior information, a Bayesian estimate of the rate in an area is shrunk towards a *local* mean, according to the rates in the neighbouring areas. A parallel can be drawn with image restoration (see Green, 1995: this volume), our goal being to reconstruct the true image (disease rates) from noisy observed data (event counts).

Empirical Bayes methods yield acceptable point estimates of the rates but underestimate their uncertainty. A direct fully Bayesian approach is rarely tractable with the non-conjugate distributions typically involved. However, Gibbs sampling has been used to simulate posterior distributions and produce satisfactory point and interval estimates for disease rates.

This application is part of the general theory of Bayesian analysis using generalized linear mixed models, developed in Breslow and Clayton (1993) and discussed in Clayton (1995: this volume).

20.2 Hypotheses and notation

Suppose the map is divided into n contiguous areas labelled $i = 1, \ldots, n$. Let $y = (y_1, \ldots, y_n)$ denote the numbers of deaths from the disease of interest during the study period. 'Expected' numbers of deaths $(e_1, \ldots, e_n)$ in each area can be calculated by applying the overall age- and sex-specific death rates, assumed to be constant during the study period, to the population at risk in the area subdivided by age and sex. Note that $e_1, \ldots, e_n$ do not reflect area-specific variations in mortality.

When the disease is non-contagious and rare, the numbers of deaths in each area are assumed to be mutually independent and to follow Poisson distributions. In each area, y_i has mean $e_i r_i$ where $r = (r_1, \ldots, r_n)$ are the unknown area-specific relative risks of mortality from the disease. The likelihood of the relative risk r_i is:

$$[y_i \mid r_i] = \exp\{-e_i r_i\} \frac{(e_i r_i)^{y_i}}{y_i!}. \tag{20.1}$$

Here the notation $[a \mid b]$ generically denotes the conditional distribution of a given b. Similarly, $[a]$ will denote the marginal distribution of a.

20.3 Maximum likelihood estimation of relative risks

The maximum likelihood estimate (MLE) of r_i is the standardized mortality ratio (SMR) for the ith area: $\hat{r}_i = y_i/e_i$, with estimated standard error $s_i = \sqrt{y_i}/e_i$.

As an illustration, we consider 3215 deaths from gall-bladder and bile-duct cancer among males in the 94 mainland French départements during the period 1971–1978 (Mollié, 1990). The SMRs in Figure 20.1 vary widely around their mean 0.98 ($s.d.$=0.27) from 0.31 for département 48, which has the smallest population size and the smallest expected number of deaths, to 1.75 for département 54, which has a tenfold larger population and a moderate expected number of deaths (Table 20.1). The standard errors of the SMRs range from 0.076 for département 75, which has one of the largest populations and the largest expected number of deaths, to 0.419 for département 90, which has one of the smallest populations and expected number of deaths (Table 20.1). Although in Figure 20.1, there is some suggestion of lower mortality in the south-west of France and higher mortality in the centre and the north-east, no clear spatial pattern emerges from the map of SMRs.

number	département	expected number of deaths	SMR	standard error of SMR	p-value
44	Loire-Atlantique	48.34	0.58	0.11	0.002
48	Lozère	6.46	0.31	0.22	0.088
54	Meurthe-et-Moselle	37.66	1.75	0.22	$< 10^{-4}$
57	Moselle	46.40	1.72	0.19	$< 10^{-4}$
62	Pas-de-Calais	73.59	1.33	0.13	0.006
73	Savoie	18.05	1.38	0.28	0.092
75	Paris	158.35	0.92	0.08	0.348
85	Vendée	28.94	0.35	0.11	$< 10^{-4}$
90	Territoire-de-Belfort	6.74	1.19	0.42	0.476

Table 20.1 *Estimates of relative risks of mortality from gall-bladder and bile-duct cancer for males in France, 1971–1978 (selected départements shown); p-value: for SMR under Poisson assumption (20.1) with $r_i = 1$.*

However, the SMR takes no account of the population size of the area, so the most extreme SMRs may be those based on only a few cases. On the other hand, p-values of tests comparing SMRs to unity are influenced by population sizes, so the most extreme p-values may simply identify areas with large population. These two drawbacks are emphasized when studying a rare disease and small areas, making the epidemiological interpretation of maps of SMRs or of p-values difficult, and even misleading.

Gall-bladder and bile-duct cancer in France has a low overall mortality rate of 1.57 per 10^5 males. Table 20.1 shows that SMR_{48} (the SMR for département 48) is small but not significantly different from unity ($p = 0.088$), while SMR_{44} is closer to unity but is significant ($p = 0.002$) because it has a much larger population ($\times 12$). Likewise SMR_{73} is large but not significant ($p = 0.092$), being based on a small population, while SMR_{62} is closer to unity but is significant ($p = 0.006$), being based on one of the

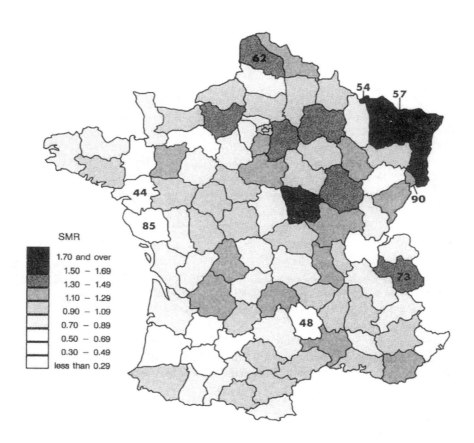

Figure 20.1 *SMRs for gall-bladder and bile-duct cancer for males in France, 1971–1978. Départements listed in Table 20.1 are labelled.*

largest populations.

For a rare disease and small areas, since individual risks are heterogeneous within each area, the variability of the average risk of the area exceeds that expected from a Poisson distribution. Extra-Poisson variation can be accommodated by allowing relative risks r_i to vary within each area. Bayesian methods can be used for this, giving smoothed estimates of relative risks. Indeed, even if the SMR is the best estimate of the rate, for each area considered in isolation, Bayesian rules produce sets of estimates having smaller squared-error loss (when $n \geq 3$) than the set of SMRs.

20.4 Hierarchical Bayesian model of relative risks

20.4.1 Bayesian inference for relative risks

Bayesian approaches in this context combine two types of information: the information provided in each area by the observed cases described by the Poisson likelihood $[y \mid r]$ and used to compute the SMRs, and prior information on the relative risks specifying their variability in the overall map and summarized by their prior distribution $[r]$.

Bayesian inference about the unknown relative risks r is based on the posterior distribution $[r \mid y] \propto [y \mid r][r]$. The likelihood function of the relative risks r given the data y is the product of n independent Poisson distributions (20.1), since the y_i can be assumed to be conditionally independent given r, and y_i depends only on r_i:

$$[y \mid r] = \prod_{i=1}^{n} [y_i \mid r_i]. \qquad (20.2)$$

The prior distribution $[r]$ reflects prior belief about variation in relative risks over the map, and is often parameterized by hyperparameters γ. The joint posterior distribution of the parameters (r, γ) is then expressed by $[r, \gamma \mid y] \propto [y \mid r][r \mid \gamma][\gamma]$. Thus the marginal posterior distribution for r given the data y is:

$$[r \mid y] = \int [r, \gamma \mid y] \, d\gamma. \qquad (20.3)$$

A point estimate of the set of relative risks is given by a measure of location of distribution (20.3): typically the posterior mean $E[r \mid y]$. However, direct evaluation of the posterior mean through analytic or numerical integration is not generally possible.

An alternative measure of location of (20.3) which is easier to compute and often used in image analysis applications (Besag, 1986, 1989) is the posterior mode or maximum *a posteriori* (MAP) estimate, which maximizes $[r \mid y]$ with respect to r. MAP estimation can be performed using penalized likelihood maximization (Clayton and Bernardinelli, 1992) and has been applied to disease mapping by Tsutakawa (1985) and Bernardinelli and

Montomoli (1992).

A Bayesian analysis employing a completely specified prior distribution $[r \mid \gamma]$ with known hyperparameters γ is seldom used in practice. The empirical Bayes (EB) approach assumes hyperparameters are unknown and drawn from an unspecified distribution. The fully Bayesian formulation comprises a three-stage hierarchical model in which the hyperprior distribution $[\gamma]$ is specified.

20.4.2 Specification of the prior distribution

Unstructured heterogeneity of the relative risks

The simplest prior assumes exchangeable relative risks, given γ:

$$[r \mid \gamma] = \prod_{i=1}^{n} [r_i \mid \gamma],$$

where prior distribution $[r_i \mid \gamma]$ is the same for each area i. It is convenient to choose for $[r_i \mid \gamma]$ a gamma distribution $\mathrm{Ga}(\nu, \alpha)$ with mean ν/α and variance ν/α^2, which is conjugate with the likelihood (20.1). Thus:

$$[r \mid \gamma] = [r \mid \alpha, \nu] = \prod_{i=1}^{n} [r_i \mid \alpha, \nu], \qquad (20.4)$$

with $[r_i \mid \alpha, \nu] \propto \alpha^\nu \, r_i^{\nu-1} \exp\{-\alpha r_i\}$. Alternatively, a normal prior distribution $\mathrm{N}(\mu, \sigma^2)$ with mean μ and variance σ^2 on the log relative risk $x_i = \log r_i$ can be used (Clayton and Kaldor, 1987), giving:

$$[x \mid \gamma] = [x \mid \mu, \sigma^2] = \prod_{i=1}^{n} [x_i \mid \mu, \sigma^2]. \qquad (20.5)$$

To allow for area-specific covariates, the independent normal prior (20.5) can easily be generalized (Tsutakawa, Shoop and Marienfeld, 1985; Clayton and Kaldor, 1987). Setting $\mu = Z\phi$, where Z is a matrix of p known covariates and $\phi = (\phi_1, \ldots, \phi_p)$ is a vector of covariate effects, we define the independent normal prior by:

$$[x \mid \gamma] = \prod_{i=1}^{n} [x_i \mid \mu_i, \sigma^2] \qquad (20.6)$$

where $\mu_i = (Z\phi)_i$.

The conjugate gamma prior (20.4) can also be generalized to accommodate covariates (Manton $et\ al.$ 1981, 1987, 1989; Tsutakawa, 1988). Gamma priors can be justified for modelling population risk processes (Manton $et\ al.$ 1981, 1987), and lead to estimates which have the best robustness properties in the class of all priors having the same mean and variance

(Morris, 1983). However, gamma priors cannot easily be generalized to allow for spatial dependence, unlike normal priors.

Spatially structured variation of the relative risks

Prior knowledge may indicate that geographically close areas tend to have similar relative risks, i.e. there exists local spatially structured variation in relative risks. To express this prior knowledge, nearest neighbour Markov random field (MRF) models are convenient. For this class of prior model, the conditional distribution of the relative risk in area i, given values for the relative risks in all other areas $j \neq i$, depends only on the relative risk values in the neighbouring areas δi of area i. Thus in this model, relative risks have a locally dependent prior probability structure. Their joint distribution is determined (up to a normalizing constant) by these conditional distributions (Besag, 1974).

Gaussian MRF models for the log relative risks specify the conditional prior distribution of x_i to be normal with mean depending upon the mean of x_j in the neighbouring areas. The usual forms of conditional Gaussian autoregression (Besag, 1974), first used for log relative risks in Bayesian mapping by Clayton and Kaldor (1987), assume that the conditional variance is constant, and hence are not strictly appropriate for irregular maps where the number of neighbours varies.

For irregular maps, intrinsic Gaussian autoregression (Besag et al., 1991; Bernardinelli and Montomoli, 1992) is more appropriate. For this, the conditional variance of x_i, given all the other log relative risks in the map, is inversely proportional to the number of neighbouring areas w_{i+} of area i. The joint prior distribution of x given the hyperparameter γ ($= \sigma^2$) is then:

$$[x \mid \gamma] = [x \mid \sigma^2] \propto \frac{1}{\sigma^n} \exp\left\{ -\frac{1}{2\sigma^2} \sum_{i=1}^{n} \sum_{j<i} w_{ij}(x_i - x_j)^2 \right\}. \qquad (20.7)$$

The mean of $[x \mid \gamma]$ is zero and its inverse variance-covariance matrix has diagonal elements w_{i+}/σ^2 and off-diagonal elements $-w_{ij}/\sigma^2$, where the w_{ij} are prescribed non-negative weights, with $w_{ij} = 0$ unless i and j are neighbouring areas, and $w_{i+} = \sum_{j=1}^{n} w_{ij}$. The simplest choice is $w_{ij} = 1$ if i and j are adjacent areas. In this case, the normal conditional prior distribution of x_i given all the other x_j and the hyperparameter σ^2, has mean and variance given by:

$$E\left[x_i \mid x_{-i}, \gamma\right] \quad = E\left[x_i \mid x_j, j \in \delta i, \sigma^2\right] \quad = \bar{x}_i$$

$$\mathrm{var}\left[x_i \mid x_{-i}, \gamma\right] \quad = \mathrm{var}\left[x_i \mid x_j, j \in \delta i, \sigma^2\right] \quad = \frac{\sigma^2}{w_{i+}}$$

where $\bar{x}_i$ denotes the mean of the x_j in areas adjacent to area i, and x_{-i}

denotes the log relative risk values in all the areas $j \neq i$.

In practice, it is often unclear how to choose between an unstructured prior and a purely spatially structured prior. An intermediate distribution on the log relative risks that ranges from prior independence to prior local dependence, called a convolution Gaussian prior, has been proposed (Besag, 1989; Besag and Mollié, 1989; Besag et al., 1991). In this prior model, log relative risks are the sum of two independent components

$$x = u + v, \tag{20.8}$$

where v is an independent normal variable (20.5) with zero mean and variance λ^2, describing unstructured heterogeneity in the relative risks; u is modelled as an intrinsic Gaussian autoregression (20.7) with conditional variances proportional to κ^2, representing local spatially structured variation; and κ^2 and λ^2 are hyperparameters contained in γ. The conditional variance of x_i given all the other x_j is the sum of the variances of the independent components u and v:

$$\text{var}\left[x_i \mid x_{-i}, \gamma\right] = \text{var}\left[x_i \mid x_j, j \in \delta i, \kappa^2, \lambda^2\right] = \frac{\kappa^2}{w_{i+}} + \lambda^2. \tag{20.9}$$

Parameters κ^2 and λ^2 control the strength of each component: if κ^2/λ^2 is close to $\bar{w}$, the average value of w_{i+}, each component has the same importance; if κ^2/λ^2 is small then unstructured heterogeneity dominates; and if κ^2/λ^2 is large then spatially structured variation dominates.

This model can be generalized to allow for covariate effects by assuming v to have mean $\mu = Z\phi$, where Z is a matrix of p known covariates measured in each area and $\phi = (\phi_1, \ldots, \phi_p)$ are their corresponding effects (Mollié, 1990; Clayton and Bernardinelli, 1992; Clayton et al., 1993).

20.4.3 Graphical representation of the model

For large models, it is often helpful to view relationships between variables in the model in the form of a *graph*, as discussed in Spiegelhalter et al. (1995: this volume) and in Clayton (1989, 1991).

Figures 20.2, 20.3 and 20.4 are graphs corresponding to the hierarchical Bayesian models discussed above. Each graph represents a set of conditional independence assumptions for one model, the models differing only in their prior on the log relative risks x. For each model, we assumed that y is independent of γ given x; this is represented on the graph by the absence of direct links between y and hyperparameters ϕ, σ^2, κ^2 or λ^2 in Figures 20.2–20.4. We also assumed that the y_i are conditionally independent given x; this is represented by the absence of links between the y_i. Finally, we asumed that y_i is independent of all other x_j ($j \neq i$), given x_i; therefore there are no direct links between y_i and x_j in Figures 20.2 and 20.3, or between y_i and u_j or v_j in Figure 20.4.

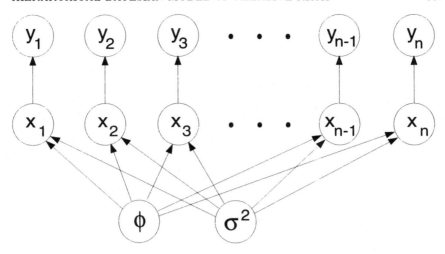

Figure 20.2 *Graphical representation of the hierarchical Bayesian model with the independent normal prior on log relative risks x.*

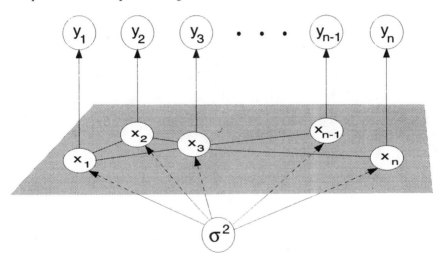

Figure 20.3 *Graphical representation of the hierarchical Bayesian model with the intrinsic Gaussian autoregressive prior on log relative risks x.*

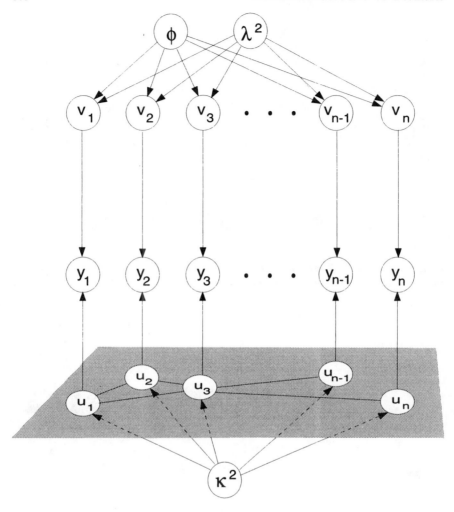

Figure 20.4 *Graphical representation of the hierarchical Bayesian model with the convolution Gaussian prior on log relative risks $x = u + v$.*

Further conditional independence assumptions depend on the prior model. For the independent normal prior (20.6), the log relative risks $\{x_i\}$ are mutually independent given $\gamma \; (= (\phi, \sigma^2))$, as indicated by the absence of links between the $\{x_i\}$, and by the directed links emanating from ϕ and σ^2 in Figure 20.2. For the intrinsic Gaussian autoregressive prior (20.7), undirected links in Figure 20.3 between the $\{x_i\}$ represent conditional dependence between log relative risks in neighbouring areas (e.g. areas 1 and 2). Absence of an undirected link between two areas (e.g. areas n and

$n - 1$) indicates they are non-neighbours, and hence their log relative risks are *a priori* conditionally independent, given σ^2. In Figure 20.4, conditional dependence between log relative risks for the convolution Gaussian prior (20.8) is indicated by links between the $\{u_i\}$.

20.5 Empirical Bayes estimation of relative risks

As noted above, a fully Bayesian analysis based on the marginal posterior distribution $[r \mid y]$ in (20.3) is often intractable. The EB idea consists in approximating $[r \mid y]$ by

$$[r \mid y, \gamma] \propto [y \mid r] [r \mid \gamma], \qquad (20.10)$$

the marginal posterior distribution for r given the data y and the hyper-parameters γ, where the unknown hyperparameters γ are replaced by suitable estimates denoted $\widehat{\gamma}$. This approximation is relevant if the distribution $[\gamma \mid y]$ is very sharply concentrated at $\widehat{\gamma}$. In general, these hyperparameter estimates are MLEs derived from information relevant to the overall map structure, contained in the marginal likelihood of γ:

$$[y \mid \gamma] = \int [y \mid r] [r \mid \gamma] \, dr. \qquad (20.11)$$

The EB estimate of the relative risks is then $E[r \mid y, \widehat{\gamma}]$, the posterior mean evaluated at $\widehat{\gamma}$.

20.5.1 The conjugate gamma prior

For the conjugate gamma prior (20.4) for relative risks, the marginal posterior distribution $[r \mid y, \gamma]$ is a product of n marginal posterior distributions $[r_i \mid y_i, \alpha, \nu]$, which are gamma distributions $\mathrm{Ga}(y_i + \nu, e_i + \alpha)$ with means given by:

$$E\left[r_i \mid y_i, \alpha, \nu\right] = \frac{y_i + \nu}{e_i + \alpha} = \omega_i \, SMR_i + (1 - \omega_i)\frac{\nu}{\alpha}, \qquad (20.12)$$

where $SMR_i = y_i/e_i$ and weight $\omega_i = e_i/(e_i + \alpha)$. Thus the posterior mean of the relative risk for the ith area is a weighted average of the SMR for the ith area and the prior mean of the relative risks in the overall map, the weight being inversely related to the variance of the SMR. As this variance is large for rare diseases and for small areas, ω_i is small and the posterior mean tends towards a global mean ν/α, thereby producing a smoothed map.

MLEs $\widehat{\alpha}$ and $\widehat{\nu}$ of the hyperparameters α and ν are obtained by maximizing the marginal likelihood $[y \mid \alpha, \nu]$, which is a product of n negative binomial distributions $[y_i \mid \alpha, \nu]$. The EB estimate of the relative risk for the ith area, $\tilde{r}_i$, is then the posterior mean $E\left[r_i \mid y_i, \widehat{\alpha}, \widehat{\nu}\right]$.

For example, for gall-bladder and bile-duct cancer in France (Mollié, 1990), $\widehat{\alpha} = 25.508$ and $\widehat{\nu} = 25.223$. The EB estimates of relative risk show much less variation than the SMRs: they vary from 0.65 for département 85 to 1.46 for département 57 (Table 20.2), with a mean of 0.99 and standard deviation 0.15. This compares to a standard deviation of 0.27 among SMRs. Comparison of Tables 20.1 and 20.2 reveals that the smoothing effect is more pronounced for départements with extreme SMRs, particularly those with small populations such as Lozère (area 48) or Savoie (area 73), than for départements with large populations such as Loire-Atlantique (44) or Pas-de-Calais (62). The map of EB estimates of relative risk (not shown) demonstrates a clear trend with lower risks in the south-west of France and higher risks in the centre and north-east.

number	département	EB estimate	Bayes estimate	95% Bayes credible interval
44	Loire-Atlantique	0.72	0.72	0.57 − 0.90
48	Lozère	0.85	0.88	0.68 − 1.13
54	Meurthe-et-Moselle	1.44	1.55	1.28 − 1.88
57	Moselle	1.46	1.63	1.34 − 1.96
62	Pas-de-Calais	1.24	1.25	1.03 − 1.50
73	Savoie	1.15	1.03	0.81 − 1.35
75	Paris	0.93	0.95	0.81 − 1.08
85	Vendée	0.65	0.69	0.53 − 0.88
90	Territoire-de-Belfort	1.03	1.21	0.92 − 1.59

Table 20.2 *Estimates of relative risks of mortality from gall-bladder and bile-duct cancer for males in France, 1971–1978 (selected départements shown); EB estimate: of the relative risk with the conjugate gamma prior; Bayes estimate: of the relative risk with the convolution Gaussian prior.*

20.5.2 Non-conjugate priors

When the prior distribution is not conjugate with the likelihood, the marginal posterior distribution $[r \mid y, \gamma]$ in (20.10) is non-standard and must be approximated to allow direct calculation of the posterior mean. For multivariate normal priors on log relative risks, Clayton and Kaldor (1987) used a multivariate normal approximation for the marginal posterior distribution of x. In addition, as the marginal likelihood (20.11) of γ is rarely tractable, its maximization requires the EM algorithm (Dempster et al., 1977), even in the simplest case of independent normal priors. Nevertheless, the posterior

mean of the log relative risk can still be expressed as a compromise between
the standard estimate and the prior mean as in (20.12), where the weight
is inversely proportional to the observed number of deaths and the prior
variance. This method has been used by Clayton and Kaldor (1987) and
Mollié and Richardson (1991) to smooth maps for rare diseases and small
areas.

20.5.3 Disadvantages of EB estimation

The use of $[r \mid y, \hat{\gamma}]$ instead of $[r \mid y]$ usually yields acceptable point estimates (Deely and Lindley, 1981). However, EB methods underestimate
the variability in r because no allowance is made for uncertainty in γ. Thus
'naive' EB confidence intervals for r based on the estimated variance of
the posterior $[r \mid y, \hat{\gamma}]$ are too narrow. To allow for uncertainty in γ, two
approaches have been proposed in the EB context: adjustments based on
the delta method (Morris, 1983) or bootstrapping the data y (Laird and
Louis, 1987). These methods have been reviewed by Louis (1991).

Another disadvantage of EB estimation is that it may not provide an
adequate description of the true dispersion in the rates. This problem has
been addressed by Devine and Louis (1994) and Devine *et al.* (1994) using
a constrained EB estimator.

20.6 Fully Bayesian estimation of relative risks

The fully Bayesian approach gives a third way to incorporate variability in
the hyperparameters γ. Fully Bayesian inference about r is based on the
marginal posterior distribution $[r \mid y]$ which, as we have already noted, is
often intractable. Working directly with $[r \mid y]$ will then require analytic approximations or numerical evaluation of integrals (Tsutakawa, 1985; Tierney and Kadane, 1986). However, MCMC methods permit samples to be
drawn from the joint posterior distribution $[r, \gamma \mid y]$ and hence from the
marginal posteriors $[r \mid y]$ and $[\gamma \mid y]$.

20.6.1 Choices for hyperpriors

Classical choices for the hyperprior distribution $[\gamma]$ generally assume independence between the hyperparameters, and may also assume non-informative priors for some hyperparameters.

Uniform hyperpriors

For the location parameters ϕ, the usual non-informative hyperprior is a
uniform distribution $U(-\infty, +\infty)$ which, although improper, does not lead
to an improper posterior distribution. However, a $U(-\infty, +\infty)$ prior for

the logarithm of scale parameter σ^2 (or κ^2 and λ^2) results in an improper posterior with an infinite spike at $\sigma^2 = 0$ (or at $\kappa^2 = 0$ and $\lambda^2 = 0$), forcing all the relative risks to be equal.

Conjugate gamma hyperprior

For independent normal or multivariate normal priors for the log relative risks, a more general class of hyperpriors for the inverse variance $\theta = \sigma^{-2}$ (or for κ^{-2} and λ^{-2}) is a conjugate gamma distributions $\mathrm{Ga}(a, b)$, where a and b are known. Setting $a = 0$ and $b = 0$ is equivalent to a $\mathrm{U}(-\infty, +\infty)$ prior on the logarithm of the variance, leading to an improper posterior as discussed above. Setting a and b strictly positive avoids the problems encountered with improper priors. In the absence of any prior information about a and b, a reasonable approach consists in choosing a vague gamma hyperprior $\mathrm{Ga}(a, b)$ for θ, with mean a/b based on the observed log-SMRs, as described below, but whose variance a/b^2 is very large so that the choice for the mean does not appreciably affect the posterior distribution.

With the independent normal prior (20.6) on the log relative risks and a gamma hyperprior $\mathrm{Ga}(a, b)$ on θ, the hyperprior mean a/b represents a prior guess θ^* for θ and the hyperprior variance $a/b^2 = \theta^*/b$ the uncertainty about this prior belief. Thus a reasonable value for θ^* is the inverse of the sample variance of the observed log-SMRs, $1/s_x^2$, and b can be set small.

With the intrinsic Gaussian autoregressive prior (20.7) on the log relative risks and a gamma hyperprior $\mathrm{Ga}(a, b)$ on θ, expression (20.9) suggests that a reasonable guess θ^* might be $1/(\bar{w}s_x^2)$, where $\bar{w}$ is the average value of w_{i+} (Bernardinelli and Montomoli, 1992).

With the convolution Gaussian prior (20.8) on the log relative risks, in the absence of prior information about the relative importance of each component u and v, it is reasonable to assume that these two components have the same strength. Then using (20.9), $2/s_x^2$ will be a reasonable prior guess ξ^* for $\xi = 1/\lambda^2$, and $2/(\bar{w}s_x^2)$ a reasonable prior guess ψ^* for $\psi = 1/\kappa^2$.

20.6.2 Full conditional distributions for Gibbs sampling

Gibbs sampling (see Gilks *et al.*, 1995: this volume) is particularly convenient for MRF models where the joint posterior distribution is complicated but full conditional distributions have simple forms and need be specified only up to a normalizing constant.

For the hierarchical Bayesian models discussed above, the joint posterior distribution of the log relative risks and hyperparameters is

$$[x, \gamma \mid y] \propto \prod_{i=1}^{n} [y_i \mid x_i] [x \mid \gamma] [\gamma]. \qquad (20.13)$$

The full conditionals required by the Gibbs sampler can be deduced from (20.13); see Gilks (1995: this volume). To readily identify those terms of (20.13) which contribute to a given full conditional distribution, it is helpful to refer to the graph of the model.

Full conditionals for x_i

Using the decomposition $[x \mid \gamma] = [x_i \mid x_{-i}, \gamma] [x_{-i} \mid \gamma]$ substituted into (20.13) and ignoring terms which do not depend on x_i, gives the full conditional distribution of x_i: $[x_i \mid x_{-i}, y, \gamma] \propto [y_i \mid x_i] [x_i \mid x_{-i}, \gamma]$. With the independent normal prior (20.6) on the log relative risks, this becomes, using (20.1):

$$[x_i \mid x_{-i}, y, \gamma] \propto \exp\left\{ y_i x_i - e_i \exp\{x_i\} - \frac{(x_i - \mu_i)^2}{2\sigma^2} \right\}, \qquad (20.14)$$

where $\mu_i = (Z\phi)_i$. Thus $[x_i \mid x_{-i}, y, \gamma]$ depends only on y_i and on the current values of ϕ and σ^2. This is clearly shown in Figure 20.2 where each node x_i is linked to nodes y_i, ϕ and σ^2.

With an intrinsic Gaussian autoregressive prior (20.7) on the log relative risks, and with the simplest choice of weighting ($w_{ij} = 1$ if i and j are neighbours, $w_{ij} = 0$ otherwise):

$$\begin{aligned} [x_i \mid x_{-i}, \gamma] &= [x_i \mid \{x_j, j \in \delta i\}, \sigma^2] \\ &\propto \frac{\sqrt{w_{i+}}}{\sigma} \exp\left\{ -\frac{w_{i+}(x_i - \bar{x}_i)^2}{2\sigma^2} \right\}, \end{aligned}$$

where $\bar{x}_i = \frac{1}{w_{i+}} \sum_{j=1}^{n} w_{ij} x_j$. Therefore, the full conditional of x_i is:

$$[x_i \mid x_{-i}, y, \gamma] \propto \exp\left\{ y_i x_i - e_i \exp\{x_i\} - \frac{w_{i+}(x_i - \bar{x}_i)^2}{2\sigma^2} \right\}. \qquad (20.15)$$

Thus the full conditional for x_i depends only on y_i, and on the current values of σ^2 and $\{x_j, j \in \delta i\}$. This can be seen clearly in Figure 20.3.

Finally, we present full conditional distributions for a convolution Gaussian prior (20.8) on the log relative risks. Substituting

$$[u, v \mid \gamma] = [u \mid \gamma] [v \mid \gamma]$$

into (20.13), and using the relationships between u, v, ϕ, λ^2 and κ^2 described in Section 20.4.2, gives:

$$[u, v, \gamma \mid y] \propto \prod_{i=1}^{n} [y_i \mid u_i, v_i] [u \mid \kappa^2] [v \mid \phi, \lambda^2] [\phi] [\lambda^2] [\kappa^2]. \qquad (20.16)$$

Using (20.1) this gives:

$$[u_i \mid u_{-i}, v, y, \gamma] \propto \exp \left\{ y_i u_i - e_i \exp \{u_i + v_i\} - \frac{w_{i+}(u_i - \bar{u}_i)^2}{2\kappa^2} \right\}$$

$$(20.17)$$

and:

$$[v_i \mid v_{-i}, u, y, \gamma] \propto \exp \left\{ y_i v_i - e_i \exp \{u_i + v_i\} - \frac{(v_i - \mu_i)^2}{2\lambda^2} \right\}, \quad (20.18)$$

where $\bar{u}_i = \frac{1}{w_{i+}} \sum_{j=1}^n w_{ij} u_j$. Thus the full conditional for u_i depends only on y_i and on the current values of $(u_j, j \in \delta i)$, κ^2, and v_i. In Figure 20.4, we see that all of these variables except v_i are directly connected to u_i. The dependence of the full conditional on v_i arises from the term $[y_i \mid u_i, v_i]$ in (20.16), which involves both u_i and v_i, as can be seen in Figure 20.4.

Distributions (20.14), (20.15), (20.17) and (20.18) are non-standard but log-concave and can be sampled by adaptive rejection sampling (Gilks and Wild, 1992; Gilks, 1995: this volume).

Full conditionals for γ

Assuming an independent normal prior (20.6) for the log relative risks x, using (20.13) and the marginal independence between ϕ and λ^2, the full conditional distribution for ϕ becomes:

$$[\phi \mid x, y, \sigma^2] \propto [x \mid \phi, \sigma^2] [\phi], \qquad (20.19)$$

and for σ^2 becomes:

$$[\sigma^2 \mid x, y, \phi] \propto [x \mid \phi, \sigma^2] [\sigma^2]. \qquad (20.20)$$

Neither depends on y because of the conditional independence of the hyperparameters and the data given x. This can be seen in Figure 20.2, where y and (ϕ, σ^2) are separated by x. With the uniform hyperprior for ϕ, the full conditional (20.19) for ϕ becomes a normal distribution with mean $(Z^T Z)^{-1} Z^T x$ and covariance matrix $\sigma^2 (Z^T Z)^{-1}$. With the gamma hyperprior $\text{Ga}(a, b)$ for $\theta = 1/\sigma^2$, the full conditional distribution (20.20) for θ becomes a $\text{Ga}(a + \frac{n}{2}, b + \frac{1}{2}(x - Z\phi)^T (x - Z\phi))$ distribution.

Assuming instead the intrinsic Gaussian autoregressive prior (20.7) for the log relative risks x, the full conditional for σ^2 becomes:

$$[\sigma^2 \mid x, y] \propto [x \mid \sigma^2] [\sigma^2]. \qquad (20.21)$$

With the gamma hyperprior $\text{Ga}(a, b)$ for $\theta = 1/\sigma^2$, the full conditional distribution (20.21) for θ becomes a $\text{Ga}(a + \frac{n}{2}, b + \frac{1}{2} \sum_{i=1}^n \sum_{j<i} w_{ij}(x_i - x_j)^2)$ distribution.

Finally, assuming the convolution Gaussian prior (20.8) for the log relative risks x, and using (20.13) and the relationships between u, v, ϕ, λ^2

and κ^2 described in Section 20.4.2, we obtain the following full conditional distributions:

$$\left[\phi \mid u, v, y, \lambda^2, \kappa^2\right] \propto \left[v \mid \phi, \lambda^2\right] [\phi], \tag{20.22}$$

$$\left[\lambda^2 \mid u, v, y, \phi, \kappa^2\right] \propto \left[v \mid \phi, \lambda^2\right] \left[\lambda^2\right], \tag{20.23}$$

$$\left[\kappa^2 \mid u, v, y, \phi, \lambda^2\right] \propto \left[u \mid \kappa^2\right] \left[\kappa^2\right]. \tag{20.24}$$

With the uniform hyperprior on ϕ, the full conditional distribution (20.22) for ϕ becomes a normal distribution with mean $(Z^T Z)^{-1} Z^T v$ and covariance matrix $\lambda^2 (Z^T Z)^{-1}$. Choosing a gamma hyperprior $\mathrm{Ga}(a_1, b_1)$ on $\xi = 1/\lambda^2$, the full conditional distribution (20.23) for ξ becomes a $\mathrm{Ga}(a_1 + \frac{n}{2}, b_1 + \frac{1}{2}(v - Z\phi)^T(v - Z\phi))$ distribution. Choosing a gamma hyperprior $\mathrm{Ga}(a_2, b_2)$ on $\psi = 1/\kappa^2$, the full conditional distribution (20.24) for ψ becomes a $\mathrm{Ga}(a_2 + \frac{n}{2}, b_2 + \frac{1}{2}\sum_{i=1}^{n}\sum_{j<i} w_{ij}(u_i - u_j)^2)$ distribution.

Standard algorithms can be used to simulate from these normal and gamma distributions.

20.6.3 Example: gall-bladder and bile-duct cancer mortality

We now return to the analysis of gall-bladder and bile-duct cancer mortality, commenced in Section 20.3. Here we use a fully Bayesian model with a convolution Gaussian prior (20.8) for log relative risks. As the empirical variance of the log-SMRs was 0.088, we set $\xi^* = 25$ as a prior guess for $\xi = 1/\lambda^2$. As $\bar{w} = 5$ for the 94 french départements, we set $\psi^* = 5$ as a prior guess for $\psi = 1/\kappa^2$. Our lack of confidence in these prior guesses were reflected in a $\mathrm{Ga}(0.0625, 0.0025)$ hyperprior for ξ and a $\mathrm{Ga}(0.0025, 0.0005)$ hyperprior for ψ. We assigned initial values: $\kappa^2 = \lambda^2 = 1$ and $x_i = 0$ for $i = 1, \ldots, n$. We performed a single run of the Gibbs sampler, with a burn-in of 100 iterations followed by 2 000 further cycles.

From the output of the Gibbs sampler, the heterogeneity component λ^2 was found to have a posterior mean of 0.0062 with a 95% Bayesian credible interval $(0.0010 - 0.0165)$, and the spatially structured component κ^2 had a posterior mean of 0.0435 with a 95% Bayesian credible interval $(0.0178 - 0.0872)$. This shows that spatial variation had a greater influence on the relative risks than unstructured heterogeneity.

The fully Bayesian estimates of relative risk in Table 20.2 show much less variation than the SMRs: they vary from 0.69 for département 85 to 1.63 for département 57, with a mean of 0.98 and standard deviation 0.17, in close agreement with the EB estimates discussed in Section 20.5. The fully Bayesian point estimates are not far from the EB point estimates, but the fully Bayesian approach additionally provides credible intervals. For départements with small population sizes, such as départements 48 and 73, the fully Bayesian estimate of the relative risks are 0.88 and 1.03, with 95% credible intervals: $(0.68 - 1.13)$ and $(0.81 - 1.35)$, neither of which excludes

1. For départements with large population sizes as départements 44 and 62, the fully Bayesian estimate of the relative risks are 0.72 and to 1.25, with 95% credible intervals: $(0.57 - 0.90)$ and $(1.03 - 1.50)$: both of which exclude 1.

The spatial structure of the relative risks is illustrated in Figure 20.5, which clearly shows increasing rates from the south-west of France to the north-east, unlike Figure 20.1.

20.7 Discussion

For a rare disease and for small areas, the Bayesian approaches overcomes the problem of overdispersion in the classical SMRs. Indeed, they smooth SMRs based on unreliable data but preserve those based on large populations, as shown in the example on gall-bladder and bile-duct cancer in Section 20.6.3. Thus Bayesian estimates of the relative risks are easier to interpret.

The EB method and the fully Bayesian method give similar point estimates of the relative risks, but the fully Bayesian approach via MCMC has the great advantage that it not only produces both point and interval estimates of the relative risks, but also permits computation of problem-specific statistics.

However, the Bayesian approaches raise the problem of choosing an appropriate prior for the relative risks: smoothing towards a global mean (conjugate gamma or independent normal) or towards a local mean (intrinsic Gaussian autoregression or convolution Gaussian). It seems that the convolution Gaussian prior gives a satisfactory intermediate prior distribution between independence and a purely local spatially structured dependence of the relative risks. Indeed, in the example on gall-bladder and bile-duct cancer, the trends from south-west to north-east in the estimates of relative risk are more distinct with the convolution Gaussian prior than with either the independent normal or the intrinsic Gaussian autoregressive prior (Mollié, 1990). Moreover, even if we do not yet have any epidemiological explanation for these trends, they have been rediscovered on a wider scale with European data (Smans et al., 1992).

It would be useful to develop other forms of prior, to take into account the presence of discontinuities in the spatial structure of the relative risks, which could be induced by natural borders such as mountains.

References

Bernardinelli, L. and Montomoli, C. (1992) Empirical Bayes versus fully Bayesian analysis of geographical variation in disease risk. *Statist. Med.*, 11, 983–1007.

Besag, J. (1974) Spatial interaction and the statistical analysis of lattice systems. *J. R. Statist. Soc.* B, **36**, 192–236.

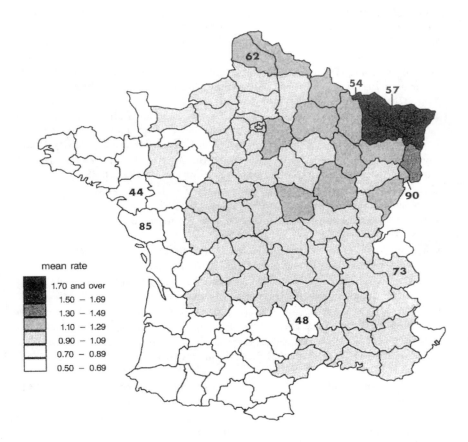

Figure 20.5 *Bayesian estimates of relative risks of mortality from gall-bladder and bile-duct cancer, for males in France, 1971–1978 (based on 2 000 iterations of the Gibbs sampler, using a convolution Gaussian prior for the log relative risks). Départements listed in Table 20.1 are labelled.*

Besag, J. (1986) On the statistical analysis of dirty pictures. *J. R. Statist. Soc.* B, **48**, 259–302.

Besag, J. (1989) Towards Bayesian image analysis. *J. Appl. Statist.*, **16**, 395–407.

Besag, J. and Mollié, A. (1989) Bayesian mapping of mortality rates. *Bull. Int. Statist. Inst.*, **53**, 127–128.

Besag, J., York, J. and Mollié, A. (1991) Bayesian image restoration, with two applications in spatial statistics. *Ann. Inst. Statist. Math.*, **43**, 1–21.

Breslow, N. and Clayton, D. (1993) Approximate inference in generalized linear mixed models. *J. Am. Statist. Ass.*, **88**, 9–25.

Clayton, D. (1989) Simulation in hierarchical models. Technical report, Department of Community Health, Leicester University.

Clayton, D. (1991) A Monte Carlo method for Bayesian inference in frailty models. *Biometrics*, **47**, 467–485.

Clayton, D. G. (1995) Generalized linear mixed models. In *Markov Chain Monte Carlo in Practice* (eds W. R. Gilks, S. Richardson and D. J. Spiegelhalter), pp. 275–301. London: Chapman & Hall.

Clayton, D. and Bernardinelli, L. (1992) Bayesian methods for mapping disease risk. In *Small Area Studies in Geographical and Environmental Epidemiology* (eds J. Cuzick and P. Elliott), pp. 205–220. Oxford: Oxford University Press.

Clayton, D. and Kaldor, J. (1987) Empirical Bayes estimates of age-standardised relative risks for use in disease mapping. *Biometrics*, **43**, 671–681.

Clayton, D., Bernardinelli, L. and Montomoli, C. (1993) Spatial correlation in ecological analysis. *Int. J. Epidem.*, **22**, 1193–1202.

Deely, J. J. and Lindley, D. V. (1981) Bayes empirical Bayes. *J. Am. Statist. Ass.*, **76**, 833–841.

Dempster, A. P., Laird N. M. and Rubin D. B. (1977) Maximum likelihood from incomplete data via the EM algorithm. *J. R. Statist. Soc.* B, **39**, 1–38.

Devine, O. J. and Louis, T. A. (1994) A constrained empirical Bayes estimator for incidence rates in areas with small populations. *Statist. Med.*, **13**, 1119-1133.

Devine, O. J., Louis, T. A. and Halloran, M. E. (1994) Empirical Bayes methods for stabilizing incidence rates before mapping. *Epidemiology*, **5**, 622–630.

Gilks, W. R. (1995) Full conditional distributions. In *Markov Chain Monte Carlo in Practice* (eds W. R. Gilks, S. Richardson and D. J. Spiegelhalter), pp. 75–88. London: Chapman & Hall.

Gilks, W. R. and Wild, P. (1992) Adaptive rejection sampling. *Appl. Statist.*, **41**, 337–348.

Gilks, W. R., Richardson, S. and Spiegelhalter, D. J. (1995) Introducing Markov chain Monte Carlo. In *Markov Chain Monte Carlo in Practice* (eds W. R. Gilks, S. Richardson and D. J. Spiegelhalter), pp. 1–19. London: Chapman & Hall.

Green, P. J. (1995) MCMC in image analysis. In *Markov Chain Monte Carlo in Practice* (eds W. R. Gilks, S. Richardson and D. J. Spiegelhalter), pp. 381–399. London: Chapman & Hall.

Laird, N. M. and Louis, T. A. (1987) Empirical Bayes confidence intervals based on bootstrap samples. *J. Am. Statist. Ass.*, **82**, 739–750.

Louis, T. A. (1991) Using empirical Bayes methods in biopharmaceutical research. *Statist. Med.*, **10**, 811–829.

Manton, K. G., Woodbury, M. A. and Stallard, E. (1981) A variance components approach to categorical data models with heterogeneous cell populations: analysis of spatial gradients in lung cancer mortality rates in North Carolina counties. *Biometrics*, **37**, 259–269.

Manton, K. G., Stallard, E., Woodbury, M. A., Riggan, W. B., Creason, J. P. and Mason, T. J. (1987) Statistically adjusted estimates of geographic mortality profiles. *J. Nat. Cancer Inst.*, **78**, 805–815.

Manton, K. G., Woodbury, M. A., Stallard, E., Riggan, W. B., Creason, J. P. and Pellom, A. C. (1989) Empirical Bayes procedures for stabilizing maps of US cancer mortality rates. *J. Am. Statist. Ass.*, **84**, 637–650.

Mollié, A. (1990) Représentation géographique des taux de mortalité: modélisation spatiale et méthodes Bayésiennes. *Thèse de Doctorat*, Université Paris VI.

Mollié, A. and Richardson, S. (1991) Empirical Bayes estimates of cancer mortality rates using spatial models. *Statist. Med.*, **10**, 95–112.

Morris, C. N. (1983) Parametric empirical Bayes inference: theory and applications (with discussion). *J. Am. Statist. Ass.*, **78**, 47–65.

Smans, M., Muir, C. S. and Boyle, P. (eds) (1992) *Atlas of Cancer Mortality in the European Economic Community*, IARC Scientific Publications **107**, pp. 70–71. Lyon: International Agency for Research on Cancer.

Spiegelhalter, D. J., Best, N. G., Gilks, W. R. and Inskip, H. (1995) Hepatitis B: a case study in MCMC methods. In *Markov Chain Monte Carlo in Practice* (eds W. R. Gilks, S. Richardson and D. J. Spiegelhalter), pp. 21–43. London: Chapman & Hall.

Tierney, L. and Kadane, J. B. (1986) Accurate approximations for posterior moments and marginal densities. *J. Am. Statist. Ass.*, **81**, 82–86.

Tsutakawa, R. K. (1985) Estimation of cancer mortality rates: a Bayesian analysis of small frequencies. *Biometrics*, **41**, 69–79.

Tsutakawa, R. K. (1988) Mixed model for analyzing geographic variability in mortality rates. *J. Am. Statist. Ass.*, **83**, 37–42.

Tsutakawa, R. K., Shoop, G. L. and Marienfeld, C. J. (1985) Empirical Bayes estimation of cancer mortality rates. *Statist. Med.*, **4**, 201–212.

21

MCMC in image analysis

Peter J Green

21.1 Introduction

Digital images now routinely convey information in most branches of science and technology. The Bayesian approach to the analysis of such information was pioneered by Grenander (1983), Geman and Geman (1984), and Besag (1986), and both fundamental research and practical implementations have been pursued energetically ever since. In broad terms, the Bayesian approach treats the recorded raw image as numerical data, generated by a statistical model, involving both a stochastic component (to accommodate the effects of noise due to the environment and imperfect sensing) and a systematic component (to describe the true scene under view). Using Bayes theorem, the corresponding likelihood is combined with a prior distribution on the true scene description to allow inference about the scene on the basis of the recorded image.

This informal and simplistic summary in fact embraces extraordinary variety, both in the style of modelling and in the type of application – from say, removal of noise and blur in electron microscopy to the recognition of objects in robotic vision. There is little common ground in this wide spectrum about which to generalize, but one feature that does stand out in any general view of image analysis literature is the prominent role played by Markov chain Monte Carlo (MCMC) as a means of calculation with image models.

The purpose of this article is to discuss general reasons for this prominence of MCMC, to give an overview of a variety of image models and the use made of MCMC methods in dealing with them, to describe two applications in more detail, and to review some of the methodological innovations in MCMC stimulated by the needs of image analysis, that may prove important in other types of application.

21.2 The relevance of MCMC to image analysis

The most obvious feature of digital images, considered as numerical data, is their very large size. Images very seldom consist of as few as 32×32 pixels (picture elements): sizes of between 256×256 and $1\,024 \times 1\,024$ are much more common. And if a time dimension is added, through consideration of 'movies' or image sequences, the volume of data available becomes very large indeed. This size factor immediately favours the use of MCMC methods in a statistical approach to image analysis since although other more conventional numerical methods sometimes provide practical routes to the calculation of point estimates of a true image, e.g. the maximum *a posteriori* (MAP) estimate, MCMC is usually the only approach for assessing the variability of such estimates.

A second key feature of image data is its spatial structure. The interplay between stochastic and spatial variation is one of the most interesting aspects of image modelling from a mathematical perspective, and to construct such models that are tractable algorithmically can be a considerable challenge. The usual solution, as in the applications in other articles in this volume, is to adopt models with a parsimonious conditional independence structure. This is the key to efficient application of MCMC methods, which are all driven by conditional probabilities.

In the case of image models, neighbours in the conditional independence graph are usually spatially contiguous pixels or picture attributes. The random field models that are used are similar, or even identical, to those studied in statistical physics: this field was the original home ground for MCMC methods (Metropolis *et al.*, 1953; Binder, 1988) and physical analogies are drawn strongly in some work in statistical image analysis (e.g. Geman and Geman, 1984). From a historical perspective, developments in the specification of image models, and in the construction of MCMC algorithms to deal with them, have proceeded in parallel.

For extra parsimony, image models are usually specified to be spatially stationary. This makes implementation of MCMC methods particularly simple: effectively the same calculations are performed repeatedly at each location in the image, so the computer code required can be very compact. This in turn explains the apparent lack of interest among researchers in statistical image analysis in using general purpose MCMC packages. It is also evident that this is fertile ground for the implementation of MCMC on parallel architecture: impressive performance is obtained in the applications described by Grenander and Miller (1994).

We have thus identified a number of reasons why MCMC is such a natural approach to computing in image analysis, and it is not surprising that these methods were first encountered in statistics in this setting, before being adapted to other problems. In some cases, in fact, not only has the style of computation been borrowed, but image models themselves have been used with little modification; this is true of applications to geographical epidemiology (Besag *et al.*, 1991; Clayton and Bernardinelli, 1992; Gilks *et al.*, 1993; Mollié, 1995: this volume), spatial archaeology (Besag *et al.*, 1991; Litton and Buck, 1993), biogeography (Högmander and Møller, 1995; Heikkinen and Högmander, 1994) and agricultural field trials (Besag and Higdon, 1993). Some of these connections are further explored in Besag *et al.* (1995).

21.3 Image models at different levels

In computer vision, a distinction is drawn between low-level analysis – for example, noise and blur removal – and high-level analysis – for example, object recognition. In between, there is a spectrum embracing such problems as tomographic reconstruction, segmentation and classification, texture modelling, etc. This variety is reflected in the range of different levels at which statistical modelling of images has been attempted. In this section, we will discuss the implications for the use of MCMC methods in three different styles of modelling. We focus on the usual problem of interest – calculation of probabilities and expectations under posterior distributions. But it is equally true that MCMC has an important role in computing with prior distributions for images.

21.3.1 Pixel-level models

Low-level image processing can be conducted with the aid of models for images that operate at the pixel level. If y denotes the recorded image, and x the description of the true scene, then x and y are large arrays, usually representing discretizations of the images onto a two-dimensional rectangular lattice, and usually of the same size. The original work of Geman and Geman (1984) and Besag (1986) had this character, and concentrated on the spatial classification problem, in which the value x_i of x at pixel i is drawn from a finite set of labels representing, say, different crop types in a remotely sensed agricultural scene. The corresponding recorded value y_i is the quantity of radiation emitted by the scene, that is recorded in the sensor as originating from pixel i on the ground. A typical form for the degradation model $p(y|x)$ to take would allow blur and independent noise, so that

$$p(y|x) = \prod_{k \in S} p(y_k | x_{\nu k}), \tag{21.1}$$

where S denotes the set of all pixels, νk the set of ground pixels influencing sensor pixel k, and for any set of pixels A, $x_A = \{x_k : k \in A\}$. In the product on the right-hand side here would be specified functions, quantifying the blur and noise. Throughout this chapter, the symbol $p(\ )$ is used generically to denote probability functions or densities. Prior information in this type of application would usually be generic, reflecting belief that adjacent pixels are likely to have the same label, for example using the Potts model

$$p(x) \propto \exp(-\beta \sum_{i \sim j} I[x_i \neq x_j]). \tag{21.2}$$

Here $i \sim j$ means that i and j are neighbours in some prescribed undirected graph with the pixels as vertices, the sum is over all such pairs of neighbours, and $I[\ ]$ is the indicator function, taking the value 1 if x_i and x_j are different, and 0 otherwise; thus the sum gives the number of neighbour pairs that have differing labels. This is a simple example of a Markov random field or Gibbs distribution. (Throughout this chapter, the symbol $\propto$ is used to denote proportionality in the first argument of the function on its left-hand side.) Combining these two model terms yields the posterior distribution

$$p(x|y) \propto p(x)p(y|x).$$

The posterior full conditional distribution (or local characteristic) of x_i is

$$p(x_i|x_{-i}, y) \propto p(x|y),$$

where $x_{-i} = \{x_j : j \neq i\}$. So from (21.1) and (21.2):

$$p(x_i|x_{-i}, y) \propto \exp(-\beta \sum_{j \in \partial i} I[x_j \neq x_i]) \prod_{k:\nu k \ni i} p(y_k|x_{\nu k}). \tag{21.3}$$

The notation ∂i means the set of neighbours of i, that is $\{j : j \sim i\}$; thus the sum above gives the number of neighbours of i labelled differently from i. The product in (21.3) is over all sensor pixels which are influenced by ground pixel i. The posterior full conditionals (21.3) are seen to take a simple form: the product of a small number of easily computed factors. The normalization to a proper distribution is easy since this is a discrete distribution over a small set of labels, and the way is clear for MCMC simulation using the Gibbs sampler (see Gilks *et al.*, 1995: this volume). This was proposed by Geman and Geman (1984), and indeed they coined this name for the sampler already known to the statistical physics community as the *heat bath* (e.g. Creutz, 1979).

Geman and Geman (1984) discuss other samplers, and simulated annealing as well as simulating at equilibrium; they also deal with a number of ramifications to the basic formulation above, including the use of edge variables that are not directly observed.

It is straightforward to amend the model to deal with the case where the true value x_i is a real number ('grey level') representing, say, the emitted radiance rather than a label. For example, we might use a pairwise difference prior

$$p(x) \propto \exp(-\beta \sum_{i \sim j} \phi(x_i - x_j)) \tag{21.4}$$

for some symmetric U-shaped function ϕ (see Besag 1986; Geman and McClure, 1987; Besag et al., 1995). The full conditional retains its simple structure, but except in very special circumstances is no longer amenable to the Gibbs sampler, as it is a density function of rather arbitrary form with an unknown normalizing constant. The most convenient MCMC sampling method then is the Metropolis–Hastings method (see Gilks et al., 1995: this volume), which is in our experience capable of quite adequate computational efficiency.

Concrete applications of such models have included gamma camera imaging (Aykroyd and Green, 1991), magnetic resonance imaging (Godtliebsen, 1989; Glad and Sebastiani, 1995), film restoration (Geman et al., 1992) and remote sensing (Wilson and Green, 1993).

Variants of these simple models have been used to deal with other low-level vision problems. Geman (1991) provides a good general account of the theory. Texture modelling is important in segmentation (the subdivision of an image into homogeneous regions) and classification (the assignment of pixels of an image to pre-assigned classes). It can be accomplished using Markov random fields of higher order than that in (21.2); modelling and associated MCMC algorithms are discussed by Cohen and Cooper (1987). Boundary detection is discussed by Geman et al. (1990), using low-level random field models with varying spatial resolutions. A general framework for the Bayesian restoration of image sequences via state-space modelling was proposed by Green and Titterington (1988), and this is amenable to approximate MCMC calculations, at least for recursive estimation.

21.3.2 Pixel-based modelling in SPECT

A particularly interesting variant applies to reconstruction from emission tomography data (PET and SPECT), based on the Poisson regression model introduced by Shepp and Vardi (1982). The novelty in tomography is the geometrical transformation between the 'body-space' in which the true scene x is located and the 'projection-space' where the recorded data y are obtained. Model (21.3) is still applicable, only the set of pixels νk is no longer a compact region surrounding k, but rather the set of all pixels in body space from which an emitted photon that is accumulated in the detected datum y_k can originate. This set can be large, which means there are many factors in (21.3) and hence relatively slow operation of MCMC methods.

In single photon emission computed tomography (SPECT), the generally accepted model states that the pixel values in the recorded image $y = \{y_t : t \in T\}$, which represent counts of detected gamma ray photons, satisfy

$$y_t \sim \text{Poisson}(\sum_{s \in S} a_{ts} x_s), \text{independently} \qquad (21.5)$$

where $x = \{x_s : s \in S\}$ denotes the true concentration of a radioisotope as a function of location in the patient, the object of inference, and $\{a_{ts}\}$ are coefficients determined by the geometry and other physical properties of the instrumentation.

Bayesian reconstruction has been discussed by Geman and McClure (1987) and Geman *et al.* (1993), using MCMC and other methods including *iterated conditional modes* (ICM) (Besag, 1986), by Green (1990) and Weir and Green (1994), using a modified EM method, and by Amit and Manbeck (1993), using deformable templates and wavelets, among others.

Here we briefly describe how to set up a Metropolis–Hastings sampler for the problem: more detail for a somewhat similar problem is given in Besag *et al.* (1995).

The prior model we shall use is related to that of (21.4), but on a log scale (since $x_s \geq 0$) and with a scale factor γ:

$$p(x) \propto \exp(-\sum_{s \sim r} \phi(\gamma(\log x_s - \log x_r)))/ \prod x_s$$

where $\phi(u) = \delta(1 + \delta) \log \cosh(u/\delta)$. The product in the denominator is a Jacobian arising from the logarithmic transformation. This expression gives absolute value and Gaussian priors as $\delta \to 0$ and ∞ respectively (resulting in $\phi(u) = |u|$ or $\frac{1}{2}u^2$). Combined with (21.5) and factoring out terms not involving x_s gives the posterior local characteristics

$$p(x_s|x_{-s}, y) \quad \propto \quad \exp\left(-\sum_{r \in \partial s} \phi(\gamma \log(x_s/x_r)) \right. \tag{21.6}$$
$$\left. + \sum_{t:a_{ts}>0} (y_t \log \sum_r a_{tr}x_r - \sum_r a_{tr}x_r) - \log x_s\right).$$

On the scale of problem presented by Green (1990), there would be around 300 terms in each of the sums over t and r above. There is, therefore, a heavy burden in computing even one value of the probability density above, although this can be partially offset by storing also $\{\mu_t = \sum_s a_{ts}x_s\}$ and updating these as x changes. The density is also of awkward, non-standard form (and is concave in neither x_s nor $\log x_s$) so the Gibbs sampler is not at all practical here.

Instead, we use a Metropolis–Hastings sampler, using a proposal in which the new value of $\log x_s$ is uniformly distributed over an interval centred at the old value. Thus the proposal density, proposing point x_s' when the current value is x_s, is

$$q_s(x_s \to x_s') \propto (x_s')^{-1}$$

on an interval $|\log(x_s'/x_s)| < c$, for some positive constant c. The corresponding acceptance function $\alpha(x, x')$ is calculated from (21.6):

$$
\begin{aligned}
\alpha(x, x') \;=\; & \min\Bigg[1, \exp\Bigg\{-\sum_{r \in \partial s} \left(\phi(\gamma \log(x_s'/x_r)) - \phi(\gamma \log(x_s/x_r))\right) \\
& + \sum_{t: a_{ts} > 0} \left(y_t \log\left\{1 + a_{ts}(x_s' - x_s)/\sum_r a_{tr} x_r\right\} \right. \\
& \left. -a_{ts}(x_s' - x_s)\right)\Bigg\}\Bigg] \;.
\end{aligned}
$$

Note that the Jacobian terms have cancelled. The sampler is applied to the pixels one-by-one in a systematic raster scan. (In some of the recent Bayesian literature, such a sampler has been called 'Metropolis-within-Gibbs'; however, this term is incorrect: the method was originally applied component-wise (Metropolis *et al.*, 1953).)

In a fully Bayesian treatment of the SPECT problem, the hyperparameters γ and δ in the prior should themselves be considered drawn from appropriate hyperpriors. This produces additional complications, as the normalization constant in the joint distribution of x and y is an intractable function of γ and δ, and has to be estimated, off-line, by simulations from the prior; this approach is followed successfully in a recent paper by Weir (1995). Alternatively, it may be appropriate to regard γ and δ as fixed, perhaps in practice chosen according to the organ being studied. For present purposes, we take γ and δ as fixed, with values chosen by trial and error.

The method just described was applied to SPECT data relating to a pelvis scan conducted at the Bristol Royal Infirmary. Data relevant to a particular horizontal slice are displayed in sinogram form in the upper left panel of Figure 21.1, and a 'filtered back-projection' reconstruction of the slice is shown in the upper right panel. This is a standard and the most commonly used method of reconstruction in current practice, and is based on standard principles of extracting a signal in the presence of noise using linear filtering (see, for example, Budinger *et al.*, 1979). Note the noisy character of this reconstruction, and the presence of radial artifacts. For alternative Bayesian reconstructions, the coefficients $\{a_{ts}\}$ were calculated using the model described in Green (1990), and the hyperparameters were set at $\gamma = 1.1$ and $\delta = 0.18$. Short pilot runs suggested that a value for the proposal spread parameter c of around 0.35 was most efficient from the point of view of minimizing the integrated autocorrelation time for estimating $E(x_s|y)$; see Sokal (1989), Green and Han (1992) and Roberts (1995: this volume). The reconstruction resulting from averaging over 2 000 sweeps of the Metropolis–Hastings sampler is shown in the lower left panel of Figure 21.1; for comparison, the lower right panel shows an approximate MAP reconstruction, under the same prior, obtained using the one-step-late algorithm of Green (1990). The posterior mean and mode differ slightly, but the two Bayesian methods have provided reconstructions of comparable quality.

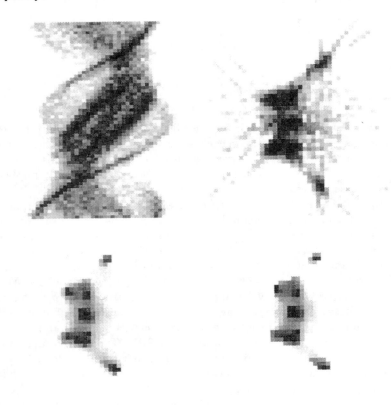

Figure 21.1 *SPECT sinogram from a pelvis scan, and three reconstructions. upper left: sinogram; upper right: filtered back-projection; lower left: posterior mean; lower right: posterior mode.*

The reconstruction obtained using MCMC was some 20 times more expensive to compute; however, the advantage of the MCMC approach is seen when we wish to assess variability. Figure 21.2 displays information relevant to a one-dimensional transect through the body-space that corresponds to a horizontal left-right line in the patient and to a vertical line in the reconstructions shown in Figure 21.1, just to the left of the centre of the diagram. All of the curves are based on the final 1 500 sweeps of the 2 000-sweep Monte Carlo run giving Figure 21.1, and are plotted on a square-root vertical scale. The solid curve is the pixelwise median from the realization, estimating the posterior median isotope concentration along the transect, and the dotted curves are the 10% and 90% quantiles. The broken lines display order-statistics-based *simultaneous* symmetric credible bounds, covering 80% of the posterior distribution for the whole transect. These were computed in the manner described in Besag *et al.* (1995).

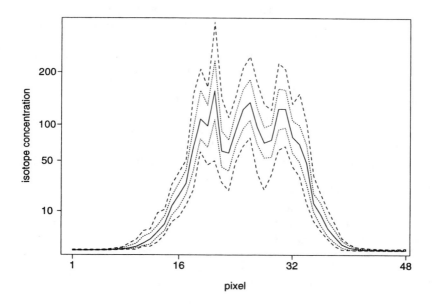

Figure 21.2 *A one-dimensional transect through a Bayesian reconstruction from a pelvis scan. Solid line: pixelwise posterior median; dotted lines: pixelwise 80% credibility bounds; broken lines: simultaneous 80% credibility bounds.*

21.3.3 Template models

The major limitation of pixel-level image models is that inference based on them can only refer to statements about the pixels of the true scene. With calculations performed by MCMC, this is not quite the restriction it might seem, since posterior probabilities of arbitrary events defined by the pixel values can be estimated. (And if a complete realization of the Markov chain is stored, such events need not even be identified until the simulation is complete!)

However, it is hardly realistic in practical terms to contemplate specifying which pixel values define, say, a platoon of tanks, and which a herd of cows. To perform inference at this higher level of objects, the role of the scene x

in the previous section must be replaced or augmented by a description of the objects that the scene contains.

Having done so, at the conceptual level, at least, nothing changes. Bayes theorem and the principle of estimating posterior expectations and probabilities by MCMC still apply. However, the details of modelling and of sampling may be very different.

One very promising route to image model building for these purposes is via deformable templates, introduced by Grenander; the original study was eventually published as Grenander *et al.* (1991). This dealt with simple images consisting of polygonal silhouettes: the black/white true silhouette is not observed directly, but only after spatial discretization and the addition of noise. Prior modelling then focuses on the polygonal boundary of the silhouette, or rather, in the deformable template paradigm, on stochastic deformations of this polygon from an ideal outline determined by the context. In early experiments, these were outlines of a human hand and of leaves.

There remains a central role for Markov random field models in this situation, incidentally. The deformation would usually be specified by its effects on the lengths and orientations of individual boundary segments of the polygon, and these effects would be modelled dependently using a simple local dependence assumption.

Deformable templates have been extended to a number of more challenging situations, including grey-level scenes, in which the recorded datum at each pixel is continuously distributed (Amit *et al.* 1991), with progress towards practical applications (Grenander and Manbeck, 1992; Ripley and Sutherland, 1990). When the number of elements defining the template or its deformation is small, the resulting image models are close to parametric rather than nonparametric in nature, and inference is less sensitive to the specification of the prior. Phillips and Smith (1994) discuss an interesting application to segmentation of facial images, and Aykroyd and Green (1991) compare performance of image analysis procedures using pixel and template based prior models. In all of these papers, MCMC methods are used.

Template models have been extended to deal with scenes containing variable numbers of objects. The appropriate parameter spaces are now unions of subspaces of varying dimension; novel MCMC methods have been developed for this situation, notably using jump-diffusion simulation (Grenander and Miller, 1994; Srivastava *et al.*, 1991). Green (1994a,b) describes an explicit class of MCMC methods using reversible Metropolis–Hastings jumps between subspaces, and applies such methods to Bayesian model determination tasks including factorial experiments and multiple change-point problems. Applications in image analysis are likely to follow.

21.3.4 An example of template modelling

We illustrate the main ideas of image analysis based on a deformable template model by describing a toy example, simplified from the silhouette problem considered by Grenander *et al.* (1991).

The template is a closed polygon, whose boundary does not intersect itself. The prior distribution forming the template model is obtained from separate local perturbations to the lengths of each boundary segment and the angles between them, the whole template then being subjected to an uniformly distributed random rotation and a translation drawn from a diffuse prior. For simplicity, we assume here that the joint density for the deformation is proportional to the product of terms for the local perturbations which have the form of normal densities. Of course, the polygon must remain closed and non-intersecting, which places a very complicated constraint on the individual perturbations. Therefore, in spite of the product form, the local perturbations are not *a priori* independent and the normalizing constraint for the joint density is completely intractable. One of the beauties of MCMC calculation is that no attention need be paid to either of these points in constructing a Metropolis–Hastings sampler, apart from checking that proposed moves satisfy the constraints.

Two simple alternative models for the likelihood $p(y|x)$ are used. In each case, the data are pixel values independently generated, depending on the polygon only according to whether the pixel centre is inside or outside the polygon. The two models used are a symmetric binary channel, with 40% error rate, and additive Gaussian noise, with standard deviation 1.5 times the difference in means.

A Metropolis–Hastings sampler was used to simulate from the resulting posterior. Each proposal consisted of a random displacement to a single vertex of the template, drawn uniformly from a small square centred at the current position. Such a move affects two boundary segment lengths and three inter-segment angles; it also changes the set of data pixel centres that are interior to the template, thus affecting a variable number of likelihood terms. The details are straightforward but notationally lengthy, so will not be further presented here.

The results from two experiments are displayed in Figures 21.3 and 21.4, showing cases with binary channel and Gaussian noise respectively. In each case, five polygons are superimposed on an image of the data to give a rough impression of the variation in the posterior; these are the states of the simulation after 200, 400, 600, 800 and 1 000 complete sweeps, each visiting every vertex once.

Figure 21.3 *Template-based restoration of a polygonal silhouette: binary channel noise with 40% error rate.*

21.3.5 Stochastic geometry models

A third approach to the issue of image modelling is one that again aims at high level image attributes, but uses established models from stochastic geometry, rather than purpose-built processes. The cleanest general approach is one in which objects are considered as realized points in a marked point process in an appropriate abstract object space. Models are specified by their densities with respect to the corresponding Poisson process. This general framework, some interesting classes of interacting object processes, and associated MCMC methods are described by Baddeley and van Lieshout (1994). Baddeley (1994) also notes that the object processes of Grenander and Miller (1994) can usefully be set in this framework.

A more specifically focused approach builds on the polygonal random field models of Arak and Surgailis (1989), and has potential for dealing with either pixellated images or point process data generated from an assumed underlying true image consisting of polygonal shapes. This is explored by Clifford and Middleton (1989), and extended by Clifford and Nicholls (1995).

All these approaches have yet to find concrete application, but they

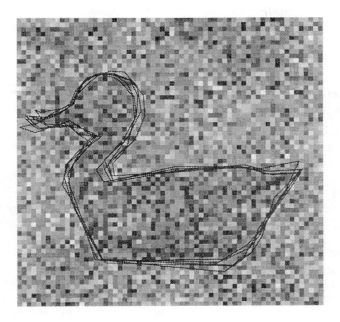

Figure 21.4 *Template-based restoration of a polygonal silhouette: Gaussian noise, with standard deviation 1.5 times the difference in means.*

certainly seem to provide a rigorous and flexible basis for high-level imaging. Again, credible alternatives to MCMC methods for computing with these models have yet to be proposed.

21.3.6 Hierarchical modelling

All of the models described so far are simple in general structure: there are just two levels, x and y, modelling of their joint description is via specified $p(x)$ and $p(y|x)$, and y is observed. Within each level, dependencies are modelled non-directionally.

In the language of graphical modelling, this is a simple chain graph model (Whittaker, 1990). Richer chain graphs have been used; for example, Phillips and Smith (1994) use hierarchical models for features in images of human faces.

Wilson and Green (1993) use a four-level model for satellite remote sensing. The four levels are the ground classification, the emitted radiance, the detected radiance and the recorded data; only the latter is observed. The merit of introducing the additional layers between the object of inference and the data is that interactions can be kept more local, and each stage of modelling reflects a separate physical process.

In this case, all the levels are discretized on the same rectangular grid, but applications for other multi-level models can be envisaged, perhaps combining template and pixel based models at different levels.

21.4 Methodological innovations in MCMC stimulated by imaging

In very recent years, the search for efficient MCMC methods in image problems has continued to stimulate new methodology, that may find use in general Bayesian inference.

A typical difficulty with single-variable updating samplers is that they mix slowly in the presence of multimodality, which can occur when the data are relatively uninformative compared to the (spatially structured) prior. This is related to the phenomenon of phase transition in statistical physics, and similar solutions have been explored to combat it.

In auxiliary variable algorithms, the variable x is augmented by additional variables, u, with an arbitrary specified conditional distribution $p(u|x)$. The joint distribution $p(x, u) = p(x)p(u|x)$ is then the target of a MCMC simulation; note that the marginal distribution of the original variables remains $p(x)$. With careful design, the resulting simulation can provide more efficient estimation than does the original process.

The first auxiliary variable method was the revolutionary algorithm designed to reduce critical slowing down in Potts models (21.2), proposed by Swendsen and Wang (1987). If β is large, MCMC using single site updating algorithms mixes very slowly. The Swendsen–Wang algorithm introduces conditionally independent, binary auxiliary bond variables u_{ij} for each neighbour pair, with $p(u_{ij} = 1|x) = 1 - \exp(-\beta I[x_i = x_j])$; here $u_{ij} = 1$ indicates presence of a bond. Consideration of the form of the conditional density $p(x|u)$ shows that this is a uniform distribution over pixel labels, subject to the constraints that clusters of pixels formed by pairs bonded together have the same label. This is equivalent to x being partitioned into uniformly labelled clusters with labels drawn independently.

The Swendsen–Wang algorithm is easily extended to some more general models and to posterior distributions, but works poorly when symmetry is lost, as in some applications to Bayesian image analysis (Higdon, 1993; Gray, 1993).

The concept of auxiliary variables extends beyond the use of binary bonds (Edwards and Sokal, 1988; Besag and Green, 1993). In general, awkward interactions are reduced in strength, or eliminated altogether, in exchange for constraints on the variables.

There are modifications of the Swendsen–Wang algorithm that control the size of clusters by modifying the distribution of the bond variables, and these can be expressed in this framework. There include explicitly limiting cluster size stochastically (Kandel *et al.*, 1989), preventing clusters growing outside deterministic blocks of pixels (Han, 1993), or having bond probabilities depend on the data (Higdon, 1993). All these approaches can be used with multiple levels of auxiliary variables to capture a multigrid effect (Kandel, *et al.*, 1989; Besag and Green, 1993), with the aim that one or more of the bond levels will generate clusters with a good chance of changing label.

Auxiliary variable methods have proved extremely effective in combating the problems of critical slowing down in calculations with statistical physics models. Their potential in statistical applications with high levels of interaction seems great, but few examples have yet materialized. To work well, the auxiliary variables need to be constructed so that either $p(x|u)$ can be simulated directly without the need for rejection methods, or there is a rapidly mixing Metropolis–Hastings algorithm for this conditional distribution.

21.5 Discussion

Both the general descriptions and the detailed examples above affirm the very valuable role of MCMC methods in computing with the models of Bayesian image analysis, in both low-level and high-level problems. As in other domains of application of modern Bayesian statistics, the availability of these methods encourages the 'model-liberation movement'; in addition the sheer size of the data structures involved means that there are really no available alternative general computational methods if variability is to be assessed.

Among MCMC methods, the Gibbs sampler has very limited use in image analysis. Only in classification problems where marginal distributions take few values, or in purely Gaussian models where simulation-based calculations are unnecessary, can the Gibbs sampler be competitive in computational efficiency. Metropolis–Hastings methods, updating pixel values or picture attributes singly or in small groups, are routine to code and efficient; the spatial stationarity usually assumed in image models cuts out any need for experimentation with proposal distributions tailored to individual variables.

Sensitivity analysis is an important aspect of any responsible statistical inference. In Bayesian image analysis, in contrast to some other applications covered by this volume, prior distributions remain highly influential even with a great volume of data (since great volume usually means a large field of view rather than replication or small variance). Thus sensitivity to the prior may be an important issue. MCMC is well attuned to sensitivity analysis, a point given particular emphasis in Besag *et al.* (1995).

On the negative side, MCMC is only available as a computational technique when a full probabilistic model is specified. This is sometimes awkward in image problems, as the field of view (especially in biological microscopy, for example) may contain very heterogeneous image structures.

Finally, there are concerns about convergence of MCMC methods. The very large dimensionality of parameter spaces in image problems makes monitoring convergence unusually difficult, while theoretical results, though limited at present, suggest that speed of convergence tends to decrease with dimensionality. The possibility of phase-transition-like behaviour generated by stochastic/spatial interactions clearly increases such concerns. However, it remains unclear to what extent these issues are important in dealing with the posterior distribution in an image problem, although they are certainly important in the prior.

Acknowledgements

This work was supported by the SERC Complex Stochastic Systems Initiative, and by the University of Washington Department of Statistics.

References

Amit, Y. and Manbeck, K. M. (1993) Deformable template models for emission tomography. *IEEE Trans. Med. Imag.*, **12**, 260–269.

Amit, Y., Grenander, U. and Piccioni, M. (1991) Structural image restoration through deformable templates. *J. Am. Statist. Ass.*, **86**, 376–387.

Arak, T. and Surgailis, D. (1989) Markov random fields with polygonal realizations. *Prob. Th. Rel. Fields*, **80**, 543–579.

Aykroyd, R. G. and Green, P. J. (1991) Global and local priors, and the location of lesions using gamma camera imagery. *Phil. Trans. R. Soc. Lond.* A, **337**, 323–342.

Baddeley, A. J. (1994) Discussion on representation of knowledge in complex systems (by U. Grenander and M. I. Miller). *J. R. Statist. Soc.* B, **56**, 584–585.

Baddeley, A. J. and van Lieshout, M. N. M. (1994) Stochastic geometry models in high-level vision. In *Statistics and Images, Volume 1* (eds K. V. Mardia and G. K. Kanji), pp. 231–256. Abingdon: Carfax.

Besag, J. E. (1986) On the statistical analysis of dirty pictures (with discussion). *J. R. Statist. Soc.* B, **48**, 259–302.

Besag, J. E. and Green, P. J. (1993) Spatial statistics and Bayesian computation (with discussion). *J. R. Statist. Soc.* B, **55**, 25–37.

Besag, J. E. and Higdon, D. M. (1993) Bayesian inference for agricultural field experiments. *Bull. Int. Statist. Inst.*, **LV**, 121–136.

Besag, J. E., York, J. C. and Mollié, A. (1991) Bayesian image restoration, with two applications in spatial statistics (with discussion). *Ann. Inst. Statist. Math.*, **43**, 1–59.

Besag, J., Green, P. J., Higdon, D. and Mengersen, K. (1995) Bayesian computation and stochastic systems. *Statist. Sci.*, **10**, 3–41.

Binder, K. (1988) *Monte Carlo Simulation in Statistical Physics: an Introduction.* Berlin: Springer.

Budinger, T., Gullberg, G. and Huesman, R. (1979) Emission computed tomography. In *Image Reconstruction from Projections: Implementation and Application* (ed G. Herman). Berlin: Springer.

Clayton, D. G. and Bernardinelli, L. (1992) Bayesian methods for mapping disease risk. In *Small Area Studies in Geographical and Environmental Epidemiology* (eds J. Cuzick and P. Elliott), pp. 205-220. Oxford: Oxford University Press.

Clifford, P. and Middleton, R. D. (1989) Reconstruction of polygonal images. *J. Appl. Statist.*, **16**, 409–422.

Clifford, P. and Nicholls, G. K. (1995) Bayesian inference for vector-based images. In *Complex Stochastic Systems and Engineering* (ed. D. M. Titterington), *IMA Conference Series*, **54**, pp. 121–139. Oxford: Oxford University Press.

Cohen, F. S. and Cooper, D. B. (1987) Simple parallel hierarchical and relaxation algorithms for segmenting non-causal Markov random fields. *IEEE Trans. Patt. Anal. Mach. Intell.*, **9**, 195–219.

Creutz, M. (1979) Confinement and the critical dimensionality of space-time. *Phys. Rev. Lett.*, **43**, 553–556.

Edwards, R. G. and Sokal, A. D. (1988) Generalization of the Fortuin–Kasteleyn–Swendsen–Wang representation and Monte Carlo algorithm. *Phys. Rev. D*, **38**, 2009–2012.

Geman, D. (1991) *Random fields and inverse problems in imaging. Lect. Notes Math.*, **1427**. Berlin: Springer.

Geman, D., Geman, S., Graffigne, C. and Dong, P. (1990) Boundary detection by constrained optimization. *IEEE Trans. Patt. Anal. Mach. Intell.*, **12**, 609–628.

Geman, S. and Geman, D. (1984) Stochastic relaxation, Gibbs distributions and the Bayesian restoration of images. *IEEE Trans. Patt. Anal. Mach. Intell.*, **6**, 721–741.

Geman, S. and McClure, D. E. (1987) Statistical methods for tomographic image reconstruction. *Bull. Int. Statist. Inst.*, **LII-4**, 5–21.

Geman, S., Manbeck, K. M. and McClure, D. E. (1993) A comprehensive statistical model for single photon emission tomography. In *Markov random fields: theory and application* (eds R. Chellappa and A. Jain), pp. 93–130. New York: Academic Press.

Geman, S., McClure, D. E., and Geman, D. (1992) A nonlinear filter for film restoration and other problems in image processing. *CVGIP: Graphical Models and Image Proc.*, **54**, 281–289.

Gilks, W. R., Richardson, S. and Spiegelhalter, D. J. (1995) Introducing Markov chain Monte Carlo. In *Markov Chain Monte Carlo in Practice* (eds W. R. Gilks, S. Richardson and D. J. Spiegelhalter), pp. 1–19. London: Chapman & Hall.

Gilks, W. R., Clayton, D. G., Spiegelhalter, D. J., Best, N. G., McNeil, A. J., Sharples, L. D. and Kirby, A. J. (1993) Modelling complexity: applications of Gibbs sampling in medicine (with discussion). *J. R. Statist. Soc.* B, **55**, 39–52.

Glad, I. K. and Sebastiani, G. (1995) A Bayesian approach to synthetic magnetic resonance imaging. *Biometrika*, **82**, 237–250.

Godtliebsen, F. (1989) *A study of image improvement techniques applied to NMR images. Dr. Ing. thesis*, Norwegian Institute of Technology.

Gray, A. J. (1993) Discussion on the meeting on the Gibbs sampler and other Markov chain Monte Carlo methods. *J. R. Statist. Soc.* B, **55**, 58–61.

Green, P. J. (1990) Bayesian reconstructions from emission tomography data using a modified EM algorithm. *IEEE Trans. Med. Imag.*, **9**, 84–93.

Green, P. J. (1994a) Discussion on representation of knowledge in complex systems (by U. Grenander and M. I. Miller). *J. R. Statist. Soc.* B, **56**, 589–590.

Green, P. J. (1994b) Reversible jump MCMC computation and Bayesian model determination. *Mathematics Research Report S-94-03*, University of Bristol.

Green, P. J. and Han, X-L. (1992) Metropolis methods, Gaussian proposals, and antithetic variables. In *Stochastic Models, Statistical Methods and Algorithms in Image Analysis* (eds P. Barone, A. Frigessi and M. Piccioni), *Lect. Notes Statist.*, **74**, pp.142–164. Berlin: Springer.

Green, P. J. and Titterington, D. M. (1988) Recursive methods in image processing. *Bull. Int. Statist. Inst.*, **LII-4**, 51–67.

Grenander, U. (1983) *Tutorial in pattern theory*. Technical report, Division of Applied Mathematics, Brown University, Providence.

Grenander, U. and Manbeck, K. M. (1992) Abnormality detection in potatoes by shape and color. Technical report, Division of Applied Mathematics, Brown University, Providence.

Grenander, U. and Miller, M. (1994) Representations of knowledge in complex systems (with discussion). *J. R. Statist. Soc.*, B, **56**, 549–603.

Grenander, U., Chow, Y. and Keenan, D. M. (1991) *Hands: a pattern theoretic study of biological shapes. Res. Notes Neur. Comp.*, **2**. Berlin: Springer.

Han, X-L. (1993) Markov Chain Monte Carlo and Sampling Efficiency. Ph.D. thesis, University of Bristol.

Heikkinen, J. and Högmander, H. (1994) Fully Bayesian approach to image restoration with an application in biogeography. *Appl. Statist.*, **43**, 569–582.

Higdon, D. (1993) Discussion on the meeting on the Gibbs sampler and other Markov chain Monte Carlo methods. *J. R. Statist. Soc.* B, **55**, 78.

Högmander, H. and Møller, J. (1995) Estimating distribution maps from atlas data using methods of statistical image analysis. *Biometrics*, **51**, 393–404.

Kandel, D., Domany, E. and Brandt, A. (1989) Simulations without critical slowing down: Ising and three-state Potts models. *Phys. Rev.* B, **40**, 330–344.

Litton, C. D. and Buck, C. E. (1993) Discussion on the meeting on the Gibbs sampler and other Markov chain Monte Carlo methods. *J. R. Statist. Soc.* B, **55**, 82.

Metropolis, N., Rosenbluth, A. W., Rosenbluth, M. N., Teller, A. H. and Teller, E. (1953) Equations of state calculations by fast computing machines. *J. Chem. Phys.*, **21**, 1087–1091.

Mollié, A. (1995) Bayesian mapping of disease. In *Markov Chain Monte Carlo in Practice* (eds W. R. Gilks, S. Richardson and D. J. Spiegelhalter), pp. 359–379. London: Chapman & Hall.

Phillips, D. and Smith, A. F. M. (1994) Bayesian faces. *J. Am. Statist. Ass.*, **89**, 1151–1163.

Ripley, B. D. and Sutherland, A. I. (1990) Finding spiral structures in images of galaxies. *Phil. Trans. R. Soc. Lond.* A, **332**, 477–485.

Roberts, G. O. (1995) Markov chain concepts related to sampling algorithms. In *Markov Chain Monte Carlo in Practice* (eds W. R. Gilks, S. Richardson and D. J. Spiegelhalter), pp. 45–57. London: Chapman & Hall.

Shepp, L. A. and Vardi, Y. (1982) Maximum likelihood reconstruction in positron emission tomography. *IEEE Trans. Med. Imag.*, **1**, 113–122.

Sokal, A. D. (1989) Monte Carlo methods in statistical mechanics: foundations and new algorithms. *Cours de Troisième Cycle de la Physique en Suisse Romande*. Lausanne.

Srivastava, A., Miller, M. I. and Grenander, U. (1991) Jump diffusion processes for object tracking and direction finding. In *Proc. 29th Annual Allerton Conference on Communication, Control and Computing*, pp. 563–570. University of Illinois.

Swendsen, R. H. and Wang, J. S. (1987) Nonuniversal critical dynamics in Monte Carlo simulations. *Phys. Rev. Lett.*, **58**, 86–88.

Weir, I. S. (1995) Fully Bayesian reconstructions from single photon emission computed tomography data. *Mathematics Research Report*, University of Bristol.

Weir, I. S. and Green, P. J. (1994) Modelling data from single photon emission computed tomography. In *Statistics and Images, Volume 2* (ed. K. V. Mardia), pp. 313–338. Abingdon: Carfax.

Whittaker, J. (1990) *Graphical Models in Applied Multivariate Statistics*. Chichester: Wiley.

Wilson, J. D. and Green, P. J. (1993) A Bayesian analysis for remotely sensed data, using a hierarchical model. *Research Report S-93-02*, School of Mathematics, University of Bristol.

22

Measurement error

Sylvia Richardson

22.1 Introduction

Errors in the measurement of explanatory variables is a common problem in statistical analysis. It is well known that ignoring these errors can seriously mislead the quantification of the link between explanatory and response variables, and many methods have been proposed for countering this. Measurement error was initially investigated in linear regression models (see for example Fuller, 1987), and recent research has led to its investigation in other regression models, motivated in particular by applications in epidemiology and other areas of biomedical research. Overviews of measurement error in epidemiology can be found in Caroll (1989), Armstrong (1990), Gail (1991), Liu and Liang (1991), Thomas *et al.* (1993) and Caroll *et al.* (1995). We too shall focus much of our discussion in the epidemiological context.

In epidemiological studies, it is rarely possible to measure all relevant covariates accurately. Moreover, recent work has shown that measurement error in one covariate can bias the association between other covariates and the response variable, even if those other covariates are measured without error (Greenland, 1980; Brenner, 1993). Apart from biasing the estimates, misspecification of explanatory variables also leads to loss of efficiency in tests of association between explanatory and response variables; this has recently been characterized in logistic regression models by Begg and Lagakos (1993).

Any method proposed for correcting parameter estimates in the presence of measurement error is dependent on some knowledge of the measurement-error process. It is often possible to build into the design of a study some assessment of the error process, either by the inclusion of a validation group, i.e. a subgroup of individuals for whom it is possible to obtain accurate measurements, or by performing repeated measurements on some of the subjects (Willet, 1989; Marshall, 1989). How best to integrate this knowledge has been the subject of much research. Existing methods for dealing with measurement error differ according to:

- the type of covariate considered (continuous or categorical);

- the assumptions made on the measurement-error process (Berkson or classical, fully parametric or not);

- the estimation framework considered (maximum likelihood, quasi-likelihood, pseudo-likelihood or Bayesian); and

- whether or not approximations are used.

For continuous covariates, methods substituting an estimate of the expectation of the unobserved covariate given the measured one (also called the 'surrogate') have been discussed by Caroll et al. (1984), Rosner et al. (1989, 1990), Whittemore (1989) and Pierce et al. (1992). Some methods make a specific assumption of small error variance (Stefanski and Caroll, 1985; Whittemore and Keller, 1988; Chesher, 1991; Caroll and Stefanski, 1990). Semi-parametric methods have been considered by Pepe and Fleming (1991), Caroll and Wand (1991), Pepe et al. (1994), Robins et al. (1994, 1995) and Mallick and Gelfand (1995). While most of the research quoted above has been concerned with models appropriate for cohort studies, the estimation of logistic regression models for case-control studies with errors in covariates has recently been elaborated by Caroll et al. (1993) using pseudo-likelihood with non-parametric estimation of the marginal distribution of the unobserved covariates.

The formulation of measurement-error problems in the framework of a Bayesian analysis using graphical models, and the associated estimation methods using stochastic simulation techniques, have recently been developed (Thomas et al., 1991; Stephens and Dellaportas, 1992; Gilks and Richardson, 1992; Richardson and Gilks, 1993a,b; Mallick and Gelfand, 1995). In this chapter, we recount this development, placing particular emphasis on two aspects: the flexibility of this approach, which can integrate successfully different sources of information on various types of measurement process; and the natural way in which all sources of uncertainty are taken account of in the estimation of parameters of interest. Outside the framework of graphical models, a Bayesian approach to logistic regression with measurement error has recently been proposed by Schmid and Rosner (1993).

The structure of the measurement-error problem in epidemiology can be formulated as follows. Risk factors (covariates) are to be related to the disease status (response variable) Y for each individual. However, for many or all individuals in the study, while some risk factors C are truly known, other risk factors X are unknown. It is sometimes possible to obtain information on the unknown risk factors X by recording one or several surrogate measures Z of X for each individual. In other situations, ancillary risk factor information, gained by carrying out surveys on individuals outside the study, but related to the study individuals by known group characteristics, are used (Gilks and Richardson, 1992). To model this general situation, we shall distinguish three submodels (following the terminology introduced by Clayton, 1992):

- a disease model, which expresses the relationship between risk factors C and X and disease status Y;

- an exposure model which describes the distribution of the unknown risk factors X in the general population, or which relates the distribution of X to ancillary information; and

- a measurement model, which expresses the relationship between some surrogate information Z and the true risk factors X, or which links the observed survey to the ancillary risk-factor information.

We shall now detail the structure of two particular epidemiological designs; the first is widely encountered in epidemiology, for example in nutritional studies, whilst the second has arisen more prominently in occupational epidemiology.

22.2 Conditional-independence modelling

22.2.1 Designs with individual-level surrogates

The structure of the three submodels described above can be characterized through the following conditional-independence assumptions:

$$
\begin{array}{rll}
\text{disease model} & [Y_i \mid X_i, C_i, \beta] & (22.1) \\
\text{measurement model} & [Z_i \mid X_i, \lambda] & (22.2) \\
\text{exposure model} & [X_i \mid C_i, \pi] & (22.3)
\end{array}
$$

where subscript i denotes the individual, and β, λ and π are model parameters. Variables in (22.1–22.3) can be scalar or vector. Equations (22.1–22.3) are called *model conditional distributions* ('model conditionals' for short). Since we work in a Bayesian framework, we require prior distributions for β, λ and π.

Conditional-independence assumptions

By asserting (22.1–22.3) as model conditionals, we imply far more than the conditional dependencies made explicit in those equations: we also imply conditional independence relationships which follow from the directed Markov assumption (Lauritzen *et al.*, 1990). This states that the joint distribution of all the variables can be written as the product of the model conditionals:

$$[\beta] \, [\lambda] \, [\pi] \, \prod_i \, [X_i \mid C_i, \pi] \, [Z_i \mid X_i, \lambda] \, [Y_i \mid X_i, C_i, \beta], \qquad (22.4)$$

where $[a]$ generically denotes the distribution of a, and $[a \mid b]$ generically denotes the conditional distribution of a given b. Thus, in particular, the following apply:

- (22.1) states that the disease status of individual i, Y_i, is only dependent on its true exposure X_i, on known covariates C_i and on unknown parameters β. We are thus in the classical case where, conditionally on the true exposure being known, the surrogate measures Z_i do not add any information on the disease status. This is a fundamental assumption made in most of the work on measurement error in epidemiology.

- (22.2) states that by conditioning on appropriately defined parameters λ and the true exposure X_i, the surrogate measures Z_i are independent among individuals. The construction of λ will be detailed in an example.

- (22.3) models the population distribution of unknown risk factors among individuals in terms of parameters π. Dependence between the different components of vector X_i can be accommodated through parameters contained in π but the risk factors X_i are assumed independent between individuals given C_i and π.

By specifying the conditional distribution of the surrogate Z given the true exposure X as in (22.2), we are placing ourselves in the Bayesian analog of what is traditionally referred to as the 'classical error model', where measurement error is independent of X. Another type of error model which has been considered in the literature is the Berkson error model, where (22.2) is replaced by $[X_i \mid Z_i, \lambda]$. With the Berkson error model, usually no model need be specified for the marginal distribution of Z.

Conditional-independence graph

An influence diagram or conditional-independence graph corresponding to (22.1–22.3), encompassing several epidemiological designs, is shown in Figure 22.1. We use squares to denote observed quantities and circles to denote unobserved quantities. See Spiegelhalter *et al.* (1995: this volume) for further discussion of conditional independence graphs. Figure 22.1 identifies six groups of individuals, grouped according to which variables are recorded. For example, for individuals in Part 1 of Figure 22.1, variables X_i, Y_i, C_i and Z_i are recorded, whilst for individuals in Part 2 only variables Y_i, C_i and Z_i are recorded. Designs which record X_i on some individuals presume the availability of a 'gold standard', i.e. an error-free method for measuring X.

Parts 1 and 4 of Figure 22.1 are validation studies. In a validation study, both X_i and Z_i are recorded on each individual, providing information

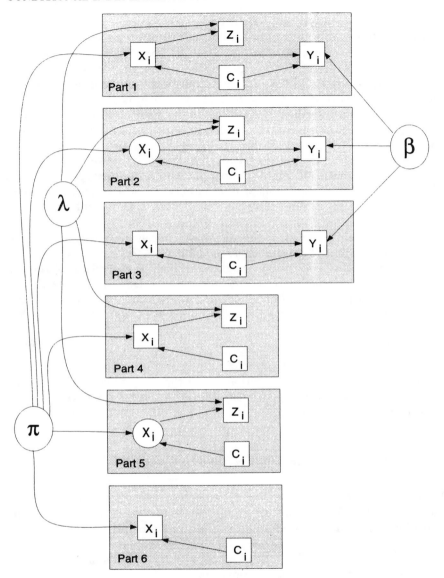

Figure 22.1 *A graph corresponding to equations (22.1–22.3).*

on the measurement-process parameters λ. A validation study is either internal, if disease status Y_i is also recorded (Part 1), or external, if there is no information on disease status (Part 4). Part 2 represents the common situation where only surrogates and disease status are known. Part 3 represents a subgroup in which only the true exposure and disease status are known. Clearly if all the individuals were in this group, there would be no measurement-error problem. In general, the number of individuals included in Parts 1 and 3 will be small compared to Part 2. Finally, Parts 5 and 6 represent 'survey' situations in which information is obtained only on surrogates or on the true exposure. This global influence diagram illustrates how information can flow from one part of the graph to another. For example, Part 6 contributes information on π which in turn provides information on X_i in Part 5, so that Part 5 can yield some information on λ.

When no gold standard is available, validation groups are ruled out and the estimation of measurement model parameters must rely on other sources of information. Designs might then include several measuring instruments with possible repeated determinations. Let Z_{ihr} denote the rth repeated measurement of instrument h for individual i. The model conditional for the measurement process (22.2) now becomes:

$$[Z_{ihr} \mid X_i, \lambda_h]. \tag{22.5}$$

This equation states that, conditionally on the true value of the covariate X_i and on λ_h, there is independence of the surrogates Z_{ihr} between repeats and between instruments.

22.2.2 Designs using ancillary risk-factor information

In occupational or environmental epidemiology, risk-factor information for each individual is often not directly available and must be obtained from ancillary, aggregate-level information, such as a job-exposure matrix in an industrial-epidemiological application. Job-exposure matrices provide information on exposures to each of many industrial agents in each of many finely subdivided categories of occupation. They are commonly constructed by industrial experts from detailed job descriptions obtained in a specially conducted survey. Thus the exposure to industrial agents of each individual in the disease study can be assessed using only his job title, by referring to a job-exposure matrix. The measurement-error model implied by this design is different from those considered above, as imprecision in exposure information provided by the job-exposure matrix must be taken into account.

Model conditionals

We consider the case of a dichotomous exposure. Let π_{jk} denote the underlying (unobserved) probability of being exposed to agent k, for individuals in job j. We assume that π_{jk} is the same in the disease study and in the job-exposure survey. Let m_{jk} denote the number of people in the job-exposure survey with job j who were considered by the experts to be exposed to agent k. The model conditional for the aggregate-level survey data $\{m_{jk}\}$ is then:

$$[m_{jk} \mid \pi_{jk}, n_j] = \text{Binomial}\,(\pi_{jk}, n_j), \qquad (22.6)$$

where n_j is the number of people with job j included in the survey. Equation (22.6) represents the measurement model in our general formulation in Section 22.1.

The unknown dichotomous exposure X_{ik} of disease-study individual i to agent k is linked to the job-exposure matrix through his job title $j = j(i)$. Since individual i is exposed to agent k ($X_{ik} = 1$) with probability $\pi_{j(i)k}$, the model conditional for exposure in the disease study is given by:

$$[X_{ik} \mid \pi_{j(i)k}] = \text{Bernoulli}\,(\pi_{j(i)k}). \qquad (22.7)$$

This represents the exposure model in our general formulation in Section 22.1. The disease model is given as before by equation (22.1).

Note that the job-exposure survey does not provide information directly on X, but rather on the prior distribution of X. We are thus in neither the classical nor the Berkson measurement-error situation. Gilks and Richardson (1992) demonstrate good performance of Bayesian modelling for analysing designs of this kind.

Suppose, in addition to the job-exposure matrix, that direct surrogate dichotomous measures Z_{ik} of X_{ik} are available for some or all disease-study individuals. These might be provided by expert assessment, as in the job-exposure survey. The previous set-up can easily be generalized to include both sources of risk-factor information. We need only specify one additional model conditional:

$$[Z_{ik} \mid X_{ik}, \delta_k], \qquad (22.8)$$

where δ_k represents misclassification parameters corresponding to errors in the coding of exposure to agent k. The measurement model now consists of both (22.6) and (22.8). The graph associated with this model is represented in Figure 22.2.

22.2.3 Estimation

Estimation of the models described above can be carried out straightforwardly by Gibbs sampling; see Gilks et al. (1995: this volume) for a general description of this method, and Gilks and Richardson (1992) and

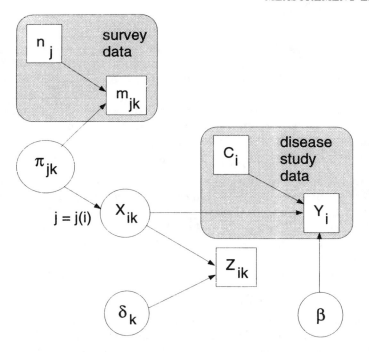

Figure 22.2 *A graph corresponding to equations (22.1) and (22.6–22.8).*

Richardson and Gilks (1993a) for computational details including full conditional distribtions for the above models. Sampling from full conditional distributions was performed using adaptive rejection sampling (Gilks and Wild, 1992; Gilks, 1995: this volume).

22.3 Illustrative examples

In this section, we present a series of examples. Our aims are to illustrate different types of measurement-error situation, to discuss the performance of our approach and to outline areas for further research. We have used simulated data sets throughout to evaluate the performance of our method of analysis.

22.3.1 Two measuring instruments with no validation group

In our first example, we present the analysis of a simulated data set reproducing a design where information on the measurement parameters is obtained through the combination of two measuring instruments.

Design set-up

Two risk factors are involved in the disease model. The first one, X, is measured with error and the second one, C, is known accurately. We consider the case of a logistic link between risk factors and disease status. Specifically, we suppose that Y_i follows a Bernoulli distribution with parameter α_i, where logit $\alpha_i = \beta_0 + \beta_1 X_i + \beta_2 C_i$. We suppose that the exposure vector (X, C) follows a bivariate normal distribution, with mean μ and variance-covariance matrix Σ. Thus the vector β in (22.1) comprises parameters $\beta_0, \beta_1, \beta_2$; and π in (22.3) comprises μ and Σ.

Concerning the measurement process, we consider the case of two measuring instruments. The first instrument has low precision but is unbiased. In contrast, the second instrument has a higher precision but is known to be biased. Since Instrument 2 has a higher precision, it is used in preference to Instrument 1 on the entire study population. However, we aim to correct the bias of Instrument 2 by including in the design a subgroup of relatively small size in which both instruments are measured, Instrument 1 being administered twice.

We suppose that the model conditional for the rth repeat of Instrument 1 is a normal distribution with mean X_i and variance θ_1^{-1}:

$$[Z_{i1r} \mid X_i, \theta_1] = \mathrm{N}(X_i, \theta_1^{-1}), \qquad r = 1, 2.$$

Here the only measurement parameter is θ_1, the precision (the inverse of the variance) of Instrument 1. Parameter θ_1 corresponds to λ_1 in (22.5). For the biased Instrument 2, the model conditional is also normal:

$$[Z_{i2} \mid X_i, \phi_2, \psi_2, \theta_2] = \mathrm{N}(\phi_2 + \psi_2 X_i, \theta_2^{-1}).$$

Here the measurement parameters are the intercept ϕ_2 and slope ψ_2 (expressing the linear relationship between the true exposure and its surrogate), and the precision θ_2. These parameters correspond to λ_2 in equation (22.5).

The study is thus designed to include two parts (i.e. two subgroups of individuals) with 1 000 individuals in Part 1 and $n = 200$ or 50 individuals in Part 2. In Part 1, only Instrument 2 has been recorded. In Part 2, Instrument 1 has been measured twice and Instrument 2 has been recorded once on each individual.

A data set was generated using 'true' values of $\beta_0, \beta_1, \beta_2, \theta_1, \phi_2, \psi_2$ and θ_2 given in Table 22.1 (column: 'true values') and with

$$\mu = \begin{pmatrix} 0.5 \\ -0.5 \end{pmatrix}, \qquad \Sigma = \begin{pmatrix} 1.02 & 0.56 \\ 0.55 & 0.96 \end{pmatrix}.$$

Thus we have simulated a situation with two detrimental risk factors X and C having relative risks 2.46 and 3.32 respectively, with a positive correlation of 0.56 between X and C. Note that Instrument 2 is three times more accurate than Instrument 1 ($\theta_2 = 3 \times \theta_1$).

parameter	true value	Gibbs sampling analysis				classical analysis with Instrum. 2	
		n=200		n=50			
		mean ±	*s.d.*	mean ±	*s.d.*	mean	± *s.e.*
β_0	-0.8	-0.81	0.32	-0.77	0.40	-0.17	0.11
β_1	0.9	1.03	0.36	0.98	0.42	0.14	0.07
β_2	1.2	1.25	0.14	1.36	0.21	1.57	0.11
θ_1	0.3	0.31	0.03	0.25	0.04	–	–
ϕ_2	0.8	0.81	0.11	0.84	0.14	–	–
ψ_2	0.4	0.44	0.09	0.45	0.14	–	–
θ_2	0.9	0.91	0.06	0.96	0.13	–	–

Table 22.1 *Gibbs sampling analysis of a design with 2 measuring instruments*

Results

Table 22.1 presents the results from a Gibbs sampling analysis of the simulated data set. Hyperparameters were chosen to specify only vague prior information (for details see Richardson and Gilks, 1993a). We have summarized marginal posterior distributions of parameters of interest by reporting posterior means and standard deviations. In the last two columns of Table 22.1, we have given classical estimates of log relative risks and their standard errors which would be obtained if the analysis relied solely on the data of Part 1, i.e. if no correction for measurement error was performed. We see that these estimates are quite hopeless and that the simulated situation is one in which measurement error, if not taken into account, substantially influences the results.

The results show that our estimation method has performed satisfactorily with all the estimated values lying within one posterior standard deviation of the values set in the simulation. As expected, the posterior standard deviation for β_2 which corresponds to the covariate C measured without error is smaller than that for β_1 which corresponds to X. Note the influence of the size n of Part 2 on all posterior standard deviations. The measurement parameters for Instrument 2 have been well estimated, even though our design did not include a validation subgroup. This highlights how information has been naturally propagated between the two measuring instruments and between the two parts of the design.

In this particular design, even though there is no gold standard, the data still contain information on ϕ_2 and ψ_2 because there is information on X_i from the repeats of the unbiased Instrument 1, information which represents in effect a 'simulated gold standard'. The size n of Part 2 is clearly crucial in this process: with a smaller validation subgroup, the results deteriorate.

22.3.2 *Influence of the exposure model*

At each step of the approach we have outlined so far, conditional distributions need to be explicitly specified in a parametric way. Whilst some of these parametric distributions arise naturally, such as the choice of a logistic model for the disease risk, other assumed distributional forms are more arbitrary. In particular, there are some cases where little is known about the distribution of the exposure X and an appropriate model for it (Thomas *et al.*, 1991).

The exposure model (22.3) we have implemented in our recent work is that of a simple Gaussian distribution. A natural generalization of the exposure model would be a mixture of Gaussians or other distributions. As a first step in that direction, we have assessed the effect of misspecifying a simple Gaussian variable for exposure when the true exposure is a mixture of Gaussians or a χ^2 distribution, as we now describe.

Simulation set-up

We consider a study with one known risk factor C and another risk factor X measured with error by a surrogate Z on 1 000 individuals. The study also includes a validation group containing 200 individuals where both X and Z are measured. As in the previous example, we assume a logistic link between risk factor and disease status. Three datasets were simulated, differing in the generating distribution for true exposures X:

$$\text{(a)} \quad X_i \sim \quad \frac{1}{2}\text{N}(-1.0, 1.0) + \frac{1}{2}\text{N}(2.0, 1.0);$$

$$\text{(b)} \quad X_i \sim \quad \frac{1}{2}\text{N}(-3.0, 1.0) + \frac{1}{2}\text{N}(4.0, 1.0);$$

$$\text{(c)} \quad X_i \sim \quad \chi_1^2.$$

In each dataset the surrogate Z_i was generated by:

$$[Z_i \mid X_i, \theta] = \text{N}(X_i, \theta^{-1}).$$

Each of the three simulated datasets was analysed by Gibbs sampling, with the assumption that the exposure model for (X, C) was misspecified as a bivariate normal distribution with mean μ and variance-covariance matrix Σ, with a vague prior distribution for μ centred around $(0.5, 0.5)$ and a Wishart prior distribution for Σ with 5 degrees of freedom and identity scale matrix.

The results are shown in Table 22.2. With a bimodal, symmetric true exposure distribution (Datasets (a) and (b)), the parameters are still adequately estimated, although there is some deterioration when the modes are well-separated (Dataset (b)). However in Dataset (c), where the true exposure distribution is skewed, misspecification has led to attenuation of the estimate of β_1, and its 95% credible interval no longer contains the true value. Hence misspecification of the exposure model can influence the regression results. Further studies of the influence of misspecification are warranted.

parameter	true value	Dataset (a) mean ± s.d.		Dataset (b) mean ± s.d.		Dataset (c) mean ± s.d.	
β_0	-0.8	-0.82	0.11	-0.72	0.13	-0.68	0.12
β_1	0.9	0.91	0.08	0.77	0.07	0.56	0.08
β_2	1.2	1.20	0.08	1.03	0.08	1.34	0.13
θ	0.9	0.87	0.08	0.87	0.09	0.86	0.08
μ_X	0.5	0.50	0.06	0.50	0.11	1.35	0.07
μ_C	0.5	0.55	0.05	0.63	0.10	1.24	0.05

Table 22.2 *Gibbs sampling analysis with misspecified exposure distribution*

Mixture models provide great flexibility in modelling distributions with a variety of shapes. The next stage in the development of our graphical model will be to employ a mixture model for the exposure distribution (22.3). Gibbs sampling analysis of mixtures is discussed by Robert (1995: this volume). A difference between our set-up and the set-up usually considered in mixture problems is that, in our case, we do not observe a fixed sample from the mixture distribution; rather, this sample (corresponding to the unknown risk factor X) is generated anew at each iteration of the Gibbs sampler. It will be thus interesting to see how convergence of the Gibbs sampler is modified by the additional level of randomness.

22.3.3 Ancillary risk-factor information and expert coding

In our last example, we illustrate how ancillary information can be combined with expert coding to provide estimates of both regression coefficients and probabilities of misclassification. This extends the work of Gilks and Richardson (1992). We first describe how the job-exposure survey data and the disease-study data were generated.

Generating disease-study data

Each of 1 000 individuals were randomly and equiprobably assigned to one of four occupations ($j = 1, \ldots, 4$). For each individual, exposure status ($X_i = 1$: exposed; $X_i = 0$: not exposed) to a single industrial agent was randomly assigned according to true job-exposure probabilities $\{\pi_j\}$:

j	π_j
1	0.9
2	0.3
3	0.5
4	0.6

Disease status ($Y_i = 1$: diseased; $Y_i = 0$: not diseased) was assigned with probability of disease α_i, where logit $\alpha_i = \beta_0 + \beta_1 X_i$.

We also supposed that, for each of the 1 000 individuals, an expert was able to code their exposure with a variable Z_i. This was generated from X_i with the following misclassification probabilities:

$$P(Z_i = 1 \mid X_i = 0) = \gamma_1; \qquad P(Z_i = 0 \mid X_i = 1) = \gamma_2.$$

Exposures X_i were thenceforth assumed unknown for each individual, so the analysis was based only on Y_i, Z_i and the job-exposure survey data, described next. Three datasets were generated:

(a) Z_i not available;

(b) $\gamma_1 = 0.2$, $\gamma_2 = 0.5$;

(c) $\gamma_1 = 0.1$, $\gamma_2 = 0.3$.

Generating the job-exposure survey data

Each of 150 individuals were assigned an occupation and an exposure status, as for the disease study individuals. The observed job-exposure matrix $\{n_j, m_j\}$ (where n_j is the number of surveyed individuals in job j and m_j is the number 'coded by the expert' as 'exposed') was compiled from these 150 individuals, and the job-exposure probabilities π_j were thenceforth assumed unknown. The job-exposure matrix was assumed to be available for each of the three datasets described above.

Analysing the simulated data

For the graphical model, we assumed a normal $N(0, 9)$ prior distribution for the regression parameter β, and almost flat priors for logit π_j, logit γ_1 and logit γ_2. We ran the Gibbs sampler for 7 000 iterations, discarding the first 100 iterations before analysis. The results are presented in Table 22.3.

By comparing the results for Datasets (b) and (c) with Dataset (a), one can clearly see that the information given by the coding of the expert leads to improved estimates of the regression coefficients, with smaller posterior standard deviations. The improvement is more marked in (c) as the misclassification probabilities are smaller than in (b). Moreover good estimates of the misclassification probabilities are also obtained.

Dataset	parameters	β_0	β_1	γ_1	γ_2
(a)	true values	-1.00	1.50	–	–
	posterior mean	-0.99	1.70	–	–
	posterior $s.d.$	0.30	0.43	–	–
(b)	true values	-1.00	1.50	0.20	0.50
	posterior mean	-1.20	1.63	0.16	0.51
	posterior $s.d.$	0.27	0.34	0.04	0.03
(c)	true values	-1.00	1.50	0.10	0.30
	posterior mean	-1.11	1.56	0.08	0.32
	posterior $s.d.$	0.18	0.24	0.04	0.03

Table 22.3 *Gibbs sampling analysis of designs with ancillary risk-factor information and expert coding*

22.4 Discussion

In this chapter, we have presented a unifying representation through conditional independence models of measurement-error problems with special reference to epidemiological applications. There are several advantages of this approach over methods previously proposed which are extensively discussed in Richardson and Gilks (1993a). Of paramount importance is its flexibility, which enables the modelling of an extensive range of measurement-error situations without resorting to artificial simplifying assumptions. This has important design implications for future studies. Now that analyses of complex measurement-error designs can be carried out successfully, there is more freedom at the design stage. An important area for future research is thus to give guidelines for complex designs.

The key to the construction of such models is the stipulation of suitable conditional independence assumptions. Careful thought has to be given to the implications of each of these assumptions in any particular context. For example, in (22.1–22.4) we have assumed independence between the Y conditional on the X, C and β. This is an appropriate assumption in chronic disease epidemiology but is likely to be violated when considering infectious diseases. Indeed, the disease status of an individual is influenced, through contagious contacts, by the disease status of other individuals. As another example, the conditional independence between repeated measures of surrogates given the true risk factor, assumed by (22.5), would not hold if there is a systematic bias in the measurement.

The approach we have developed is fully parametric. The influence on regression parameter estimates of misspecification of the measurement error or exposure distributions gives cause for concern. The use of flexible mixture distributions is a natural way to relax the fully parametric set-up. This approach has been taken by Mallick and Gelfand (1995). By using a mixture of cumulative beta distribution functions to model unknown cumulative distribution functions, they formulate a semi-parametric Bayesian approach to a measurement-error problem, implemented through a single-component Metropolis–Hastings algorithm.

References

Armstrong, B. G. (1990) The effects of measurement errors on relative risk regression. *Am. J. Epidem.*, **132**, 1176–1184.

Begg, M. D. and Lagakos, S. (1993) Loss in efficiency caused by omitting covariates and misspecifying exposure in logistic regression models. *J. Am. Statist. Ass.*, **88**, 166–170.

Brenner, H. (1993) Bias due to non-differential misclassification of polytomous confounders. *J. Clin. Epidem.*, **46**, 57–63.

Caroll, R. J. (1989) Covariance analysis in generalized linear measurement error models. *Statist. Med.*, 8, 1075–1093.

Caroll, R. J. and Stefanski, A. (1990) Approximate quasi-likelihood estimation in models with surrogate predictors. *J. Am. Statist. Ass.*, **85**, 652–663.

Caroll, R. J. and Wand, M. P. (1991) A semiparametric estimation in logistic measurement error models. *J. R. Statist. Ass.* B, **53**, 573–585.

Caroll, R. J., Gail, M. H and Lubin, J. H. (1993) Case-control studies with errors in covariates. *J. Am. Statist. Ass.*, **88**, 185–199.

Carroll, R. J., Ruppert, D. and Stefanski, L. A. (1995) *Non-linear Measurement Error Models*. London: Chapman & Hall.

Caroll, R. J., Spiegelman, C., Lan, K. K. G., Bailey, K. T. and Abbott, R. D. (1984) On errors in variables for binary regression models. *Biometrika*, **71**, 19–26.

Chesher, A. (1991) The effect of measurement error. *Biometrika*, **78**, 451–462.

Clayton, D. G. (1992) Models for the analysis of cohort and case-control studies with inaccurately measured exposures. In *Statistical Models for Longitudinal Studies of Health* (eds J. H. Dwyer, M. Feinleib and H. Hoffmeister), pp. 301-331. Oxford: Oxford University Press.

Fuller, W. A. (1987) *Measurement Error Models*. New York: Wiley.

Gail, M. H. (1991) A bibliography and comments on the use of statistical models in epidemiology in the 1980s. *Statist. Med.*, **10**, 1819–1885.

Gilks, W. R. (1995) Full conditional distributions. In *Markov Chain Monte Carlo in Practice* (eds W. R. Gilks, S. Richardson and D. J. Spiegelhalter), pp. 75–88. London: Chapman & Hall.

Gilks, W. R. and Richardson, S. (1992) Analysis of disease risks using ancillary risk factors, with application to job-exposure matrices. *Statist. Med.*, 11, 1443–1463.

Gilks, W. R. and Wild, P. (1992) Adaptive rejection sampling for Gibbs sampling. *Appl. Statist.*, **41**, 337–348.

Gilks, W. R., Richardson, S. and Spiegelhalter, D. J. (1995) Introducing Markov chain Monte Carlo. In *Markov Chain Monte Carlo in Practice* (eds W. R. Gilks, S. Richardson and D. J. Spiegelhalter), pp. 1–19. London: Chapman & Hall.

Greenland, S. (1980) The effect of misclassification in the presence of covariates. *Am. J. Epidem.*, **112**, 564–569.

Lauritzen, S. L., Dawid, A. P., Larsen, B. N. and Leimer, H. G. (1990) Independence properties of directed Markov fields. *Networks*, **20**, 491–505.

Liu, X. and Liang, K. J. (1991) Adjustment for non-differential misclassification error in the generalized linear model. *Statist. Med.*, **10**, 1197–1211.

Mallick, B. K. and Gelfand, A. E. (1995) Semiparametric errors-in-variables models: a Bayesian approach. Technical report, Imperial College, London University.

Marshall, J. R. (1989) The use of dual or multiple reports in epidemiologic studies. *Statist. Med.*, **8**, 1041–1049.

Pepe, M. S. and Fleming, T. R. (1991) A non-parametric method for dealing with mismeasured covariate data. *J. Am. Statist. Ass.*, **86**, 108–113.

Pepe, M. S., Reilly, M. and Fleming, T. R. (1994) Auxiliary outcome data and the mean score method. *J. Statist. Plan. Inf.*, **42**, 137–160.

Pierce, D. A., Stram, D. O., Vaeth, M. and Schafer, D. W. (1992) The errors in variables problem: considerations provided by radiation dose-response analyses of the A-bomb survivor data. *J. Am. Statist. Ass.*, **87**, 351–359.

Richardson, S. and Gilks, W. R. (1993a) Conditional independence models for epidemiological studies with covariate measurement error. *Statist. Med.*, **12**, 1703–1722.

Richardson, S. and Gilks, W. R. (1993b) A Bayesian approach to measurement error problems in epidemiology using conditional independence models. *Am. J. Epidem.*, **138**, 430–442.

Robert, C. P. (1995) Mixtures of distributions: inference and estimation. In *Markov Chain Monte Carlo in Practice* (eds W. R. Gilks, S. Richardson and D. J. Spiegelhalter), pp. 441–464. London: Chapman & Hall.

Robins, J. M., Hsieh, F. and Newey, W. (1995) Semiparametric efficient estimates of a conditional density with missing or mismeasured covariates. *J. R. Statist. Soc. B*, **57**, 409–424.

Robins, J. M., Rotnitzky, A. and Zhao, L. P. (1994) Estimation of regression coefficients when some regressors are not always observed. *J. Am. Statist. Ass.*, **89**, 846–866.

Rosner, B., Spiegelman, D. and Willett, W. C. (1989) Correction of logistic regression relative risk estimates and confidence intervals for systematic within-person measurement error. *Statist. Med.*, **8**, 1051–1069.

Rosner, B., Spiegelman, D. and Willett, W. C. (1990) Correction of logistic regression relative risk estimates and confidence intervals for measurement error: the case of multiple covariates measured with error. *Am. J. Epidem.*, **132**, 734–745.

Schmid, C. H. and Rosner, B. (1993) A Bayesian approach to logistic regression models having measurement error following a mixture distribution. *Statist. Med.*, **12**, 1141–1153.

Spiegelhalter, D. J., Best, N. G., Gilks, W. R. and Inskip, H. (1995) Hepatitis B: a case study in MCMC methods. In *Markov Chain Monte Carlo in Practice* (eds W. R. Gilks, S. Richardson and D. J. Spiegelhalter), pp. 21–43. London: Chapman & Hall.

Stefanski, L. A. and Caroll, R. J. (1985) Covariate measurement error in logistic regression. *Ann. Statist.*, **13**, 1335–1351.

Stephens, D. A. and Dellaportas, P. (1992) Bayesian analysis of generalised linear models with covariate measurement error. In *Bayesian Statistics 4* (eds J. M. Bernardo, J. O. Berger, A. P. Dawid and A. F. M. Smith), pp. 813–820. Oxford: Oxford University Press.

Thomas, D. C., Gauderman, W. J. and Kerber, R. (1991) A non-parametric Monte Carlo approach to adjustment for covariate measurement errors in regression problems. Technical report, Department of Preventive Medicine, University of Southern California.

Thomas, D., Stram, D. and Dwyer, J. (1993) Exposure measurement error: influence on exposure-disease relationship and methods of correction. *Annual Rev. Pub. Health, 14*, 69–93.

Whittemore, A. S. (1989) Errors in variables regression using Stein estimates. *Am. Statist.*, **43**, 226–228.

Whittemore, A. S. and Keller, J. B. (1988) Approximations for regression with covariate measurement error. *J. Am. Statist. Ass.*, **83**, 1057–1066.

Willett W. (1989) An overview of issues related to the correction of non-differential exposure measurement error in epidemiologic studies. *Statist. Med.*, **8**, 1031–1040.

23

Gibbs sampling methods in genetics

Duncan C Thomas

W James Gauderman

23.1 Introduction

The essential feature of genetic studies is that they involve related individuals. Relationships between individuals within a family produce correlations in their outcomes that render standard analysis methods of epidemiology (which assume independence) inappropriate. This has motivated much of the recent work on methods of analysis for correlated outcomes. To the geneticist, however, these correlations are viewed not as a nuisance, but rather as the central piece of information for determining whether genes are involved. Indeed, it is the pattern of correlations between different types of relatives that must be examined, so that methods such as frailty models (Clayton, 1991), which assume a common degree of correlation between all members of a family, are not helpful. In this chapter, we discuss the standard likelihood approach to analysis of genetic data and show how Gibbs sampling can help overcome some of the formidable computational problems.

23.2 Standard methods in genetics

23.2.1 Genetic terminology

We begin with a brief introduction to the genetic terminology that will be used throughout this chapter. Mendel, in his First Law in 1866, proposed that discrete entities, now called *genes*, are passed from parents to offspring. Genes occur in pairs and are located on strands of DNA called *chromosomes*. A normal human being has 22 paired chromosomes and two unpaired sex chromosomes.

Mendelian genes can occur in different forms, or *alleles*. If there are only two possible alleles, as is commonly assumed for unobservable disease genes, we say the gene is *diallelic* and typically represent the forms by the letters a and A. A *genotype* consists of a pair of alleles, one inherited from the mother and one from the father. A person is *homozygous* if his genotype is aa or AA and is *heterozygous* if his genotype is aA. The observable expression of a genotype is called the *phenotype*. For example, the genotypes aa, aA, and AA may correspond to the phenotypes: unaffected, affected, and affected, respectively. This is an example of a *dominant trait*, in which an individual needs only one copy of the disease susceptibility allele (A) to become affected. If two copies of A are needed to become affected, the trait is called *recessive* while if each genotype produces a different phenotype, the trait is *codominant*. A *fully penetrant* trait is one whose expression is completely determined by the genotype at a particular chromosomal position or *locus*. Many traits are affected not only by genotype but also by other factors, such as age or environmental exposures, and are called *partially penetrant*. Traits may also depend on a *polygene*, i.e. the cumulative effect of many genes, each with a small influence on the trait.

The population frequencies of each genotype are determined by the population allele frequencies. If the frequency of allele A is q, then under 'Hardy–Weinberg equilibrium' the population frequencies of genotypes AA, aA, and aa are q^2, $2q(1 - q)$, and $(1 - q)^2$, respectively. However, the probability distribution of a given individual's genotype is determined solely by the genotypes of his parents. Under mendelian inheritance, each parent passes to each of his offspring one of his two alleles with probability $\frac{1}{2}$. Thus, for example, if the mother has genotype aa and the father aA, their children will have either genotype aa or aA, each with probability $\frac{1}{2}$. The population distribution of a polygene, relying on the Central Limit Theorem, is normal; the polygene distribution for a given individual is also normal with mean and variance depending on parental values.

Linkage analysis is used to determine the chromosomal position of a disease gene in relation to one or more marker genes at known locations. Consider a disease locus with alleles a and A, and a second (marker) locus with alleles b and B; there are then ten possible joint genotypes: $ab|ab$, $ab|aB$, $aB|aB$, $ab|Ab$, $ab|AB$, $aB|Ab$, $aB|AB$, $Ab|Ab$, $Ab|AB$, $AB|AB$. Here the vertical bar separates alleles appearing on different chromosomal strands, so for example $aB|Ab$ means that aB appears on one strand, and Ab appears on the other. During meiosis (a stage of sperm and egg development), the chromosomal strands can crossover, break, and recombine. For example, if a parent with genotype $ab|AB$ passes alleles aB to an offspring, a *recombination* between the disease and marker locus has occurred. The probability of recombination, θ, is roughly proportional to the genetic distance between the two loci. Thus, if the disease and marker locus are close (tightly linked), θ will be close to zero, while if the two loci are far apart,

θ will be close to $\frac{1}{2}$.

23.2.2 Genetic models

Let y_i denote the phenotype for subject $i = 1, \ldots, I$, which may depend on measured risk factors x_i, and unobserved major mendelian genes G_i, polygenes z_i, or other shared factors. Letting P denote a probability, a genetic model is specified in terms of two submodels, (1) a penetrance model $P(y|G, x, \Omega)$, specifying the relationship between phenotypes and genotypes, and (2) a genotype model $P(G, z|\Theta)$, describing the distribution of genotypes and polygenes within families, where Ω and Θ are vectors of parameters to be estimated. In *segregation analysis*, one aims to determine whether the pattern of disease in families is consistent with one or more major genes, polygenes, or other mechanisms (e.g. environmental influences) and to estimate the parameters of the model using only information on the distribution of phenotypes in the families and on the family structures. If there is evidence of a major gene, linkage analysis is then used to identify its chromosomal location in relation to one or more genetic markers $m^{(\ell)}$ at loci $\ell = 1, \ldots, L$.

In the penetrance model, we assume that the phenotypes y_i of individuals are conditionally independent given their genotypes, i.e., that the modelled genotypes fully account for any dependencies within families. The particular form of model will depend on the nature of the outcome y under study, which might be continuous, discrete, or a censored survival time. In this chapter, we will limit attention to late-onset disease traits, characterized by a dichotomous disease status indicator d (1=affected, 0=not affected) and an age variable t, which is the age of disease onset if $d=1$ or the last known unaffected age if $d=0$. In this case, we model penetrance in terms of the hazard function (age-specific incidence rate) $\lambda(t)$. The penetrance for an unaffected individual is the probability of surviving to age t free of the disease, $S(t) = \exp[-\Lambda(t)]$ where $\Lambda(t) = \int_0^t \lambda(u)\, du$, and the penetrance for an affected individual is the density function $\lambda(t)S(t)$, evaluated at the observed age at onset. One model that might be considered is the proportional hazards model

$$\lambda(t, G, x, z) = \lambda_0(t) \exp\{\beta x + \gamma.\mathrm{dom}(G) + \eta z + \ldots\} \qquad (23.1)$$

where $\lambda_0(t)$ is the baseline disease risk function, β, γ, and η are regression coefficients to be estimated, and '...' indicates the possibility of including additional age- or gene-environment interaction terms. The function $\mathrm{dom}(G)$ translates the genotype G into a value in the range $[0, 1]$ depending on the assumed mode of inheritance. For example, in a dominant model $\mathrm{dom}(G)=0$ for $G = aa$ and $\mathrm{dom}(G)=1$ for $G = aA, AA$. For notational simplicity, we let phenotype y denote the pair (d, t) and Ω denote the entire set of penetrance parameters $(\lambda_0(.), \beta, \gamma, \eta, \ldots)$.

The genotype model is decomposed into a series of univariate models, one for each individual. Based on the way genetic information is passed from parents to offspring, subjects may be ordered in such a way that their genotype depends only on the genotypes of individuals appearing previously; thus

$$P(G|\Theta) = P(G_1|\Theta) \prod_{i=2}^{I} P(G_i|G_1, \ldots, G_{i-1}; \Theta). \qquad (23.2)$$

Although the genotype model may include alternative forms of dependency among family members, we will restrict attention to major mendelian genes. The factors in the product consist of marginal genotype probabilities for subjects without parents in the data (founders), and parent-to-offspring genotype transmission probabilities for nonfounders.

As described in Section 23.2.1, the marginal (population) distribution of a major disease gene depends only on q, the frequency of the high risk allele A. The probability that a parent with genotype $G = aa$, aA or AA transmits allele A to an offspring is $\tau_G = 0$, $\frac{1}{2}$or 1, respectively. Thus, for a major gene model, the factors of (23.2) have the form

$$P(G_i|G_1, \ldots, G_{i-1}; q, \tau) = \begin{cases} P(G_i|q) & \text{for founders} \\ P(G_i|G_{M_i}, G_{F_i}; \tau) & \text{for nonfounders} \end{cases},$$
$$(23.3)$$

where M_i denotes the mother of individual i, F_i denotes the father, and τ denotes the set of transmission probabilities. Allowing the τ to be free (instead of 0, $\frac{1}{2}$ or 1) provides a general alternative model to strict mendelian inheritance, which still allows for familial aggregation based on some unobservable discrete characteristic, known as an *ousiotype*. In linkage analysis, G_i will denote the joint genotype at the disease and marker loci. The parameters Θ include the allele frequencies q_ℓ at each locus ℓ, transmission parameters τ, and recombination fractions θ_ℓ between loci ℓ and $\ell + 1$.

The decomposition of the joint distribution (23.2) could be extended to include polygenes. As mentioned in Section 23.2.1, polygenes are normally distributed in the population, and without loss of generality can be treated as having zero mean and unit variance. Since half of one's polygene is inherited from the mother and the other half from the father, it follows that the conditional distribution of z_i is normal with mean that depends on parental values, and so (23.2) for a polygenic model has the form

$$P(z_i|z_1, \ldots, z_{i-1}) = \begin{cases} N(0, 1) & \text{for founders} \\ N\left(\frac{z_{M_i} + z_{F_i}}{2}, \frac{1}{2}\right) & \text{for nonfounders} \end{cases}, \qquad (23.4)$$

(where $N(a, b)$ denotes a normal distribution with mean a and variance b). Hence, in a pure polygenic model there are no parameters in Θ to be estimated.

23.2.3 Genetic likelihoods

A genetic likelihood is formed by summing (and/or integrating) over all possible combinations of genotypes (and/or polygenes) in each family. Assuming a single major-gene model, the likelihood is given by:

$$
\mathcal{L}(y; \Omega, \Theta) \;=\; \sum_G P(y|G; \Omega) P(G; \Theta)
$$

$$
\;=\; \sum_G \prod_{i=1}^{I} P(y_i|G_i; \Omega)\, P(G_i|G_1, \ldots, G_{i-1}; \Theta), \quad (23.5)
$$

where individuals are assumed to be numbered from the top of the pedigree.

The summation in (23.5) involves 3^I terms, although many of these terms will have zero probability under mendelian laws. Systematic enumeration of all these possibilities would be impossible for moderate sized pedigrees, even on modern high-speed computers. The standard approach to calculation of this likelihood involves a recursive algorithm due to Elston and Stewart (1971), known as *peeling*. Starting at the bottom of the pedigree, one successively stores $P(y_{i+1}, \ldots, y_I|G_i)$ and uses these stored values in the next step of the recursion to compute

$$
P(y_i, \ldots, y_I|G_1, \ldots, G_{i-1}) =
$$
$$
\sum_{G_i} P(y_i|G_i)\, P(G_i|G_1, \ldots, G_{i-1})\, P(y_{i+1}, \ldots, y_I|G_i). \quad (23.6)
$$

In practice, this term is evaluated for persons who connect two subpedigrees, i.e. persons who are both an offspring and a parent in the pedigree. The final summation (over G_1) is over one of the founders at the top of the pedigree and produces the likelihood value. Although the Elston–Stewart algorithm makes it feasible to compute likelihoods for even very large pedigrees, it is still quite time-consuming and certain restrictions apply. Exact calculations are not possible for pedigree structures with a high degree of inbreeding, and for certain types of models, particularly mixed major gene/polygene models and linkage models with multi-allelic markers. Faced with these difficulties, there is considerable interest in Gibbs sampling and other MCMC methods as a means of facilitating these calculations by sampling from the domain of possible genotypes rather than systematically enumerating all possibilities.

The likelihood given in (23.5) applies only to families sampled at random from the population. For studying rare diseases, families are generally ascertained through *affected probands*, i.e. persons with disease (*cases*) are sampled from the population and their respective family members are then included in the sample. In this case, the likelihood conditional on ascertainment must be computed,

$$\mathcal{L}(y; \Omega, \Theta, \text{Asc}) = \frac{P(y, \text{Asc}|\Omega, \Theta)}{P(\text{Asc}|\Omega, \Theta)} = \frac{P(\text{Asc}|y)\mathcal{L}(y; \Omega, \Theta)}{\sum_y P(\text{Asc}|y)\mathcal{L}(y; \Omega, \Theta)} \qquad (23.7)$$

where 'Asc' denotes the event that the observed families were ascertained.

In general, the denominator of (23.7) involves a daunting double summation over all possible combinations of both phenotypes and genotypes, but simplifications are possible in special situations. In *simplex single ascertainment*, the probability that any case in the population is a proband is small, in which case the denominator is proportional to

$$\prod_{\text{probands}} P(d_i = 1) = \prod_{\text{probands}} \sum_{G_i} P(d_i = 1|G_i; \Omega)P(G_i; \Theta),$$

(Cannings and Thompson, 1977). In *simplex complete ascertainment*, the probability that a case is a proband equals one, so all families with at least one case are ascertained. For this scheme, the denominator becomes

$$\prod_{\text{families}} \left\{ 1 - \prod_i P(d_i = 0) \right\},$$

where the inner product is over members within a family.

Because of the complexity of the calculations, parameter estimation is particularly daunting, involving repeated evaluation of the likelihood at various points of the parameter space. For this reason, geneticists tend to adopt relatively simple penetrance models involving a small number of free parameters, typically ignoring environmental risk factors and using simple parametric models for age at onset. Furthermore, any correction other than for single ascertainment can nearly double the required computing time. These problems are accentuated in linkage analysis, owing to the far greater number of possible genotypes to be considered in the summation in (23.5). The standard practice in linkage analysis is to fix all parameters, except the recombination fraction θ, at values obtained from previous segregation analyses, and treat them as known with certainty. This approach is somewhat justified by the approximate orthogonality between θ and the other parameters, but an analysis that allows for the uncertainty in these parameters would be more appropriate. In this chapter, we discuss how Gibbs sampling can also be used for parameter estimation, either from a likelihood or a Bayesian perspective.

23.3 Gibbs sampling approaches

We first consider Gibbs sampling of genotypes with a fixed set of parameters and show how this approach can be used in approximating the likelihood, first for segregation analysis with major genes and polygenes, and then for linkage analysis. We then consider the use of Gibbs sampling for parameter estimation.

23.3.1 Gibbs sampling of genotypes

Segregation analysis

Figure 23.1 illustrates the directed conditional independence graph for a typical pedigree. The genotypes of the founders at the top of the graph are determined by the population allele frequency q. The genotype of individual i, shown in the middle of the pedigree, depends on his parents' genotypes and on inheritance parameters θ, τ. The genotype of individual i, together with the genotype of his spouse S_{ij}, determine the genotypes of his J_i offspring $O_{ij}, j = 1, \ldots, J_i$; where $J_i = 3$ in Figure 23.1. His phenotype y_i is determined by his genotype, together with the penetrance model parameters Ω. (Figure 23.1 also shows a possible further dependence of the phenotype on observed risk factors x_i and unobserved polygene z_i.) The full conditional probability of individual i's genotype can therefore be decomposed by repeated application of Bayes formula as

$$P(G_i|G_{-i}, y_i; \Omega, \Theta) \quad \propto \quad P(y_i|G_i; \Omega)\, P(G_i|G_{M_i}, G_{F_i}; \Theta)$$
$$\times \prod_{j=1}^{J_i} P(G_{O_{ij}}|G_i, G_{S_{ij}}; \Theta) \qquad (23.8)$$

where G_{-i} consists of the genotypes for all subjects except i. See Spiegelhalter *et al.* (1995: this volume) for further discussion of full conditional distributions arising from directed conditional independence graphs. The second factor in (23.8) will take one of the two forms given in each of (23.3) or (23.4), depending on whether the subject's parents are in the dataset.

This basic formulation of the full conditional (23.8) also applies if major gene G_i is replaced by polygene z_i. In mixed models, one could consider each component one at a time, updating the major gene conditional on the current assignment of polygenes and then *vice versa*. For a major gene with two alleles, one would evaluate (23.8) for $G_i = aa, aA, AA$ and then randomly pick one of these choices with the corresponding probabilities. In this manner, one continues through the entire pedigree to complete one scan, each individual being updated conditional on the current assignment of genotypes for all other pedigree members (only parents, offspring, and spouses being relevant). Since the calculations for each individual are extremely simple, it is possible to complete a large number of scans in a small amount of computer time, depending only linearly on the size of the dataset. However, because individuals are updated conditional on the current assignment of their neighbours, the resulting Markov chain from such 'person-by-person' Gibbs sampling can have high autocorrela-

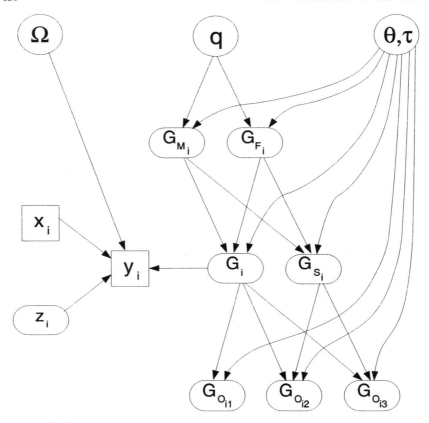

Figure 23.1 *Directed conditional independence graph for pedigree data: boxes represent observed phenotypes, ovals represent genotypes, and circles represent parameters.*

tion, with larger pedigrees requiring more Gibbs samples to adequately explore the genotype space. Nevertheless, the Markov chain is irreducible (see Roberts, 1995: this volume) under very weak conditions that are met in any segregation analysis of practical importance, the crucial factor being the assumption that there are only two alleles at the disease locus (Sheehan and Thomas, 1992). Alternatives involving simultaneous updating of entire genotype vectors will be described below and issues of convergence will be discussed further in Section 23.3.3.

For polygenes, one is faced with the problem of sampling from the continuous distribution

$$P(z_i|y_i, z_{-i}) \propto P(y_i|z_i)P(z_i|z_{-i}). \qquad (23.9)$$

The first factor can be approximated by $\exp[\eta z_i(d_i - \Lambda_i)]$ where $\Lambda_i =$

$\Lambda(t_i)/\exp(\eta z_i)$, using (23.1). Since the polygene of subject i depends only on the polygenes of his parents, spouse(s), and offspring, the second factor of (23.9) can be written

$$P(z_i|z_{-i}) \propto P(z_i|z_1, \ldots, z_{i-1}) \prod_{j=1}^{J_i} P(z_{O_{ij}}|z_i, z_{S_{ij}}). \qquad (23.10)$$

Each of these factors is normal with form given in (23.4), and so

$$P(z_i|z_{-i}) = \begin{cases} N\left(\dfrac{\sum_j (z_{O_{ij}} - \frac{1}{2}z_{S_{ij}})}{(1+\frac{1}{2}J_i)}, \dfrac{1}{(1+\frac{1}{2}J_i)}\right) & \text{for founders} \\[4mm] N\left(\dfrac{z_{M_i}+z_{F_i}+\sum_j (z_{O_{ij}} - \frac{1}{2}z_{S_{ij}})}{(2+\frac{1}{2}J_i)}, \dfrac{1}{(2+\frac{1}{2}J_i)}\right) & \text{for nonfounders} \end{cases}$$

$$(23.11)$$

The full conditional in (23.9) is thus approximately normal, with mean $\bar{z}_i + \eta(d_i - \Lambda_i)v_i$ and variance v_i, where $\bar{z}_i$ and v_i are the mean and variance of the normal distributions in (23.11). This is an adequate approximation provided the disease is rare; otherwise, it can be used as a proposal density in a Metropolis–Hastings step, or adaptive rejection sampling methods (Gilks, 1992) can be used; see Gilks (1995: this volume).

Linkage analysis

The same basic approach to sampling genotypes can also be applied in linkage analysis. Thomas and Cortessis (1992) describe the necessary extension to (23.8) for sampling the joint disease and marker genotype when the marker locus has only two possible alleles. For each of the ten joint genotypes (see Section 23.2.1), the full conditional probability is computed using (23.8), where the vector of genotype parameters Θ now includes the recombination fraction θ. The resulting multinomial is easily sampled.

With more than two alleles at the marker locus or with multiple marker loci, this approach can quickly become unwieldy in terms of both computer time and array storage. Two alternatives are to sample locus-by-locus rather than jointly and to sample grandparental sources rather than joint genotypes; the two techniques can also be combined (Wang and Thomas, 1994). Locus-by-locus sampling proceeds as described above for mixed models: first one updates disease genotypes conditional on the current assignment of marker genotypes; then one updates each marker locus, conditional on the disease locus and any other marker loci. At each cycle, one could do a complete scan across individuals one locus at a time, or one could update each locus in turn one individual at a time. Either way, the amount of computing is reduced from $\prod N_\ell^2$ genotypes per person to $\sum N_\ell^2$, where N_ℓ is the number of alleles at locus ℓ. Grandparental source sampling works by distinguishing which grandparent was the source of each allele in a person's genotype. Let $s_\ell = (s_\ell^M, s_\ell^F)$ denote a pair of indicator variables each taking the values M or F depending on whether the allele inherited from mother or father derived from the corresponding grandmother or grandfather. There are thus only 4^L possible values s can take across all L loci jointly, or $4L$ possibilities to be considered if one proceeds locus-by-locus. Sampling by grandparental source constitutes a reparameterization of the genotype model so that sources, not genotypes, are the variables of interest. The full conditional probabilities for the sources are a simple function of the recombination fractions and the penetrances for the corresponding genotypes. One therefore samples a vector of grandparental sources, and then assigns the genotypes accordingly.

In addition to these computational issues, a new conceptual problem arises in linkage analysis with marker loci that have more than two alleles and some untyped individuals (persons for whom marker genotypes are not available): there is no longer any guarantee that the resulting Markov chain is irreducible. To see this, consider the ABO blood group locus and suppose two offspring are typed as AB and OO, and the two parents are untyped. Given the offsprings' genotypes, one parent must have genotype AO and one must have BO. If the mother is initially assigned AO, then the father must be assigned BO and on the next iteration, the mother can only be reassigned to AO. Although the reverse ordering is a legal state (i.e. has positive probability), it can never be reached using person-by-person Gibbs sampling, and so the two possible configurations represent 'islands' in the genotype space, and the Markov chain is reducible. Since some islands may support linkage to the disease locus and others not, one would be led to incorrect inferences about the posterior distribution of the recombination fraction θ.

Several approaches have been proposed to deal with this problem. One is to dispense with person-by-person Gibbs sampling in favour of updating groups of individuals jointly. Updating marriages jointly would solve the simple example described above, but more complex counter-examples can be constructed. Sampling all genotypes jointly for each family would guarantee irreducibility in any pedigree structure, but this requires use of the peeling algorithm which we are trying to avoid. The utility of this approach arises in Gibbs sampling of parameters, discussed in Section 23.3.2.

One could run a large number of Gibbs samplers independently, each one starting from a randomly selected configuration, and then average their values. There are two difficulties with this approach. First, there may be a large number of islands and without careful analysis of each pedigree, one would have no idea how many series would be needed to represent adequately all the possibilities. More seriously, not all islands are equally likely and there is no obvious way to compute these probabilities for weighting the different series. A promising approach for initializing genotypes (Wang, 1993) involves peeling one locus at a time, followed by random selection of genotypes at that locus using the peeled probabilities, each peeling being conditional on the genotypes selected at the previous locus. In this way, initial configurations can be selected that will at least approximately represent the marginal island probabilities.

Sheehan and Thomas (1992) suggested 'relaxing' the marker penetrances (replacing zero penetrance probabilities by $\rho > 0$) in such a way that the resulting Markov chain would be irreducible. The resulting sample would then include genotypes that were incompatible with the observed marker phenotypes. Such configurations would simply not be included in tabulating the posterior distribution. Sheehan and Thomas show that the resulting Markov process retains the equilibrium distribution of the correctly specified model. They also discuss the trade-off in the selection of the relaxation parameter between the proportion of samples that must be rejected and the ease with which the process is able to jump from one island to another. A variant of this approach would be to relax the transmission probabilities (τ) rather than the penetrances; to date, no comparison of the relative efficiency of the two approaches has been reported. See Gilks and Roberts (1995: this volume) for further discussion of this method.

The most promising approach to this problem incorporates both of the previous ideas in an algorithm known as *Metropolis-coupled Markov chain* (Lin, 1993; Geyer, 1992). Several Gibbs samplers are run in parallel, one constrained to the true model (but therefore not necessarily irreducible), the others containing relaxed penetrances or transmission probabilities which guarantee irreducibility (without necessarily generating legal states). According to a predetermined schedule, one considers two of the chains, r and s, and proposes swapping the current states of their genotype vectors $G^{(r)}$ and $G^{(s)}$, accepting the swap with probability given by the Hastings (1970) odds ratio,

$$p_{rs} = \min\left[1, \frac{P(G^{(r)}|\Theta^{(s)})\,P(G^{(s)}|\Theta^{(r)})}{P(G^{(r)}|\Theta^{(r)})\,P(G^{(s)}|\Theta^{(s)})}\right]. \tag{23.12}$$

The real chain thereby becomes irreducible and retains its equilibrium distribution under the true model. Multiple chains with a range of relaxations can be effective in speeding up the mixing across islands. See Gilks and Roberts (1995: this volume) for further discussion.

23.3.2 Gibbs sampling of parameters

In addition to (or instead of) sampling genotypes, Gibbs sampling can be used to sample model parameters. In this case, one treats the current assignment of genotypes as known and samples model parameters from their full conditionals, which are much simpler than the likelihood given in (23.5) because they do not involve summation over all possible genotypes. The parameters of a genetic model fall naturally into two groups, those in the penetrance model and those in the genotype model.

Consider first the parameters $\Omega = (\lambda_0(.), \beta, \gamma, \eta)$ of the penetrance model (23.1). The full conditional is proportional to $P(\Omega)P(y|G, z, x; \Omega)$, i.e. the product of a prior distribution and the likelihood conditional on the assigned genotypes and polygenes (and ascertainment). Again, one can procede component by component, updating the regression coefficients β, γ and η (jointly or singly) conditional on $\lambda_0(.)$ and then vice-versa. In moderate sized datasets, the regression coefficients have approximately normal conditional likelihoods. One might simply adopt independent flat priors and sample from the normal density with mean the MLE and variance the inverse of the observed information. Alternatively, some form of rejection sampling can also be used if there is any concern about the adequacy of the asymptotic normal approximation. Although finding the MLE may require several Newton–Raphson iterations, a single iteration will provide an adequate approximation for use as a guess density in rejection methods. Updating of the baseline hazard $\lambda_0(.)$ can be done by treating it as an independent increments gamma process, as described by Clayton (1991), i.e. sampling an independent gamma deviate at each time interval k with shape parameter given by the number of events D_k in that interval and scale parameter $1/\sum_i t_{ik} \exp(\beta x_i + \gamma.\mathrm{dom}(G_i) + \eta z_i)$ where t_{ik} is the length of time subject i was at risk during interval k.

The parameters $\Theta = (q, \theta)$ of the genotype model (23.2) depend on the configuration of alleles in the current assignment of genotypes. For the disease allele frequency q, let F_A denote the number of A alleles among the F founders. The likelihood for F_A is Binomial$(2F, q)$, and so the full conditional for q is a beta distribution, assuming a beta prior. The prior can be either be informative or uninformative, as described by Gauderman and Thomas (1994). Updating allele frequencies for polymorphic markers (i.e. markers with more than two alleles) follows similar principles using multinomial and Dirichlet distributions. For θ, the sufficient statistics are the number of recombination events (R) among H doubly heterozygous parents; R is then distributed as Binomial(H, θ). The full conditional distribution for θ is a beta distribution proportional to this binomial likelihood, if an uninformative prior is chosen. However, one may wish to choose an informative beta prior truncated to the range $(0, 0.5)$, as described by Faucett et al. (1993), in which case the full conditional will be a beta distribution truncated on the same range.

As described in Section 23.2.3, valid parameter estimation requires allowance for the ascertainment scheme. Gibbs sampling greatly simplifies this process by treating the genotypes as known when updating the parameters. Hence, one need consider only the probability of ascertainment as a function of model parameters, summing over possible phenotypes but not over genotypes. Under simplex single ascertainment, sampling allele frequencies is achieved by simply subtracting the number of A alleles in probands from F_A, reducing F by the number of probands, and proceeding as described above. For sampling penetrance parameters, one can simply omit the probands. Under complete ascertainment, the proper correction is to divide the sampling distributions described above by

$$\prod_{\text{families}} \left[1 - \prod_i P(d_i = 0|G_i) \right],$$

where the inner product is over subjects within a family. A more complex ascertainment scheme is described in the example in Section 23.5.

One can combine Gibbs sampling of genotypes with Gibbs sampling of parameters in the obvious fashion. There is no need to do the two operations in sequence, however. For optimal efficiency, the more highly autocorrelated series should be updated more often. Since sampling of genotypes often requires much less computation and since the genotype series tends to be highly autocorrelated, one might perform 10 genotype updates for each parameter update. Alternatively, Kong *et al.* (1992) suggested sampling the genotype vector jointly at each iteration using the peeling algorithm (Ploughman and Boehnke, 1989), and then Gibbs sampling new parameter values conditional on the assigned genotypes. In this way, the realizations of the genotypes are independent of previous genotype assignments, conditional on the parameter values. Also, the supremum of the plot of the peeled likelihoods against the corresponding values of any particular parameter provides an estimate of the profile likelihood. In contrast, Thompson and Guo (1991) fix the parameter vector to some value, say (Ω_0, Θ_0), and Gibbs sample only over genotypes. They periodically evaluate the likelihood over a grid of parameter values and, by averaging over Gibbs samples, obtain an estimate of the likelihood ratio function relative to (Ω_0, Θ_0). This approach is described in further detail in Section 23.4.

23.3.3 Initialization, convergence, and fine tuning

For an irreducible Markov chain, the Gibbs sampler is assured of converging to its equilibrium distribution no matter where it is started, but convergence time will be greatly shortened by starting at a reasonable place. This is more of an issue for sampling genotypes than for sampling parameters. After choosing initial values for the parameters, a technique known as 'gene dropping' is used to assign initial genotypes. This process entails first drawing a random selection of genotypes for founders from $P(G_i|y_i, q)$ and then proceeding down through the pedigree, sampling from $P(G_i|y_i, G_{M_i}, G_{F_i})$ for nonfounders. However, for pedigrees with missing marker data at the upper generations, even arriving at a legal assignment can be difficult. Two approaches that have been helpful are to incorporate peeled probabilities in gene dropping or to use relaxed penetrances, starting all subjects as heterozygotes and running Gibbs sampling until the chain becomes legal

(Wang and Thomas, 1994). Since model parameters will typically be used in initializing genotypes, convergence to likely genotype configurations is expedited by an initial selection of parameters that is close to the MLE.

Having achieved a legal initial configuration, the Gibbs sampler must be allowed to run for some time before the equilibium distribution is attained, and then for enough subsequent iterations to characterize adequately that distribution. The number of iterations required for convergence depends on the complexity of the problem and can range from under 100 to several thousand. The multiple chains approach of Gelman and Rubin (1992) has been particularly useful for monitoring convergence. Diagnostics for assessing convergence are discussed in more detail in Raftery and Lewis (1995) and Gelman (1995) in this volume.

Existence of local maxima is major determinant of the number of iterations required. Even if the Markov chain is irreducible, there can exist genotype configurations that may be quite unlikely, but once sampled can persist for a long time. An example might be a situation where there are no unaffected carriers of the disease allele or no affected non-carriers: the genetic relative risk parameter γ then diverges to infinity and the baseline risk $\lambda_0(t)$ to zero, thereby perpetuating this situation in subsequent Gibbs iterations. A promising approach to speeding up movement around the parameter space is the *heated Metropolis algorithm* (Lin, 1993). Letting $f(g)$ denote the true conditional distribution which one would use in Gibbs sampling, one selects a new genotype g' from the heated distribution $h(g) \propto f(g)^{1/T}$ where $T > 1$, and then either accepts g' with probability

$$\min\left\{1, \left[\frac{f(g')}{f(g)}\right]^{1-\frac{1}{T}}\right\}$$

or retains the previous assignment g. This device facilitates large movements across the genotype space that would be quite unlikely by ordinary Gibbs sampling, while maintaining the equilibrium distribution; see Gilks and Roberts (1995: this volume).

23.4 MCMC maximum likelihood

Although Gibbs sampling is more directly applied in a Bayesian setting, it can be used by the non-Bayesian as a device to evaluate complex likelihoods by sampling (see Geyer, 1995: this volume). In this section, we explore two basic approaches to this idea.

As described above, Thompson and Guo (1991) perform Gibbs sampling over genotypes with a fixed set of parameter values. To illustrate the ideas, consider a single parameter ω and let ω_0 denote the null value. Then the likelihood can be written as

$$
\begin{aligned}
\mathcal{L}(\omega) &= P(y|\omega) = \sum_G P(y, G|\omega) \\
&= \sum_G \frac{P(y, G|\omega)}{P(y, G|\omega_0)} P(G|y, \omega_0) P(y|\omega_0) \\
&= \mathcal{L}(\omega_0) \sum_G LR_G(\omega : \omega_0) P(G|y, \omega_0) \qquad (23.13)
\end{aligned}
$$

where $LR_G(\omega : \omega_0)$ denotes the contribution of genotype vector G to the likelihood ratio $LR(\omega : \omega_0)$. Gibbs sampling at the fixed parameter ω_0 generates K random samples from the marginal distribution of G given y and ω_0, so one can estimate the likelihood ratio by a simple arithmetic mean of the sampled likelihood ratio contributions,

$$
\hat{LR}(\omega : \omega_0) = \frac{1}{K} \sum_{k=1}^{K} LR_{G^{(k)}}(\omega : \omega_0), \qquad (23.14)
$$

where $G^{(k)}$ denotes the value of G at the kth iteration of the Gibbs sampler. Thus, for each $k = 1, \ldots, K$, one evaluates the genotype-specific likelihood ratio contributions over a grid of ω values and accumulates their sums. Since this evaluation may be computationally more intense than generating the genotype samples, and since the genotype samples will tend to be highly autocorrelated, it may be more computationally efficient to do this only for a subset of the sampled genotypes.

This procedure provides an efficient estimate of the likelihood function for ω in the neighbourhood of ω_0, which can be used for constructing score tests. However, it will be increasingly unstable for distant values of ω, since genotype configurations which are likely under distant values are unlikely to be sampled under ω_0. To remedy this, one might run several independent Gibbs samplers over a range of values for ω_0 spanning the region of interest and then take a weighted average of the estimates. Geyer (1992) has described a method for doing this, called *reverse logistic regression*; see Gilks and Roberts (1995: this volume).

An alternative is to sample genotypes and parameters jointly, as described above, using flat priors. One can then estimate the posterior distribution of parameters as

$$
P(\omega|y) = \sum_G P(\omega|G, y) P(G|y). \qquad (23.15)
$$

But joint Gibbs sampling of genotypes and parameters generates G from its marginal distribution given only y. Hence, one can estimate the posterior distribution of ω by a simple arithmetic mean of the conditional distributions,

$$\hat{P}(\omega|y) = \frac{1}{K} \sum_{k=1}^{K} P(\omega|G^{(k)}, y) \qquad (23.16)$$

as described by Gelfand and Smith (1990). By using flat priors on the parameters, the posterior density is then proportional to the likelihood. This approach will provide an efficient estimate of the likelihood in the vicinity of the MLE, and is preferred for finding point and interval estimates and for Wald tests. However, if the MLE is far from the null value ω_0, this approach may produce inefficient likelihood ratio tests of the null hypothesis.

The latter approach for estimating likelihoods can be combined with importance sampling methods by using an informative prior to focus the sampling over the entire range of parameter values of interest. Suppose one adopts $h(\omega)$ as the prior distribution, and uses (23.16) to obtain the posterior distribution. The likelihood function can then be recovered as

$$\hat{L}(\omega) = \frac{1}{K} \frac{1}{h(\omega)} \sum_{k=1}^{K} P(\omega|G^{(k)}, y). \qquad (23.17)$$

The challenge is then to choose $h(\omega)$ so as to minimize the Monte Carlo variance of $\hat{L}(\omega)$.

23.5 Application to a family study of breast cancer

Epidemiologic studies have long recognized family history as a risk factor for breast cancer, with familial relative risks generally in the 2–3 fold range, but apparently stronger at young ages. Segregation analyses have demonstrated that a dominant major gene with an estimated allele frequency of 0.3% may be responsible (Claus et al., 1991), and linkage analyses have provided strong evidence of linkage to genes on chromosome 17q (Hall et al., 1992; Easton et al., 1993). The genetic relative risk appears to be strongly age dependent, declining from 98 at young ages to 5 at older ages (Claus et al., 1991). Nongenetic factors are also involved, including reproductive history, radiation, benign breast disease, diet, and exogenous estrogens. There may also be important gene-environment interactions.

To investigate the joint effects of genetic and environmental factors, a population-based family study has been conducted by Dr Robert Haile at UCLA (Goldstein et al., 1987, 1989). This involves ascertainment of premenopausal, bilateral breast-cancer cases from population-based cancer registries in Los Angeles County and Connecticut, and from large hospitals in Montreal and Quebec City that ascertain 95% of all cases in southern Quebec. Each proband was sent a family history questionnaire. Risk-factor and diet-history questionnaires were then sent to the proband and all surviving first-degree female relatives. Families with one or more cases in close

relatives of the proband were extended and blood samples and questionnaire data were obtained from all surviving members of these families that
were potentially informative for linkage. By October 1992, 464 probands
had been ascertained and 70 families totalling 964 individuals had two or
more cases. The latter set of *multiplex* families have been analysed by the
Gibbs sampling approach to segregation analysis (Thomas *et al.*, 1993), using an ascertainment correction based on single ascertainment of probands
followed by complete ascertainment of multiple case families. Marker data
for the CMM86 locus at 17q21 were analysed by standard methods (Haile
et al., 1993) and Gibbs sampling methods (Wang, 1993) for a subset of 35
of the most informative families for whom blood samples were available.

The proportional hazards model (23.1) was used initially, including as
covariates education, benign breast disease, use of oral contraceptives, and
two indicator variables for data-collection center. Major-gene and polygenic models were considered, separately and in combination. There was
no evidence for a polygenic component, but strong evidence of a major
gene. Dominant, codominant, and recessive modes of inheritance could not
be distinguished, but a dominant model was used for subsequent analyses,
based on previous literature. The estimated genetic relative risk, e^γ, was 20
(i.e. carriers of the disease allele have 20 times the risk of developing breast
cancer compared to noncarriers) with 95% confidence limits from 8 to 98,
and the estimated gene frequency, q, was 1.2% with 95% confidence limits
from 0.4% to 2.2%. In all analyses, samples were obtained from the full conditional distributions of model parameters and genotypes as described in
Section 23.3. Uniform prior distributions were utilized so that the posterior
distribution was proportional to the likelihood.

The familial relative risk, comparing the cohort of family members (subtracting the two cases required for ascertainment) to population rates
shows a marked decline with age (Figure 23.2), suggesting that the proportional hazards model may be inappropriate. As an alternative, an age-
dependent genetic relative risk model was fitted, replacing γ by γ_k for
five-year age intervals, and using the population rates $\lambda^*(t)$ as constraints
(i.e. $\lambda^*(t) = \sum_G \lambda_0(t) \exp(\gamma.\mathrm{dom}(G)) P(G|q)$). Even with this constraint,
the γ_k were quite unstable, owing to the possibility of genotype vectors
with no unaffected gene carriers or no affected noncarriers in certain age
intervals. To overcome this problem, the γ_k were smoothed with an intrinsic autoregressive process (Besag, 1974). The resulting smoothed curve is
shown in Figure 23.3, declining from about 200 at age 20 to about 10 above
age 65.

A recurring problem with family data is missing data, particularly for
family members who are long dead. In epidemiology, the conventional approach is to omit such individuals, assuming the data are missing completely at random (MCAR). However, even if the MCAR assumption were
plausible, this approach would not be possible in genetic analysis, because

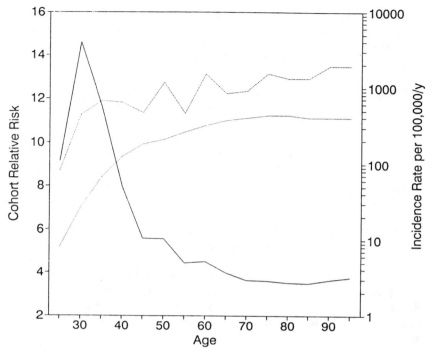

Figure 23.2 *Age-dependent familial risk in the UCLA breast cancer family study.*
Upper broken line: cohort incidence rates; lower broken line: population incidence
rates; solid line: relative risks.

subjects with missing data are members of a pedigree and this would mean
excluding the entire pedigree. Gibbs sampling provides a natural solution
by imputing values for the missing data at each iteration, sampling from
their full conditional distribution given the available data. This might include other information on the same subject (e.g. date of birth, sex, vital
and disease status) and the distribution of the variable among other family
members. Regression coefficients are then updated treating the imputed
values as known. This strategy produces unbiased estimates of the regression coefficients under the weaker assumptions that the data are missing
at random given the modelled covariates and the imputation model is correctly specified. However, estimates of the *genetic* relative risk are robust to
misspecification of the covariate imputation model, since in a proportional
hazards model the posterior distribution of genotypes does not depend on
covariates (except trivially through differential survival if the disease is not
rare).

Interactions between oral contraceptives and the major gene were examined but no evidence of departures from the multiplicative model were

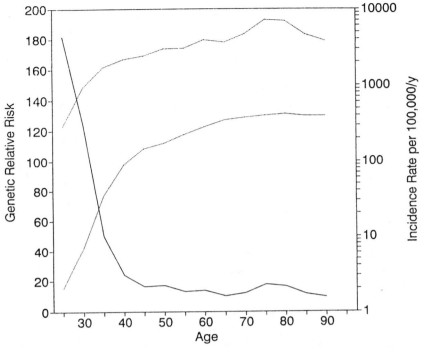

Figure 23.3 *Age-dependent genetic risk in the UCLA breast cancer family study. Upper broken line: carrier incidence rates; lower broken line: non-carrier incidence rates; solid line: relative risks.*

found. Addition of such interactions considerably increased the variances of the estimates of the component main effects, indicating instability. Simulation studies (Gauderman and Thomas, 1994) indicate that interactions involving an unmeasured major gene need to be quite strong to be detectable by segregation analysis. Power for detecting a gene-environment interaction is somewhat increased by including a tightly linked marker, but is still far less than if the disease gene could be observed.

23.6 Conclusions

Gibbs sampling has been shown to provide a feasible means of solving complex genetic problems, including large or complex pedigrees for which peeling is infeasible, and complex models involving many parameters. Simulation studies have demonstrated that it provides unbiased and efficient estimates and valid significance tests, and results that are comparable to standard likelihood methods for relatively simple problems. Whether it is computationally more efficient for problems that are amenable to standard methods remains to be seen, but probably its real appeal will be for the more complex problems for which no other method is feasible, or for testing the validity of the approximations that must otherwise be used.

Numerous problems remain to be explored before it can be recommended for routine use, including criteria for judging how many series and iterations are needed for convergence and for adequately exploring the parameter space, more efficient means for initializing the process, methods of accelerating it, and means for dealing with reducibility and local maxima.

References

Besag, J. (1974) Spatial interaction and the statistical analysis of lattice systems. *J. R. Statist. Soc.* B, **36**, 192–236.

Cannings, C. and Thompson E.A. (1977) Ascertainment in the sequential sampling of pedigrees. *Clin. Genet.*, **12**, 208–212.

Clayton, D. G. (1991) A Monte Carlo method for Bayesian inference in frailty models. *Biometrics*, **47**, 467–486.

Claus, E. B., Risch, N. J. and Thompson, W. D. (1991) Genetic analysis of breast cancer in the Cancer and Steroid Hormone Study. *Am. J. Hum. Genet.*, **48**, 232–242.

Easton, D. F., Bishop, D. T., Ford, D., Crockford, G. P. and the Breast Cancer Linkage Consortium (1993) Genetic linkage analysis in familial breast and ovarian cancer: results from 214 families. *Am. J. Hum. Genet.*, **52**, 678–701.

Elston, R. and Stewart, J. (1971) A general model for the analysis of pedigree data. *Hum. Hered.*, **21**, 523–542.

Faucett, C., Gauderman, W. J., Thomas, D. C., Ziogas, A. and Sobel, E. (1993) Combined segregation and linkage analysis of late onset Alzheimer's disease in Duke families using Gibbs sampling. *Genet. Epidem.*, **10**, 489–494.

Gauderman, W. J. and Thomas, D. C. (1994) Censored survival models for genetic epidemiology: a Gibbs sampling approach. *Genet. Epidem.*, **11**, 171–188.

Gelfand, A. E. and Smith, A. F. M. (1990) Sampling based approaches to calculating marginal densities. *J. Am. Statist. Ass.*, **85**, 398–409.

Gelman, A. (1995) Inference and monitoring convergence. In *Markov Chain Monte Carlo in Practice* (eds W. R. Gilks, S. Richardson and D. J. Spiegelhalter), pp. 131–143. London: Chapman & Hall.

Gelman, A. and Rubin, D. B. (1992) Inference from iterative simulation using multiple sequences (with discussion). *Statist. Sci.*, **7**, 457–511.

Geyer, C. J. (1992) Practical Markov chain Monte Carlo (with discussion). *Statist. Sci.*, **7**, 473–511.

Geyer, C. J. (1995) Estimation and optimization of functions. In *Markov Chain Monte Carlo in Practice* (eds W. R. Gilks, S. Richardson and D. J. Spiegelhalter), pp. 241–258. London: Chapman & Hall.

Gilks, W. R. (1992) Derivative-free adaptive rejection sampling for Gibbs sampling. In *Bayesian Statistics 4* (eds J. M. Bernardo, J. O. Berger, A. P. Dawid and A. F. M. Smith), pp. 641–649. Oxford: Oxford University Press.

Gilks, W. R. (1995) Full conditional distributions. In *Markov Chain Monte Carlo in Practice* (eds W. R. Gilks, S. Richardson and D. J. Spiegelhalter), pp. 75–88. London: Chapman & Hall.

Gilks, W. R. and Roberts, G. O. (1995) Strategies for improving MCMC. In *Markov Chain Monte Carlo in Practice* (eds W. R. Gilks, S. Richardson and D. J. Spiegelhalter), pp. 89–114. London: Chapman & Hall.

Goldstein, A., Haile, R. W. C., Marazita, M. L. and Pagannini-Hill, A. (1987) A genetic epidemiologic investigation of breast cancer in families with bilateral breast cancer. I. Segregation analysis. *J. Nat. Cancer Inst.*, **78**, 911–918.

Goldstein, A. M., Haile, R. W., Spence, M. A., Sparkes, R. S. and Paganinihill, A. (1989) A genetic epidemiologic investigation of breast cancer in families with bilateral breast cancer. II. Linkage analysis. *Clin. Genet.*, **36**, 100–106.

Haile, R. W., Cortesis, V. K., Millikan, R., Ingles, S., Aragaki, C. C., Richardson, L., Thompson, W. D., Paganni-Hill, A. and Sparkes, R. S. (1993) A linkage analysis of D17S74 (CMM86) in 35 families with premenopausal bilateral breast cancer. *Cancer Res.*, **53**, 212–214.

Hall, J. M., Friedman, L., Guenther, C., Lee, M. K., Weber, J. L., Block, D. M. and King, M. C. (1992) Closing in on a breast cancer gene on chromosome 17q. *Am. J. Hum. Genet.*, **50**, 1235–1242.

Hastings, W. K. (1970) Monte Carlo sampling methods using Markov chains and their applications. *Biometrika*, **57**, 97–109.

Kong, A., Frigge, M., Cox, N. and Wong, W. H. (1992) Linkage analysis with adjustment for covariates: a method combining peeling with Gibbs sampling. *Cytogenet. Cell Genet.*, **59**, 208–210.

Lin, S. (1993) Achieving irreducibility of the Markov chain Monte Carlo method applied to pedigree data. *IMA J. Math. Appl. Med. Biol.*, **10**, 1–17.

Ploughman, L. M. and Boehnke, M. (1989) Estimating the power of a proposed linkage study for a complex genetic trait. *Am. J. Hum. Genet.*, **43**, 543–551.

Raftery, A. E. and Lewis, S. M. (1995) Implementing MCMC. In *Markov Chain Monte Carlo in Practice* (eds W. R. Gilks, S. Richardson and D. J. Spiegelhalter), pp. 115–130. London: Chapman & Hall.

Roberts, G. O. (1995) Markov chain concepts related to sampling algorithms. In *Markov Chain Monte Carlo in Practice* (eds W. R. Gilks, S. Richardson and D. J. Spiegelhalter), pp. 45–57. London: Chapman & Hall.

Sheehan, N. and Thomas, A. (1992) On the irreducibility of a Markov chain defined on a space of genotype configurations by a sampling scheme. *Biometrics*, **49**, 163–175.

Spiegelhalter, D. J., Best, N. G., Gilks, W. R. and Inskip, H. (1995) Hepatitis B: a case study in MCMC methods. In *Markov Chain Monte Carlo in Practice* (eds W. R. Gilks, S. Richardson and D. J. Spiegelhalter), pp. 21–43. London: Chapman & Hall.

Thomas, D. C. and Cortesis, V. (1992) A Gibbs sampling approach to linkage analysis. *Hum. Hered.*, **42**, 63–76.

Thomas, D. C., Gauderman, W. J. and Haile, R. (1993) An analysis of breast cancer segregation using Gibbs sampling. *Technical Report 55*, Department of Preventive Medicine, University of Southern California.

Thompson, E. A. and Guo, S. W. (1991) Evaluation of likelihood ratios for complex genetic models. *IMA J. Math. Appl. Med. Biol.*, **8**, 149–169.

Wang, S. J. (1993) A Gibbs sampling approach for genetic linkage parameter estimation. Ph.D. dissertation. University of Southern California, Los Angeles.

Wang, S. J. and Thomas, D. C. (1994) A Gibbs sampling approach to linkage analysis with multiple polymorphic markers. *Technical Report 85*, Department of Preventive Medicine, University of Southern California.

24

Mixtures of distributions: inference and estimation

Christian P Robert

24.1 Introduction

24.1.1 Modelling via mixtures

Although many phenomena allow direct probabilistic modelling through classical distributions (normal, gamma, Poisson, binomial, etc.), the probabilistic structure of some observed phenomena can be too intricate to be accomodated by these simple forms. Statistical remedies are either to acknowledge the complexity of probabilistic modelling by choosing a non-parametric approach, or to use more elaborate parameterized distributions. The first approach does not force the data into a possibly inappropriate representation, but is only mildly informative and requires large sample sizes for its validation. For example, the representation of a kernel approximation to a density involves a term for each data point, and rarely permits basic inferential procedures such as hypothesis testing or interval estimation. The parametric alternative may appear vague because of its breadth, but this is equally its strength. It relies on standard densities $f(x|\theta)$ as a functional basis, approximating the true density $g(x)$ of the sample as follows:

$$g(x) \simeq \hat{g}(x) = \sum_{i=1}^{k} p_i f(x|\theta_i), \qquad (24.1)$$

where $p_1 + \ldots + p_k = 1$ and k is large (Dalal and Hall, 1985; Diaconis and Ylvisaker, 1985). The right-hand side of (24.1) is called a *finite mixture distribution* and the various distributions $f(x|\theta_i)$ are the *components of the mixture.* Note that the usual *kernel density* estimator,

$$\hat{g}(x) = \frac{1}{nh} \sum_{i=1}^{n} K\left(\frac{x - x_i}{h}\right),$$

also appears (at least formally) as a particular mixture of location-scale versions of the distribution K, where the number of components is equal to the number of observations, the location parameters are the observations, and the scale parameter is constant.

Mixture components do not always possess an individual significance or reality for the phenomenon modelled by (24.1), as we will see in a character-recognition example. Rather they correspond to particular zones of support of the true distribution, where a distribution of the form $f(x|\theta)$ can accurately represent the local likelihood function. There are also situations where the mixture components are interpretable, as in discrimination and clustering. Here the number of components, or the originating component of an observation, is of explanatory interest, as in the star cluster example described below.

Mixtures of distributions can model quite exotic distributions with few parameters and a high degree of accuracy. In our opinion, they are satisfactory competitors to more sophisticated methods of nonparametric estimation, in terms of both accuracy and inferential structure. Even when the mixture components do not allow for a model-related interpretation, the various parameters appearing in the final combination are still much easier to assess than coefficients of a spline regression, local neighbourhood sizes of a histogram density estimate, or location-scale factors of a wavelet approximation. Moreover, subsequent inferences about the modelled phenomenon are quite simple to derive from the original components of the mixture, since these distributions have been chosen for their tractability.

24.1.2 A first example: character recognition

Figure 24.1 illustrates the modelling potential of mixture distributions: two-component normal mixture densities accurately fit histograms of different shapes. The data corresponding to these histograms have been analysed in Stuchlik *et al.* (1994) and the modelling of these histograms is part of a Bayesian character recognition system. In this approach, handwritten words are automatically decomposed into characters which are then represented by a vector of 164 characteristics describing concavities and other geometric features of the pixel maps corresponding to the characters. A training sample of identified characters is analysed by creating pseudo-independence through a principal components analysis, and then applying mixture models independently to each principle component. Some of the principle components provide fairly regular histograms which can be estimated by standard distributions. Others, like those produced in Figure 24.1, are multimodal and call for more elaborate distributions. For the present data, two-component normal mixtures were enough to represent accurately all the multimodal histograms. Interpretation of mixture components in this context is difficult, as the mixture models vary over the principle components of a given character of the alphabet, and the principal components themselves are difficult to interpret. The estimated normal mixtures in Figure 24.1 were obtained using the MCMC algorithm of Mengersen and Robert (1995), described in Section 24.3.4. Having obtained estimates of the distributions of each character, $\hat{g}_a, \ldots, \hat{g}_{z,}$, a new observation x can be analysed: for each letter i the probability that x corresponds to i is

$$\frac{q_i \hat{g}_i(x)}{\sum_{\ell=a}^Z q_\ell \hat{g}_\ell(x)},$$

where the q_ℓ are the overall frequencies of letters in the dictionary.

24.1.3 Estimation methods

It is natural to try to implement a Bayesian estimation of the parameters $\{p_i, \theta_i\}$ of the mixture, as this approach automatically integrates learning processes and parameter updating, and is usually more reliable than other statistical techniques. In some situations, such as a mixture of two normal distributions, the main competing estimator, namely the maximum likelihood estimator, does not exist whatever the sample size, as pointed out in Lehmann (1983). This is because, for a given sample $x_1, \ldots, x_n$ from

$$\sum_{i=1}^k p_i f(x|\theta_i),$$

there is usually a positive probability that one component $f(x|\theta_i)$ generates none of the observations $\{x_j\}$, i.e. that the sample brings no information on this particular component. This phenomenon does not necessarily prevent the derivation of a maximum likelihood estimator when the likelihood is bounded, but it contributes to its instability. This peculiar identifiability problem of mixture models motivates the use of Bayesian methods to produce acceptable estimators or to devise tests about the number of mixture components.

We refer the reader to Titterington et al. (1985) for a theoretical and practical survey of estimation methods in mixture settings. Here we merely point out that, besides the maximum likelihood approach, the *moment method* was favoured for a while, following its introduction by Pearson (1894), but was later abandoned because of its limited performance.

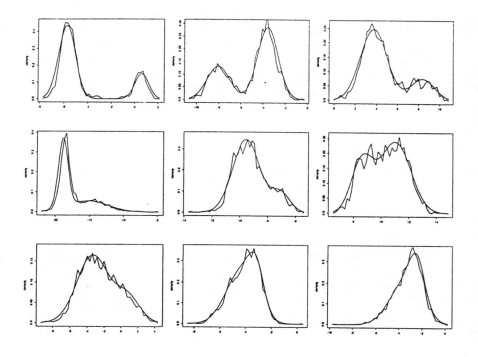

Figure 24.1 *Results from mixture models of distributions of some character components in a character recognition system (Source: Stuchlik et al., 1994).*

24.1.4 Bayesian estimation

Ironically, although Bayes estimators for mixture models are well-defined (so long as prior distributions are proper), it is only recently that implementation of the Bayesian paradigm has been made practical. Bayesian mixture estimation is definitely an 'after Gibbs' field. We now explain why this is the case and refer the reader to Titterington *et al.* (1985) for 'before Gibbs' approximation techniques to Bayesian procedures.

Assume, with little loss of generality, that the distributions $f(x|\theta)$ forming the functional basis are all from the same exponential family,

$$f(x|\theta) = h(x)e^{\theta \cdot x - \psi(\theta)}, \qquad (24.2)$$

where θ is a real-valued vector. This family allows a *conjugate prior* for θ:

$$\pi(\theta|y, \lambda) \propto e^{\theta \cdot y - \lambda \psi(\theta)} \qquad (24.3)$$

where y is a real-valued vector of constants of the same length as θ, and λ is a scalar constant (see for example Berger, 1985, or Robert, 1994). A conjugate prior for $(p_1, \ldots, p_k)$ is the Dirichlet distribution $D(\alpha_1, \ldots, \alpha_k)$, which has density

$$\pi^D(p_1, \ldots, p_k) \propto p_1^{\alpha_1 - 1} \cdots p_k^{\alpha_k - 1},$$

where $p_1 + \ldots + p_k = 1$. The Dirichlet distribution is a generalization of the Beta distribution $Be(\alpha_1, \alpha_2)$.

The posterior distribution of $(p_1, \ldots, p_k, \theta_1, \ldots, \theta_k)$, given the sample $(x_1, \ldots, x_n)$, can be written

$$[p_1, \ldots, p_k, \theta_1, \ldots, \theta_k | x_1, \ldots, x_n] \propto$$

$$\pi^D(p_1, \ldots, p_k) \prod_{i=1}^{k} \pi(\theta_i | y_i, \lambda_i) \prod_{j=1}^{n} \left(\sum_{i=1}^{k} p_i f(x_j | \theta_i) \right), \qquad (24.4)$$

where $[a|b]$ generically denotes the conditional distribution of a given b. Note that in (24.4), each component parameter θ_i is assigned a distinct prior via unique constants y_i and λ_i in (24.3).

Equation (24.4) involves k^n terms of the form

$$\prod_{i=1}^{k} p_i^{\alpha_i + n_i - 1} \pi(\theta_i | y_i + n_i \bar{x}_i, \lambda_i + n_i),$$

where $n_1 + \ldots + n_k = n$ and $\bar{x}_i$ denotes the average of n_i of the n observations $\{x_j\}$. While the posterior expectation of $(p_1, \ldots, p_k, \theta_1, \ldots, \theta_k)$ can be written in closed form (Diebolt and Robert, 1990), the computing time required to evaluate the k^n terms of (24.4) is far too large for the direct Bayesian approach to be implemented, even in problems of moderate size. For example, a simple two-component normal mixture model cannot be estimated in a reasonable time for 40 observations. It is therefore not surprising that it took roughly one century to evolve from Pearson's (1894) approach to a truly Bayesian estimation of mixtures, given the intractable analytic expansion of the Bayes estimator. Bayesian estimation of mixtures can now be carried out straightforwardly using Gibbs sampling. This approach might have been introduced sooner from MCMC methods developed by Metropolis *et al.* (1953) or Hastings (1970), or by extending the EM algorithm (Dempster *et al.*, 1977) to the Bayesian setting, but it is the developments of Tanner and Wong (1987) and Gelfand and Smith (1990) that led the way to the MCMC estimation of mixtures in Diebolt and Robert (1990), Verdinelli and Wasserman (1991), Lavine and West (1992), Diebolt and Robert (1994), Escobar and West (1995) and Mengersen and Robert (1995).

The following sections motivate and describe the Gibbs sampler for a general exponential family setting, and discuss some parameterization issues in the normal case in terms of stabilization of the algorithm and noninformative priors. A specific parameterization of normal mixtures is used to test the number of components in a mixture model, following Mengersen and Robert (1995). Finally, some extensions to infinite mixture models are considered.

24.2 The missing data structure

Consider a sample $x_1, \ldots, x_n$ from the mixture distribution

$$g(x) = \sum_{i=1}^{k} p_i f(x|\theta_i), \qquad (24.5)$$

assuming $f(x|\theta_i)$ is from an exponential family as in (24.2). It is possible to rewrite $x \sim g(x)$ under the hierarchical structure

$$x \sim f(x|\theta_z)$$

where z is an integer identifying the component generating observation x, taking value i with prior probability $p_i, 1 \le i \le k$. The components of (24.5) may or may not be meaningful; it is a feature of many Gibbs implementations that artificial random variables are introduced to create conditional distributions which are convenient for sampling. The vector $(z_1, \ldots, z_n)$ is the *missing data* part of the sample, since it is not observed. The above decomposition is also exploited in applications of the EM algorithm to mixture distributions (Celeux and Diebolt, 1985); see Diebolt and Ip (1995: this volume) for a description of the EM algorithm.

Were the missing data $z_1, \ldots, z_n$ available, it would be much easier to estimate the mixture model (24.5). From (24.5)

$$[x_1, \ldots, x_n | z_1, \ldots, z_n] = \prod_{j:z_j=1} f(x_j|\theta_1) \ldots \prod_{j:z_j=k} f(x_j|\theta_k) \qquad (24.6)$$

which implies that

$$[p_1, \ldots, p_k, \theta_1, \ldots, \theta_k | x_1, \ldots, x_n, z_1, \ldots, z_n] \propto p_1^{\alpha_1 + n_1 - 1} \ldots p_k^{\alpha_k + n_k - 1}$$
$$\pi(\theta_1 | y_1 + n_1 \bar{x}_1, \lambda_1 + n_1) \ldots \pi(\theta_k | y_k + n_k \bar{x}_k, \lambda_k + n_k), \qquad (24.7)$$

where now

$$n_i = \sum_j I(z_j = i) \qquad \text{and} \qquad n_i \bar{x}_i = \sum_{j : z_j = i} x_j,$$

and $I(.)$ is the indicator function taking the value 1 when its argument is true, and 0 otherwise. This conditional decomposition implies that the conjugate structure is preserved for a given allocation/missing data structure $z_1, \ldots, z_n$ and therefore that simulation is indeed possible conditional on $z_1, \ldots, z_n$.

The above decomposition is also of considerable interest from a theoretical point of view. In the implementation of Gibbs sampling described below, samples for the missing data z_j and the parameters (p, θ) are alternately generated, producing a missing-data chain and a parameter chain, both of which are Markov. The finite-state structure of the missing-data chain allows many convergence results to be easily established for it, and transferred automatically to the parameter chain (Diebolt and Robert, 1993, 1994; Robert, 1993a). These properties include geometric convergence, ϕ-mixing and a central limit theorem (see Geyer, 1992, or Tierney, 1994). Practical consequences can also be derived in terms of convergence control since, as shown in Robert (1993a), renewal sets can easily be constructed on the missing-data chain. Tierney (1995: this volume) provides an introduction to some of these theoretical concepts.

The introduction of the z_i is not necessarily artificial, although the algorithm works similarly whether it is natural or not. In some cases, the determination of the posterior distribution of these indicator variables is of interest, to classify the observations with respect to the components of the mixture. This is the case for the astronomical data described in Section 24.3.3.

Missing data structures can also be exhibited in more complex models, such as qualitative regression models (logit, probit, etc.) (Albert and Chib, 1993) or models for censored and grouped data (Heitjan and Rubin, 1991). In these contexts, the missing data are usually naturally specified in relation to the form of the model, and the implementation of the Gibbs sampler is easily derived. For example, consider a *probit regression model*, where

$$P(x_i = 1) = 1 - P(x_i = 0) = \Phi(\alpha^T v_i)$$

where Φ is the cumulative normal distribution function and v_i is a vector of regressors. The observation x_i can then be interpreted as the indicator that an unobserved $N(\alpha^T v_i, 1)$ variable x_i^* is smaller than 0, namely $x_i = I(x_i^* < 0)$. If the missing data $x_1^*, \ldots, x_n^*$ were available, the estimation of

α would be simple, since the likelihood function would then be a normal likelihood. See Diebolt and Ip (1995: this volume) for an application of stochastic EM to probit regression.

24.3 Gibbs sampling implementation

24.3.1 General algorithm

Once a missing data structure is exhibited as in (24.6) above, the direct implementation of Gibbs sampling is to run successive simulations from (24.7) for the parameters of the model and from $[z_1, \ldots, z_n | p_1, \ldots, p_k, \theta_1, \ldots, \theta_k, x_1, \ldots, x_n]$ for the missing data. Provided conjugate prior distributions for the parameters are used, the simulation is usually straightforward:

Algorithm 1

Step 1. Simulate

$$\theta_i \quad \sim \quad \pi(\theta_i | y_i + n_i \bar{x}_i, \lambda_i + n_i), \quad (i = 1, \ldots, k)$$
$$(p_1, \ldots, p_k) \quad \sim \quad D(\alpha_1 + n_1, \ldots, \alpha_k + n_k)$$

Step 2. Simulate

$$[z_j | x_j, p_1, \ldots, p_k, \theta_1, \ldots, \theta_k] = \sum_{i=1}^{k} p_{ij} I(z_j = i), \qquad (j = 1, \ldots, n)$$

with

$$p_{ij} = \frac{p_i f(x_j | \theta_i)}{\sum_{t=1}^{k} p_t f(x_j | \theta_t)}, \quad (i = 1, \ldots, k).$$

Step 3. Update n_i and $\bar{x}_i$, $(i = 1, \ldots, k)$.

The number of iterations necessary to reach convergence of the Gibbs sampler cannot be evaluated simply in a general setting. See Gelman (1995) and Raftery and Lewis (1995) in this volume; see also Robert (1993a) for the use of renewal theory on the missing data chain. For the mixture models of one and two-dimensional data considered below, 5 000 iterations were 'enough' in the sense that a substantial increase in the number of iterations did not usually perturb values of ergodic averages.

A deeper difficulty in implementing *Algorithm 1* is the existence of computational *trapping states*. Despite the theoretical irreducibility of the chain, the algorithm becomes effectively trapped when one of the components of the mixture is allocated very few observations. This difficulty is addressed and resolved below, along with an additional problem related to noninformative priors, when we consider parameterization issues.

The next section illustrates the above issues in a multivariate normal setting where all parameters are unknown. The resulting algorithm is then applied to real data from an astronomy experiment. Before this, we briefly consider a discrete setting.

24.3.2 Extra-binomial variation

A *logit model* differs from the above probit model by the form of the dependence on the regressors v_i:

$$P(x_i = 1) = \frac{\exp\left\{\alpha^{\mathrm{T}} v_i\right\}}{1 + \exp\left\{\alpha^{\mathrm{T}} v_i\right\}}. \tag{24.8}$$

Extra-binomial variation may also be present, jeopardizing the application of (24.8). Instead of abandoning logit modelling, a *mixture of logit models* can be used to capture the extra-binomial variability. We have therefore the following structure:

$$P(x_i = 1 | v_i) = p_1 \frac{\exp\left\{\alpha_1^{\mathrm{T}} v_i\right\}}{1 + \exp\left\{\alpha_1^{\mathrm{T}} v_i\right\}} + \ldots + p_k \frac{\exp\left\{\alpha_k^{\mathrm{T}} v_i\right\}}{1 + \exp\left\{\alpha_k^{\mathrm{T}} v_i\right\}}, \tag{24.9}$$

with $p_1 + \ldots + p_k = 1$. *Algorithm 1* may be applied to the estimation of (24.9), by partitioning the sample into k groups at each step and updating the distributions of the α_j separately on each group. For a conjugate prior distribution of the form

$$[\alpha_j | y_j, \lambda_j] \propto \frac{\exp\left\{\alpha_j^{\mathrm{T}} y_j\right\}}{\prod_{i=1}^{n} (1 + \exp\left\{\alpha_j^{\mathrm{T}} v_i\right\})^{\lambda_j}},$$

the simulation of the α_j at *Step 1* of *Algorithm 1* can be done either by a Metropolis–Hastings step (Robert, 1993b) or by adaptive rejection sampling (Gilks and Wild, 1992).

24.3.3 Normal mixtures: star clustering

Consider a sample $x_1, \ldots, x_n$ from the normal mixture model

$$x \sim \sum_{i=1}^{k} p_i \mathrm{N}_v(\mu_i, \Sigma_i),$$

where $\mathrm{N}_v(.,.)$ denotes a v-dimensional multivariate normal distribution. The corresponding conjugate prior distributions are, for $(i = 1, \ldots, k)$:

$$\begin{aligned}
(p_1, \ldots, p_k) &\sim \mathrm{D}(\alpha_1, \ldots, \alpha_k), \\
\mu_i &\sim \mathrm{N}_v(\xi_i, \tau_i \Sigma_i), \\
\Sigma_i^{-1} &\sim \mathrm{W}_v(r_i, W_i),
\end{aligned}$$

where $\mathrm{W}_v(r, W)$ denotes the Wishart distribution. (A Wishart random matrix A has density proportional to

$$|A|^{\frac{r-(v+1)}{2}} \exp\{-\mathrm{trace}(AW^{-1})/2\}$$

on the space of symmetric positive semi-definite $v \times v$ matrices. The Wishart distribution $W_1(r, W)$ reduces to the gamma distribution $\text{Ga}(r/2, W^{-1}/2)$.) *Step 1* of *Algorithm 1* then simulates from the full conditionals

$$
\begin{aligned}
(p_1, \ldots, p_k) &\sim \text{D}(\alpha_1 + n_1, \ldots, \alpha_k + n_k), \\
\mu_i &\sim \text{N}_v(\xi_i^*, \tau_i^* \Sigma_i), \\
\Sigma_i^{-1} &\sim \text{W}_v(r_i^*, W_i^*),
\end{aligned}
$$

where n_i is the number of observations currently allocated to the ith component,

$$
\xi_i^* = \frac{n_i \bar{x}_i + \xi_i \tau_i^{-1}}{n_i + \tau_i^{-1}}, \quad r_i^* = r_i + n_i, \quad \tau_i^* = \frac{\tau_i}{\tau_i n_i + 1},
$$

$$
W_i^* = \left(W_i^{-1} + S_i + \frac{n_i}{\tau_i n_i + 1} (\bar{x}_i - \xi_i)(\bar{x}_i - \xi_i)^{\text{T}} \right)^{-1}
$$

and

$$
S_i = \sum_{j \,:\, z_j = i} (x_j - \bar{x}_i)(x_j - \bar{x}_i)^{\text{T}}, \qquad (i = 1, \ldots, k).
$$

Step 2 derives the missing data z_j $(1 \leq j \leq n)$ from a simulated standard uniform random variable u_j, as follows: $z_j = i$ if $p_{j1} + \ldots + p_{j(i-1)} < u_j \leq p_{j1} + \ldots + p_{ji}$, where

$$
p_{ji} \propto p_i \exp \left\{ -(x_j - \mu_i)^{\text{T}} \Sigma_i^{-1} (x_j - \mu_i)/2 \right\} |\Sigma_i|^{-\frac{1}{2}}. \tag{24.10}
$$

Evaluation of the normalizing coefficient in (24.10) can be avoided for large k by replacing the simulation in *Step 2* with a Metropolis–Hastings step.

Figures 24.2, 24.3 and 24.4 illustrate the behaviour of the above Gibbs sampler for $v = 2$ dimensions on a set of data analysed in Robert and Soubiran (1993). Each dot on the graph represents the angular and radial speeds of a given star in the plane of the galaxy. The purpose behind the probabilistic modelling is to examine a mixture structure with $k = 2$ mixture components, corresponding to an astronomical hypothesis of *formation in bursts*. As described in detail in Robert and Soubiran (1993), this hypothesis assumes that stars are created in clusters when wandering matter has accumulated to a critical mass in a particular region of space. Thus stars which originate from the same cloud of matter form clusters which can be identified while the stars are young. This is a setting in which the missing data $z_1, \ldots, z_n$ are meaningful, z_i indicating the formation cloud of star i. Figure 24.2 provides the dataset along with contours

$$
(x - \hat{\mu}_i)^{\text{T}} \hat{\Sigma}_i^{-1} (x - \hat{\mu}_i) = 1
$$

of $\text{N}_2(\hat{\mu}_i, \hat{\Sigma}_i)$, for $i = 1, 2$, representing 39% coverage regions for the angular and radial speed of a star, given its formation cloud. Here $\hat{\mu}_i$ and $\hat{\Sigma}_i$ are

empirical Bayes estimates of μ_i and Σ_i, obtained using the above implementation of *Algorithm 1* with data-based hyperparameter values: $\xi_i = \bar{x}$
and $W_i = 2S^{-1}/(r_i - 3)$, where $\bar{x}$ and S are the empirical mean and covariance matrix of the whole sample. The above setting for W_i implies a
prior expectation for Σ_i of $S/2$. Constants τ_i and r_i were set to 1 and 6
respectively, to reduce prior information. This example is typical of situations where noninformative priors cannot be used, since improper priors
on (μ_i, Σ_i) lead to improper posterior distributions: the absence of prior
information can only be overcome by empirical-Bayes approximations.

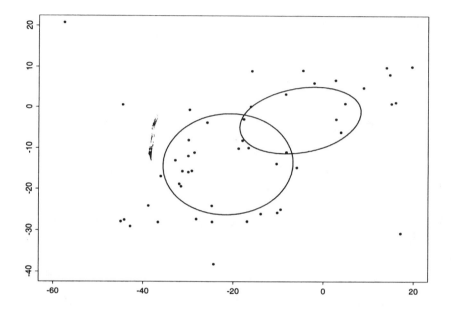

Figure 24.2 *Data set and 39% coverage sets of estimated components of a mixture
model for star speeds (Source: Robert and Soubiran, 1993).*

The values of the estimates given in Table 24.1 were obtained after 5 000
iterations. As noted above, this choice is rather *ad hoc*, but larger numbers
of iterations produced negligible differences. Figures 24.3 and 24.4 illustrate
the clustering ability of the Gibbs sampler: for each star, probabilities of
belonging to the first and second components can be estimated. These probabilities are represented in Figures 24.3 and 24.4 by line segments directed
towards the mean of each component, with lengths proportional to the
probabilities. Note the strong attraction of the leftmost component, since
all observations located left of the larger ellipse have higher probabilities
of belonging to this component. We emphasize that sample sizes of 51 are

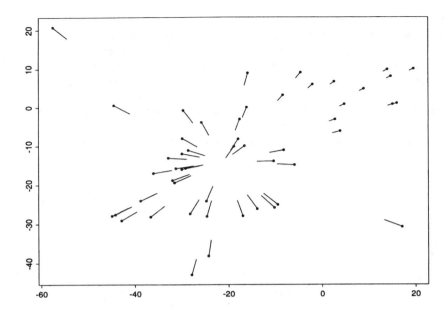

Figure 24.3 *Representing probabilities of originating from the first (leftmost) component of Figure 24.2.*

very difficult to manage using other inferential methods (see Soubiran *et al.*, 1991).

24.3.4 Reparameterization issues

As noted above and in Bernardo and Girón (1988), the extension of Bayesian estimation techniques to the noninformative case is impossible in some cases because the identifiability deficiency exposed in Section 24.1.3 also prohibits the use of improper priors on the parameters θ_i. This problem cannot be avoided by using vague but proper priors, since the simulated chains become increasingly unstable and tend to converge to trapping states as the

	μ_1	μ_2	σ_1^2	σ_2^2	σ_{12}	p
first component	-21.2	-13.9	207.9	152.0	7.1	0.64
second component	-5.1	-3.1	170.8	77.6	41.3	0.36

Table 24.1 *Estimates of the components of a two-dimensional mixture model for the star speed dataset of Figure 24.2*

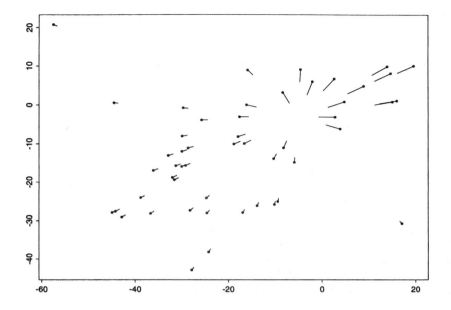

Figure 24.4 *Representing probabilities of originating from the second (rightmost) component of Figure 24.2.*

variance hyperparameters are increased. A reason for this is the choice of parameterization of the distribution (24.5): the parameters of each compon-ent of the mixture are distinct in both prior and sampling distributions. When few observations are allocated to a given component, the fit of this component to these observations is tight, and in practice the Gibbs sampler cannot escape this local mode of the posterior distribution. If however the parameters of the different components were somehow linked together, the influence of such extreme allocations might be substantially reduced. Con-sider therefore the following parameterization of a two-component normal mixture, as introduced in Mengersen and Robert (1995):

$$pN(\mu, \tau^2) + (1 - p)N(\mu + \tau\theta, \tau^2\sigma^2) \qquad (24.11)$$

with the restriction $\sigma < 1$ to ensure identifiability of the parameters. This representation is clearly equivalent to the more common one of Sec-tion 24.3.3, but it creates a strong link between both components. In (24.11), μ and τ represent location and scale parameters for the first com-ponent, while θ and σ represent deviations in location and scale for the second component relative to the first. An equivalent 'centred' formulation of (24.11) is also possible:

$$pN(\mu - \tau\theta, \tau^2/\sigma^2) + (1-p)N(\mu + \tau\theta, \tau^2\sigma^2).$$

Since the location and scale parameters, μ and σ, appear in both components, the restriction on proper priors partially vanishes and we can propose prior distributions of the type

$$[\mu, \theta, \tau, \sigma, p] \propto \frac{1}{\tau}e^{-\theta^2/2\zeta}I(\sigma \in [0,1])I(p \in [0,1]), \qquad (24.12)$$

which are then mildly informative and depend on a single parameter ζ. The influence of this parameter is limited and variations of ζ in the range $[1,4]$ do not produce significant modifications in the resulting estimates. This approach stabilizes the Gibbs sampling algorithm since the trapping states observed in the original parameterization disappear. The algorithm is able to escape from states where one of the two components is attributed one observation at most. This is not the case with the parameterization used in Diebolt and Robert (1994), where a regeneration procedure is required.

The Gibbs sampler in *Step 1* of *Algorithm 1* can be readily adapted for the new parameterization (24.11) and prior (24.12) (Mengersen and Robert, 1995):

Algorithm 2

Step 1. Simulate

$$p \sim \text{Be}(n_1 + 1, n_2 + 1);$$

$$\theta \sim N\left(\frac{(\bar{x}_2 - \mu)}{\tau}\left[1 + \frac{\sigma^2}{n_2\zeta}\right]^{-1}, \frac{\sigma^2}{n_2 + \sigma^2\zeta^{-1}}\right);$$

$$\mu \sim N\left(\frac{n_1\bar{x}_1 + \sigma^{-2}n_2(\bar{x}_2 - \tau\theta)}{n_1 + n_2\sigma^{-2}}, \frac{\tau^2\sigma^2}{n_1\sigma^2 + n_2}\right);$$

$$\sigma^{-2} \sim \text{Ga}\left(\frac{n_2 - 1}{2}, \frac{s_2^2 + n_2(\bar{x}_2 - \mu - \tau\theta)^2}{2\tau^2}\right)I(\sigma^{-2} > 1);$$

$$\tau^{-2} \sim \text{Ga}\left(\frac{n}{2}, \frac{1}{2}\left(s_1^2 + n_1(\bar{x}_1 - \mu)^2 + \sigma^{-2}s_2^2 + \frac{n_2(\bar{x}_2 - \mu)^2}{n_2\zeta + \sigma^2}\right)\right).$$

Steps 2 and *3* remain as in *Algorithm 1*.

Moreover, the location-scale parameterization can be extended to a larger number of components quite simply since we need only consider

$$p_1N(\mu, \tau^2) + (1-p_1)p_2N(\mu + \tau\theta_1, \tau^2\sigma_1^2) +$$
$$(1-p_1)(1-p_2)p_3N(\mu + \tau\theta_1 + \tau\sigma_1\theta_2, \tau^2\sigma_1^2\sigma_2^2) + \ldots +$$
$$(1-p_1)\ldots(1-p_{k-1})N(\mu + \tau\theta_1 + \ldots + \tau\sigma_1\ldots\sigma_{k-2}\theta_{k-1},$$
$$\tau^2\sigma_1^2\ldots\sigma_{k-1}^2)$$

with the extended identifiability constraint $\sigma_1 < 1, \ldots, \sigma_{k-1} < 1$ and the following generalization of (24.12):

$$[\mu, \theta_1, \ldots, \theta_{k-1}, \tau, \sigma_1, \ldots, \sigma_{k-1}, p_1, \ldots, p_{k-1}] \propto \frac{1}{\tau} \exp\left\{-\sum_{i=1}^{k-1} \theta_i^2/2\zeta\right\}$$

$$I(\sigma_1 \in [0,1]) \ldots I(\sigma_{k-1} \in [0,1]) I(p_1 \in [0,1]) \ldots I(p_{k-1} \in [0,1]).$$

The following example shows that the reparameterization technique can easily be extended to situations other than a normal mixture.

24.3.5 Extra-binomial variation: continued

For the logit mixture model described in Section 24.3.2, the noninformative prior $[\alpha_1, \ldots, \alpha_k] \propto 1$ is not acceptable, since no allocations to a given component i_0 of the mixture brings no information about the parameter α_{i_0}. A more 'restrictive' parameterization can partially alleviate this problem. For example, we can replace (24.9) by

$$P(x_i = 1|v_i) \quad = p_1 \frac{\exp\left\{\alpha_1^{\mathrm{T}} v_i\right\}}{1 + \exp\left\{\alpha_1^{\mathrm{T}} v_i\right\}} + (1 - p_1) p_2 \frac{\exp\left\{\alpha_2^{\mathrm{T}} v_i\right\}}{1 + \exp\left\{\alpha_2^{\mathrm{T}} v_i\right\}} + \ldots +$$

$$(1 - p_1) \ldots (1 - p_{k-1}) \frac{\exp\left\{\alpha_k^{\mathrm{T}} v_i\right\}}{1 + \exp\left\{\alpha_k^{\mathrm{T}} v_i\right\}},$$

under the identifying restriction that $||\alpha_i|| > ||\alpha_{i+1}||$, where $||\alpha_i||$ denotes $\alpha_i^{\mathrm{T}} \alpha_i$. In this case, a manageable prior is

$$[\alpha_1, \ldots, \alpha_k] \propto e^{-||\alpha_1||^2/2\xi^2},$$

with large ξ. The sample contains sufficient information on some of the α_i for the corresponding posterior distribution to be proper.

24.4 Convergence of the algorithm

The approach advocated in *Algorithm 1* is a special case of the Data Augmentation algorithm proposed by Tanner and Wong (1987) while *Algorithm 2* follows the fully conditional decomposition of Gelfand and Smith's (1990) Gibbs sampling. The latter approach is less powerful in term of convergence properties and speed (see Liu *et al.*, 1994, and Diebolt and Robert, 1994) but both guarantee ergodicity for the generated Markov chain and thus the validity of the resulting approximation. We can also suggest the alternative of Metropolis–Hastings perturbations in delicate cases, namely random or periodic replacements of the exact full conditional distributions by Metropolis–Hastings substitutes with larger variability, to increase excursions of the chain in the whole parameter space and, presumably, the speed of convergence to the stationary distribution; see Gilks and Roberts (1995: this volume). Mykland *et al.* (1995) also recommend this 'hybrid' algorithm to achieve theoretical conditions for the implementation of renewal theory as a convergence control.

24.5 Testing for mixtures

Since mixture models are often used in situations where individual components are meaningless, it is important that methods are available for assessing model adequacy in terms of the number of mixture components. The number of mixture components is a key ingredient of the modelling, akin in a weak way to the bandwidth parameter in kernel density estimation. Methods for determining the number of mixture components are still under development. Here we discuss just one possibility: a test for the presence of a mixture (Mengersen and Robert, 1995). For other approaches related to model choice, see Gelfand (1995), Raftery (1995), Gelman and Meng (1995), George and McCulloch (1995) and Phillips and Smith (1995) in this volume.

We assume that the primary aim of the analysis is prediction. Suppose we consider using $h(x)$ instead of $g(x)$ to describe the sampling distribution of the data. The inadequacy of $h(x)$ can be assessed in terms of a distance between the two distributions. The most common choice of distance metric between probability densities is *entropy distance* (or Kullback–Leibler divergence),

$$\text{ED}[g, h] = \int \log(g(x)/h(x))g(x)dx. \qquad (24.13)$$

To test for the presence of a mixture model, we propose comparing a two-component normal mixture to the closest single-component model, i.e. a normal distribution $N(\mu, \sigma^2)$, where 'closest' is in the sense of entropy distance. Thus, for given μ_1, μ_2, σ_1^2, and σ_2^2, we choose μ and σ to minimize

$$\text{ED}\left[\, p N(\,.\mid \mu_1, \sigma_1^2) + (1-p)N(\,.\mid \mu_2, \sigma_2^2)\,,\ N(\,.\mid \mu, \sigma^2)\,\right]. \qquad (24.14)$$

Equation (24.14) is minimized when $\mu = p\mu_1 + (1-p)\mu_2$ and $\sigma^2 = p\sigma_1^2 + (1-p)\sigma_2^2 + p(1-p)(\mu_1 - \mu_2)^2$, and this is called the *projection* of the two-component model onto the one-component model. If the distance (24.14) is smaller than a given bound α, with α small, parsimony dictates that a mixture is not warranted. While (24.14) cannot be computed explicitly, Mengersen and Robert (1995) proposed Laplace approximations to this distance, and find that $\alpha = 0.1$ is quite satisfactory for one-dimensional normal mixtures. The corresponding value of ζ in (24.12) can then be derived by imposing the constraint that the prior probability of a 'null hypothesis' H_α is 0.5, where H_α asserts that entropy distance (24.14) is less than α.

We illustrate the behaviour of this testing procedure on another astronomical dataset shown in Figure 24.5. This particular star sample is known to be homogeneous for astrophysical reasons, as discussed in Robert and Soubiran (1993). The testing procedure described above is restricted to one-dimensional settings, so it must be applied separately to each of the velocity axes. The resulting probabilities are thus partial indicators of the homogeneity of the sample since heterogeneity can be due to another axis. However, tests in other directions (i.e. using convex combinations of the two components) give posterior probabilities of $H_{0.1}$ that are intermediate between the two values 0.61 and 0.94 in Table 24.2.

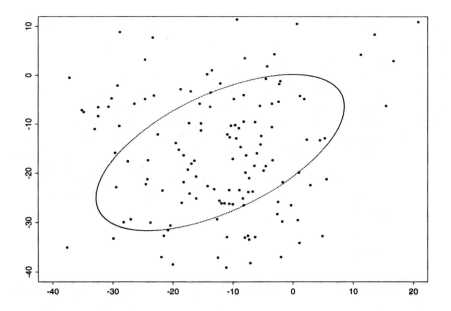

Figure 24.5 *A homogeneous dataset for star speeds, showing a 39% coverage set under a single-component normal model.*

Since a Gibbs sampling algorithm is available and produces a Markov chain $\{\theta^{(t)}\}$, it is straightforward to check at each iteration whether the entropic distance (24.13) (between the the two-component mixture at that iteration and its one-component projection) is smaller or larger than α, by computing

$$\omega_t = I \left(\mathrm{ED} \left[p^{(t)} \mathrm{N}(.|\mu_1^{(t)}, \sigma_1^{2(t)}) + (1 - p^{(t)}) \mathrm{N}(.|\mu_2^{(t)}, \sigma_2^{2(t)}) , \right. \right.$$

$$\left. \left. \mathrm{N}(.|\mu^{(t)}, \sigma^{2(t)}) \right] \leq \alpha \right) ,$$

	μ	τ	θ	σ	p	probability
first axis	-13	12	0.32	0.57	0.75	0.94
second axis	-12	11	-1.30	0.67	0.57	0.61

Table 24.2 *Componentwise tests of homogeneity for the star speed data in Figure 24.5. The values in the final column are the posterior probabilities that (24.14) is smaller than $\alpha = 0.1$. The parameters values for $\mu, \tau, \theta, \sigma$ and p were estimated using Algorithm 2 and with $\zeta = 4$ in (24.12)*

where $\mu^{(t)}$ and $\sigma^{(t)}$ are the projections defined above. The average of the ω_t then provides the posterior probability that the distance between the normal mixture and the closest normal distribution is smaller than α.

Extensions to larger numbers of components, i.e. tests of k component mixtures versus $k + 1$ component mixtures are more delicate to implement for lack of a closed form expression for the k component mixture closest to a $k + 1$ normal mixture. More classical approaches based on Bayes factors for both models can be derived; see Raftery (1995: this volume).

An entropic distance seems much easier to use in discrete situations, since it can always be obtained as a finite sum. Nonetheless, the related bound α on (24.14) is quite dependent on the nature of the problem and, in particular, on the cardinality of the state space, although bounds on the entropic distance are still infinite. See Dupuis (1995) for an illustration in a capture-recapture setting.

24.5.1 Extra-binomial variation: continued

For the logit mixture model in Section 24.3.2, the main problem is the determination of the number of components k. We suggest a simple upwards procedure to test $k = 1$ versus $k = 2$; $k = 2$ versus $k = 3$; $\ldots$, until the more elaborate level is rejected. The $k - 1$ versus k test can be based upon the entropy

$$
\frac{1}{n} \sum_{i=1}^{n} \log \left(\frac{p_1 \frac{\exp\{\alpha_1^T v_i\}}{1+\exp\{\alpha_1^T v_i\}} + \ldots + p_k \frac{\exp\{\alpha_k^T v_i\}}{1+\exp\{\alpha_k^T v_i\}}}{q_1 \frac{\exp\{\beta_1^T v_i\}}{1+\exp\{\beta_1^T v_i\}} + \ldots + q_{k-1} \frac{\exp\{\beta_{k-1}^T v_i\}}{1+\exp\{\beta_{k-1}^T v_i\}}} \right)
$$
$$
\times \left(p_1 \frac{\exp\{\alpha_1^T v_i\}}{1 + \exp\{\alpha_1^T v_i\}} + \ldots + p_k \frac{\exp\{\alpha_k^T v_i\}}{1 + \exp\{\alpha_k^T v_i\}} \right),
$$

where $(q_1, \ldots, q_{k-1}, \beta_1, \ldots, \beta_{k-1})$ is the projection of the k component estimated mixture logit model on a $k - 1$ component mixture logit model, that is, $(q_1, \ldots, q_{k-1}, \beta_1, \ldots, \beta_{k-1})$ is chosen to minimize the entropic distance between the $k - 1$ and k component models. Unfortunately, this minimization is rather complex and requires numerical or stochastic minimization procedures. For example, the method of prior feedback (Robert, 1993b), which approximates the maximum likelihood estimator as a sequence of Bayes estimators, could be used.

24.6 Infinite mixtures and other extensions

24.6.1 Dirichlet process priors and nonparametric models

As noted in Section 24.1, a major purpose of mixture modelling is to replace most traditional nonparametric techniques. It may be that some situations call for an infinite number of components or, alternatively, for a continuous mixture of known parameterized distributions. Although we feel that a proper modelling by a finite mixture should be adequate in many cases, we briefly mention these models and related results here because they follow approximately the same features as regular mixture models and very similar MCMC algorithms apply for their estimation.

For a parameterized model $x \sim f(x|\theta)$, Escobar and West (1995) introduce *Dirichlet process hyperpriors* on the prior distribution $\pi(\theta)$. The consequence of this modelling is to allow in theory for an arbitrary number of components in the mixture, while limiting in practice the actual number of these components. Even though this approach depends on the selection of smoothing parameters and an average prior distribution, it nevertheless appears as a satisfactory Bayesian alternative to nonparametric estimation methods and can be extended to other situations such as nonparametric regression (Müller *et al.*, 1995).

A Dirichlet process is a distribution on distributions. Suppose π is a distribution, distributed according to a Dirichlet process: $\pi \sim \mathrm{DP}(\alpha, \pi_0)$, where π_0 is a probability distribution and $\alpha > 0$. A Dirichlet process has the property that $(\pi(A_1), \ldots, \pi(A_v))$ is distributed according to a regular Dirichlet distribution $\mathrm{D}(\alpha\pi_0(A_1), \ldots, \alpha\pi_0(A_v))$ for any partition $A_1, \ldots, A_v$ of the parameter space, where $\pi_0(A_i)$ denotes the probability contained in a set A_i, under the distribution π_0. Instead of embarking any further on the complex description of a Dirichlet process, we describe below its Gibbs implementation in a normal setting, as in Escobar and West (1995).

Consider x_i distributed as $\mathrm{N}(\theta_i, \sigma_i^2)$ $(1 \leq i \leq n)$ with $(\theta_i, \sigma_i^2) \sim \pi$ and π distributed as a Dirichlet process $\mathrm{DP}(\alpha, \pi_0)$. The prior distribution π is thus considered as a 'point' in the infinite-dimensional space of probability distributions, centred around the distribution π_0 with a degree of concentration α around π_0. The influence of this choice of π on the data is that the marginal distribution of x (marginalizing over all possible π) is a mixture of normal distributions, with a random number of components ranging from 1 up to n. Note that the assumption on the distribution of the x_i is not restrictive at all since the means and variances can differ for every observation. The fact that the number of components can be as high as the sample size is a reflection of this lack of constraints on the model. Another important consequence of this modelling is that the prior full conditional distributions of the (θ_i, σ_i^2) can be expressed as follows

$$[\,(\theta_i, \sigma_i^2)\mid(\theta_j, \sigma_j^2)_{j\neq i}\,] = \alpha(\alpha + n - 1)^{-1}\pi_0(\theta_i, \sigma_i^2) +$$

$$(\alpha + n - 1)^{-1}\sum_{j\neq i} I\left((\theta_i, \sigma_i^2) = (\theta_j, \sigma_j^2)\right). \tag{24.15}$$

The decomposition (24.15) demonstrates the moderating effect of the Dirichlet process prior, as new values of (θ, σ^2) only occur with probability $\alpha/(\alpha + n - 1)$. A similar conditional distribution can be obtained *a posteriori*: for observations $x_1, \ldots, x_n$,

$$[\,(\theta_i, \sigma_i^2)\mid(\theta_j, \sigma_j^2)_{j\neq i},\ x_1, \ldots, x_n\,] = q_{i0}\pi_0(\theta_i, \sigma_i^2|x_i) +$$

$$\sum_{j\neq i} q_{ij} I\left((\theta_i, \sigma_i^2) = (\theta_j, \sigma_j^2)\right), \tag{24.16}$$

where $q_{i0} + \sum_{j\neq i} q_{ij} = 1$ and

$$q_{i0} \propto \alpha \int e^{-(x_i-\theta_i)^2/2\sigma_i^2}\sigma_i^{-1}\pi_0(\theta_i, \sigma_i^2)d\theta_i d\sigma_i^2,$$

$$q_{ij} \propto e^{-(x_i-\theta_j)^2/2\sigma_j^2}\sigma_j^{-1} \quad (i\neq j).$$

The features of the conditional distributions (24.16) are similarly interesting, as they imply that (θ_i, σ_i^2) is a new parameter with probability q_{i0} and is equal to another parameter otherwise. Therefore, once the Gibbs sequence of simulations from (24.16) has been run for all i, and k different values of (θ_i, σ_i^2) have been obtained, the actual distribution of the sample $(x_1, \ldots, x_n)$ is indeed a mixture of k normal distributions, although the generating component of each observation x_i is well-known. This feature makes quite a difference with the usual mixture approach, as the number of components varies at each simulation step. If we are estimating the distribution of a future x, the Gibbs approximation to the (predictive) density f is to simulate a sample $\{\theta_{n+1}^{(t)}; \sigma_{n+1}^{2(t)};\ t = 1, \ldots, T\}$ of size T from $\pi(\theta, \sigma^2|x_1, \ldots, x_n)$ by simulating successively $\{\theta_i^{(t)}, \sigma_i^{2(t)}; 1 \leq i \leq n\}$ according to (24.16), and $(\theta_{n+1}^{(t)}, \sigma_{n+1}^{2(t)})$ according to

$$\pi(\theta_{n+1}, \sigma_{n+1}^2) = [\,(\theta_{n+1}, \sigma_{n+1}^2)\mid\{\theta_i^{(t)}, \sigma_i^{2(t)}; 1 \leq i \leq n\}\,]$$
$$= \alpha(\alpha + n)^{-1}\pi_0(\theta_{n+1}, \sigma_{n+1}^2) +$$
$$(\alpha + n)^{-1}\sum_{j=1}^{n} I\left((\theta_{n+1}, \sigma_{n+1}^2) = (\theta_j, \sigma_j^2)\right).$$

The predictive density can then be estimated by

$$\frac{1}{T} \sum_{t=1}^{T} \pi(\theta_{n+1}^{(t)}, \sigma_{n+1}^{2(t)}) \tag{24.17}$$

and is therefore of the same order of complexity as a kernel density estimator, since it involves formally T terms. The sample $(\theta_{n+1}^{(t)}, \sigma_{n+1}^{2(t)})$ will include a few values simulated according to $\pi_0(\theta_{n+1}, \sigma_{n+1}^2)$ and most of the values in $\{\theta_i^{(t)}, \sigma_i^{2(t)}; 1 \le i \le n\}$, which are also simulated according to π_0 but with replications. Improvements upon this direct implementation of the Dirichlet process prior are suggested in Escobar and West (1995), such as a derivation of the average number of components in the distribution of $\{\theta_i, \sigma_i^2; 1 \le i \le n\}$. However, the choice of hyperparameter values is quite crucial to the performance of the resulting estimator.

24.6.2 Hidden Markov models

The second extension we discuss in this overview of mixture inference concerns *dependent observations*. When there is a Markov structure underlying the observations, the model is called a *hidden Markov model*. As before, the observations x_i are generated according to a mixture model (24.1), but now the missing data are states of a Markov chain $z_1, \ldots, z_n$ with state space $\{1, \ldots, k\}$, where z_i is an integer identifying the underlying state of observation x_i. Thus, given z_i:

$$x_i \sim f(x|\theta_{z_i})$$

and $z_i = i$ with a probability $p_{z_{i-1}i}$, which depends on the underlying state of the previous observation, z_{i-1}. Marginalizing over the missing data $\{z_i\}$ produces the mixture structure. Gibbs sampling treatment of hidden Markov models is quite similar to the methods described above, except that the simulation of the hidden state z_i must take into account its dependence on z_{i-1} and z_i. Therefore, *Step 2* in *Algorithm 1* is replaced by simulation from

$$[\, z_i \mid z_{i-1}, z_{i+1}, x_i, P, \theta_1, \ldots, \theta_k \,] \propto p_{z_i z_{i+1}} p_{z_{i-1} z_i} f(x_i|\theta_{z_i}),$$

for $i = 1, \ldots, n$, where P is the transition matrix comprising the probabilities p_{ij}, and *Step 1* remains roughly the same, with independent Dirichlet distributions on the rows of P. Despite the small additional complexity due to the dependence, the probabilistic properties of the Gibbs sampler for mixture models still hold and ensure proper convergence of the algorithm (see Robert *et al.*, 1993, for details).

References

Albert, J. H. and Chib, S. (1993) Bayesian analysis of binary and polychotomous response data. *J. Am. Statist. Ass.*, **88**, 669–679.

Berger, J. O. (1985) *Statistical Decision Theory and Bayesian Analysis.* New York: Springer-Verlag.

Bernardo, J. M. and Girón, F. J. (1988) A Bayesian analysis of simple mixture problems. In *Bayesian Statistics 3* (eds J. M. Bernardo, M. H. DeGroot, D. V. Lindley and A. F. M. Smith), pp. 67–78. Oxford: Oxford University Press.

Celeux, G. and Diebolt, J. (1985) The SEM algorithm: a probabilistic teacher algorithm derived from the EM algorithm for the mixture problem. *Comput. Statist. Quart.*, **2**, 73–82.

Dalal, S. R. and Hall, W. J. (1983) Approximating priors by mixtures of natural conjugate priors. *J. R. Statist. Soc.*, B, **45**, 278–286.

Dempster, A., Laird, N. and Rubin, D. B. (1977) Maximum likelihood from incomplete data via the EM algorithm (with discussion). *J. R. Statist. Soc.*, B, **39**, 1–38.

Diaconis, P. and Ylvisaker, D. (1985) Quantifying prior opinion. In *Bayesian Statistics 2* (eds J. M. Bernardo, M. H. DeGroot, D. V. Lindley, A. F. M. Smith), 163–175. Amsterdam: North-Holland.

Diebolt, J. and Ip, E. H. S. (1995) Stochastic EM: methods and application. In *Markov Chain Monte Carlo in Practice* (eds W. R. Gilks, S. Richardson and D. J. Spiegelhalter), pp.259–273. London: Chapman & Hall.

Diebolt, J. and Robert, C. P. (1990) Estimation of finite mixture distributions through Bayesian sampling (Parts I and II). *Rapports Techniques 109, 110*, LSTA, Université Paris 6.

Diebolt, J. and Robert, C. P. (1993) Discussion on the meeting on the Gibbs sampler and other Markov chain Monte Carlo methods. *J. R. Statist. Soc.*, B, **55**, 73–74.

Diebolt, J. and Robert, C. P. (1994) Estimation of finite mixture distributions through Bayesian sampling. *J. R. Statist. Soc.*, B, **56**, 163–175.

Dupuis, J. A. (1995) Bayesian estimation of movement probabilities in open populations using hidden Markov chains. *Biometrika*, (to appear).

Escobar, M. D. and West, M. (1995) Bayesian density estimation and inference using mixtures. *J. Am. Statist. Ass.*, **90**, 577–588.

Gelfand, A. E. (1995) Model determination using sampling-based methods. In *Markov Chain Monte Carlo in Practice* (eds W. R. Gilks, S. Richardson and D. J. Spiegelhalter), pp.145–161. London: Chapman & Hall.

Gelfand, A. E. and Smith, A. F. M. (1990) Sampling based approaches to calculating marginal densities. *J. Am. Statist. Ass.*, **85**, 398–409.

Gelman, A. (1995) Inference and monitoring convergence. In *Markov Chain Monte Carlo in Practice* (eds W. R. Gilks, S. Richardson and D. J. Spiegelhalter), pp.131–143. London: Chapman & Hall.

Gelman, A. and Meng, X.-L. (1995) Model checking and model improvement. In *Markov Chain Monte Carlo in Practice* (eds W. R. Gilks, S. Richardson and D. J. Spiegelhalter), pp.189–201. London: Chapman & Hall.

George, E. I. and McCulloch, R. E. (1995) Stochastic search variable selection. In *Markov Chain Monte Carlo in Practice* (eds W. R. Gilks, S. Richardson and D. J. Spiegelhalter), pp. 203–214. London: Chapman & Hall.

Geyer, C. (1992) Practical Markov chain Monte Carlo. *Statist. Sci.*, 7, 473–482.

Gilks, W. R. and Roberts, G. O. (1995) Strategies for improving MCMC. In *Markov Chain Monte Carlo in Practice* (eds W. R. Gilks, S. Richardson and D. J. Spiegelhalter), pp. 89–114. London: Chapman & Hall.

Gilks, W. R. and Wild, P. (1992) Adaptive rejection sampling for Gibbs sampling. *Appl. Statist.*, 41, 337–348.

Hastings, W. K. (1970) Monte Carlo sampling methods using Markov chains and their applications. *Biometrika*, 57, 97–109.

Heitjan, D. F. and Rubin, D. B. (1991) Ignorability and coarse data. *Ann. Statist.* 19, 2244–2253.

Lavine, M. and West, M. (1992) A Bayesian method for classification and discrimination. *Canad. J. Statist.*, 20, 451–461.

Lehmann, E. L. (1983) *Theory of Point Estimation*. New York: Wiley.

Liu, J., Wong, W. H. and Kong, A. (1994) Correlation structure and convergence rate of the Gibbs sampler with applications to the comparison of estimators and sampling schemes. *Biometrika*, 81, 27–40.

Mengersen, K. L. and Robert, C. P. (1995) Testing for mixture via entropy distance and Gibbs sampling. In *Bayesian Statistics 5* (eds J. O. Berger, J. M. Bernardo, A. P. Dawid, D. V. Lindley and A. F. M. Smith). Oxford: Oxford University Press, (in press).

Metropolis, N., Rosenbluth, A. W., Rosenbluth, M. N., Teller, A. H. and Teller, E. (1953) Equations of state calculations by fast computating machines. *J. Chem. Phys.*, 21, 1087–1092.

Müller, P., Erkanli, A. and West, M. (1995) Curve fitting using Dirichlet process mixtures. *Biometrika*, (to appear).

Mykland, P., Tierney, L. and Yu, B. (1995) Regeneration in Markov chain samplers. *J. Am. Statist. Ass.*, 90, 233–241.

Pearson, K. (1894) Contribution to the mathematical theory of evolution. *Phil. Trans. Roy. Soc.*, A, 185, 71–110.

Phillips, D. B. and Smith, A. F. M. (1995) Bayesian model comparison via jump diffusions. In *Markov Chain Monte Carlo in Practice* (eds W. R. Gilks, S. Richardson and D. J. Spiegelhalter), pp. 215–239. London: Chapman & Hall.

Raftery, A. E. (1995) Hypothesis testing and model selection. In *Markov Chain Monte Carlo in Practice* (eds W. R. Gilks, S. Richardson and D. J. Spiegelhalter), pp. 163–187. London: Chapman & Hall.

Raftery, A. E. and Lewis, S. M. (1995) Implementing MCMC. In *Markov Chain Monte Carlo in Practice* (eds W. R. Gilks, S. Richardson and D. J. Spiegelhalter), pp. 115–130. London: Chapman & Hall.

Robert, C. P. (1993a) Convergence assesments for Markov Chain Monte-Carlo methods. *Working Paper 9248*, Crest, Insee, Paris.

Robert, C. P. (1993b) Prior feedback: A Bayesian approach to maximum likelihood estimation. *Comput. Statist.*, 8, 279–294.

Robert, C. P. (1994) *The Bayesian Choice*. New York: Springer-Verlag.

Robert, C. P. and Soubiran, C. (1993) Estimation of a mixture model through Bayesian sampling and prior feedback. *TEST*, **2**, 323–345.

Robert, C. P., Celeux, G. and Diebolt, J. (1993) Bayesian estimation of hidden markov models: a stochastic implementation. *Statist. Prob. Lett.*, **16**, 77–83.

Soubiran, C., Celeux, G., Diebolt, J. and Robert, C. P. (1991) Estimation de mélanges pour de petits échantillons. *Revue de Statistique Appliquée*, **39**, 17–36.

Stuchlik, J. B., Robert, C. P. and Plessis, B. (1994) Character recognition through Bayes theorem. *Working Paper URA CNRS 1378*, University of Rouen.

Tanner, M. and Wong, W. (1987) The calculation of posterior distributions (with discussion). *J. Am. Statist. Ass.*, **82**, 528–550.

Tierney, L. (1994) Markov chains for exploring posterior distributions (with discussion). *Ann. Statist.*, **22**, 1701–1762.

Tierney, L. (1995) Introduction to general state-space Markov chain theory. In *Markov Chain Monte Carlo in Practice* (eds W. R. Gilks, S. Richardson and D. J. Spiegelhalter), pp. 59–74. London: Chapman & Hall.

Titterington, D. M., Smith, A. F. M. and Makov, U. E. (1985) *Statistical Analysis of Finite Mixture Distributions*. New York: Wiley.

Verdinelli, I. and Wasserman, L. (1991) Bayesian analysis of outlier problems using the Gibbs sampler. *Statist. Comput.*, **1**, 105–117.

25

An archaeological example: radiocarbon dating

Cliff Litton
Caitlin Buck

25.1 Introduction

Although perhaps not widely appreciated, archaeology provides statisticians with a rich source of challenging problems. Many of the fields of archaeological study which stand to benefit most, however, require development of non-standard, often complex, models which cannot be analysed using off-the-peg statistical techniques. This is clearly demonstrated by the wide range of specific case studies given in Fieller (1993), covering areas as varied as the study of migration patterns of prehistoric horses, Bronze Age woodland management and Roman field systems.

By virtue of the way in which archaeologists develop interpretations of the data they observe, they are constantly learning by experience and updating their ideas in the light of new evidence. Consequently, in addition to the development of complex models, they also require *a priori* knowledge to be included in the statistical investigation. Archaeologists are thus naturally attracted to the Bayesian view of statistics, although they generally lack detailed knowledge of its implementation. Over the last ten years, archaeological appreciation of the benefit of explicit inclusion of prior information has grown and a number of case studies have been published; see Litton and Buck (1995) for a review. The statistical methods involved include, among others, spatial analysis, multivariate mixtures and change-point analysis.

An essential feature of almost every archaeological investigation is the desire to obtain both absolute and relative chronological information and possibly the most widely adopted technique for doing this is radiocarbon dating. This application area is of fundamental importance for archaeological research and will be the focus of this chapter. In the next section, we give a brief overview of the salient scientific background to radiocarbon dating. Subsequently, we develop the basic statistical model used and show how it is applied, using MCMC methods, to a variety of real archaeological problems.

25.2 Background to radiocarbon dating

We provide here only a brief description of the major features of radiocarbon dating and refer the reader to Bowman (1990) for a more detailed account. Carbon occurs naturally in the form of three isotopes, carbon-12, carbon-13 and carbon-14 (denoted by ^{12}C, ^{13}C and ^{14}C respectively). Now ^{14}C, which forms about one part per million million of modern carbon, is radioactive and so it decays with time. Moreover ^{14}C is continually being formed in the upper atmosphere by the interaction of cosmic rays with nitrogen atoms. After formation, ^{14}C (along with ^{12}C and ^{13}C) is, as a result of photosynthesis, eventually taken into the food chain. Whilst alive, plants and animals constantly exchange carbon with the naturally replenishing reservoir. Upon death, however, no more new carbon is taken up and consequently the amount of radioactive carbon (^{14}C) begins to decay. In other words, during the lifetime of an organism, the proportion of ^{14}C to ^{12}C and ^{13}C remains constant within its cells, but after its death this proportion decreases. The decomposition of the ^{14}C (like other radioisotopes) obeys the law of radioactive decay which may be expressed as

$$M = M_0 \, e^{-\lambda x},$$

where M is the amount of ^{14}C remaining after time x, M_0 is the initial unknown amount of ^{14}C present upon death and λ is the decay rate (the reciprocal of the meanlife). If we can assume that the ^{14}C concentration in the atmosphere and living organic material is constant through time, then, provided we can estimate M relative to a standard M_0, the time elapsed since the material died can be estimated by

$$x = \lambda^{-1}\log(M_0/M).$$

However, due to changes in atmospheric conditions as a result of sun spot activity and the like, the ^{14}C 'equilibrium' level fluctuates slightly from year to year. Thus M_0 in the above equation varies depending upon when the organism died and hence there is the need to calibrate the radiocarbon age, x, onto the calendar time-scale.

Consider a sample of organic material that died at calendar date θ BP, where BP denotes 'before present'. For radiocarbon dating purposes, 'present' is taken as 1950 AD; for example, since 0 AD does not exist, 35 BC is equivalent to 1984 BP. Associated with this unknown calendar date is a unique 'radiocarbon age', related to the amount of ^{14}C currently contained in the sample; we represent 'radiocarbon age' by $\mu(\theta)$. Due to the nature of the samples and the chemical and physical techniques used, the *experimentally derived* values available for $\mu(\theta)$ are not totally accurate or precise. They provide an observation, x, which is a realization of a random variable, X, and is an estimate of $\mu(\theta)$. Conventionally, X is assumed to be normally distributed: $P(X|\theta, \sigma) = N(\mu(\theta), \sigma^2)$, where P generically denotes a probability density and σ is the laboratory's quoted error on x; see Bowman (1990) for details on how laboratories assess σ.

The need to calibrate provided the impetus for the radiocarbon community to undertake world-wide collaboration during the 1970s and 1980s. The result was the internationally agreed high precision calibration curve (Stuiver and Pearson, 1986; Pearson and Stuiver, 1986; Pearson et al., 1986) which was established by determining the ^{14}C content of timber samples already accurately dated on the calendar time-scale using dendrochronology (the science of dating wood samples using their annual growth rings: see Baillie, 1982). The calibration curve so obtained consists of estimates of the radiocarbon age corresponding to samples whose calendar ages are at approximately twenty-year intervals from the present day back through time to about 6000 BC. Because of the atmospheric changes in ^{14}C levels, the calibration curve is non-monotonic and exhibits a significant number of wiggles (see Figure 25.1).

Now, a suitable working approximation is to express $\mu(\theta)$ in a piece-wise linear form

$$
\mu(\theta) = \begin{cases} a_0 + b_0\theta & (\theta \leq t_0) \\ a_\ell + b_\ell\theta & (t_{\ell-1} < \theta \leq t_\ell, \; \ell = 1, 2, \ldots, L) \\ a_{L+1} + b_{L+1}\theta & (\theta > t_{L+1}) \end{cases}
$$

where t_ℓ are the knots of the calibration curve, $L+1$ is the number of knots, and a_ℓ and b_ℓ are assumed to be known constants which ensure continuity at the knots. Assuming a uniformly vague prior for θ, the posterior density of θ is given by

$$
P(\theta|x) \propto \begin{cases} \exp -\frac{(x - a_0 - b_0\theta)^2}{2\sigma^2} & (\theta \leq t_0) \\ \exp -\frac{(x - a_\ell - b_\ell\theta)^2}{2\sigma^2} & (t_\ell - 1 < \theta \leq t_\ell, \; \ell = 1, 2, \ldots, L) \\ \exp -\frac{(x - a_{L+1} - b_{L+1}\theta)^2}{2\sigma^2} & (\theta > t_{L+1}) \end{cases} \cdot
$$

Incidentally, the most well-known and well-used computer program for the calibration of a radiocarbon determination (see Stuiver and Reimer, 1993) calculates this posterior although nowhere is Bayes mentioned! Typically, because of the wiggles in the calibration curve, the posterior density is not symmetric and is often multimodal (see Figure 25.2) which makes interpretation difficult.

Calibrating and interpreting a single radiocarbon determination provides

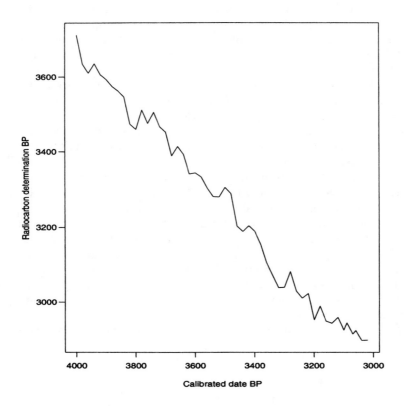

Figure 25.1 *A section of the radiocarbon calibration curve.*

archaeologists with minimal information. Commonly archaeologists wish to use radiocarbon dating to provide absolute dates for events whose relative chronology is already at least partially understood. As a result, they are positively encouraged by funding bodies and laboratories to submit for dating collections of samples for which they possess *a priori* relative temporal information, even if nothing is known of their exact calendar dates. For example, samples may be taken from several different layers in a prehistoric midden (rubbish heap). Where such layers clearly lie one above the other, archaeologists describe this as a *stratigraphic sequence*. Although the exact dates of such layers are unknown *a priori*, the order of the calendar dates is known and is based upon which layers underlie which in the stratigraphic sequence. It is in the handling of such groups of related samples that MCMC methods have proved extremely useful. This will be illustrated

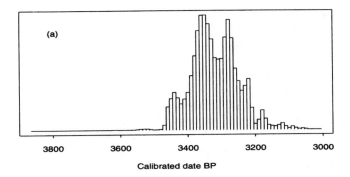

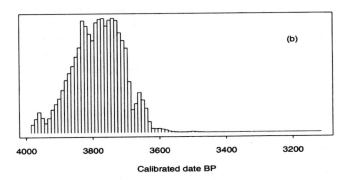

Figure 25.2 *Two calibrated date distributions, illustrating non-symmetry and multimodality; in (a) $x = 3\,500$, $\sigma = 60$ and in (b) $x = 3\,100$, $\sigma = 60$.*

later in this chapter using real examples.

25.3 Archaeological problems and questions

Provided that samples for radiocarbon dating are selected with extreme care, so that only material from short-lived organisms is used, each radiocarbon determination provides a date at death for that one particular organism. Archaeologists are not usually, however, specifically interested in obtaining a date of death for the sampled organism they submit, but for the *event* which caused its death (for example, the harvesting of corn, the cutting of reeds for thatch or the burning of twigs on a fire). As a result, archaeologists usually refer to the date of an event which they equate to the date of death of a particular organic sample they have obtained. Given a collection of radiocarbon determinations, commonly posed archaeological questions include, when did an event occur, what is the time period between successive events, what is the likely order of the events, and so on.

In situations where archaeologists are able to identify sophisticated relationships between events, the questions they pose of statisticians become more complex. An example of this occurs when archaeologists subdivide a timespan into *phases*. Each phase is usually seen as a period of time during which a particular type of archaeological activity took place. Such a phase might be defined by the presence of an individual ceramic type or style, for example Bronze Age beakers are a ceramic type found widely in Europe. Consequently, if radiocarbon determinations are available which relate to several phases, questions will arise about the relationship between the phases: Do they overlap?; Do they abutt?; Is there a hiatus between them?; and so on.

In terms of prior information, what do the archaeologists know? In a few cases, there may be information about the time period between successive events. More commonly, prior information takes the form of chronological orderings (or partial orderings) of events or phases which result in constraints on the parameters. Such constraints are extremely hard to handle without MCMC methods.

25.4 Illustrative examples

25.4.1 Example 1: dating settlements

Buck *et al.* (1994) describe the application of Bayesian methods to the analysis of fifteen radiocarbon determinations (corresponding to fifteen archaeological events) from the early Bronze Age settlement of St Veit-Klinglberg, Austria. The primary objective of excavating this site was to investigate the organization of a community believed to be associated with the production and distribution of copper. A further objective was to obtain a series of radiocarbon determinations which would allow the placement of the site in its true temporal location on the calendar scale. During the excavation, archaeologists therefore collected samples for radiocarbon dating in the hope that they would provide an estimate of the absolute date of the development of copper production in the region, and a foundation for the dating of Early Bronze Age domestic pottery. With the exception of some recent dendrochronological work, there are still few absolute dates for the early Bronze Age in central Europe. The dating of such settlements is particularly problematical because they lack the metalwork which has been used to establish typological chronologies, and their domestic pottery (which is often used to provide such information) is not distinctive enough to allow fine chronological divisions.

At St Veit-Klinglberg, however, the radiocarbon determinations were collected in the light of archaeological information (based upon stratigraphic evidence) which was used, *a priori*, to partially order the calendar dates of ten of the fifteen events. St Veit-Klinglberg exhibited complex depositional patterns as a consequence of its sloping location on a hill spur: five of the fifteen events were from deposits which had suffered erosion and could not, therefore, reliably be placed in the sequence. So, letting θ_i denote the calendar date BP of event i, Buck *et al.* (1994) provide the archaeological stratigraphic information in the form of the following inequalities:

$$\theta_{758} > \theta_{814} > \theta_{1235} > \theta_{358} > \theta_{813} > \theta_{1210},$$

$$\theta_{493} > \theta_{358},$$

$$\theta_{925} > \theta_{923} > \theta_{358}$$

and

$$\theta_{1168} > \theta_{358}.$$

where the subscripts are the archaeological location numbers allocated during excavation. At this site, primary archaeological interest focuses upon the dates of these ten events, with secondary interest in adding the five unordered events to the same sequence. In addressing the first of these, we note that nothing is known *a priori* about the time period between the ten events. Therefore, their θ_i are assumed to have a jointly uniform improper prior subject to the above inequalities. Thus the prior for θ (the θ_i for all 15 events) may be expressed as

$$I_A(\theta) = \begin{cases} 1 & \theta \in A \\ 0 & \theta \notin A \end{cases}$$

where A is the set of values θ satisfying the above inequalities. By Bayes theorem, the joint posterior of θ is given by

$$P(\theta|\mathbf{x}) \propto \left\{ \prod_i \exp\left\{ -\frac{(x_i - \mu(\theta_i))^2}{2\sigma_i^2} \right\} \right\} I_A(\theta).$$

Note here that each radiocarbon determination, x_i, has its own standard deviation, σ_i, reported by the radiocarbon laboratory.

Were it not for the constraints provided by the archaeological prior information, each θ_i would be *a posteriori* independent and the marginal information could be easily evaluated numerically without the need for MCMC methods. Indeed, even with these constraints, evaluation of the marginal posteriors of interest is not impossible using conventional numerical integration techniques. However, implementation of the Gibbs sampler when order constraints are present is simple and this is the approach selected by Buck *et al.* (1994). See Gilks *et al.* (1995: this volume) for an introduction to Gibbs sampling.

At each iteration of the Gibbs sampler, we need to evaluate the posterior density for the θ_i: this is proportional to

$$\exp\left\{-\frac{(x_i - \mu(\theta_i))^2}{2\sigma_i^2}\right\}\mathrm{I}_{(c_i, d_i)}(\theta_i)$$

where

$$\mathrm{I}_{(c_i, d_i)}(\theta_i) = \begin{cases} 1 & c_i < \theta_i < d_i \\ 0 & \text{otherwise} \end{cases}.$$

Note that c_i and d_i possibly depend upon the values of the other θ_i. For example, consider the sample from archaeological location 358 at St Veit-Klinglberg. The calendar date for this event is restricted to be earlier than the date of event 813, but later than events 1235, 493, 923 and 1168. Thus, $c_i = \theta_{813}$ and $d_i = \min\{\theta_{1235}, \theta_{493}, \theta_{923}, \theta_{1168}\}$.

In Figure 25.3, we give plots of the posterior probability of θ_{358}. Figure 25.3(a) shows the calendar date obtained when the radiocarbon determination from archaeological location 358 is calibrated alone, without the *a priori* information provided by the known order of the events. In contrast, Figure 25.3(b) shows the calendar date arising from the radiocarbon determination from location 358 when it is calibrated in conjunction with all the other determinations and the above archaeological constraints are accounted for. The 95% highest posterior density region for the distribution in Figure 25.3(a) is 3825 to 3785, 3765 to 3745 and 3735 to 3395 BP. This is disjoint because of the non-monotonic nature of the calibration curve. The 95% highest posterior density region for the distribution in Figure 25.3(b) is, however, continuous and is 3535 to 3385 BP. We note that there is a considerable difference in the highest posterior density regions depending upon whether the archaeological information is included or ignored. In fact, when the archaeological information is included, the region is shorter by about 200 years, since early calibrated dates are precluded by the archaeological ordering information. Histogram-like plots similar to those in Figure 25.3, together with highest posterior density regions, have become standard tools for reporting the strange, multimodal posteriors that frequently occur in radiocarbon dating. For obvious reasons, the use of conventional summary statistics such as means, standard deviations, modes, etc. is generally avoided.

Returning to the archaeological question of secondary interest, the Gibbs sampler also provides posterior probabilities of the temporal location of the five deposits which could not be stratigraphically related to the other ten. Such information offers the archaeologists greatly enhanced insight into the chronological relationships of several parts of the site that were not directly linked by archaeological stratigraphy.

Until the advent of MCMC methods, archaeologists were not offered such probabilistic determinations. As a consequence, assessments and interpretations were often made on the basis of mean or modal values which, given

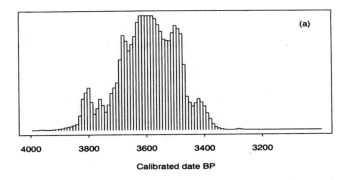

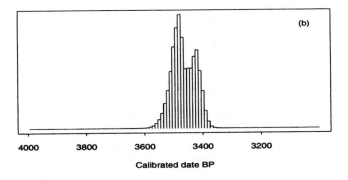

Figure 25.3 *The posterior probability of θ_{358} at St Veit-Klinglberg (a) when the stratigraphic information is ignored and (b) when it is included.*

the nature of the distributions, was clearly misguided and often actually misleading.

25.4.2 Example 2: dating archaeological phases

Zeidler *et al.* (1996) report the analysis of radiocarbon determinations from 16 sites in the Jama River Valley in Ecuador. As a result of a recent archaeological survey in the Jama River Valley, seven major archaeological phases have thus far been defined spanning about 4 000 years. Initially, one major site was excavated (San Isidro) in the upper valley which, through its deep stratigraphy, led to a 'master' ceramic sequence. Over several subsequent seasons of study, the archaeologists undertook further excavation, correlated the information between the various sites studied and selected suitable organic samples found on these sites to be submitted for radiocarbon dating. Some 37 radiocarbon determinations were made on samples taken from 16 different archaeological sites whose occupation spanned the whole period of interest, with each sample being clearly and unambiguously assigned to one, or other, of the seven phases.

Despite the importance of the Jama River Valley as the territory of the Jama-Coaque culture (a series of complex chiefdoms with an elaborate ceramic tradition spanning some 2 000 years in the pre-Columbian history of Ecuador), temporal information about the Jama River Valley has until recently been limited and poorly understood. Consequently, the primary use of the radiocarbon determinations was in estimating the beginning and ending dates for each of the seven phases. So let α_j and β_j represent the beginning and ending dates (measured BP) of phase j ($j = 1, 2, \ldots, 7$). Note that, since the dates are measured in years BP, we have $\alpha_j > \beta_j$. Let n_j be the number of samples assigned to the jth phase, let x_{ij} be the ith radiocarbon determination obtained from phase j and let σ_{ij} be the corresponding quoted laboratory error. Similarly, let θ_{ij} represent the calendar date of the ith radiocarbon sample in the jth phase. Since nothing is known about its date within the phase, we assume that

$$P(\theta_{ij}|\alpha_j, \beta_j) = \mathrm{U}(\beta_j, \alpha_j),$$

where $\mathrm{U}(a, b)$ denotes a uniform distribution on the interval a to b.

Several lines of investigation were utilized in establishing the archaeological phasing including archaeological stratigraphy (which in this region is interspersed with layers of volcanic ash) and detailed study of ceramic typologies. Careful study of the various sources of archaeological information enabled the archaeologists to identify clearly interrelationships between the seven phases and even to provide partial orderings for them. Consequently, they were able to offer prior information about the α_j and β_j in the form of inequalities:

$$\alpha_1 > \beta_1 \geq \alpha_2 > \beta_2 \geq \alpha_3 > \beta_3 \geq \alpha_4 > \beta_4$$

$$\alpha_4 > \alpha_5 > \alpha_6$$

$$\beta_4 > \beta_5 > \beta_6$$

and

$$\beta_6 = \alpha_7.$$

The relationships between the other α_j and β_j (except $\alpha_j > \beta_j$) were unknown *a priori* and information about them was sought *a posteriori*. Let $\theta = \{\theta_{ij}, i = 1, \ldots, n_j, j = 1, \ldots, 7\}$, $\alpha = \{\alpha_j, j = 1, \ldots, 7\}$, $\beta = \{\beta_j, j = 1, \ldots, 7\}$. Assuming a jointly uniform improper prior for α_j and β_j, subject to the above inequalities, the joint posterior density of θ, α and

β is given by

$$P(\theta, \alpha, \beta | x) \propto I_A(\alpha, \beta) \prod_{j=1}^{7} \left\{ (\alpha_j - \beta_j)^{-n_j} \prod_{i=1}^{n_j} z_{ij} I_{(\beta_j, \alpha_j)}(\theta_{ij}) \right\},$$

where, in this case, A is the set of α and β satisfying the above inequalities,

$$z_{ij} = \exp \left\{ -\frac{(x_{ij} - \mu(\theta_{ij}))^2}{2\sigma_{ij}^2} \right\}$$

and

$$I_A(\alpha, \beta) = \begin{cases} 1 & (\alpha, \beta) \in A \\ 0 & \text{otherwise} \end{cases}$$

As with the St Veit-Klinglberg example, it is not impossible to evaluate the posterior densities of the α_j and β_j using quadrature methods. Indeed, this was demonstrated by Naylor and Smith (1988) for a similar example which involved dating the boundaries between successive ceramic phases from the Iron Age hillfort at Danebury, England. However, considering the Jama River Valley problem from an MCMC viewpoint, we note that the partial ordering information offers no real difficulty since the θ parameters can be sampled from a restricted range as in the previous example and the full conditionals of the α and β parameters are fairly easy to write down. For instance, since phases 4 and 5 may or may not overlap but phase 3 must precede phase 4, the full conditional density for α_4 is proportional to $(\alpha_4 - \beta_4)^{-n_4} I_{(c,d)}(\alpha_4)$ where $c = \max\{\alpha_5, \theta_{i4}; i = 1, \ldots, n_4\}$ and $d = \beta_3$. Thus, α_4 can be sampled using an inversion method. The situation for β_6 is slightly more complicated, since phases 6 and 7 abut one another (i.e. $\beta_6 = \alpha_7$) and phase 5 is known to have ended before phase 6 ended (i.e. $\beta_5 > \beta_6$). The full conditional for β_6 is, therefore, proportional to

$$(\alpha_6 - \beta_6)^{-n_6} (\beta_6 - \beta_7)^{-n_7} I_{(c,d)}(\beta_6)$$

where

$$c = \max(\theta_{i7}; \ i = 1, \ldots, n_7)$$

and

$$d = \min(\beta_5, \theta_{i6}; \ i = 1, \ldots, n_6).$$

See Gilks (1995: this volume) for further details on constructing and sampling from full conditional distributions.

Zeidler *et al.* (1996) provide all such conditionals and use a Gibbs sampling scheme to obtain the posterior densities for the α and β parameters. These posterior densities are given in the form of histogram-like plots and highest posterior density regions, as for the posteriors in the St Veit-Klinglberg example. In addition, archaeological interest in the Jama River Valley focuses not only on the beginning and ending dates of the phases, but on the lengths of each of the phases. The posterior distribution of each phase length are easy to compute within the Gibbs sampler framework, since the time between the α_j and β_j can be evaluated at each iteration.

Following the presentation of posterior information about the dates of the phase boundaries and the lengths of the phases, Zeidler *et al.* (1996) consider the sensitivity of their posterior statements to changes in the prior information. This was felt to be particularly important since the calendar dates obtained were likely to be very dependent on the validity of the archaeological information used to define the phases. The authors place a great deal of emphasis on this point and are content that the radiocarbon evidence and the archaeological prior knowledge are in broad agreement.

To summarize, the calendar dates provided for the phase boundaries offer a reliable and specific calendar-based chronology for the Jama River Valley. This can now profitably be compared with other, more generalized, chronologies established by other workers for other parts of Ecuador. This in turn will enable greater understanding of the development of ancient cultures in Ecuador and further afield.

25.4.3 Example 3: accommodating outliers

One of the major difficulties in using radiocarbon dating techniques is that they are extremely prone to outliers. In fact Baillie (1990) demonstrated that, for radiocarbon determinations made in the 1970s and early 1980s, something like a third of the results had true dates outside the 2σ ranges produced by the radiocarbon laboratories. Since that time considerable efforts have been made to improve the accuracy of laboratory methods but, as with any sophisticated scientific process, erroneous or misleading radiocarbon results will still arise for a variety of reasons, some as a consequence of mistakes by the archaeologists and some by the laboratories themselves. Thus an important requirement of any statistical analysis of radiocarbon data is the ability to incorporate a means of identifying outliers.

Christen (1994) proposes the following Bayesian methodology, based on the Gibbs sampler, for identifying outliers. Suppose that we have n determinations with the jth determination, x_j, having reported standard deviation σ_j corresponding to unknown calendar date θ_j. For simplicity of explanation we suppose that $P(\theta_j | \alpha, \beta) = \mathrm{U}(\beta, \alpha)$ where α and β are unknown. The basic idea is that x_j is an outlier if x_j needs a shift δ_j on the radiocarbon scale to be consistent with the rest of the sample. Formally we have

$$P(X | \theta_j, \phi_j, \delta_j) = \mathrm{N}\left(\mu(\theta_j) + \phi_j \delta_j, \; \sigma_j{}^2\right),$$

where

$$\phi_j = \begin{cases} 1 & \text{if } x_j \text{ needs a shift} \\ 0 & \text{otherwise} \end{cases}.$$

Thus an error in the radiocarbon dating process of the jth determination will be modelled by a shift in the true radiocarbon age $\mu(\theta_j)$. (An alternative more drastic precaution would be to inflate the standard deviation of all the observations.) In many situations, it is reasonable to assume a vague prior $P(\delta_j)$ for δ_j. Let q_j denote the prior probability that a given determination is an outlier. In practice, since accuracy of radiocarbon laboratories has recently improved, taking $q_j = 0.1$ seems reasonable although possibly somewhat pessimistic.

Suppose that prior information about α and β can be expressed in the form of conditional densities $P(\alpha|\beta)$ and $P(\beta|\alpha)$ respectively. Then the full conditionals for use with the Gibbs sampler are

$$P(\theta_j|\cdot) \propto z_j I_{(\beta,\alpha)}(\theta_j),$$
$$P(\alpha|\cdot) \propto (\alpha - \beta)^{-n} I_{(d,\infty)}(\alpha) P(\alpha|\beta),$$
$$P(\beta|\cdot) \propto (\alpha - \beta)^{-n} I_{(0,c)}(\beta) P(\beta|\alpha),$$
$$P(\delta_j|\cdot) \propto z_j P(\delta_j),$$
$$P(\phi_j|\cdot) \propto q_j^{\phi_j}(1 - q_j)^{1-\phi_j} z_j,$$

where $c = \min(\theta_i; \ i = 1, \ldots, n)$; $d = \max(\theta_i; \ i = 1, \ldots, n)$;

$$z_j = \exp\left\{ -\frac{(\mu(\theta_j) - (x_j - \phi_j\delta_j))^2}{2\sigma_j^2} \right\};$$

and the '$\cdot$' in the conditions denotes 'all other variables'; see Gilks (1995: this volume). The full conditionals for θ_j, α and β can be readily sampled as in earlier examples and, moreover, the sampling of δ_j and ϕ_j is quite straightforward (a normal and a Bernoulli respectively). Thus, the Gibbs sampler is quite simple to implement in this, and other, situations and allows us to compute the posterior probability that y_j is an outlier. Further details with illustrative examples are to be found in Christen (1994).

25.4.4 Practical considerations

In summary, it is now possible to address a range of interesting and challenging problems associated with the calibration of radiocarbon determinations that were, prior to the advent of MCMC methods, relatively difficult and needed highly specialized and sophisticated software. We should, however, consider one or two points relating to the practicalities of these approaches. For event-based problems such as the example in Section 25.4.1, convergence of the Gibbs sampler is rapid and run times (even when using computer code that is not optimized in any way) are commonly less than an hour for around 10 000 iterations on a SUN workstation. For more complex models involving ordered phases such as the example in Section 25.4.2, since the parameters are highly correlated, convergence is slower and typical run times are between six and twelve hours for around 100 000 iterations. Generally, outlier analysis takes a matter of minutes. However, even a day's computer time is insignificant when compared with the time taken in excavation, obtaining radiocarbon determinations, post excavation interpretation and publication of final reports. For larger sites, this process commonly takes several years!

25.5 Discussion

It is clear that there is a wide range of archaeological problems associated with radiocarbon dating, most of which can only realistically be approached from a Bayesian perspective. Even in the case of single determinations, the multimodality of the likelihood function makes the use of maximum-likelihood based techniques inappropriate and nonsensical. Having said this, due to the nature of the available prior information, evaluating posteriors of interest using quadrature methods requires specialist software not widely available and often not easy to use. The advantage of MCMC methods in this context is their intuitive appeal and their ease of implementation. Moreover, the simple iterative nature of the algorithms is easily understood by archaeologists and consequently they are well received. In fact, we are aware of at least one radiocarbon laboratory which is developing its own software based on these ideas.

The implications for archaeology are enormous. For the first time, archaeologists have a framework in which to combine a range of sources of information which can be readily implemented. As more illustrative case studies are published using these techniques, popularity will undoubtedly grow. Of course, as with any statistical technique, statisticians should bear in mind that without clear guidelines, misuse and abuse will almost certainly occur.

In the examples described above, the archaeological information available is very precise. The nature of archaeological investigation is, however, such that relative chronological information is often stated much less clearly. Hence future models will need to allow for this and will be more complex. Moreover, it is not hard to imagine that situations will arise where the nature of the archaeology is such that we need to combine events, phases and outliers. All this, we believe, can be easily encompassed within the MCMC framework. In this chapter, we have placed all the emphasis on the application of MCMC methods to radiocarbon dating and, indeed, this is the archaeological application area that has received most attention from Bayesian statisticians in the last few years. As stated in the introduction, however, there are already very many applications of model-based statistics to archaeological data interpretation. A good number of these stand to benefit from the use of MCMC-based Bayesian interpretations over the next few years.

Acknowledgements

We are grateful to Adrian Smith and Sue Hills for introducing us to the Gibbs sampler methodology and for their continuing interest. In addition, we wish to acknowledge the large number of colleagues from around the world who have collaborated with us in the investigation of archaeological problems, given generously of their time and have consistently encouraged us to pursue the work reported here. In particular, we are indebted to Andrés Christen whose presence in our research team over the last few years has added greatly to our motivation and enthusiasm. Finally, the second named author is supported by the UK Science and Engineering Research Council (reference GR/G142426).

References

Baillie, M. G. L. (1982) *Tree-ring Dating and Archaeology*. London: Croom Helm.

Baillie, M. G. L. (1990) Checking back on an assemblage of published radiocarbon dates. *Radiocarbon*, **32**, 361–366.

Bowman, S. (1990) *Radiocarbon Dating*. London: British Museum Publications.

Buck, C. E., Litton, C. D. and Shennan, S. J. (1994) A case study in combining radiocarbon and archaeological information: the early Bronze Age settlement of St Veit-Klinglberg, Land Salzburg, Austria. *Germania*, **72**, 427–447.

Christen, J. A. (1994) Summarising a set of radiocarbon determinations: a robust approach. *Appl. Statist.*, **43**, 489–503.

Fieller, N. R. J. (1993) Archaeostatistics: old statistics in ancient contexts. *The Statistician*, **42**, 279–295.

Gilks, W. R. (1995) Full conditional distributions. In *Markov Chain Monte Carlo in Practice* (eds W. R. Gilks, S. Richardson and D. J. Spiegelhalter), pp. 75–88. London: Chapman & Hall.

Gilks, W. R., Richardson, S. and Spiegelhalter, D. J. (1995) Introducing Markov chain Monte Carlo. In *Markov Chain Monte Carlo in Practice* (eds W. R. Gilks, S. Richardson and D. J. Spiegelhalter), pp. 1–19. London: Chapman & Hall.

Litton, C. D. and Buck, C. E. (1995) The Bayesian approach to the interpretation of archaeological data. *Archaeometry*, **37**, 1–24.

Naylor, J. C. and Smith, A. F. M. (1988) An archaeological inference problem. *J. Am. Statist. Ass.*, **83**, 588–595.

Pearson, G. W. and Stuiver, M. (1986) High-precision calibration of the radiocarbon time scale, 500–2500 BC. *Radiocarbon*, **28**, 839–862.

Pearson, G. W., Pilcher, J. R., Baillie, M. G. L., Corbett, D. M. and Qua, F. (1986) High-precision ^{14}C measurements of Irish oaks to show the natural ^{14}C variations from AD 1840–5210 BC. *Radiocarbon*, **28**, 911–34.

Stuiver, M. and Pearson, G. W. (1986) High-precision calibration of the radiocarbon time scale, AD 1950–500 BC. *Radiocarbon*, **28**, 805–838.

Stuiver, M. and Reimer, P. J. (1993) Extended ^{14}C data base and revised CALIB 3.0 ^{14}C age calibration program. *Radiocarbon*, **35**, 215–230.

Zeidler, J. A., Buck, C. E. and Litton, C. D. (1996) The integration of archaeological phase information and radiocarbon results from the Jama River Valley, Ecuador: a Bayesian approach. *Latin Amer. Antiq.*, (in press).

Index